ALLE · ZEIT · WACH · 1842 · SJ

Werner Vogel

Glass Chemistry

Second Edition

With 357 Figures

Springer-Verlag
Berlin Heidelberg New York
London Paris Tokyo
Hong Kong Barcelona Budapest

Prof. Dr. Werner Vogel

Otto-Schott-Institut
Friedrich-Schiller-Universität Jena
Fraunhoferstr. 6
07743 Jena

The first edition has been published
by The American Ceramic Society, Inc. in 1985

Translation of the German third edition published by
Springer-Verlag, 1992, ISBN-13:978-3-642-78725-6
Translated by N. Kreidl and M. Lopes Barreto

ISBN-13:978-3-642-78725-6 e-ISBN-13:978-3-642-78723-2
DOI: 10.1007/978-3-642-78723-2

Library of Congress Cataloging-in-Publication Data
Vogel, Werner, 1925-[Glaschemie. English] Glass chemistry/Werner Vogel, -- 3rd ed./translated by N. Kreidl and M. Lopes Barreto. Includes bibliographical references and index.
ISBN-13:978-3-642-78725-6
1. Glass- -Analysis. I. Title.

Production: PRODUserv Springer Produktions-Gesellschaft, Berlin
Typesetting Macmillan India Ltd., Bangalore
SPIN: 100055748 51/3020-543210-Printed on acid-free paper

Foreword

Glass chemistry represents a relatively young branch of chemistry. It comprises a boundary area between chemistry, physics, mineralogy, biology, and, recently, medicine as well. It is difficult to present this area with the proper balance and well-placed accents.

No doubts chemistry and also physics – the latter particularly in its technological aspects – occupy a central position in the treatment of glass problems.

Glass technology is not treated within the framework of this book, although a glass chemist or physicist certainly cannot manage without sufficient background in glass technology, nor a glass technologist without one in glass chemistry. In the area of glass technology, excellent older texts are available, such as those by R. Guenther, F. Tooley, J. J. Kitaigorodski, etc.

Glass chemistry represents a special branch of high-temperature chemistry in which chemical processes of transformation of matter take place. They have, during the past 20 years, led to results hardly believed possible before.

The importance of glass and silicate chemistry for mankind is going to increase considerably at the turn of this century. This is because, once the classical sources of raw materials such as oil, natural gas, and ores begin to be exhausted as a basis for materials development and production, the supply of raw materials for the glass and silicate industry will remain indefinitely.

This book has been conceived as a text for university-level students of glass science and industry. Its content is largely the core of a course at the Friedrich-Schiller-University in Jena, offered since 1961 and updated annually. In some cases, the book goes beyond educational needs and thus might be useful to those already active in glass science or production.

The introductory sections of the book have been guided by the impulses emerging in Jena from Otto Schott, pioneer of modern glass research. Special attention has been paid to the interdisciplinary cooperation of Ernst Abbe, Otto Schott and Carl Zeiss – a physicist, a chemist and a technologist – which was instigated by Ernst Abbe and is still an example to be followed. The main thrust is the elucidation of the relation of composition, structure, and properties. Only thorough understanding in this respect will lead from the empiricism which has governed glass investigation for so long toward a firm foundation for systematic development and production control.

Of the many modern methods applied in contemporary studies of the structure of glasses, nuclear magnetic resonance spectroscopy and electron microscopy have been singled out for more detailed coverage. These two complementary methods have significantly advanced glass research during the past 20 years. While nuclear magnetic resonance provided closer insight into atomic interaction and the coordination of ions, electron microscopy revealed structure-property relations in glasses over a range orders of magnitude larger than those accessible through magnetic resonance. Many examples demonstrate the utilization of both methods in practice.

The section on nuclear magnetic resonance spectroscopy was written by Professor P. J. Bray, Physics Department, Brown University, Providence, Rhode Island, USA. Professor Bray is known worldwide as an expert in the field of nuclear magnetic resonance and as an outstanding teacher. I thank him for his collaboration.

The first english edition has been published by the American Ceramic Society. Special thanks are due to Norbert J. Kreidl for translating, editing, and updating the book. No better translator than Prof. Kreidl, who is also a glass scientist and teacher, could have been found anywhere in the world for this project.

The 2nd English edition of "Chemistry of Glass" is a translation of the 3rd German edition by Springer-Verlag. It has been considerably expanded and enhanced, especially as concerns, e.g. electron-optical techniques in glass research, vitreous silica, gel glasses, metal glasses, optical high-performance glasses, bioglass ceramics for medical purposes, etc. I should like to thank several assistants and scientists for the support they gave me in the writing of the new manuscript. I should specifically like to name Dr. Seeber, Dr. sc. Stachel. Dr. sc. Ehrt, Dr. Müller, Dr. sc. Bürger and Dr. Völksch. As in the past, L. Horn and Ms. Keinert were always ready to help in drawing figures and processing photographs. I also extend my thanks to Ms. Kraft for typing the manuscript. All the aforementioned and others have played a substantial role in the creation of this book.

The present 2nd English edition of "Chemistry of Glass" was translated by M. Lopes Barotto and N. J. Kreidl. Professor Kreidl is one of the world's most renown glass researchers. Thus the quality of the translation of the specific terminology is guaranteed, and the book may therefore serve as a work aid in the German-English translation of technical texts concerning glass.

Regrettably, Professor Kreidl passed away shortly after his 90th birthday on 12 July 1994. I should like to express my deep gratitude for the great help he afforded me during the past 30 years. I shall keep him in fond memory.

The author would also like to thank Springer-Verlag for the fine appearance of all my books on glass. Special thanks are due to Ms. Maas for her support and cooperation in producing this book.

Jena, July 1994 Werner Vogel

Contents

1 **Historical Development of Glass Chemistry** 1

1.1 The Beginnings of Glass Research 1
1.2 History of the Chemistry of Optical Glass 2
1.2.1 Ernst Abbe and Otto Schott 2
1.2.2 Carl Zeiss and the Zeiss Foundation 11
1.2.3 The Development of New Optical Glasses after 1939 15
1.3 History of Technical Glass 19

2 **Freezing of a Melt to a Vitreous Solid** 22

2.1 Fusion and Crystallization. General 22
2.2 Significant Differences Between Crystalline and Non-crystalline (Glassy) Solids 24
2.3 Standard Viscosity Temperatures for Solidification of Glasses 30
2.4 Annealing of Optical Glass 31

3 **Structural Elements of Silicates** 34

3.1 The SiO_4 Tetrahedron as the Basic Building Block of Silicates 34
3.2 Building Units of Natural Crystalline Silicates 36

4 **Classical Theories of Glass Structure** 41

4.1 Glass Structure According to Tammann (since 1903) 41
4.2 Glass Formation According to Goldschmidt 41
4.3 The Zachariasen–Warren Network Theory 42
4.4 Extension of the Network Theory by Dietzel 45
4.5 Additional Concepts Supplementing the Network Theory 48
4.6 Lebedev's Crystallite Theory 50
4.7 Further Development of the Crystallite Theory 54
4.8 Kinetic Theory 54

5 **Methodology in Glass Research** 57

5.1 Structure of Liquids and Melts 58
5.2 The Nuclear Magnetic Resonance Method as Applied to Glass Research (P.J. Bray) 59

5.2.1 Introduction 59
5.2.2 Basic NMR Theory 60
5.2.3 Dipolar Interaction. 61
5.2.4 Chemical Shift 62
5.2.5 Quadrupole Interaction 64
5.3 Electron Microscopy 71
5.3.1 Introduction 71
5.3.2 Relations Between Light and Electron Microscopy 71
5.3.3 Imaging and Preparation of Samples 72
5.3.3.1 Direct Penetration of the Sample by Electrons. 73
5.3.3.2 Carbon Replica Method after Bradley 75
5.3.3.3 Further Development of the Experimental Technique of Bradley's Carbon Replica 77
5.3.3.4 Treatment of Glass Surfaces Prior to Replication 82
5.3.4 The Scanning Electron Microscope and Electron Microprobe. 85

6 Microphase Separation 92

6.1 Early History 92
6.2 Electron Microscopy Evidence for Immiscibility Phenomena in Glasses 93
6.3 Theoretical Treatment 98
6.3.1 Thermodynamics of Phase Separation. General (Kortüm). 98
6.3.1.1 Conditions of Equilibrium and Stability. 98
6.3.1.2 Derivation of Stability Conditions for a Binary Mixed Phase. 99
6.3.1.3 Characterization of the Regions of Immiscibility in Binary and Ternary Systems. 101
6.3.2 Thermodynamics of Immiscibility in Glasses 103
6.3.3 Kinetics of Immiscibility in Glasses 105
6.4 Experimental Evidence 109
6.4.1 Functional Change of Microphases 109
6.4.2 Multiple Phase Separation 111
6.4.3 Shells around Microphases. 112
6.4.4 Droplet Agglomeration after Secondary Phase Separation. 114
6.4.5 Composition of Microphases and Distribution of Heavy Metal Ions. 116
6.4.6 General Conclusions on Immiscibility Behavior and Microstructure 119
6.4.7 Control of Phase Separation. 120

7 **Structure and Properties of Colorless Glasses** 123

7.1 Silica Glass 123
7.2 Alkali Silicate Glasses 128
7.2.1 The Mixed-Alkali Effect 132
7.3 Alkaline Earth and Alkali-Alkaline Earth Silicate Glasses 135
7.4 Borate and Borosilicate Glasses 137
7.4.1 Binary Alkali Borate Glasses. The "Boron Anomaly" 138
7.4.1.1 Temperature Dependence of the Boric Acid Anomaly 139
7.4.1.2 Tendency Toward Immiscibility 141
7.4.1.3 Present State of Interpretations of the "Boron Anomaly" 144
7.4.2 Borosilicate Glasses 145
7.4.2.1 The Ternary System Na_2O–B_2O_3–SiO_2 145
7.4.2.2 Vycor-Type Glasses 149
7.4.2.3 Pyrex-Type Glasses 151
7.5 Glasses of High Lead Content 157
7.5.1 Glass Formation in Lead-Containing Systems 157
7.5.2 Phase Separation in Glasses Containing Lead 160
7.5.3 Structurally Conditioned Coloration of High-Lead Silicate Glasses 161
7.6 Phosphate Glasses 163
7.6.1 Structure of Phosphate Glasses 163
7.6.2 Phase Separation in Pure Phosphate Glasses 166
7.7 Tellurite Glasses 166
7.7.1 Glass-Formation Range and Optical Properties of Tellurite Glasses 166
7.7.2 Structure of Tellurite Glasses 172
7.8 Beryllium Fluoride Glasses – "Model Glasses" 174
7.8.1 Theoretical Discussion of "Model Glasses" 174
7.8.2 Ranges of Glass Formation in BeF_2 Model Systems. Properties of These Glasses 176
7.8.2.1 Density Plots 177
7.8.2.2 Refractive Index Plots 178
7.8.3 Phase Separation in Pure BeF_2 Glasses 180
7.8.4 Fluoride Glasses Free of Beryllium 181
7.8.5 Fluorophosphate Glasses 183
7.9 Zirconium Fluoride Glasses 184
7.9.1 Glass Formation, Structure and Properties 184
7.10 Germanate Glasses 187
7.10.1 Glass Formation from GeO_2 and Germanate Melts 187
7.10.2 Structure and Properties 188

7.11 Glasses Containing Arsenic Oxide . . . 189
7.11.1 Glass Formation . . . 189
7.11.2 Structure and Properties of Glasses of High Arsenic Oxide Content . . . 192
7.12 Glasses Containing Antimony Oxide . . . 193
7.12.1 Glass Formation and Some Important Properties . . . 193
7.12.2 Structure . . . 194
7.13 Glasses Containing Bismuth Oxide . . . 194
7.14 Limited Glass Formation in Systems of Exclusively Scientific Interest . . . 195
7.14.1 Titanate Glasses . . . 195
7.14.2 Vanadate Glasses . . . 196
7.14.3 Nitrate Glasses . . . 197
7.14.4 Carbonate Glasses and Glasses Based on $ZnCl_2$ . . . 199
7.14.5 Oxyhalide Glasses . . . 199
7.14.6 Oxynitride Glasses . . . 199
7.14.7 Oxycarbonate Glasses . . . 200
7.14.8 High-H_2O Glasses . . . 200
7.15 Metal Glasses . . . 200
7.16 Vitreous Carbon . . . 202
7.17 The Sol-Gel Method for Production of Glasses and Glass Ceramics . . . 203
7.17.1 Introduction . . . 203
7.17.2 The Alkoxide Sol-Gel Method . . . 203
7.17.3 The Silica Hydrosol Process . . . 204
7.17.4 The Ormocer Method . . . 205
7.17.5 The Importance and Application of Gel Glasses . . . 206

8 New Optical High-Performance Glasses . . . 208

8.1 Fundamental Principles of the Dispersion Behaviour of Glasses . . . 208
8.2 Change of the Dispersion with the Introduction of Additional Absorption Centers . . . 211
8.3 Optical Glasses with Unusual Partial Dispersions . . . 212
8.4 Athermal Optical Glasses . . . 216
8.5 Non-linear Refraction . . . 221
8.6 Prerequisites on the Raw Material for the Production of Optical Glasses . . . 222

9 Structure and Properties of Colored Glasses . . . 223

9.1 General . . . 223
9.2 Absorption of Colorless Base Glasses . . . 224
9.3 Glasses Colored by Ions . . . 226
9.3.1 Dependence of Absorption on Network-Former . . . 227
9.3.2 Dependence of Absorption on Modifiers . . . 228

9.3.3 Dependence of Absorption on the Valency of the Chromophore 229
9.3.4 Dependence of Absorption on the Coordination Number of the Chromophore 230
9.3.4.1 Coordination Change Due to Change in Chromophore Concentration 230
9.3.4.2 Coordination Change of Chromophore Due to Concentration Change of Network-Modifier 230
9.3.4.3 Coordination Change of Chromophore Due to Changed Network-Former 230
9.3.5 Problems of Interpretation 231
9.3.6 Technologically Important Chromophores and Selected Transmission Curves 231
9.4 Striking Glasses 235
9.4.1 Composition, Preparation, and Absorption Behavior 235
9.4.2 Base Glass Structure and Coloring Mechanism in "Striking" Glasses 236
9.4.3 Coloring Mechanism in Striking Glasses 241
9.4.4 Related Glasses with Other Chromophores 245
9.5 Glasses Colored by Metal Colloids (Ruby Glasses) 246
9.5.1 Composition, Fabrication, and Absorption Behavior 246
9.5.2 Structure and Coloring Mechanism in True Ruby Glasses 247
9.5.3 Silver Stain 249
9.6 IR-Absorbing Glasses (Heat-Absorbing Glasses) 250
9.6.1 Application of Heat-Absorbing Glasses and Absorption Behavior of Glasses Containing Fe^{2+} and Fe^{3+} Ions 250
9.6.2 Development, Production, and Properties of Heat-Absorbing Glasses 251
9.7 IR-Transmitting Glasses 254
9.7.1 IR-Transmission of Solids 254
9.7.2 IR-Transmission of Germanate, Tellurite, and Aluminate Glasses 257
9.7.3 IR-Transmitting Chalcogenide Glasses 258
9.7.3.1 Arsenic Sulfide Glasses 260
9.7.3.2 Other Chalcogenide Systems 261
9.8 Opacified Glasses 263
9.8.1 Mechanisms of Opacification 263
9.8.2 History and Classification of Opacified Glasses 266
9.8.3 Phosphate Opal Glasses 266
9.8.4 Fluorine Opal Glasses 272
9.8.5 Opal Glasses Based on SnO_2, TiO_2, ZrO_2, CeO_2, ZnO, and Other Compounds 274
9.8.6 Light Scattering and Color of Microdisperse Two-Phase Glasses 275

10 **Crystallization of Glasses** 280

10.1 General 280
10.2 Theoretical Considerations 281
10.2.1 Homogeneous Nucleation 281
10.2.2 Heterogeneous Nucleation 283
10.2.3 Crystal Growth 283

10.3 Crystallization as a Defect in Glass 287
10.4 Controlled Crystallization 290
10.4.1 Principles of Controlled Crystallization 290
10.4.2 Pioneering Developments at Corning Glass Works 295
10.4.2.1 Glass Ceramics with Minimal Coefficients of Thermal Expansion 295
10.4.2.1.1 Composition, Production, and Application 295
10.4.2.1.2 Structure and Properties 296

10.4.2.2 Machinable Glass Ceramics 300
10.4.2.2.1 General, Composition, Production 300
10.4.2.3 New Mica-Containing Glass Ceramics 302
10.4.2.4 Chain Silicate Glass Ceramics 302
10.4.2.5 Strengthening of a Special Glass by the Chemcor Process 302

10.4.3 Fundamental Investigations in the Development of Glass-Ceramics at the Otto Schott Institute of the Friedrich Schiller University in Jena 304
10.4.3.1 Nucleation and Crystallization Kinetics of a Base Glass from the $MgO–Al_2O_3–SiO_2$ System 305
10.4.3.2 Doping of the Base Glass with 11.2 mol % Fluorine Ions 306
10.4.3.3 High-Strength Glass Ceramics Containing Spinel 307
10.4.3.4 Single Doping of the Base Glass with 2–10 mol % TiO_2 (Ti_2O_3) also Leads to High Strength Glass Ceramics 311
10.4.3.5 Double Doping of the Base Glass with 11.2 mol % F and 5.2 mol % Na_2O Yields Machinable Glass Ceramics 313
10.4.3.6 Ferrimagnetic Glass Ceramics 324

10.4.4 Development of Bioglass Ceramics for Medicine 330
10.4.4.1 Introduction 330
10.4.4.2 Development of Bioglass Ceramics, Present State, Requirements and Targets 331
10.4.4.3 Development of Biocompatible and Machinable Glass Ceramics 332
10.4.4.4 Development of Bioactive Glass Ceramics 335

10.5 Bioactive, Piezoelectric, Phosphate Glass Ceramics Free of Silica 344

10.5.1 Development Trends of Phosphate Glass Ceramics 344
10.5.2 Structure and Crystallization Behavior of Phosphate Glasses 345
10.5.3 Development of Pure Biophosphate Glass Ceramics 346
10.5.4 Animal Experiments at the Academy of Medicine of Dresden on the Intergrowth Between Phosphate Glass-Ceramic Implants and Bones 352
10.5.5 Clinical Tests of the New Bioglass Ceramics on Humans 353
10.5.6 Summary and Outlook 358
10.6 Sintered and Special Glass Ceramics 360
10.6.1 Sintered Glass Ceramics 360
10.6.2 Special Glass Ceramics 361

11 The Strength of Glass 363

11.1 Theoretical Strength 363
11.2 Effective Strength: Attempts at Theoretical and Practical Explanations 364
11.2.1 Theoretical Concepts Regarding the Strength of Glass. 364
11.2.2 Experimental Investigations 365
11.2.2.1 A Demonstration of Griffith Flaws. 365
11.2.2.2 Strength After Elimination of Crude Surface Defects 368
11.2.2.3 Fatigue 368
11.2.2.4 Aging 368
11.2.3 The Strength of Glass Fibers 368
11.3 Strengthening Methods in Practice. 369
11.3.1 Tempering. 369
11.3.2 Compound Glass 369
11.3.3 Silicon-Organic Coatings 370
11.3.4 Dealkalizing 370
11.3.5 Ion Exchange: "Chemical Strengthening" 371
11.3.6 Multiple Layer Glass 371

12 Interaction Between High Energy Radiation and Glass 373

12.1 General Considerations 373
12.2 Photosensitive Glasses Based on the Formation of Metal Colloids 373
12.3 Photosensitive Glasses Based on Partial Crystallization in Lithium and Barium Silicate Systems. 375
12.3.1 Composition and Preparation 375
12.3.2 Structure, Properties, and Microprocesses 375
12.3.3 Special Properties and Applications, Photoform, Photoceram 376
12.3.3.1 Photoform 377
12.3.3.2 Photoceram 378

12.4 Dosimeter Glasses . . . 378
12.5 Photochromic Systems and Glasses . . . 381
12.5.1 Requirements of Photochromic Systems . . . 381
12.5.2 Combination of Photochromic Organic Compounds and Glass . . . 381
12.5.3 Inorganic Photochromic Glasses . . . 382
12.5.3.1 Development and Application . . . 382
12.5.3.2 Photochromic Glasses Activated by Rare Earths . . . 383
12.5.3.3 Borosilicate Glasses Doped with Silver Halides . . . 385
12.5.3.4 Borosilicate Glasses Doped with Silver Molybdate and Tungstate . . . 391
12.5.3.5 Borosilicate Glasses Doped with Copper or Cadmium Halides . . . 392
12.5.3.6 Thermally Darkening Photochromic Glasses ("TDPC") . . . 391
12.6 Laser Glasses . . . 392
12.6.1 Introduction . . . 392
12.6.2 Light Absorption and Light Emission in Solids . . . 392
12.6.3 The Solid Laser . . . 394
12.6.3.1 Laser Principle – Oscillator – Optical Pumps . . . 394
12.6.3.2 Mode of Operation of Lasers . . . 395
12.6.3.3 Properties of a Solid Laser Material . . . 396
12.6.4 Efficiency Increase by Sensitization . . . 399
12.6.5 Applications of Lasers . . . 399
12.7 Radiation Protection and Radiation-Resistant ("Protected") Glasses . . . 402
12.8 Transmission Changes of Colored Glasses under γ Irradiation . . . 404
12.9 Solarization . . . 405

13 Survey of the Physical Basis of Some Glass Properties . . . 408

13.1 Introduction . . . 408
13.2 Refraction of Light, Dispersion and Abbe's Value . . . 408
13.3 Density . . . 410
13.4 Molar Refraction . . . 411
13.5 Thermal Expansion . . . 413
13.6 Viscosity . . . 414
13.7 Strain . . . 418
13.8 Surface Tension . . . 421
13.9 Heat Conductivity, Specific Heat . . . 423
13.10 Electrical Conductivity . . . 424

References . . . 426
Index . . . 451

1 Historical Development of Glass Chemistry

1.1 The Beginnings of Glass Research

The history of glass is quite old, that of its scientific exploration very young. During formation of the earth, highly siliceous melts of rocks occasionally froze to natural glasses such as obsidian. Water, dissolved homogeneously under high pressure, evaporated rapidly if such melts penetrated the surface, e.g., in a volcanic eruption; and, when cooled quickly, the foaming melt became a solid natural glass foam, since high viscosity prevented the escape of bubbles. Modern industrial technologies may be considered variations of this basic process.

Glass was first produced by man about 4000 years ago, apparently in furnaces used in the ancient art of pottery, as, for example, in ancient Egypt. As with most materials, such as bronze or iron, the first application was to jewelry. However, the favorable shaping properties of glass were soon utilized in other fields, such as containers, windows, and lenses. Yet, during the long first period of application to life and progressing technology, the nature of glass remained unexplored.

For a long time, too, systematic studies of the relation of composition to properties and the development of new glasses remained inhibited by the inability to produce glass of sufficient homogeneity. For this reason, the first significant progress waited until about 1800, when Guinand and Fraunhofer devised special stirring methods. Thus, Fraunhofer was able to determine that a glass containing lead refracted and dispersed light quite differently than a glass containing lime.

Apart from his desire to further vary glass compositions, he recognized the necessity and precondition to exactly measure the refractive index of glasses at specific wavelengths. Thus he became the inventor of the spectrometer and provided a foundation for the entire field of optics. The correlations between glass composition and light refraction could be systematically and scientifically studied for the first time. Fraunhofer predominantly worked with seven elements in his melting experiments (not accounting for unintentional trace elements) which were, apart from oxygen, silicon, sodium, calcium, potassium, aluminum, lead and iron. The English pastor Harcourt and the famous physicist Stokes went much beyond the seven oxides of Fraunhofer for their choice of raw material for their glassy melts. They consequently laid the foundation of a "chemistry of glasses" in 1834. It is astonishing that Harcourt could as early as 1871 introduce 20 new elements in glassy melts. They included Li, Be, Mg, Sr, Ba, Zn, Cd, As, Sb, Sn, Tl, W, Mo, V, Ti, B, P and F. This was not only the beginning of glass chemistry, but also the first step anticipating later optical glassmaking, since a platinum crucible was used to avoid contamination. Hydrogen was developed in a lead bomb, led via a washing bottle and copper tubing to a spiral platinum point, where it was burned. The platinum crucible was suspended on platinum wires above the platinum nozzle and turned by clockwork to ensure uniform heating.

Not only could the melting temperature be varied within large limits with this arrangement but the use of the platinum crucible eliminated the otherwise uncontrollable contamination of the melt by crucibles based on fireclay. Unfortunately, Harcourt did not live to see the success of his more than 35 years of labor. The homogeneity of his 166 prisms did not suffice to measure dispersion with the required accuracy, and some of the glasses were hygroscopic and unstable. Yet, qualitative comparison suggested future production of telescopes free from chromatic aberration, particularly by combinations with glasses in which SiO_2 would be replaced by B_2O_3 or P_2O_5.

Apart from laying the foundations for a "chemistry of glasses," Harcourt also discovered the glass forming properties of boric and phosphoric acid. Research in this field was also done before, during and after Harcourt's work in France and in Germany. At the University of Jena the famous Chemist Döbereiner had attempted to introduce large amounts of barium, and later strontium, into glass as early as 1879. It was the German poet Goethe who strived for the establishment of a scientific glass foundry in Jena. He wrote to Döbereiner on the 28th of March 1829: "Your Honour has, by sending me the "strontian glass" samples, aroused in me the wish to contribute something for the further advancement of this great discovery. It would be of importance here to investigate the refraction and scattering properties of this glass. Should you be so inclined, to assist the court mechanic Körner in experiments of this kind with your helpful guidance, I would be more than inclined to meet the necessary modest expenses in order to, in my opinion, enjoy the result." These experiments led to the scientific knowledge that it is possible to alter considerably the properties of a glass by the addition of new elements. These experiments did not have an immediate practical significance.

1.2 History of the Chemistry of Optical Glass [2–8]

1.2.1 Ernst Abbe and Otto Schott

The decisive chapter in the history of optical glass as a practical product was shaped by two people: chemist Otto Schott and physicist Ernst Abbe, both at Jena, and their fortuitous collaboration.

Ernst Abbe was born on January 23, 1840 in Eisenach, Thüringia. He studied physics, mathematics and philosophy in Jena and later settled in Göttingen where he earned his doctorate. After some years as an assistant in Göttingen and lecturer in Frankfurt, he returned to Jena to qualify as a university lecturer. He became professor of mathematics and physics. When Abbe first

Fig. 1.1. Ernst Abbe. Born January 23, 1840 in Eisenach, Thuringia; died January 14, 1905 in Jena

met Otto Schott, he was already Director of the Observatory in Jena and was very famous for his work in the area of microscopy in collaboration with the University's optician and microscopist Carl Zeiß.

Jena owes to the trio of Abbe-Schott-Zeiß: the formation of the large concerns "Zeiss" and "Schott," the growth of the University and of the whole town. Ernst Abbe was not only an eminent University professor and researcher. This survey would be incomplete if one did not indicate the social attitude, unique in its time, of the trio Abbe-Schott-Zeiss which finally led to the Zeiss-Schott foundation. The foundation established relations for the workers of both enterprises which could not be found anywhere else in the world at the time.

Ernst Abbe died on January 14, 1905 in Jena. He was responsible for having recruited Otto Schott for cooperative work and having convinced him to move to Jena.

Otto Schott was born on December 17, 1851 in Witten, the sixth child of a large family. He was the descendant of a glassmaking family. His great grandfather Johann Schott (1730–1800) and also his grandfather Anton Schott (1767–1832) were glassmakers in a glass factory in the town of Harberg (Vogesen). The father Simon Schott (1809–1874), also a glassmaker in Harberg, later moved to Witten and founded a window glass plant with three businessmen of Witten. He purposely encouraged his son's career as a scientist, recognizing the need to supplement practical expertise by fundamental knowledge. Otto Schott studied at the Technical College of Aachen and at the Universities of Würzburg and Leipzig. He received his doctorate in 1875 from the University of Jena. The title of his thesis was "Contributions to the Theory and Practice of Glassmaking" and Fig. 1.3 shows the cover page.

Fig. 1.2. Otto Schott. Born December 17, 1851 in Witten, The Ruhr; died August 27, 1935 in Jena

Contributions to the Theory and Practice of Glassmaking

Glassmaking seldom was the topic of scientific investigations until now. The journal "The Glass Factory," which was demanded by those interested in this field, has already shed some light on the manufacturing of glasses, however more exact investigations are still very much required. I have tried to examine from a chemical and technological standpoint the use of raw materials and also some glassmaking processes, in particular melting and the transition from glass to liquid. This scientific reflection was based on independent observations in practice and in the laboratory as well as on the literature available and the opinions of specialists.

Visits to English, Scottish, French and Spanish glass factories appear to have also been of great significance. They have essentially increased his knowledge and his aptitude to make clear decisions. When Otto Schott began his first melting experiments in his father's sheet glass factory without the knowledge of the long forgotten experiments of Harcourt or Döbereiner, he had absolutely not thought about optical glass. As a pyrochemist he was more interested in the chemistry of high-temperature melts. He wrote in his first voluminous scientific paper (1880): "A general and methodical study of melting processes, which would sweep the entire inorganic world was not yet attempted. We are still missing a great deal in this field before we can be successful in confidently determining the reactions on the basis of concrete laws, as is the case for aqueous solvents at usual temperatures." Schott's first investigations did not involve glasses but salt melts. He actually first investigated the solution behavior of chlorides, fluorides, sulfates and carbonates in melting cooking salt. He deduced

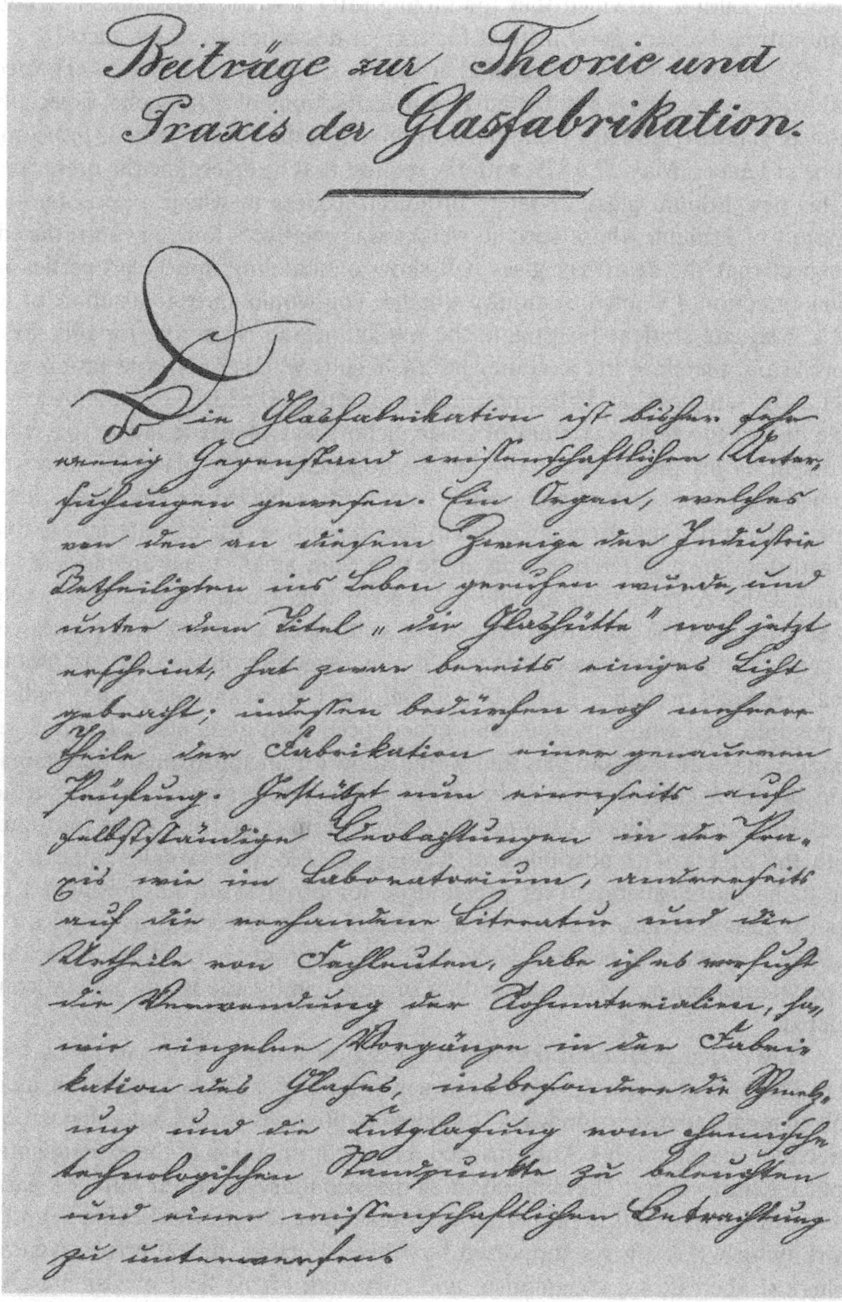
Beiträge zur Theorie und Praxis der Glasfabrikation.

Die Glasfabrikation ist bisher sehr wenig Gegenstand wissenschaftlicher Untersuchungen gewesen. Ein Organ, welches von den an diesem Zweige der Industrie Betheiligten ins Leben gerufen wurde, und unter dem Titel „die Glashütte" noch jetzt erscheint, hat zwar bereits einiges Licht gebracht; indessen bedürfen noch mehrere Theile der Fabrikation einer genaueren Prüfung. Gestützt nun einerseits auf selbstständige Beobachtungen in der Praxis wie im Laboratorium, andererseits auf die vorhandene Literatur und die Urtheile von Fachleuten, habe ich es versucht die Anwendung der Rohmaterialien, sowie einzelne Vorgänge in der Fabrikation des Glases, insbesondere die Schmelzung und die Entglasung vom chemisch-technologischen Standpunkte zu beleuchten und einer wissenschaftlichen Betrachtung zu unterwerfen.

Fig. 1.3. Title page of Otto Schott's thesis

that silicates, borates and phosphates could also form homogeneous melts in a similar fashion, provided that the mixing ratio was correctly chosen. Schott demonstrated a very good instinct for correct deduction from few facts.

Döbereiner especially dedicated himself to strontium, thus Schott expected extraordinary properties in the current investigations of glass melts, especially those of glasses containing lithium. He approached the already famous professor Abbe in Jena on May 27, 1879, with the request that he determine the properties of his new lithium glass: "I lately produced a glass in which a considerable amount of lithium, whose specific weight was relatively low, was introduced. I expect that the described glass will show outstanding optical properties in some direction. I wanted to inquire whether you would agree to examine or to let a graduate student investigate the refraction and scattering for this glass sample and therefore to ascertain from the results whether my expectations are met." The samples that Abbe immediately investigated proved to be unmeasurable due to too high a content of striae. Schott was therefore facing the same problem as his predecessors, the low homogeneity of the glass. He however immediately realised that homogeneity could be achieved for his glassy melts with sufficiently long stirring. He finally found a proper stirrer material capable of withstanding high temperatures in the thin white small stoneware pipes of the Dutch Tobacco pipes (which could be found in each grocer's store). Abbe wrote to Schott about the numerous samples he sent which were now well processed: "I consider it a great success that you have managed to obtain melt samples in small crucibles of such a quality that a complete optical analysis of the product is possible. Feil who is famous and experienced with glass melts, has not yet supplied me with melt samples that would allow for an approximate determination of the average scattering let alone a reliable assessment of the partial scattering such as I have obtained with your samples and still hope to obtain with the others. The possibility of making useable melt samples appears to me to be the most important prerequisite for progress in the production of optical glasses because it allows for methodical analysis. As long as samples are needed in a quantity between 60 and 80 pounds in order to obtain a useable experimental prism, systematic testing of new combinations can not be considered."

However a great disappointment for Schott followed this enormous success. The optical properties of Schott's new lithium glass had changed considerably in regard to dispersion when compared to previous glasses but in the wrong direction! In spite of this, Abbe insisted on the continuation of the experimental optical glass foundry. The elimination of the secondary spectrum with new glass compositions and optical properties was only one of the motives behind Schott's work (which was always supported by Abbe). Further optical errors such as spherical aberrations, astigmatism and curvature of the field of vision could similarly be reduced with new optical glasses.

It is characteristic of Otto Schott that he instinctively made the correct deductions as if "he could see into the heart of the glass." He also always attempted to reduce his results to fundamental laws and to base his work on these. He once wrote the following to Abbe:

"Investigations about atomic refraction were concluded in recent years which they supplied the proof that every element has a definite coefficient for the refraction of light. The refraction can be calculated in advance from these constants. The same behavior may exist in inorganic chemistry. I will first send you boric acid anhydride, phosphoric acid . . . for measurement."

This attempt, especially for boric acid, led to the glass works Schott & Gen. in Jena.

The most decisive work really began once Schott moved to Jena in 1881. He systematically tried to add all the accessible elements to his glassy liquids always taking into account the enormous world of the minerals. After only a few weeks, Abbe already wrote the following to Schott:

"Your melt specimens offer a diversity in the grading of optical character that could scarcely be expected with the uniformity of the ones known until now. The versatility of phosphoric acid is fabulous."

In a relatively quick succession the "glass chemistry laboratory" grew into a "technical glass institute" and finally into the Glass Works Schott & Gen. There were of course still numerous obstacles to overcome such as the technical handling of bigger melts, the heating system and furnace problems and the scale up from experimental results into production. The work had to be redone from the start in many cases.

Whereas Fraunhofer and Harcourt had to capitulate when they faced these problems, Schott solved them for the first time and led the way to future scientific fundamental glass research.

The first 15 kg melts produced caused extraordinary difficulties. Melts which until then were produced in small laboratory crucibles, often slightly crystallized in 15 kg quantities due to the much lower cooling rate after casting. They were no longer glassy after conventional processing.

Schott was even forced to interrupt the work on the new borate and phosphate glasses and to concentrate again exclusively on silicate glasses. Schott's tenacity and cleverness were shown by the fact that he did not capitulate when faced with difficulties but rather mastered them.

Now, what were the fundamental problems which led physicist Abbe, chemist Schott and microscopist Zeiss to cooperative work? A general and short summary of the development of optical image systems is in order here.

The telescopes of Galileo (1564–1642) consisted of lenses of one single glass type, crown glass, a (Na_2O–CaO–SiO_2) glass. The designation "crown glass" originates from the English because the glass was first blown to a crown-like image shape before flattening to the final shape. The colored borders obtained with these lenses gave rise to the demand for color free images, that is to make the lenses achromatic.

Newton (1643–1727) interpreted this defect and established the theory of astigmatism. He thought the correction of the color defect was impossible.

Around the end of the 17th century, flint glass, an alkali lead silicate glass, was discovered and produced in England. This designation also originates from the English referring to the fact that the most pure and colorless lead silicate glasses had been produced with the addition of flint rock as a silica source. Flint rock was found in great quantities on the southern english coast.

In 1729 Hall already produced the first relatively colour free lenses, or achromatic telescope objectives, by combining flint glass with crown glass. The apochromatic lenses of the London optician Dollond (1706–1765) became world famous. The basic problem facing Abbe and Schott was the elimination of the so-called "secondary spectrum" in lens systems, in particular in microscope systems. What is to be understood by "secondary spectrum"?

Since the refractive index varies with the wavelength for monochromatic light, white light is decomposed into the colors of the spectrum by going through a glass prism. The so called color scattering or dispersion varies however with glasses that are differently assembled (see Fig. 1.4). Lenses rather than prisms are predominantly present in optical instruments and systems. One can however imagine every lense as assembled from an infinite number of prisms (see Fig. 1.5). It is thus understandable that the dispersion of white light will also affect every lense.

By the use of lenses in optics one predominently desires to guide the light rays and thus obtain a true and sharp image of objects. However this is very often hindered by the unfavorable dispersion behavior of many glasses. Figure 1.6 shows how the rays emerging from a convergent lens illuminated with parallel monochromatic light are reassembled at the focal point. When the same lens is illuminated with parallel white light, the emerging rays reassamble at

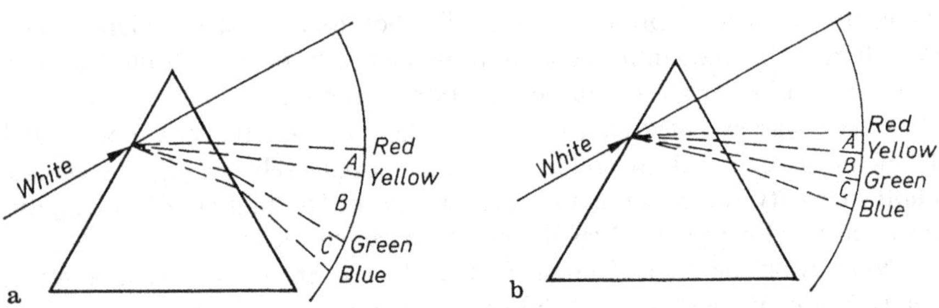

Fig. 1.4. Ray trace in a crown and a flint prism. Difference in dispersion of white light. **a** Flint glass prism; **b** crown glass prism

Fig. 1.5. Lens combination of converging and diverging lens. Each lens may be imagined as built up from an infinity of small prisms

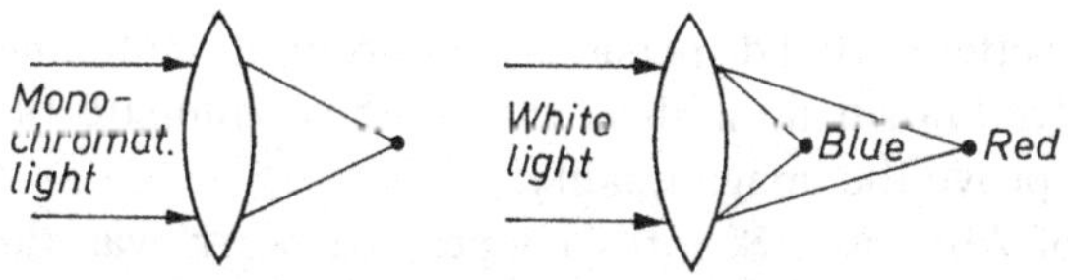

Fig. 1.6. Ray trace of monochromatic and white light in a converging lens

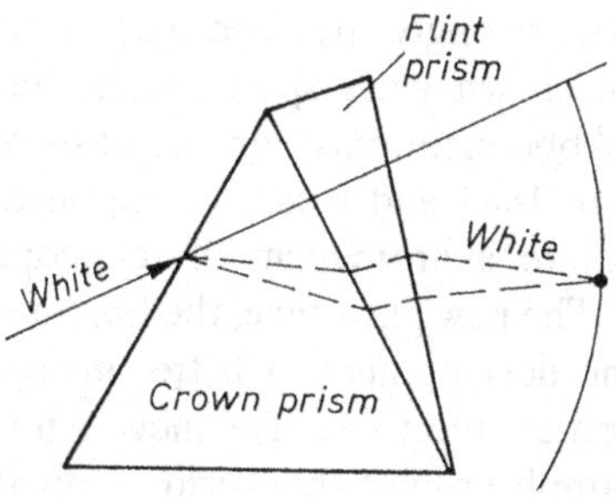

Fig. 1.7. Ray trace of white light in a combination of crown and flint prisms

different focal points for different wavelengths of the light. Sharp images therefore emerge in entirely different planes, or distances behind the lens, for light of different wavelengths. In other words: at a given distance behind the lens, numerous blurred images overlie a sharp image. This can be counteracted by using a combination of lenses, for example, a convergent crown glass lens and a divergent flint glass lens (see Fig. 1.5). The path of the light is thus the same as for the prism combination shown in Fig. 1.7.

Theoretically the dispersion of light in the crown glass prism should be compensated by the inverse addition of a flint glass prism. All rays should join again into white light. This is already possible through the addition of a flint glass prism which has a much smaller refractive angle than the crown glass because of the higher dispersion of flint glasses. The light path therefore changes direction after passing through the prism or lens combinations and objects can thus be reproduced. One decisive fact has however not been mentioned yet it concerns scientific problem facing Abbe and Schott at the beginning of their cooperation.

As shown in Fig. 1.4, the ratio of the refractive indexes for light of various wavelengths is not the same for crown and flint glass prisms. The border rays for red and blue can be completely rejoined by the proper choice of the angle of the flint glass prism used to correct the crown glass prism. For the entire spectrum, however, this would be possible only when the relative dispersion in the crown and flint prisms were identical (from A:B:C). This is however not the case. Not all colors thus join at the same focal point (see Fig. 1.6). Residual colors therefore border the image. The color of the image contour can be changed simply by raising or lowering the sample plate of the microscope, in other words by changing the focal plane. A fully sharp position is therefore not possible under these conditions. This phenomenon is known as "secondary spectrum." The elimination of the "secondary spectrum" is only possible if the partial dispersions of two glasses with very different mean dispersions (crown and flint glasses) were completely equal. Fraunhofer already recognized this. He did not succeed however in obtaining better partial dispersions in appropriate combinations of the seven glass oxides known until then and to thus eliminate the "secondary spectrum" and improve the image quality.

The first decisive result of Abbe and Schott's cooperative work was the discovery of borosilicate glasses. The introduction of boric acid to flint glasses known at the time produced a contraction of the strongly stretched part of the flint glass spectrum from blue to violet. The crown glass spectrum is thus approximated in terms of relative dispersion. Abbe mentioned, on the basis of the experimental melts which already existed in 1881 and which he supplied, that he considered the problem of the absolute achromatism of telescope objectives or of the secondary spectrum solved. The new glass type, the borosilicate glasses, had a decisive significance for the development of better microscopes, telescopes, binoculars and photographic objectives. The new Abbe microscope, free of the color defect, was a triumph around the world. A great frontier had been passed and new possibilities opened for the natural sciences and medicine. Robert Koch's identification of the pathogens of tuberculosis, cholera and malaria would never have been possible without the result of the cooperation of Abbe, Schott and Zeiß.

The elimination of the secondary spectrum by producing glasses with new compositions and optical properties was only one of the motives behind Ot to Schott's work which was always encouraged by Ernst Abbe. Further optical defects, such as spherical aberration, astigmatism and the curvature of the field of vision could similarly be reduced with new optical glasses. These successes would however not have been possible had Otto Schott not posessed the talent to quickly scale up laboratory results to production. He also had discovered a ceramic material which withstood attack from a liquid melt much better than the analogous materials known at the time and thus allowed the striae problem to be solved. Otto Schott also had the merit of making the "Siemens regenerative gas furnace" suitable for the glass industry, thus allowing higher temperatures to be reached. Schott's process of casting liquid glass into preheated molds is still known today as the Jena method. The first optical glass works of the world were established in 1884 in Jena. Schott's new optical glasses definitely refuted the opinion (which was described as the "law of the iron line" and was never broken before) that all glasses possess the same relative dispersion (Newton) or that the dispersion uniformly changes with the refractive index (see Fig. 1.8). Schott's new optical glasses possessed dispersion and special partial dispersion values which almost fully corrected the "secondary spectrum."

Schott's discovery of borosilicate glasses must be described as his greatest achievement not only because a revolutionary effect in the field of optics but they had also found applications as technical glasses in the chemical industry and in daily life.

Table 1.1 gives the basic composition of the most important optical glass families produced in the Schott & Gen Glassworks in Jena until 1939. Almost all of Schott's optical glasses developed in order to adjust for special dispersion ratios contain boric acid. The different special glass types played a special role. (see Table 1.1 Special glasses). Apart from the development of glass

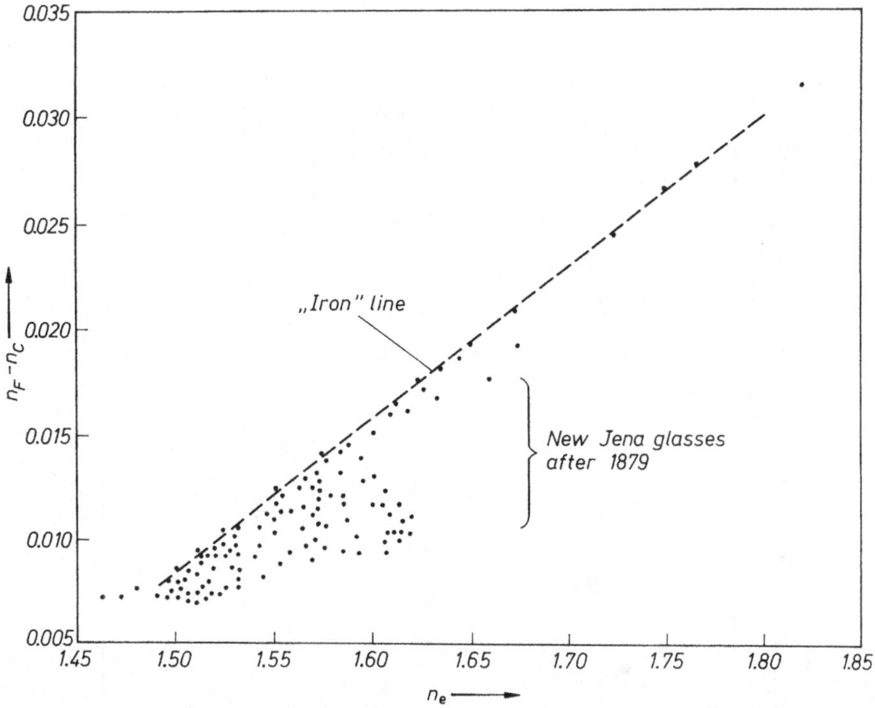

Fig. 1.8. Dependence of the refractive index n_e on n_c–n_f shown by the so-called "iron-clad lines." The new optical glasses of Otto Schott (after 1879) are the first to deviate from these lines

compositions near the limit of glass formation limiting production to small sizes, a wide range of glass formation in new systems such as those containing phosphoric acid were discovered.

In relatively quick succession, the great number of Jena optical glasses were born: borosilicate crowns (BK), fluor crowns (FK), phosphate crowns (PK), barium crowns (BaK), barium flints (BaF) and light barium flints (BaLF), the dense and extra dense crowns (SK and SSK), heavy phosphate crowns (PSK), dense barium flints (BaSF), dense flints (SF) and other optical glasses with extreme properties (see Fig. 1.9). Practically all of Schott's crown glasses are made from alkali borosilicate base glasses to which special oxides were added in order to modify their optical properties (as shown in Table 1.1). Until approximately 1939 the enlargement of the range of optical glasses was due to Schott's achievements.

1.2.2 Carl Zeiß and the Zeiß Foundation

Carl Zeiß was born on September 11th 1816 in Weimar, the fifth of twelve children. His father was an artist wood turning artisan. He completed his

Table 1.1. Jena optical glasses 1884–1939. Concentrations in mass %

Type	SiO_2	B_2O_3	$Na_2O + K_2O$	CaO	BaO	Al_2O_3	F	ZnO	TiO_2	PbO	Sb_2O_3
Borosilicate crowns	50 ... 70	10 ... 20	10 ... 20	+	+	+					
Fluor crowns	40 ... 60	10 ... 30	10 ... 20			5 ... 20	5 ... 10				
Phosphate crowns	40 ... 75	10 ... 20	10 ... 20		+	+	+	+			
Barium crowns	40 ... 60	+	5 ... 10		15 ... 30	+		5 ... 15			
Zinc crowns	50 ... 70	+	5 ... 20	+		+		5 ... 25			
Dense barium crowns	30 ... 50	5 ... 20	+	+	30 ... 40	+		+	+		
Extra dense barium crowns	30 ... 40	5 ... 10	+		30 ... 50	+		+		+	
Heavy phosphate crown	40 ... 60	10 ... 20	5 ... 15		20 ... 30	+					
Light barium crowns	60 ... 70	10 ... 20	+		5 ... 15			+			
Crown flints	50 ... 70	+	5 ... 20				+	+	+	5 ... 20	
Barium flints	30 ... 50	5 ... 10	0 ... 10	+	10 ... 40	+		+	+	5 ... 20	
Dense flints	25 ... 50		+			+				50 ... 70	
Dense barium flints	30 ... 45	5 ... 10	5 ... 10	+	10 ... 40	+		+	+	10 ... 40	
Light flints	50 ... 60		5 ... 15	+	+					30 ... 40	
Light barium flints	45 ... 60		5 ... 15		10 ... 25			5 ... 15		5 ... 15	
Extra light flints	50 ... 60		5 ... 15	+		+		+		20 ... 30	
Short flints	40 ... 55	10 ... 20	5 ... 10			+		+			10 ... 25

Special Glasses	SiO_2	B_2O_3	P_2O_5	Al_2O_3	MgO	$Na_2O + K_2O$	BaO	PbO	ZnO
Special phosphate crown		+	+ +	+	+	+			
Special heavy phosphate crown		+	+ +	+			+		
Special short flint		+ +		+		+	+	+	+
Speical dense flint				+		+		+ +	

\+ This component is contained in this particular glass.
\+ + Contained as a major constituent.

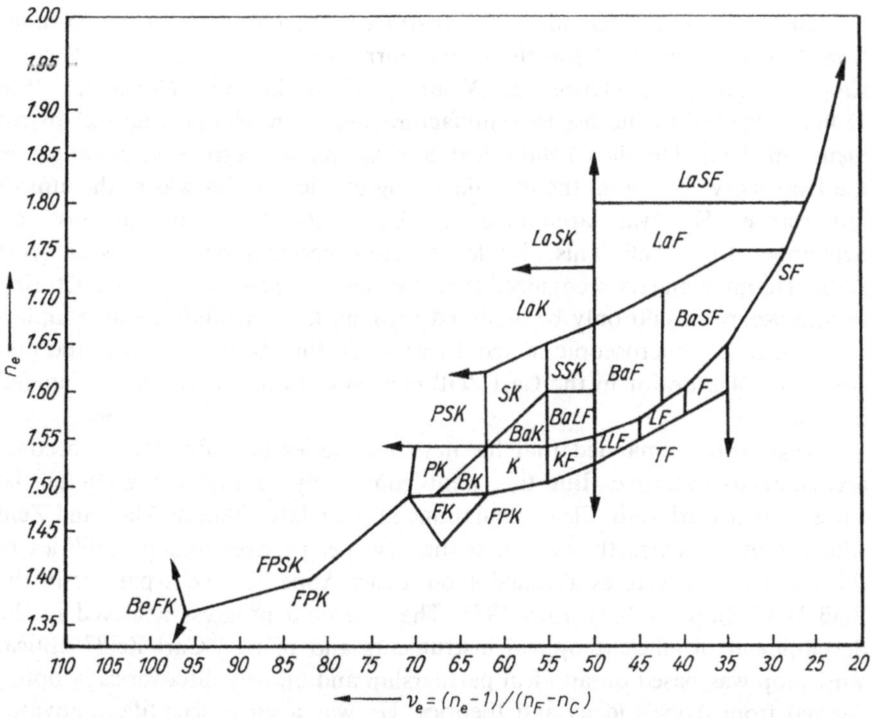

Fig. 1.9. Delineation of various optical glass types in the n_d–v_d diagram

Fig. 1.10. Carl Zeiß. Born September 11, 1816 in Weimar; died December 3, 1888 in Jena

apprenticeship as a mechanic with Friedrich Körner who also received funds from Goethe. The usual travels of the journeymen led him to optical workshops in Stuttgart, Darmstadt, Vienna and Berlin. On November 19th 1846 he received the license to manufacture and sell mechanical optical instruments in Jena. He thus established a close relationship with scientists of the University, for whom the manufacturing of microscopes was of the utmost importance. He was appointed as University technical specialist on September 10th 1860. This also led to close cooperation work with Ernst Abbe. He immediately recognized that decisive progress in the manufacturing of microscopes could only be achieved through fundamental scientific understanding of the microscopic image. Ernst Abbe thus became in 1866 the first scientific collaborator in the Carl Zeiß mechanical optical workshop founded in 1846.

Ernst Abbe demanded that his new discoveries be published to become accessible to everyone. But this went completely against the grain of the entrepreneur Carl Zeiß. Clear regulations of the relationship of Abbe and Zeiß who absolutely correctly estimated the new perspectives and possibilities of this cooperation were established soon. Ernst Abbe became a partner in the Zeiß Workshop on May 15th 1875. The all-round progress achieved in the development of efficient optical instruments and of the "Carl Zeiß" optical workshop was based on an ideal partnership and on new discoveries in optics derived from Abbe's ideas and theories. He was a great scientific innovator who also won Otto Schott's cooperative work. He once declared "Carl Zeiß and I would have had to limit ourselves to half of our agenda if Otto Schott hadn't joined. He was an inventive and energetic man whose dim view of his own profession led him to us at the beginning of the eighties." Ernst Abbe and Carl Zeiß equally participated in the foundation of the Schott Glass Works in 1884. The triumph that the trio of physicist, chemist and technical specialist henceforth enjoyed around the world with their new microscopes (developed with new glasses and scientific calculations) can be accounted for by their comradely cooperation, which still serves as a model for us today.

Carl Zeiß died on December 3rd 1888 in Jena. The Carl Zeiß foundation was quickly established by Ernst Abbe on May 19th 1889. It created social conditions for its members and the workers which were unique in its time. Abbe once said: "To be social means to work, and to work for the society. As long as we are living in capitalist age and as long as an entrepreneur is forced by the economic system to produce capitalistically, a profit must be made. When this is not the case, his managers have only worked and led poorly. The enterprise only becomes social if putting its net profit, like the foundation, at the disposal of the state, that is of culture. If all industries would do so, the surplus of means for cultural aims would be such that a large part of the social missions of a better state, than the present, would be solved without much else." The Zeiß Works and half of the Schott Works joined the foundation on June 30th 1891. In 1919 the entire Schott Works joined.

1.2.3 The Development of New Optical Glasses after 1939

A patent was registered by G. W. Morey (Kodak Company, USA) in 1936 in the USA, Germany, England, France and Switzerland. The patented compositions were completely different of the optical glass compositions known at the time (see Table 1.2) [6–8]. The amounts of the glass-former B_2O_3 were considerably reduced while the oxides of the rare elements La, Th, Zr, Nb, Ta, W, Mo etc. were introduced as main glass components. Abbe had introduced the so called Abbe ratio $\nu_e = (n_e - 1)/(n_F - n_C)$, which is also described as the relative reciprocal dispersion, for the characterization of the optical properties of a glass. The glass' basic optical properties or its performance are clearly characterized when the Abbe ratio is plotted against the refractive index of the glass for the wavelength of the e-line.

The optical positions obtained with the new glasses opened up a completely new region in the $\nu_e - n_e$ plot (see Table 1.2 and Fig. 1.11). A low viscosity and a tendency to crystalise as well as an unusual aggressiveness towards every ceramic crucible material did not allow these glasses to be produced in large

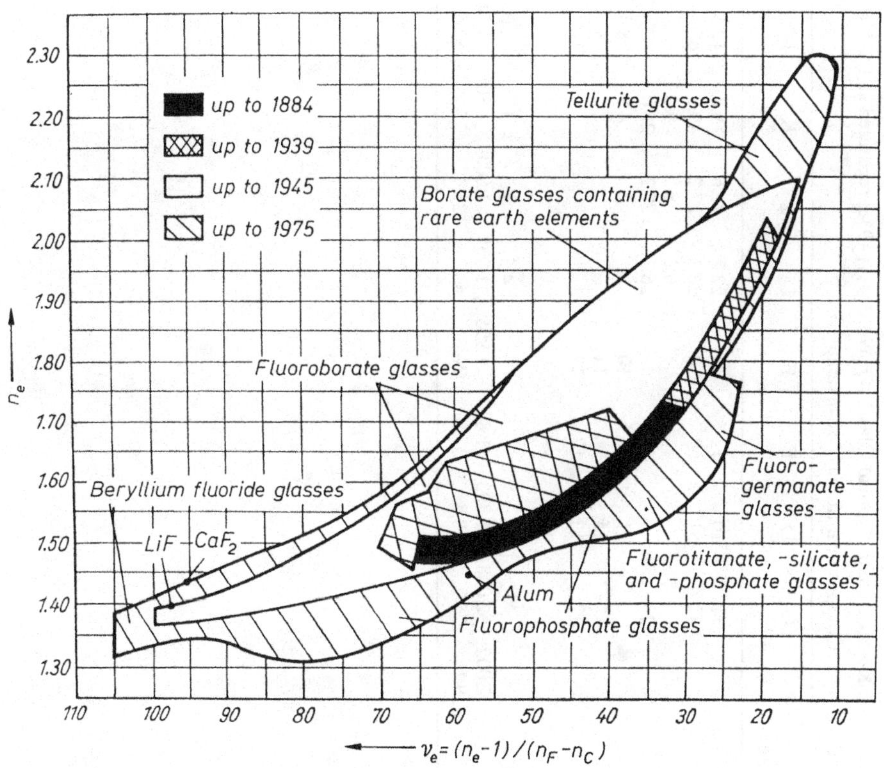

Fig. 1.11. State of development of optical glass, characterizing an optical glass by its location in the n_e–υ_e diagram, as proposed by Abbe

Table 1.2. Basic Kodak patent "Optical Glass" (G. W. Morey) 1936, registered as DRP 691 356, Brit. pat. 462 304, US pat. 2 206 081, French pat. 810 442, and Swiss pat. 206 664. Composition in mass % (examples)

Glass Component	A	C	D	E	F	I	J	K	L	M	P	R	S	T	U	V
B_2O_3		12	10	12.5	14	22.7	22.6	9.8	12.5	16.6	20	33.4	40	26	22	20
La_2O_3		42	51	60	50	26.3	27.2	49.2	60	37.5	36	22.2	60	33	28	28
ThO_2						20.2	8.5	9.8		16.7	16	22.2		41	14	14
Ta_2O_5	50	28				26.3	27.2	23.0		29.2	28	22.2			28	28
ZrO_2		6	11	8.5	10			4.9	8.5							
WO_3			15		14											
TiO_2	50	12	13	17	12				17							
Li_2O															8	10
Na_2O				2												
$Na_2B_4O_7$						4.5	4.5	3.3	2.0							
n_d	2.002	2.008	1.893	1.842	1.995	1.805	1.800	1.809	1.848	1.898	1.85	1.7175	1.7227	1.7667	1.8119	1.8037
ν_d	19.1	25.2	38.8	35.5	26.6	40.3	38.4	35	32.5	36.9	42	53.5	54.1	51.4	41.2	42.4

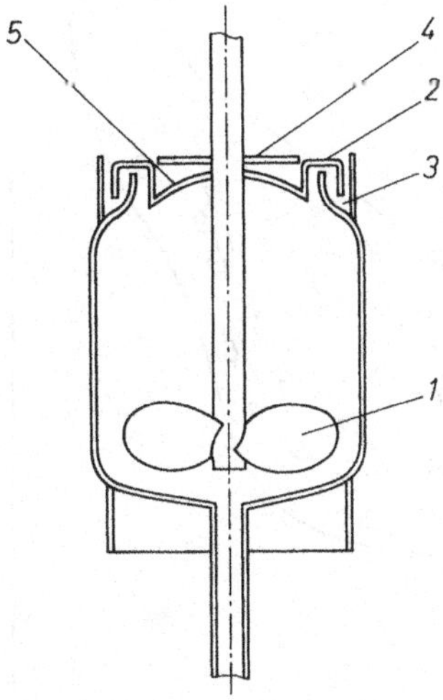

Fig. 1.12. Inductively heated platinum melt crucible with lid used for the production of thin running, easily vaporizable and easily crystallizable optical glasses. The lid (*2* and *5*) and washer (*4*) are firmly attached to the agitator (*1*) in the glass (*3*)

sizes initially. This period of development of new optical glasses was characterized by the introduction of entirely new melting and forming techniques including electrical heating. For the first time the production of the new glasses at a technical scale had been made possible.

Figure 1.12 shows a platinum melt crucible (used today) which is inductively heated and hermetically closed, and is used especially for the production of easily vapourizable, fluid and easily crystallizable optical glasses with extreme optical properties [9]. The lid with lining goes into the glass stirrer which holds the liquid glass. The stirrer has the same composition as the glass to be produced. While the first decisive progress in the sector of optical glasses, during the Jena developments, had been reached with a relatively small number of known elements (see Table 1.1), the second development period initiated by Morey's fundamental USA patent is known for the use of almost exclusively new elements.

The chemical industry plays a fundamental role in the development pace since it has to produce the completely new raw materials at a satisfactory purity level and at reasonable prices. Only once these requirements were met, could the second decisive development step of optical glass be initiated. The large number of Jena optical glasses can be clearly reduced in terms of their composition to the two types: crown, (low dispersion, $\nu_e > 50$) and flint (high dispersion, $\nu_e < 50$) (the crown family essentially consists of alkali borosilicate basic glasses with barium and zinx oxide as the main component). The main feature of almost all new high refractive index glass types is the basic binary glass system La_2O_3–B_2O_3 (see Table 1.2). The main component lanthanum is therefore directly used in the nomenclature of the new types lanthanum crown and lanthanum flint as was the case for barium in Schott's barium crown (BaK) and barium flint (BaF) or for zinc in Schott's zinc crown (ZnK). ThO_2 and Ta_2O_5 are other main components of the new high refractive index glasses (see Fig. 1.11). The most different new developments in optical glass during the final years of

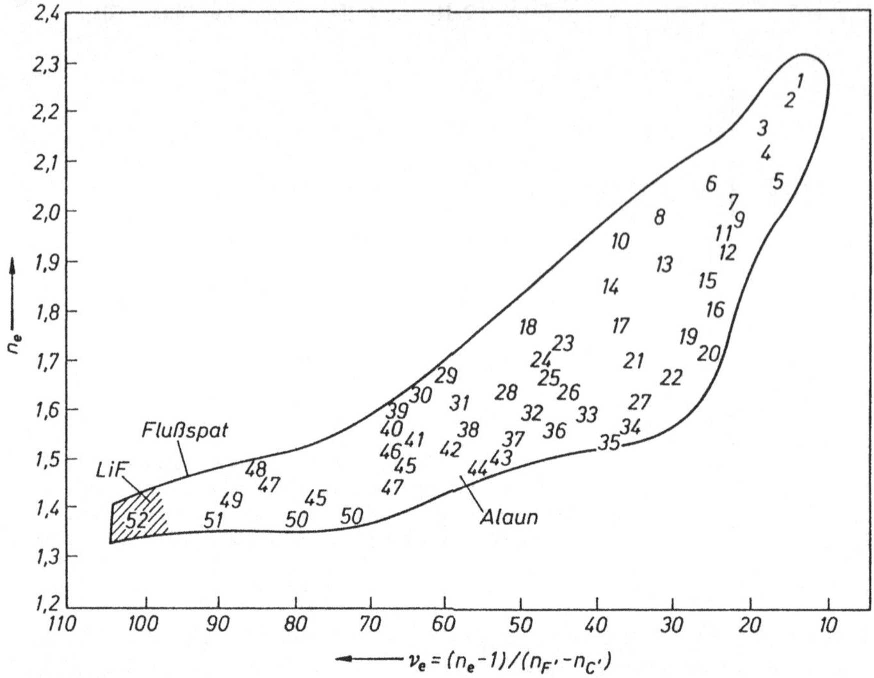

1 TeO_2-PbBr_2
2 $TeO_2-PbO-PbOCl_2$
3 $TeO_2-La_2O_3$
4 $TeO_2-Al_2O_3-MgO-SrO-BaO-ZnO$
5 $PbO-CdO-ZnO-B_2O_3$
6 $TeO_2-Ba(PO_3)_2-Pb(PO_3)_2$
7 $La_2O_3-TiO_2-Ta_2O_5-B_2O_3$
8 $La_2O_3-ThO_2-Ta_2O_5-B_2O_3$
9 $PbO-SiO_2$
10 $ThO_2-Gd_2O_3-Ta_2O_5-Nb_2O_5-B_2O_3$
11 $La_2O_3-TiO_2-ZrO_2-WO_3-B_2O_3$
12 $PbO-SiO_2$
13 $La_2O_3-TiO_2-B_2O_3$
14 $La_2O_3-Nb_2O_5-Ta_2O_5-ThO_2-B_2O_3$
15 $Sb_2O_3-Al_2O_3-SiO_2$
16 $PbO-SiO_2$
17 $ZnO-MoO_3-WO_3-CdO-B_2O_3$
18 $La_2O_3-ThO_2-CdF_2-B_2O_3$
19 $PbO-SiO_2$
20 $NaF-TiO_2-GeO_2$
21 $Na_2O-BaO-PbO-B_2O_3-SiO_2$
22 $PbO-SiO_2$
23 $Li_2O-ZrO_2-La_2O_3-Ta_2O_5-B_2O_3$ $-(SrO, CaO, BaO)$
24 $BaO-B_2O_3-SiO_2-La_2O_3-ThO_2$
25 $La_2O_3-ThO_2-BaO-CaO-MgO$ $-B_2O_3$
26 $Na_2O-BaO-PbO-B_2O_3-SiO_2$
27 $Na_2O-BaO-PbO-B_2O_3-SiO_2$
28 $BaO-B_2O_3-SiO_2$
29 $CaF_2-La_2O_3-B_2O_3$
30 $CaF_2-BaO-B_2O_3$
31 $BaO-B_2O_3-SiO_2$
32 $Na_2O-BaO-ZnO-PbO-SiO_2$
33 $Na_2O-PbO-SiO_2$
34 $NaPO_3-Sr(PO_3)_2-K_2TiF_6$
35 $KF-PbF_2-AlF_3-NaPO_3-Mg(PO_3)_2$ $-Al(PO_3)_3$
36 $Na_2O-PbO-SiO_2$
37 $Na_2O-K_2O-CaO-SiO_2$
38 $Na_2O-B_2O_3-SiO_2$
39 $Na_2O-BaO-B_2O_3-SiO_2$
40 $BaO-B_2O_3-P_2O_5-Al_2O_3$
41 $Na_2O-B_2O_3-SiO_2$
42 $Na_2O-K_2O-CaO-SiO_2$
43 $KF-AlF_3-K_2TiF_6-Sb_2O_3-PbO$ $-B_2O_3-SiO_2$
44 $NaF-KF-PbF_2-BeF_2-Pb(PO_3)_2$
45 $Na_2O-B_2O_3-SiO_2-Al_2O_3-F$
46 $BaO-B_2O_3-P_2O_5-(Al_2O_3)$
47 $LiF-NaF-KF-ZnF_2-MgF_2-BaF_2$ $-PbF_2-AlF_3-LaF_3-ThF_4-Al(PO_3)_3$
48 $CaF_2-AlF_3-MgF_2-SrF_2-BaF_2$ $-Me^{II}(PO_3)_2-Me^{III}(PO_3)_3$
49 $NaF-LaF_3-CeF_3-BeF_2-Be(PO_3)_2$ $-CePO_4-La(PO_3)_3$
50 $NaF-BaF_2-BeF_2-NaAlF_6$ $-Al_2(SiF_6)_3-NaPO_3$
51 $NaF-BaF_2-BeF_2-NaPO_3$
52 $KF-AlF_3-BeF_2-CaF_2-SrF_2$ $-BaF_2-CdF_2-LaF_3-CeF_3-ThF_4$

Fig. 1.13. Glasses of the mentioned base glass system lead to the corresponding position in the n_e–ν_e Abbe diagram

World War II in Germany and later after 1945 in Japan, were more or less derived from Morey's patent.

Figure 1.11 shows the advances in the development of optical glasses characterising them with refractive index and dispersion in Abbe's diagram. Until approximately 1884, optical crown and flint glasses were exclusively based on the components Na_2O/K_2O, CaO, PbO and SiO_2. During Schott's glass developments, B_2O_3, BaO, ZnO, TiO_2, Al_2O_3, F, P_2O_5, MgO and Sb_2O_3 were added to the list. Figure 1.11 shows the rapid advance until approximately 1945 through the use of almost all the elements of the periodic table in particular La_2O_3, ThO_2, Nb_2O_5, Ta_2O_5, WO_3, ZrO_2, CdO and MoO_3. Further decisive progress in the enlargement of the $\nu_e - n_e$ diagram was achieved only with the beryllium fluoride, germanate and tellurite glass types.

The main glass systems which compose the optical glasses in question are shown in the $\nu_e - n_e$ diagram in Fig. 1.13 for comparison with Figs. 1.9, 1.11, 1.13 [8]. Any further development of optical glasses by introducing new elements are limited since most possibilities have been exhausted. Further developments were therefore addressing just properties such as the improvement of chemical resistance through the application of protective coatings, unusual dispersion properties and athermal properties.

1.3 History of Technical Glass

The history of technical glass is closely associated with that of optical glass. After all practically all of Schott's technical glasses originated from the investigations of glass forming systems responding Abbe's suggestions for the improvement of the resolution of optical instruments. The production volume of technical glasses became much larger than that of optical glasses and, in turn, encouraged the production of optical glass on a large scale.

As already mentioned, Schott's decisive step was the use of the glass forming boric and phosphoric acids. He wanted to eventually combine all three, silica, boric acid and phosphoric acid in one melt and to obtain new optical properties by varying them. However it quickly became evident, as Schott wrote:" . . . that SiO_2 and B_2O_3 were compatible, while P_2O_5 exhibits hostile behavior. While SiO_2 and B_2O_3 as a 'glass forming double acid accept bases in any proportions, the addition of P_2O_5 almost always causes milky opalescence, apparently because phosphates are as insoluble in the melt as oil in water."

This fundamental investigation if glass melts along with the recognition that almost every mixture with other oxides $SiO_2 + B_2O_3$ glass formation, was later pursued by Otto Schott in every direction. Thus the new borosilicate glasses were developed. Schott's discovery must be described as most important as far as the consequences for the whole field of technical glass and the growth of the Jena glass involves.

At that time the change from petroleum lamps to gas lighting was hindered by the fact that almost all lamp cylinders bursted due to the strong temperature changes occurring with the lighting and cooling of the lamp. Schott solved the problem with his thermally resistant cylinder glass. Thus nothing hindered the promotion of gas lighting. The old lamp cylinder glass was a Na_2O–CaO–SiO_2 glass while the new cylinder glass was a thermally resistant borosilicate glass.

	Old Cylinder Glass	Schott Cylinder Glass
SiO_2	72	66
CaO	10	-
Na_2O	18	6
B_2O_3		24
Sb_2O_3		4

Similarly a pressing demand for thermally and chemically resistant glass for laboratory ware came from the chemical laboratories. Whereas earlier apparatus glass bursted with too rapid heating because of a high coefficient of expansion, the new "Jena Apparatus Glass," also known as "Jena Glass," stood up to such pressures. It additionally showed an excellent hydrolytic resistance. No laboratory and chemical enterprise is conceivable today without the "Jena Glass." This development originally was based on the borosilicate glasses produced for optical applications. Further developments of technical borosilicate glasses are shown in Table 1.3. This table also shows that the technical borosilicate glasses of Pyrex type originated from the "Jena Apparatus Glass 20." It is this type of glass which is produced world wide today and which has the largest significance.

A further milestone in Schott's successful career was the discovery of the Jena Normal Glass for thermometers right at the beginning of his work in Jena. Glass thermometers are still used today for most physical temperature measure-

Table 1.3. Technical borosilicate glasses

Composition in Mass %	Jena Apparatus Glass 20	Jena Duran	Jena Rasothern	Pyrex Glass (USA)
SiO_2	76	74	78	80.8
B_2O_3	7	14	12.5	12.0
Na_2O	6.5	4.5	5.5	4.3
BaO	4.0	3.0	—	—
Al_2O_3	4.5	3.5	3.0	2.2
Fe_2O_3, CaO, As_2O_3, F, K_2O	Traces	Traces	Traces	Traces
Thermal expansion $\alpha \cdot 10^7\,\text{grd}^{-1}$	$\alpha_{20-100} = 46$	$\alpha_{20-100} = 38$	$\alpha_{20-100} = 33$	$\alpha_{20-300} = 33$
T_g in °C	550	534	527	560

ments. The earlier glass thermometers were afflicted with a large measurement error which could no longer be accepted in exact scientific research. The mercury line of the thermometer was observed to no longer return to its starting point once the thermometer had been used for higher temperature measurements. The glass had expanded with heating and did not return to its initial state with cooling. The glass sphere which holds the mercury remained expanded. Since the mercury found more room in the glass sphere, the mercury line does not read 0 °C in ice water but reads negative values. The error was as high as 0.7 °C heating to 100 °C and considerably larger upon higher heating. The error also gradually decreased upon storage until eventually a reading of 0 °C for ice water could be obtained. Since used thermometers were rarely kept stored for a long period, the measurement error had to be accepted.

After a few melting experiments, Otto Schott had discovered that the simultaneous presence of sodium and potassium in the old thermometer glass was responsible for the ice point depression. The "Jena Normal Glass 16^{III}" was therefore developed and showed an insignificant ice-point depression of $8/100^{th}$ of a degree Celsius. The borosilicate "Jena Thermometer Glass 59^{III}" followed and exhibited a still smaller expansion and an ice-point depression of $4/100^{th}$ of a degree Celsius. The latter can also be used for temperature measurements of up to 500 °C, therefore with dark red embers. Both thermometer glasses are still in standard use today. The so called "Supremax Glass 2950^{III}" was developed later and exhibits even slightly better values than the 16^{III} and the 59^{III}. Table 1.4 gives an overview of the development of thermometer glass.

A whole series of Schott's pioneering discoveries could still be quoted such as glasses with increased UV transmission, compound glasses and glasses with special electrical properties among others. In many cases, such as in the case of pyrex glass, the level of performance attained was so high, that further significant improvements are quite unlikely.

Table 1.4. Development of Jena thermometer glasses

Composition in mass %	Jena normal glass 16^{III} (1985)	Jena thermometer glass 59^{III}	Jena thermometer glass 2950^{III} (supremax)
SiO_2	67	72	53
B_2O_3	2	12.0	10
Na_2O	14.5	11.0	
CaO	7.0		5.0
ZnO	7.0		
Al_2O_3	2.5	11.0	21.0
MgO			10.0
Zero point depression in °C	0.08	0.04	0.03
Area of applicability to °C	360	510	625

2 Freezing of a Melt to a Vitreous Solid

2.1 Fusion and Crystallization. General

At a certain temperature (the melting point, T_m), vibrating lattice elements of a crystalline solid will no longer return to sites bonded to their neighbors. The crystal lattice collapses and the crystal loses its shape; it changes to the liquid state, i.e., it melts. Unless additional energy is supplied, sufficient to remove all elements from lattice sites, the system maintains equilibrium: As many elements as enter the melt return to the remaining orderly lattice. In this state of equilibrium, a crystal will neither grow nor melt completely.

If, however, additional thermal energy is supplied to the system at the melting point, this energy will be used exclusively to remove further elements from their sites while the temperature remains constant at T_m (Fig. 2.1). The internal energy is increased by what is called the heat of fusion. If a liquid freezes, the reverse process takes place. At the melting (or freezing) point T_m, the entire heat of fusion consumed in the melting process is released as heat of crystallization.

The freezing of a melt to a non-crystalline (vitreous, glassy, amorphous) solid represents a deviation from this "normal" (thermodynamically required) freezing process. Only in a special case of solidification is the lowest energy not attained.

Numerous observations suggest that a liquid will solidify to a glass, if the mobility of the elements that should constitute the lattice is sufficiently restricted to hinder their orderly arrangement in the form of nuclei, during the time available in the cooling process.

In silicate glasses, for example, this impediment is caused by the rapid increase in viscosity on cooling, due to the tendency of silicate groups to form 3-dimensional networks. In organic compounds, polymerization and crosslinking induce such increases in polymerization and lack of crystallinity. It is, for instance, difficult to crystallize viscous sugar solutions. If a steep viscosity-temperature relation is associated with a melting (freezing) temperature in the low viscosity region, glass formation is hindered, and it has become customary to call such glasses "fragile" (e.g. zirconium fluoride glasses) (Angell [10]).

Time plays a decisive role in the formation of criteria of glass formation. The length of time a melt dwells in a temperature range favoring nucleation or growth of crystals is important. For this reason, glass formation depends critically on the rate of cooling. According to the prevailing theories of glass formation, therefore, any material (gaseous or liquid), not just typical "glass-formers," can be frozen to a non-crystalline (vitreous, glassy, amorphous) solid,

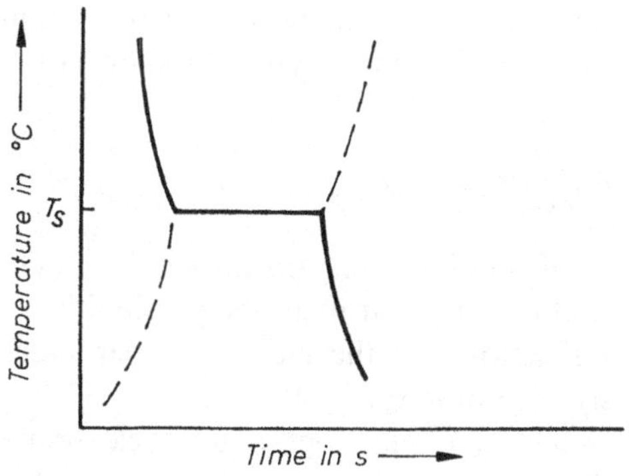

Fig. 2.1. Cooling and melting curves for simple solids

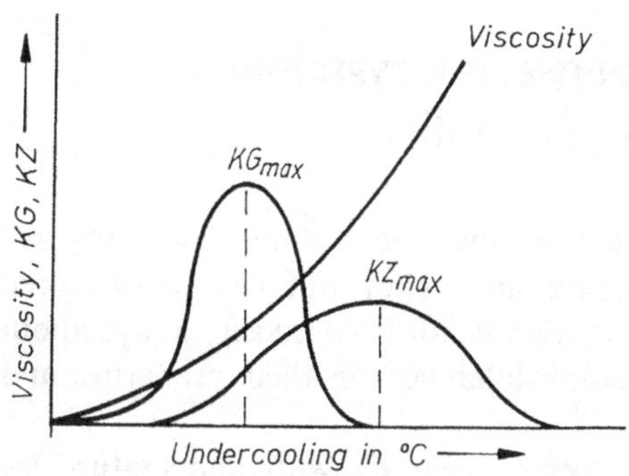

Fig. 2.2. Dependence of nucleation, crystal growth, and viscosity on undercooling of a melt which solidifies to a glass easily. As a rule, the maximum for nucleation is somewhat broader than that for growth. (K Z nucleation K G groth).

provided that the material is cooled quickly enough to suppress nucleation and growth of crystals. In fact, the recent attainment of cooling rates of 10^6 K·s^{-1} and more has allowed us to obtain small volumes of wires or ribbons and even fluid metallic melts (metal glasses) in the glassy state.

The first investigations on cooling behavior and glass formation were carried out by Tammann [11–13]. According to his now classical investigations, the behavior of the melt (crystallization or glass formation) depends on two parameters: (1) the number of "nuclei" of crystals forming in unit time (nucleation frequency, I), and (2) the linear growth rate in cm·s^{-1} (u) of these nuclei.

Both factors depend on the magnitude of undercooling ($T_m - T$), first increasing under an increasing (thermodynamic) driving force below the melting point and then decreasing at lower temperatures due to kinetic hindrance as the viscosity increases. Plots of I and u vs temperature thus have maxima, and are similar but not equal. Usually the maximum for I occurs at a lower temperature and is broader (Fig. 2.2). Of course, in this case a large separation of the I and u curves favors glass formation, since in the higher T range favoring growth u, not enough nuclei are available to grow. Conversely, for closeness of I and u maxima, easy crystallization is to be expected.

Dietzel [14] later proposed to introduce the inverse of the maximum crystal growth rate, KG_{max}, as a measure of the tendency to glass formation of a melt:

$$\text{Tendency to glass formation} = \frac{1}{KG_{max}}$$

This first attempt at evaluating glass formation is very insufficient in light of today's knowledge of the glassy state and in particular of all the possibilities of influencing glass formation by various actions on the melt. This particular problem remains the subject of intensive exploration.

Tammann also recognized that impurity ("inhomogeneous nucleation") may radically affect crystallization. Thus, purified water may be undercooled to $-30\,°C$.

2.2 Significant Differences between Crystalline and Non-crystalline (Glassy) Solids

Both crystals and glasses are solids; that is, constituent atoms essentially will no longer respond to stress by changing sites. They will just be displaced within their ranges of vibration. The common factor is a viscosity above about 10^{13} Pa. However, there are fundamental differences in their properties and behavior.

If we inspect the change of the specific heat C_p with temperature for crystalline and vitreous solids of equal composition, the features illustrated in Fig. 2.3 are observed. Regardless of whether melts had originated from a crystalline or vitreous solid, above the melting point T_m they are identical and show the identical decrease of C_p with temperature.

If at T_m the melt freezes to a crystal, C_p discontinuously decreases to the much lower value of the crystalline solid. On further cooling, a much less pronounced linear decrease of C_p follows until, near absolute zero temperature, the Debye T^3 relation holds.

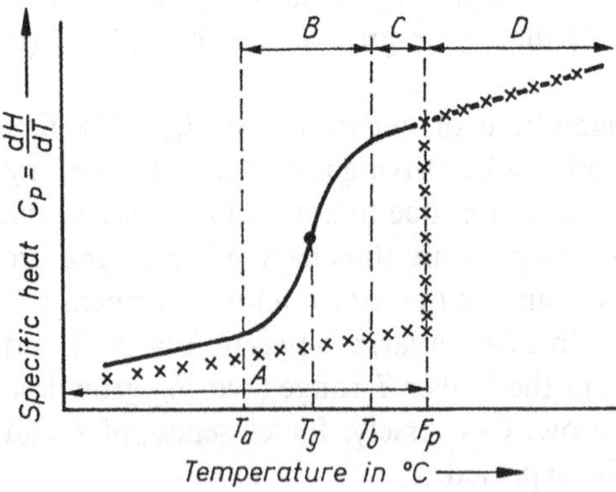

Fig. 2.3. Change of specific heat in the range of high temperatures for crystalline and vitreous solids. F_p melting point (liquidus) of crystals; T_g transition temperature; B ($T_a - T_b$) softening interval (plastic range); C ($F_p - T_b$) undercooled liquid glass melt; A crystalline solid; D liquid melt

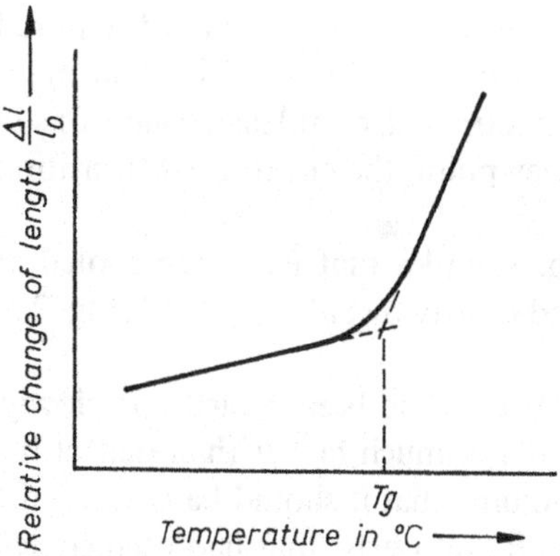

Fig. 2.4. Determination of the transition temperature of a glass by measuring the change of length in temperature dependence

If, however, at T_m crystallization is avoided, further cooling first shows a continuation in the decrease of C_p, the melt representing an "undercooled liquid."

However, at a temperature T_b, C_p begins to drop at an increasing slope through an inflection at temperature T_g down to a temperature T_a, where it approaches the value of the crystalline solid. During this interval, the melt solidifies to a glass, attaining a viscosity of about $10^{13.5}$ at T_a. On further cooling, C_p values remain generally near but slightly above those for the crystalline solid, decreasing at a slightly greater rate. This shows clearly that the glass is in a higher energy, metastable state compared to the crystal. The range T_b to T_a represents the glass transition[1] (the "softening range" of glass technologists). The inflection point is named "glass (transition) temperature," usually designated T_g. Although it will be shown that T_g depends on thermal history, it is widely used in science as well as technology to characterize noncrystalline solids in the same manner as T_m characterizes crystalline solids. This requires standardization and/or disclosure of the heat treatment used. In the glass-transition interval, viscosity increases from about 10^8 to $10^{13.5}$.

Similar curves as for C_p (Fig. 2.3) are obtained for other properties such as the coefficient of expansion, ($\alpha = (1/v)(\partial v/\partial T)$), volume refractive index, electrical resistance, etc. The conventional method for the determination of T_g is the measurement of the change in length of a glass sample at a continuous standardized rate of heating (Fig. 2.4), but various other methods are used as well.

In the transition range, decisive changes in microstructure occur which take increasing amounts of time as the lower temperature limit is approached. For this reason, all glass properties in this range depend on temperature and time, so that different heating and cooling rates, or generally any variations in time and temperature (thermal history), will induce variations in property values. This is the main reason that T_g varies with thermal history. Comparisons can be

[1] In contrast to the German use of the term "transformation" for this range. American usage limits the term "transformation" to true solid phase changes.

made only if conditions are standardized. A widely accepted standard required cooling at 1 °C·min^{-1} through the transition interval, followed by heating at 5 °C·min^{-1}. Otherwise, the sample should be retreated, or at least, conditions of the experiment reported. From a scientific viewpoint, the entire glass transition rather than T_g characterizes the glass.

The difference in T_g values obtained for samples that had been cooled at different cooling rates had been demonstrated clearly by Salmang [15] (Fig. 2.5) long ago.

Curve A shows the elongation of a sample that had been cooled very slowly. The glass had been able to shrink to a volume not much higher than that of the crystal, which is the metastable equilibrium volume that it should have at T_a, the lower end of the transition. However, in curve B (for a glass that had been cooled much faster), the time did not suffice for contraction, so that on reheating an interval of shrinkage is observed when the glass approaches its metastable equilibrium volume. A glass fiber permits extreme rates of cooling which result in curve C, showing that reheating provokes a considerable shrinkage. If such specimens were used to determine T_g values, they might differ by more than 100 °C.

Only a T_g value obtained under standard conditions can therefore be considered as the transition point of the glass. The more or less strongly quenched glasses B and C respectively shrink during the course of the reheating process involved in the determination of T_g. These shrinkages are caused by the freezing of a voluminous structure during the quenching process of the aforementioned glasses. The structure elements of the glass are rearranged into denser and lower energy sites and packings as the glass is reheated in the transition region where decisive structure and property changes are possible. This rearrangement leads to shrinkages. An increase in length however will in all cases occur with further increases in temperature.

Rötger has more closely investigated the processes which occur while heating and cooling glass in or near the transition region and has reached

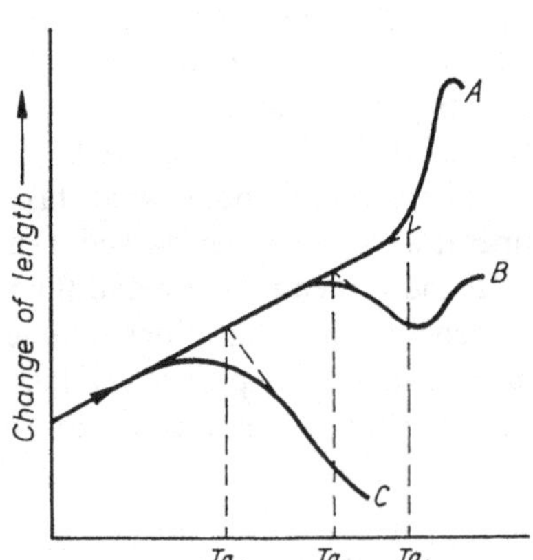

Fig. 2.5. Change in length of a borosilicate glass with increasing temperature (after Salmang *A* slowly cooled, bulk: *B* rapidly cooled, bulk; *C* extremely rapidly cooled (6 μm fiber)

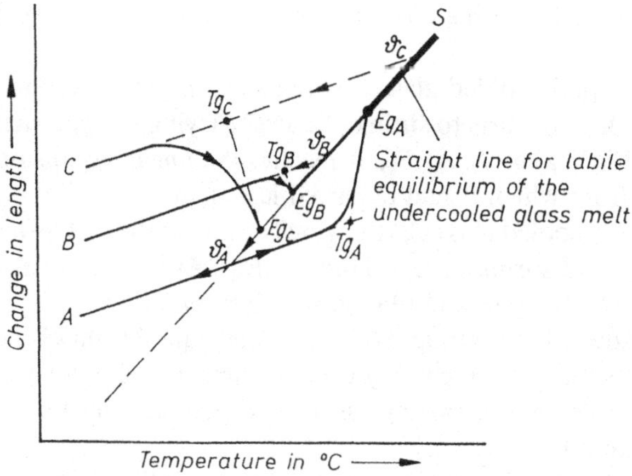

Fig. 2.6. "Salmang" curves for the identical glass after different undercooling (Rötger *A* slow cooling; *B* rapid cooling; *C* extremely rapid cooling [16, 17])

fundamental clarifications [16, 17]. Salmang's three curves (Fig. 2.5) representing the different change of length for glass samples quenched at different rates, must lie above each other in the order A, B, C from bottom to top, as rectified by Rötger. The curves are parallel in the lower region at lower temperatures. They can at times coincide at higher temperatures (Fig. 2.6). This fact is easily understood since for a quickly quenched glass sample the structure which freezes is more voluminous than the one of a glass which is more slowly cooled. In the latter cases the sample continues to contract and becomes denser. Different length values are hence observed at room temperature. Therefore the three curves cannot merge into each other (ending in one point). It is necessary at this point to consider the indication of a small correction which changes nothing to the fundamental interpretation of Salmang. The correction is indeed necessary since Salmang's observation of the very different changes in length occuring in glasses cooled at different rates appeared in the technological literature in the form of Fig. 2.5.

Rötger also clarified other features observed on glasses that had been cooled at different rates. Above T_a (Fig. 2.3), the undercooled liquid shows linear shrinkage (top of curve, Fig. 2.6). If cooling proceeds slowly (case A), this course continues to point v_A, where the melt begins to solidify. At this point, the rate of change in length changes direction (v_A to A). The smaller decrease is then typical of a solid. If the sample is cooled more rapidly (B) or extremely rapidly (C), freezing starts at higher temperatures: v_B and v_C, respectively. In this manner three curves S-v_C-C, S-v_B-B, and S-v_C-A describe shrinkage. On reheating, essentially different curves are obtained.

If glass A, originally cooled very slowly, is reheated, the direction of the low-temperature volume-change curve is maintained beyond v_A to T_{gA}, where a steep expansion toward E_{gA} follows. For this sample, obviously, T_{gA} would be reported as T_g of this glass. Above E_{gA}, i.e., above the glass transition, further expansion is that of the liquid melt, corresponding to the standardized procedure.

If, however, the more rapidly cooled glass B is reheated, it expands along the "solid" curve until T_{gB} where it starts to shrink ("thaws"), trying to approach the lower volume obtained in regime A. At E_{gB} it reaches the linear expansion mode above its transition from which it had departed less than A.

For the extremely rapidly cooled glass C, "thawing" occurs at the still lower value $T_{gC}(T_{gC} < T_{gB} < T_{gA})$, and shrinking continues to E_{gC}. At that point the linear expansion characteristic for the liquid labile equilibrium follows.

All three glasses tend towards the straight line for labile equilibrium of the undercooled melt upon reheating, although they were quenched differently and had achieved different structures for the same glass melt. They reach this line at different temperatures (t values).

The critical observer will at this point ask a fundamental question. Why are the cooling and heating curves of a glass not identical? There is a clear answer:

(a) cooling and heating curves, aside from some insignificant hysteresis phenomena, are identical when cooling and heating rates are the same,

(b) the cooling curve (more precisely the equilibrium straight line) will be crossed upon reheating as in the case of A at n_A, when the heating rate of the glass is greater than the cooling rate,

(c) when the heating rate is smaller than the cooling rate, as in cases B and C, a deviation from the cooling curve occurs upon reheating, already before reaching the equilibrium straight line at T_{gB} and T_{gC}.

The understanding of these somewhat complex phenomena is critical for the annealing and tempering processes which are important for all glass production. An incomplete or superficial analysis of the various behaviors of a glass melt in the transition region is no longer sufficient today for modern glass production.

It has often been suggested that specific structures are frozen upon cooling a glass melt, or that the packing of the building blocks of the glass played a role. It has even been proposed that glass represents a third state of matter. Thus other critical differences between the vitreous and other states shall be shown in this section.

X-ray diffraction offers means of investigating structure and has already quite early brought an enormous progress in this area. It therefore comes as no surprise that it would also be used to understand the structure of glasses. In what follows, other critical differences between crystalline and noncrystalline (glassy) solids are indicated briefly.

Figure 2.7 shows the dependence of the intensity of X-ray diffraction as a function of glazing angle ($\sin \theta/\lambda$) for gases, liquids, glasses, and crystals. For gases, diffraction is very intensive at small angles and decreases continuously with $\sin \theta/\lambda$. For liquids, diffraction at low angles remains high but is much lower than for gases, and broad diffuse maxima appear. For crystals, narrow sharp intensity maxima appear for small as well as large angles. Finally, the curve for the glasses resembles that for liquids more than that for crystals. One more difference between the glassy and crystalline states is demonstrated in this

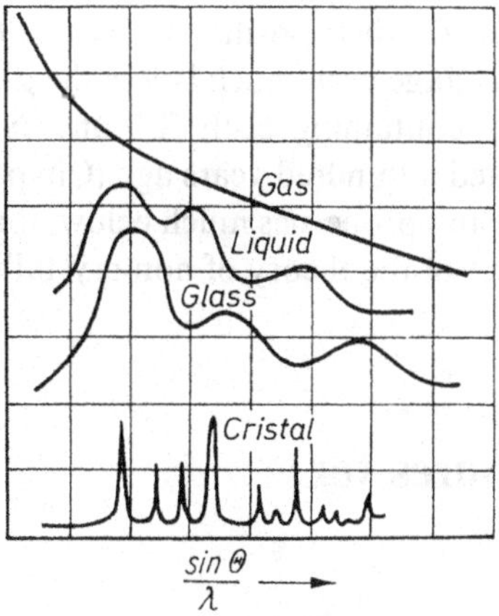

Fig. 2.7. Distribution of the intensity of scattered X rays in dependence on the grazing angle for substances in various states of aggregation

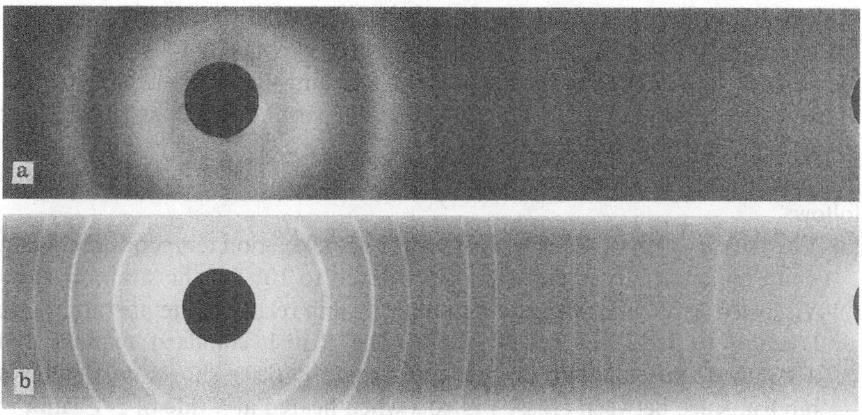

Fig. 2.8. Debye-Scherrer X-ray patterns of glass (**a**) and crystal (**b**)

manner. In Fig. 2.8 the Debye–Scherrer pattern is compared for a glass (a) and a crystal (b). Figures 2.7 and 2.8 atleast demonstrate the general conclusion that in liquids and glasses a short-range order of the glass lattice sites is found while no long range order exists. The interpretation is the presence in crystals of both short-range and long-range order. The glassy and crystalline solids, which are both forms of the solid state, show critical differences in their behavior as was clearly demonstrated by the two examples treated in this section, the behavior with temperature changes and the Debye– Sherrer X-ray patterns. More detailed information about the structure of glasses is much more difficult to obtain than for crystals, although much progress is being made (Wong–Angell [18]) in all sorts of spectroscopy. Before beginning with this topic, it is important to discuss certain viscosity temperatures which have been standardized worldwide to serve workers concerned with the development and production of glasses.

If mobile ions are present in a rigid network, their positions affecting the entire structure and properties of glasses can change even much below the glass transition. A change in thermometer glasses containing both K^+ and Na^+ affecting performance is observed and prevented a hundred years ago (Chapt. 1 pp. 20–21). But right now changes of structure and properties much below T_g are increasingly recognized as an important aspect of the theory of non-crystalline solids (Pye and Lacourse [19]).

2.3 Standard Viscosity Temperatures for Solidification of Glasses

The temperatures chosen in the preceding section to describe the process of freezing to a glass are connected with structural features and associated with discontinuities in property curves. It is, however, equally important to map the entire viscosity-temperature range in relation to glassmaking and forming

Figure 2.9 [20] shows the derivation of certain standard viscosity temperatures from certain steps in glass production permitting the specification of values most favorable for each step.

Definitions for these and other points, as well as measuring specifications, follow:

(a) Viscosity $\eta = 10^{13.5}$ Pa·s strain point (15-h relaxation temperature). At this point an artificially strained glass is relaxed to 10% of the strain in 15 h.

(b) Viscosity = 10^{12} Pa·s annealing point (15-min relaxation temperature). Relaxation to 10% occurs in 15 min. The British standard requires that a strained rod of 25 mm length and 5 mm diameter shows no significant birefringence between crossed Nicols when heated at a rate of 2 °C·min^{-1}.

(c) Viscosity = $10^{10.3}$ Pa·s: dilatometric softening point. The temperature at which dilatation stops in a standard dilatometer.

(d) Viscosity ≈ 10^7 Pa·s: "Z_1" temperature. The temperature at which a filament of about 0.3 mm diameter is elongated 1 mm/min when loaded with 9.81×10^3 Pa, on heating at a rate of 10 °C·min^{-1} in a constant furnace heating zone of 40–50 mm.

(e) Viscosity $10^{6.6}$ Pa·s: softening point. The temperature at which a filament of 23.5 cm length and a 0.55–0.77 mm diameter elongates by 1 mm·min^{-1} under its own weight.

(f) Viscosity = 10^4 Pa·s.: flow point.

In a specific instrument, a glass filament of 0.65 mm diameter held under tension, is heated abruptly at one point and the time to reach a known elongation is measured. Determination of the flow point is carried out practically only in the USA.

(g) Viscosity = 10^3 Pa·s: working point. In the United States, determined by a rotation viscometer. In the German industry predominantly the penetration, point E_g according to Dietzel and Brückner and the transition point T_g according to the DIN (German Industry Norm) 52324 are determined.

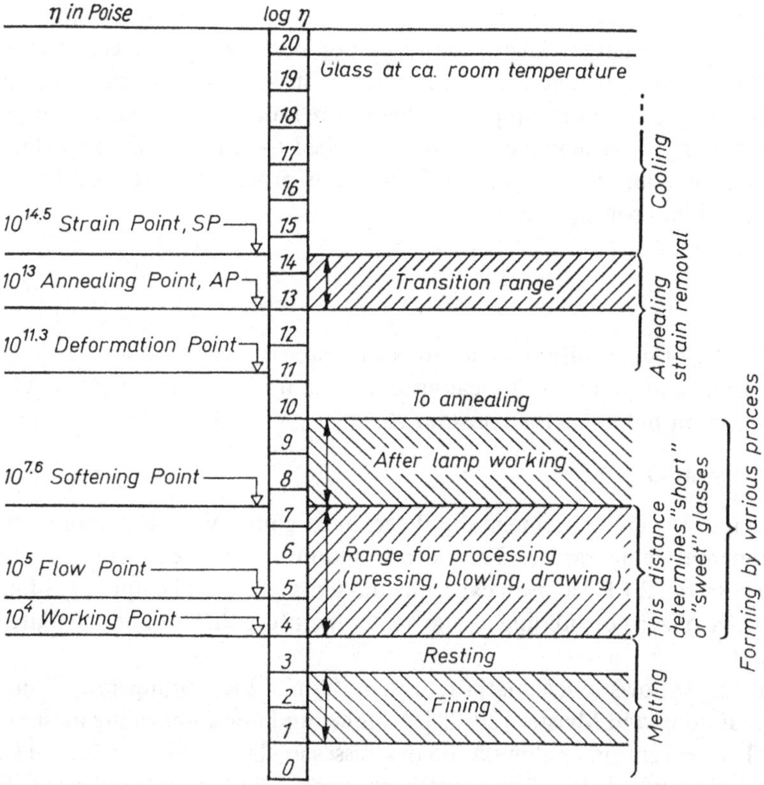

Fig. 2.9. Fix points in the entire viscosity range (after Lindig [20]). 1 poise = 10^{-1} Pa·s

(h) Viscosity $\approx 10^{11}$ Pa·s: penetration ("einsink") point (Dietzel et al. [21]). The temperature at which a PtRh (20 Rh–80 Pt) rod, diameter 0.5 mm, length 24.0 cm, weighing 0.902 g, requires 1 min to penetrate 2 cm (used in Germany).

(i) Viscosity $\eta \approx 10^{12}$–$10^{13.5}$: T_g (according to German Standard TGL 94-06009). The sample is cooled to room temperature at 1 °C·min^{-1}, reheated to determine elongation at 5 °C·min^{-1}; T_g is defined as the point at which the elongation becomes discontinuous (Fig. 2.4).

2.4 Annealing of Optical Glass

The term "annealing" generally refers to the elimination of stresses present in the glass.

Temporary stresses originate in a homogeneous glass due to too rapid cooling through the glass transition region leaving insufficient time for constituents to occupy lowest energy sites. If the glass is subsequently heated to near the transition point and annealed in a controlled fashion, the aforementioned stresses can be eliminated. This is however not possible with latent stresses produced for example by inhomogeneities such as striae. A very small change in refractive index, affecting optical performances, are induced by variations in the annealing process. These changes can be controlled to obtain desirable values. Temperature, time and birefringence influence this process as described by the Adams and Williamson equation:

$$\frac{1}{\Delta n} - \frac{1}{n_0} = A \cdot t \tag{2.1}$$

where n_0 is the original birefringence, n is the birefringence at time t, t is the annealine time and A is the annealing constant for the specific glass. The following relation holds for the constant A:

$$\log A = M_1 T - M_2 \tag{2.2}$$

where M_1 and M_2 are experimentally determined constants and T represents the experimental temperature. The corresponding cooling constant can be calculated with Eq. 2.1 for a given experimental temperature. The time to obtain a birefringence of zero by holding at a given temperature thus can be calculated from Eq. 2.1.

Figure 2.10 shows a schematic representation of the cooling process according to Günther and Meister [20, 22]. A higher and lower annealing limit are observed. The prevention or elimination of stresses in glasses can be achieved by holding the glass which has either just been poured or already cooled, at its higher annealing limit for a determined amount of time. At this annealing limit, as stresses are quickly eliminated.

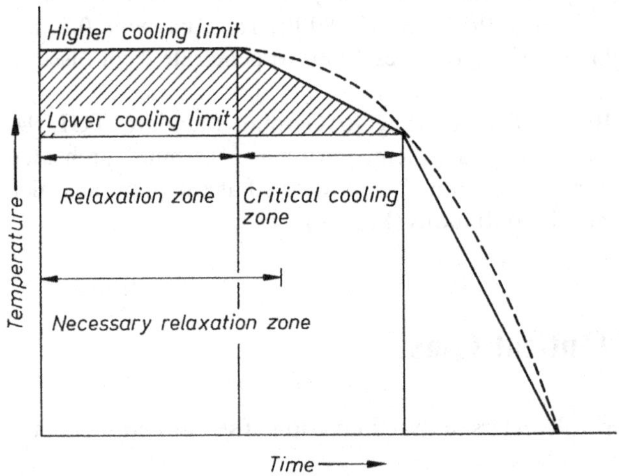

Fig. 2.10. Schematic representation of the theoretically calculated and the practically adjustable cooling curve of a glass according to Günther and Meister [22]. --- theoretical cooling curve; . . . experimental cooling curve

After this treatment the lower annealing limit must be approached very slowly. This often means a rate of 2 °C per minute or considerably less. At the lower limit, the possibility of eliminating stresses still exists. The formation of new stresses is also avoided by very slow cooling in the critical annealing zone. The glass can subsequently be cooled to room temperature relatively quickly without danger of breaking. Figure 2.10 shows the theoretical and experimental cooling curves of a glass. In practice, the curve is limited by the actual rate of cooling of the oven. However modern developments as for example air circulation have led to quite considerable progress in this field.

3 Structural Elements of Silicates

When one considers the overwhelming use of silicate glasses in the world-wide production of glasses, an urgent need for the thorough understanding of the structural elements of crystalline silicates is seen. In this respect one deals with one of two comparisons generally mobilized to describe the structure and behavior of glasses: the liquid and the crystalline states of matter. The crystallography of silicates encompasses so vast a body of information that, in the context of glass chemistry, the treatment must be limited to the most necessary foundations. For more information, refer to specialized literature sources (see, for example, Hinz [23]).

3.1 The SiO_4 Tetrahedron as the Basic Building Block of Silicates

The basic building block of all silicates is the SiO_4 tetrahedron, i.e., a structural unit with a silicon atom in the center of four surrounding oxygen atoms. All four oxygen atoms simultaneously touch the silicon atom and their external coordinating partners. The bond angles between two oxygen atoms are 109° 28′; the Si–O bond distance is 0.160 nm (1.60 Å), and the edge length of the SiO_4 tetrahedron is 0.262 nm (2.62 Å) (Fig. 3.1).

The origin of this polyhedron and its aggregations, which will be discussed later, are best understood from the consideration of the electronic configuration of silicon and the resulting bonding character.

The electronic structure of silicon is $1s^2 2s^2 2p^6 3s^2 3p^2$. Relatively little energy is required for the conversion of the ground-state $3s^2 3p^2$ to four unpaired electrons $3s^1 3p^3$ (Fig. 3.2), the spherical s orbital and the three extended p orbitals of which are hybridized to four equivalent orbitals pointing toward the corners of a tetrahedron (Fig. 3.3). In the case of oxygen being the coordination partner, this leads to the stable structural element (SiO_4).

The bond between silicon and oxygen in this tetrahedron is most likely a mixed bond (Fig. 3.4) (Brill et al. [24]). Pauling's [25] Si–O value of 0.162 nm (1.62 Å) calculated under the assumption that all three possible bond types occur, agrees quite well with the experimental value of about 0.160 nm (1.60 Å) found for the modifications of crystalline SiO_2. Similarly, the bond angles agree with those of the model. For purely covalent σ bonding, an angle of 90° would be expected, for pure π bonding one of 180°. Experimental values for Si–O–Si in silicates are between 130° and 140°.

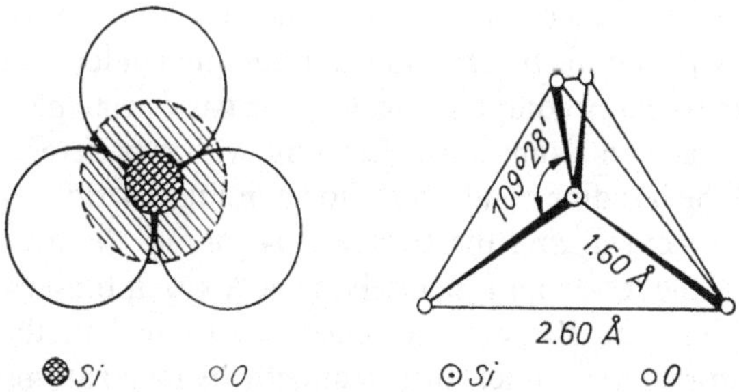

a b

Fig. 3.1. The $[SiO_4]$ tetrahedron mode I. **a** $[SiO_4]$ tetrahedron consisting of oxygen and silicon ions. The simply cross-hatched oxygen lies above or below the plane of the drawing; **b** $[SiO_4]$ tetrahedron (only centers of charge of ions are marked) with bond angles and ionic distance indicated

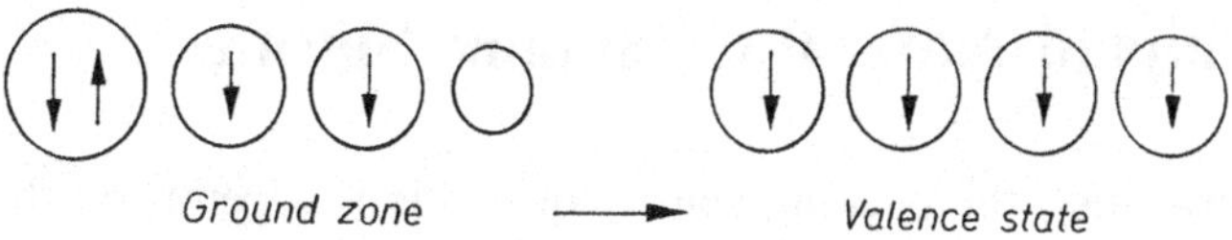

Fig. 3.2. Electronic states of Si and resulting ground and valence states

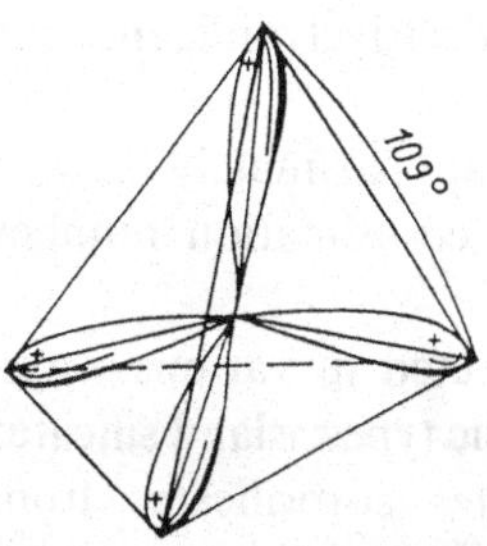

Fig. 3.3. sp^3 hybrid orbitals of Si

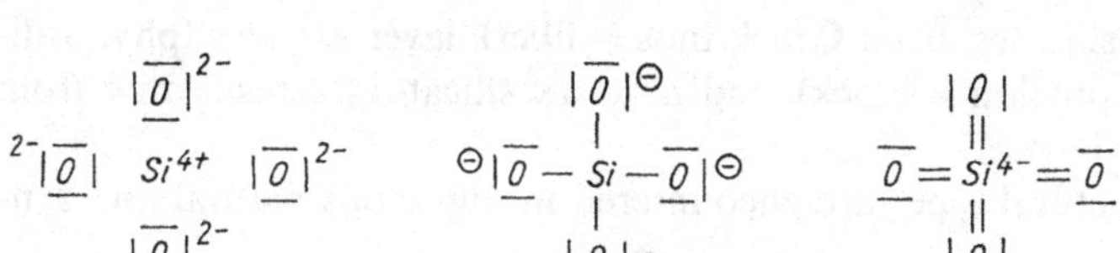

a b c

Fig. 3.4. Boundary conditions of bonding in the $[SiO_4]$ complex. **a** Polar; **b** isovalent; **c** double bonding

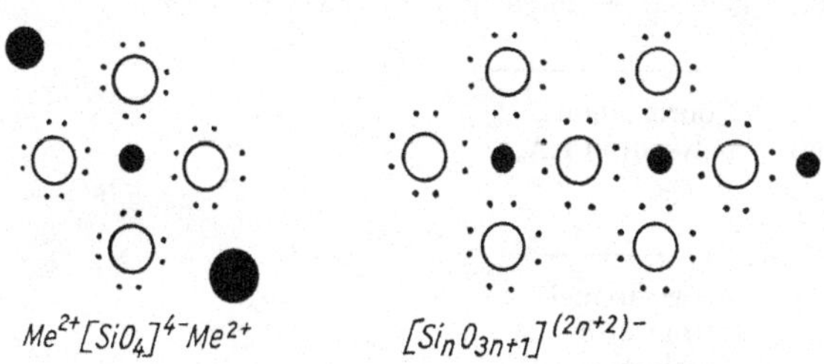

a b

Fig. 3.5. Electron octet formation in silicates. **a** By accepting of electrons from adjacent metal ions; **b** by linking of tetrahedra via bridging oxygen

A deeper knowledge of the electronic configuration and the bond type of the SiO_4 tetrahedron is a prerequisite for better understanding its tendency to polymerize and, therefore, form so many complex silicates. The basic principle is the complete saturation of the oxygen atoms with electrons, with a preference for octet shells. On the basis of the bonding conditions shown in Fig. 3.4, this can be realized by either (a) metals neutralizing the complex by joining the SiO_4 complex, or (b) a connection of the tetrahedra themselves via "oxygen bridges" forming octets of electrons (Fig. 3.5). If giant molecules are formed in this manner, it only remains to saturate terminal tetrahedra directly by metal atoms.

3.2 Building Units of Natural Crystalline Silicates

Pauling [25] has shown that the packing density of two ions depends on the relations of their radii. A large number of experimental data led to the establishment of empirical rules. Table 3.1 shows how, on exceeding certain cation:anion radius ratios, coordination numbers change from 3 to 4, from 4 to 6, from 6 to 8, and from 8 to 12. The resulting coordination polyhedra decisively influence the crystal structure of a compound.

In the case of the Si–O coordination, a relation of approximately 0.3 was obtained, which represents the stability region of the coordination number 4 (tetrahedron).

The tetracoordinate polyhedra SiO_4 can be connected in various ways. A lucid survey (Hinz [23]) distinguishes the following basic types: island silicates (nesosilicates, from Greek nesos = island), group silicates (sorosilicates, from Greek soros = group), ring silicates (cyclosilicates, from Greek kyklos = circle), chain silicates (inosilicates, from Greek inos = fiber), layer silicates (phyllosilicates, from Greek phyllon = block), and network silicates (tectosilicates, from Greek tektoneia = lattice).

The basic structural types are encountered in numerous natural and synthetic silicates.

Table 3.2 gives a typical example for each kind. Table 3.3 lists the resulting Si:O relations and the formulas for the anion complexes resulting from various

Table 3.1. Dependence of coordination numbers on the ionic radius ratios in crystals

Radius ratio $r_K:r_A$ as a boundary condition for structure formed	Coordination number (CN)	Coordination polyhedron KA_n
$\geqslant 0.155$	3	Planar triangle
$\geqslant 0.225$	4	Tetrahedron
$\geqslant 0.415$	6	Octahedron
$\geqslant 0.732$	8	Cube
$\geqslant 1.0$	12	Densest packing of spheres

Table 3.2. The most important types of silicate structures

Type	Linking of $[SiO_4]$ tetrahedra	Building group	Example
Island silicates	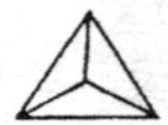	$[SiO_4]^{4-}$	Phenacite $Be_2[SiO_4]$ Willemite $Zn_2[SiO_4]$ Forsterite $Mg_2[SiO_4]$
Group silicates	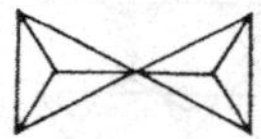	$[Si_2O_7]^{6-}$	Thortveitite $(Sc, Y)_2[Si_2O_7]$
Ring silicates 3-ring	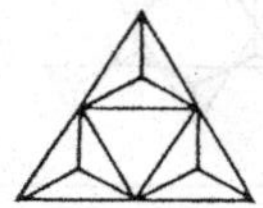	$[Si_3O_9]^{6-}$-Ring	β-wollastonite $Ca_3[Si_3O_9]$
6-ring	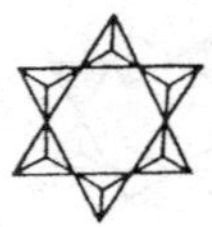	$[Si_6O_{18}]^{12-}$-Ring	Beryl $Al_2Be_3[Si_6O_{18}]$
Chain silicates Single chain	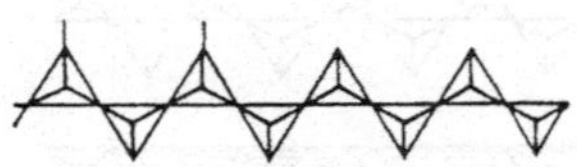	$[SiO_3]_x^{2-}$-chain	Enstatite $Mg_2[Si_2O_6]$ Spodumene $LiAl[Si_2O_6]$
	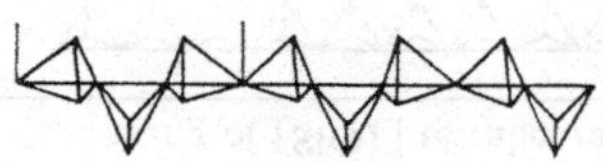	$[SiO_3]_x^{2-}$-chain	α-wollastonite $Ca_3[Si_3O_9]$

Table 3.2. (Continued)

Type	Linking of $[SiO_4]$ tetrahedra	Building group	Example
Double chain	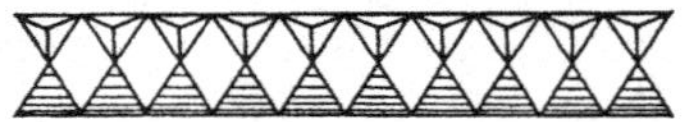	$[Si_2O_5]_x^{2-}$-chain	Sillimanite $Al[AlSiO_5]$ ($Al^{[6]}[Al^{[4]}SiO_5]$)
Double binary chain	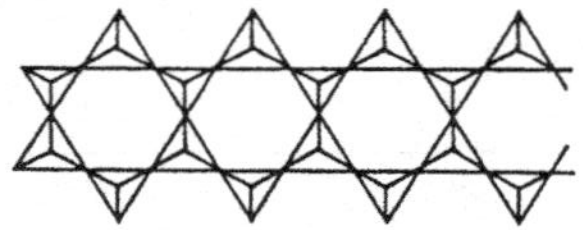	$[Si_4O_{11}]_x^{6-}$-chain	Tremolite $Ca_2Mg_5[(OH_1F)]$
Double teriary chain	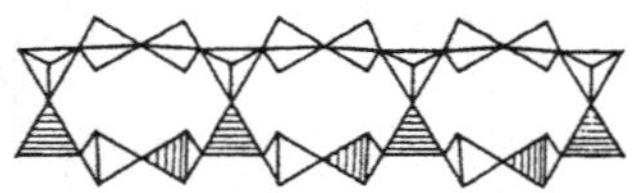	$[Si_6O_{17}]_x^{10-}$-chain	Xonolith $Ca_6[(OH)_2/Si_6O_{17}]$
Layer silicates Single layer	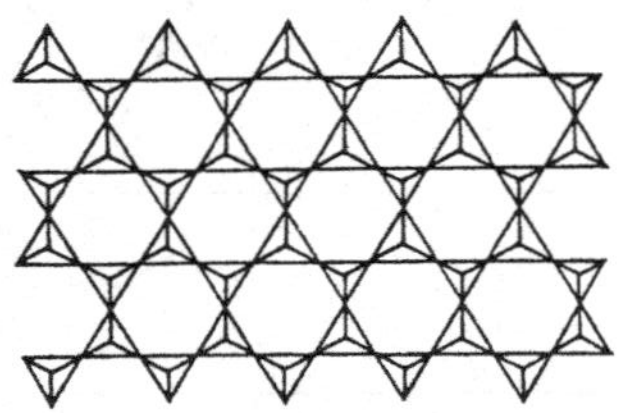	$[Si_2O_5]_{xx}^{2-}$-layer	Talcum $Mg_3[(OH)_2/Si_4O_{10}]$ Montmorillonite $Al_2[(OH)_2/Si_4O_{10}]$
Double layer	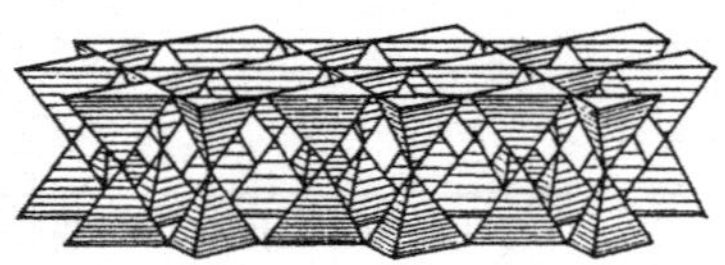		$CaAl_2Si_2O_8$(hex.)

Network silicates	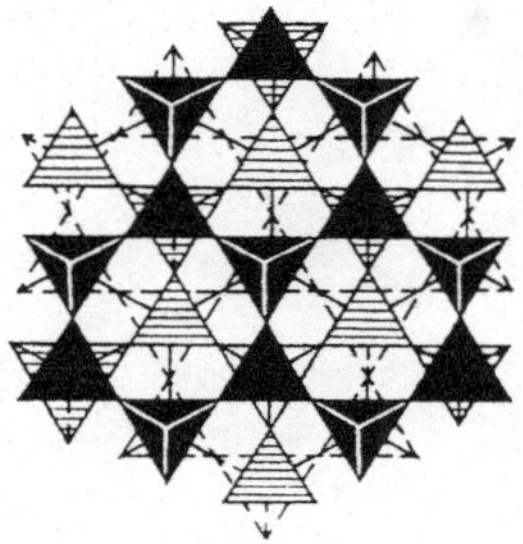	$[SiO_2]_{xxx}$ ($2SiO_2 = Si[SiO_4]$) Three-dimensional network	High-cristobalite Low-cristobalite High-tridymite Low-tridymite High-quartz Low-quartz
	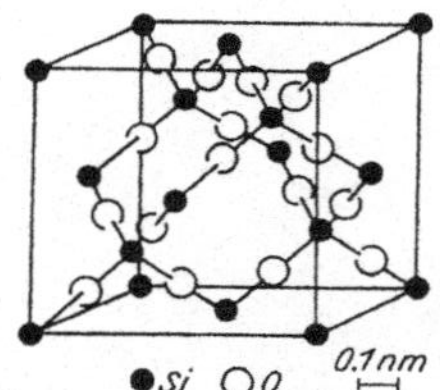 Structure of α-cristobalite (High cristobalite) a	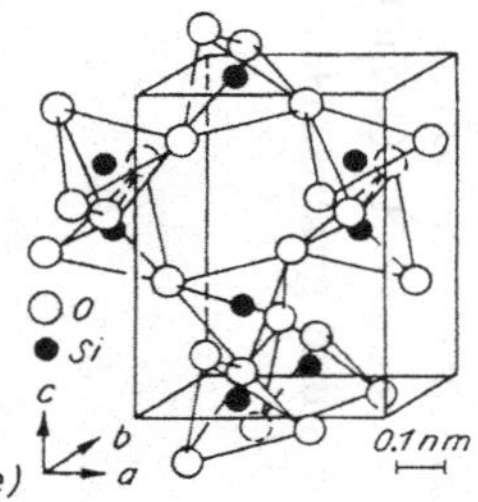 Structure of β-cristobalite (Low cristobalite) b	
	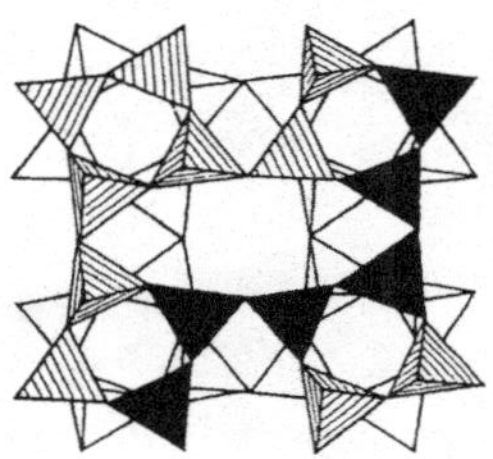		Orthoclase-$KAlSi_3O_8$ plagioclase (mixed crystals between albite $NaAlSi_3O_8$ and anorthite $CaAl_2Si_2O_8$)

Table 3.3. Silicate anion complexes and their Si:O ratios

Type	Formula anion of complex	Si (or Si + Al):O Ratio	Number of bridging oxygens
Island silicates	$[SiO_4]^{4-}$	1:4	0
Group silicates	$[Si_2O_7]^{6-}$	1:3.5	1
Ring silicates	$[SiO_3]^{2-}$	1:3	2
Chain silicates	$[SiO_3]^{2-}_{\infty}$ [a]	1:3	2
Layer silicates	$[Si_2O_5]^{2-}_{\infty\infty}$	1:2.5	3
Network silicates	$[SiO_2]_{\infty\infty\infty}$	1:2	4

[a]Subscripts ∞ mean 1-, 2-, or 3-dimensional linkage.

extensive connections. A systematic classification by anion complex required the development of X-ray diffraction as a tool for structural investigation by Laue and collaborators (since 1912). Until then silicates were classified from a morphological viewpoint and oxide formulas such as $CaO \cdot SiO_2$ were used. The elucidation of numerous silicated structures by X-ray diffraction is due to Machatschki [26], Bragg [27], and their collaborators. The now conventional formulation, e.g., $Ca_3(Si_3O_9)$ instead of $CaO \cdot SiO_2$, tells much more about the degree of polymerization of anion groups. Nevertheless, the classification of silicates remains an open field.

4 Classical Theories of Glass Structure

4.1 Glass Structure According to Tammann (since 1903) [11–13]

Tammann, the first investigator concerned with the constitution of glasses, regarded them as strongly undercooled liquids. To a certain degree this is in agreement with later interpretations using X-ray diffraction data. At that time the structures of liquid crystals and of liquids of Bernal, Stewart, or Frenkel type (which will be discussed later) were not yet known. Tammann's model of the freezing-in of a structure quite similar to that of the liquid was quite general, but did represent a good start.

Tammann also was the first to investigate low-melting-point model glasses, organic glasses obtained by rapid cooling of molten piperidine, salicin, phenolphthalein, sugars, etc. These investigations resulted in fundamental information about nucleation and crystallization of glasses (see Chapter 2, Section 2.1: Fusion and Crystallization).

4.2 Glass Formation According to Goldschmidt [28]

Goldschmidt, who is considered the founder of modern crystal chemistry, was in a position to derive empirical rules for glass formation from his studies. Just as in the case of crystal structure, relations of ionic size were to play a decisive role. He postulated a ratio of from 0.2 to 0.4 of radius-of-cation (r_c) to radius-of-anion (r_a) as a condition of glass formation. This condition is fulfilled in the case of SiO_2, B_2O_3, P_2O_5, etc. (Table 4.1).

Later, a convincing confirmation of this prediction was obtained by the discovery of the formation of BeF_2 glass, which indeed satisfied these specific

Table 4.1. Radius Ratios for Typical Glass-Formers

Compound	Radius Ratio ($r_k : r_A$)
SiO_2	$r_{Si}:r_O = 0.39$ Å[a] $: 1.4$ Å ≈ 0.28
B_2O_3	$r_B:r_O = 0.20$ Å $: 1.4$ Å ≈ 0.15
P_2O_5	$r_P:r_O = 0.34$ Å $: 1.4$ Å ≈ 0.25
GeO_2	$r_{Ge}:r_O = 0.44$ Å $: 1.4$ Å ≈ 0.31
BeF_2	$r_{Be}:r_F = 0.34$ Å $: 1.36$ Å ≈ 0.25

[a] 1 Å = 10^{-8} cm

conditions. Moreover, the same laws and dependencies should be valid for the formation of a glass structure as for crystal chemistry. Predominently qualitative interpretations of measurable glass properties were attempted at that time. Only at the beginning of the twenties and thirties of our century structure hypotheses appeared which became more or less successful bases for glass development and research. e.g. the dependence of packing density on the state of polarisation and deformation of constituent ions.

4.3 The Zachariasen–Warren Network Theory

The network hypothesis proposed by Zachariasen [29] and reinforced by Warren's [30] X-ray diffraction represented a leap forward in the knowledge about the structure of glasses. For instance, in the case of SiO_2 glass the same tetrahedra which are uniformly connected in crystalline silicates (Fig. 4.1a, Bragg) are connected irregularly (Fig. 4.1b). The formation of a 3-dimensional disordered infinite network from building units of low coordination number (e.g., 4 in SiO_4 tetrahedra, or 3 in BO_3 triangles) explained convincingly the enormous increase in viscosity on cooling.

According to the Zachariasen-Warren network theory, the following rules are valid for the formation of low-order 3-dimensional networks, i.e., the glass

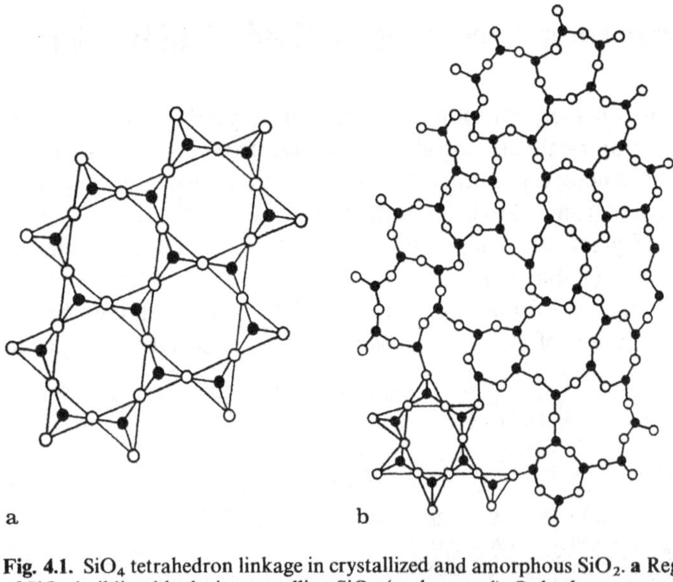

a b

Fig. 4.1. SiO_4 tetrahedron linkage in crystallized and amorphous SiO_2. **a** Regularly ordered netting of SiO_4 building blocks in crystalline SiO_2 (rock crystal). Only three oxygens of the SiO_4 building groups are pictured. The fourth oxygen lies alternatively above or below the plane of the drawing (after Zachariasen [29] and Warren [30]); **b** disordered netting of SiO_4 building blocks on SiO_2 glass (molten rock crystal) (after Zachariasen [22] and Warren [30]) formation of simple compounds such as SiO_2, B_2O_2, P_2O_5, GeO_2, As_2S_3, BeF_2:

(a) An oxide or compound tends to form a glass if it easily forms polyhedral groups as smallest building units.
(b) Polyhedra should not share more than one corner.
(c) Anions (e.g., O^{2-}, S^{2-}, F^-) should not bind more than two central atoms of a polyhedron. Thus, in simple glasses anions form bridges between two polyhedra.
(d) The number of corners of polyhedra must < 6.
(e) At least three corners of a polyhedron must connect with neighbor polyhedra.

Figure 4.2 shows examples of a 3-dimensional connection of building units in the case of BeF_2 and As_2S_3 glasses.

If large cations are introduced into such simple glasses, i.e., in their low-order networks – for instance, by melting SiO_2 and Na_2O or CaO – bridges are broken. The oxygen (or other anion like S^{2-}, F^-) coming along with the large cation will assume the site at the free end of the separated tetrahedron, while the large cation itself is accommodated in the larger cavity originating in the severance of the network at the site.

The theory assumes that the local severance of the network and the placement of large cations in the cavities of the network are distributed statistically Fig. 4.3.

Zachariasen classified the cations in a glass as follows:

(a) Network-formers, such as Si, B, P, Ge, As, Be (with F), etc., with a coordination number of generally 3 or 4.
(b) Network-modifiers, such as Na, K, Ca, Ba, etc., with coordination numbers generally $\geqslant 6$.
(c) Intermediates may either reinforce the network (coordination number 4) of further loosen the network (coordination numbers 6–8), but cannot form a glass per se.

At least for conventional glasses, a large number of properties can be explained or predicted by the Zachariasen–Warren concept [29, 30]. As an increase in

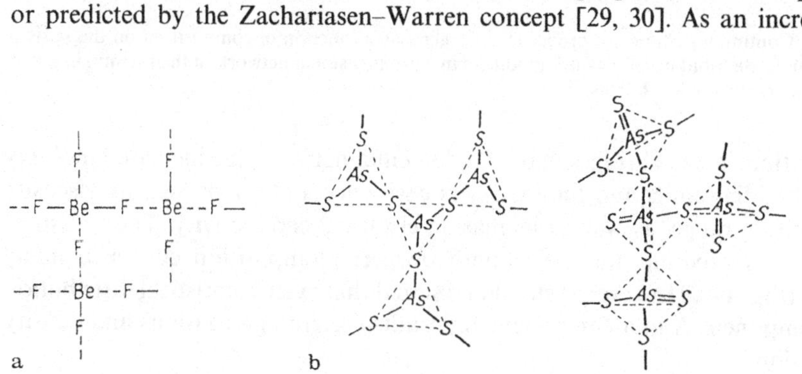

Fig. 4.2. Possibilities of spatial netting of basic building elements in BeF_2 and As_2S_3 glasses. **a** BeF_2 glass (building element BeF_4 tetrahedron); **b** As_2S_3 glass (building element AsS_3 planar triangle); As_2S_3 glass (building element AsS_4 tetrahedron)

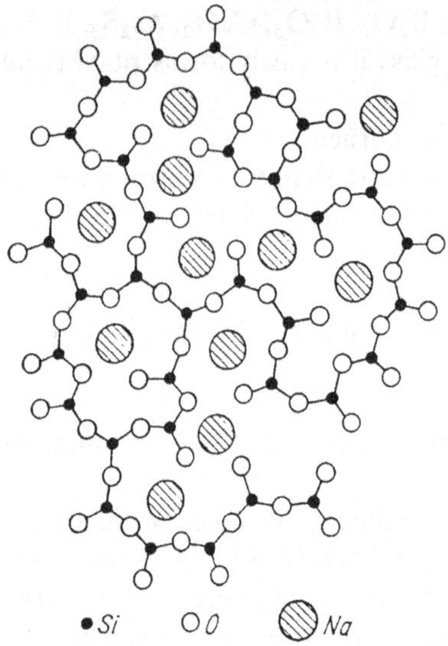

Fig. 4.3. Arrangement of atoms (or ions) in a sodium silicate glass. When Na_2O is incorporated, the large sodium ions are after rupture of oxygen bridges, located in the new, larger cavities (after Zachariasen [29] and Warren [30])

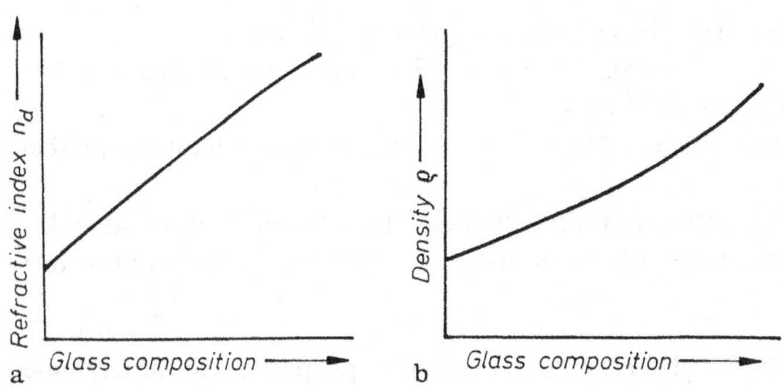

Fig. 4.4. Continuous change of properties in a glass as a function of composition on the basis of a continuous distribution of network modifiers in a 3-dimensional network of the surrounding base glass. **a** Refractive index; **b** density

large cations breaks bridges in the fundamental network, the increased mobility of the building units and the modifiers easily account for decreasing viscosity and melting range, as well as increasing electrical conductivity. The statistical distribution accounts for the smooth property-composition curves in many glasses (Fig. 4.4). However, it should be stated that even a statistical distribution of a component A in a component B permits certain aggregations and density fluctuations.

The problem of distribution of components in crystals or glass has been studied by Laves [31] by considering, for example, possible distribution of 50% black and 50% white squares (Fig. 4.5). He concluded that, in a mixed crystal for

Fig. 4.5. Distribution of black and white squares (Laves [31]. **a** Statistical; **b** better than; **c** worse than statistical

example, the distribution is close to the statistical one. In the case of aggregation, such as Fig. 4.5a or 4.5c, one speaks of clustering.

The network theory of Zachariasen and Warren will not apply to all glass systems.

(a) A few phenomena, such as discontinuities of property-composition relations, indicate the influence of considerable changes in short-range order in some systems.

(b) Under more stringent cooling procedures, an ever-increasing number of unconventional melt and vapor systems can be obtained as glasses in the absence of conformity to the Zachariasen rules, and network formation.

The kinetic theories of glass formation (see pp. 54–56) have become more satisfactory in accommodating all known glass systems, but provide less guidance in detail than the structural theories within the confines of their validity. It is for this reason that the Zachariasen–Warren theory remains applicable in many cases.

4.4 Extension of the Network Theory by Dietzel

Dietzel [32] for the first time extended Goldschmidt's considerations of the role of size and polarizability of constituent ions to that of their charges. In this manner he included the influence of the interaction of forces between cations and anions during solidification of a melt. Applying the fundamental relation of physics for the attraction (or repulsion) P between electrical charges Q at distance a

$$P = QQ/a^2$$

to anions and cations we obtain:

$$K = \frac{Z_c Z_a e^2}{(r_c + r_a)^2} = \frac{Z_c Z_a e^2}{a^2}$$

where K describes, in the first approximation, the interacting forces; Z_c, Z_a are the valencies of cation and anion; e is the elementary charge; r_c, r_a are the radii of cation and anion, and $a = r_a + r_a$.

Dietzel introduced the term "field strength" F

$$F = Z_c/a^2$$

to characterize the effect of a single cation, assuming that the anion is oxygen ($a = r_c + r_{O_2-}$). For nonoxide glasses, other values for a have to be introduced.

Table 4.2 illustrates the fact that Zachariasen's classification of ions as network-formers, network-modifiers, and intermediates may be clearly related to their field strength: For network-modifiers $F \approx 0.1$ to 0.4; for network-formers $F \approx 1.4$ to 2, larger by nearly one order of magnitude; for intermediates $F \approx 0.5$ to 1.0. The concept of field strength made it possible for Dietzel to propose rules for the behavior of a cooling melt which could not be explained on the basis of the Zachariasen–Warren hypotheses alone.

Pauling's concept of electronegativity [25] leads to an analogous classification. We limit ourselves to Dietzel's field strength, which has been applied more frequently in glass technology.

Table 4.2. Classification of cations according to their field strength (Dietzel)

Element	Valence Z	Ionic Radius (for CN = 6 r in Å [a]	Most Frequent Coordination Number CN	Ionic Distance for Oxides a in Å	Field Strength at Distance of O^2 Ions Z/a^2	Function in Glass Structure
K	1	1.33	8	2.77	0.13	
Na	1	0.98	6	2.30	0.19	
Li	1	0.78	6	2.10	0.23	
Ba	2	1.43	8	2.86	0.24	
Pb	2	1.32	8	2.74	0.27	Network-modifier $Z/a^2 \approx 0.1 \ldots 0.4$
Sr	2	1.27	8	2.69	0.28	
Ca	2	1.06	8	2.48	0.33	
Mn	2	0.91	6	2.23	0.40	
Fe	2	0.83	6	2.15	0.43	
Mn	2	0.83	4	2.03	0.49	
Mg	2	0.78	6	2.10	0.45	
			4	1.96	0.53	
Zr	4	0.87	8	2.28	0.77	
Be	2	0.34	4	1.53	0.86	Intermediate $Z/a^2 \approx 0.5 \ldots 1.0$
Fe	3	0.67	6	1.99	0.76	
			4	1.88	0.85	
Al	3	0.57	6	1.89	0.84	
			4	1.77	0.96	
Ti	4	0.64	6	1.96	1.04	
B	3	0.20	4	1.50	1.34	
Ge	4	0.44	4	1.66	1.45	
Si	4	0.39	4	1.60	1.57	Network-former $Z/a^2 \approx 1.5 \ldots 2.0$
P	5	0.34	4	1.55	2.1	
B	3	0.20	3		1.63	

[a] 1 Å = 10^{-8} cm

For example, on cooling a binary silicate melt, two cations compete for their oxygen environment to achieve densest packing. If the field strength of both cations is approximately equal, demixing will usually occur, leading to two pure oxide phases. If, however, field strengths differ, the oxygen ions will be used predominantly for the densest packing around the cation with a higher field strength. The cation with a lesser field strength must be satisfied with a higher coordination number vis-a-vis the anion complex formed by the other cation (e.g., $(SiO_4)^{4-}$). In this case, compound formation and crystallization are favored. According to Dietzel, stable compounds require a difference of field strength ΔF between two cations of a binary system exceeding 0.3. With a further increase, the number of compounds, but also the tendency to glass formation, increases. Melts of binary systems with ΔF of the constituent ions exceeding 1.33 usually solidify as glasses.

Silica, B_2O_3, and P_2O_5 are the best glass-formers. The Zachariasen Warren conditions for glass formation are also met for any of their combinations. Yet combinations of two of these oxides in one melt show a decrease in the tendency of glass formation or immiscibility.

Melts within the system SiO_2-B_2O_3 solidify as glasses in all proportions, with some region of immiscibility. Melts within the systems SiO_2–P_2O_5 and B_2O_3-P_2O_5, however, solidify in most proportions to crystalline solids or, on rapid quenching, as phase-separated glasses.

In this connection, Otto Schott's early observation should be recalled: "While those two "$(SiO_2 + B_2O_3)$" as a glass forming double acid may accept bases in any proportions, phosphorus entering the system will almost always effect milky turbidity, apparently because the glass salts of phosphoric acid are insoluble like oil in water. . . "

This contradiction to Zachariasen's concept is easily explained by Dietzel's results:

The field strength of

$$B^{3+} \text{ in } (BO_3) = 1.63$$

$$Si^{4+} \text{ in } (SiO_4) = 1.57$$

$$P^{5+} \text{ in } (PO_4) = 2.1$$

The differences in field strength ΔF shown in Table 4.3 make the crystalline solidification of the systems P_2O_5-B_2O_3 and P_2O_5-SiO_2 plausible since they exceed the limit of 0.3.

Yet these differences are not large enough to encourage stable compounds. The tendency to segregate to single oxides is apparent in the compounds $B(PO_4)$ and $Si(P_2O_7)$. The compound borophosphate used to be called "boron phosphorus oxide" and its structure remained unclear for a long time. In the similar case of "silicyl phosphate" the oxygen ions are seized, for satisfying its own coordination requirements, by the phosphorus ion which has the higher field strength, forcing (or "demoting") the silicon ion to accept a higher coordination number than the usual one of four.

Table 4.3. Difference in field strength of cations in binary glasses for chief glass-formers SiO_2, B_2O_3, and P_2O_5

Oxide glass system	Difference in field strength	Solidification behavior of melt
Si–B	0.06	Glass formation
B–P	0.47	Precipitation of crystalline BPO_4
P–Si	0.53	Precipitation of crystalline $SiO_2 \cdot P_2O_5$ and $3\ SiO_2 \cdot P_2O_5$

It seems that this does not yet occur in the melt, because unmixing is observed in some regions and/or on fast cooling. In the SiO_2-B_2O_3 system, unmixing or phase separation should actually occur at a ΔF value of 0.06 according to Dietzel. Relatively good glassformation is observed however throughout this system. After all, the success of Schott in the development of new optical and technical glasses is based on the combination of SiO_2 and B_2O_3. Thus the interpretations of Dietzel and further developments of the network theory appeared at the time to contain an unexplainable "hole". However this "hole" only seems to exist as will be proved later.

It is instructive to cite Dietzel's [33] comments regarding the applicability of the field-strength concept to the mixed-alkali effect (which he considers just a special case of a more general "mixed-ion" effect). "Mixed-alkali effect" is a term used to describe extreme property maxima or minima in pseudobinary systems in which the ratio of two alkalis is varied (e.g., $x\ Na_2O \cdot (1-x)\ K_2O \cdot 2SiO_2$). Dietzel's abbreviated comments were:

"If the difference ΔF in field strength F of two ions ($= Z/a^2$) increases, the mixed-alkali ("ion") effect increases, more ternary compounds form, and immiscibility decreases. Stable couplings like Li–$\overset{\text{O}}{\underset{\text{Si}}{|}}$–K will decrease the diffusion coefficient and conductivity."

4.5 Additional Concepts Supplementing the Network Theory

Smekal [34] further considered polarization and deformation phenomena influencing the interaction of the forces based on the field strength of the ions. He relates glass formation to the types of chemical bonding.

Figure 4.6 is a schematic illustration of the transition from pure ionic to pure atomic bonding with changing polarity. Figure 4.7 shows the distribution of electron density for various types of bonding.

According to Smekal, glasses are typically characterized by mixed bonding. Materials with pure ionic or homeopolar bonding melt at relatively low temperatures, have low hardness, and crystallize easily. Materials with mixed

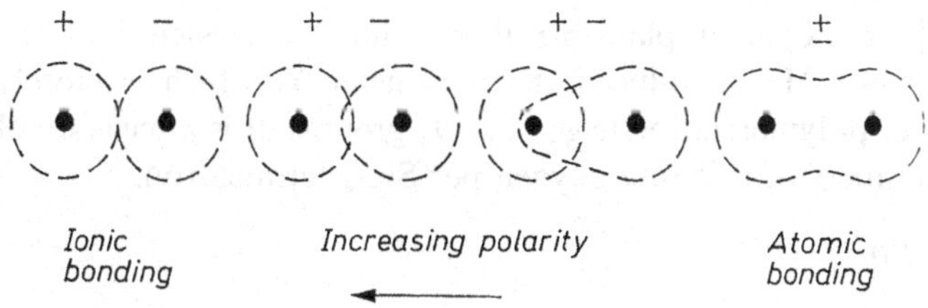

Fig. 4.6. Schematic of the transition from pure ionic to pure atomic bonding with increasing polarity

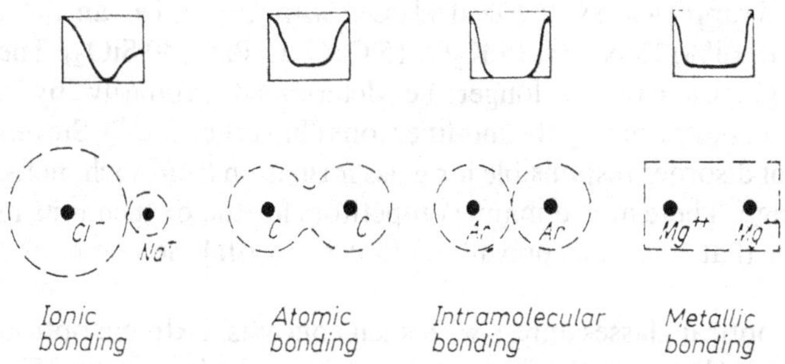

Fig. 4.7. Schematic of electron density distribution for different bonding types

bonding, for example quartz, have high hardness, high melting temperature, high viscosity in the melting range, and solidify easily to a glass.

They all show conchoidal fracture as well. According to Smekal, bonding is about 50% homeopolar in rock crystal, 44% in B_2O_3, 60% in Al_2O_3, and mixed with heteropolar bonding. In the BeF_2 "model" glasses first prepared by Goldschmidt, the homeopolar shares is only about 21% (Mackenzie [35], Pauling [25]). What is to be expected in terms of greater insight into the glass structure on the basis of their weaker bonding will be discussed later.

Weyl postulates [36] what has become known as the screening theory, which may be conceived as a theory of complexes as applied to glasses. It stresses the residual valence forces promoted by polarization and deformation of coordination groups such as SiO_4, BO_3, PO_4, or BeF_4 to irregular networks. If a highly charged cation is screened completely, such as S by F in SF_6 polymerization, glass formation will not take place. In a silicate melt with sufficiently high network-modifier concentration, the silicon ions will be saturated in coordination; isolated, electron-neutral SiO_4 complexes will form; and crystals will precipitate.

If, however, the concentration of network-modifiers, (e.g., Na_2O) causes a deficiency in oxygen ions, a complete coordinative saturation of the silicon ions is possible only by the joint utilization of certain oxygen ions, the bridging oxygen ions. Thus, the degree of polymerization, and therefore glass formation, will be determined by the ratio Si:O. Weyl's screening theory was mobilized for simple atomistic explanations of the mechanical properties and the strength of glasses.

Stevels [37–39] developed a plausible theory for the existence of the so-called "invert glasses." He introduced the parameter Y, which is closely related to the degree of polymerization (e.g., of SiO_4 groups). It is a measure of the mean possible numbers of bridging oxygen per SiO_4 tetrahedron.

$$Y = 8 - 2(O:Si)$$

It is 4 for $O:Si = 2$: a 3-dimensional network exists; 3 for $O:Si = 2.5$ ($Na_2O \cdot 2SiO_2$): the tetrahedra are connected at three corners; 2 for $O:Si = 3$($Na_2O \cdot SiO_2$): corresponding to 2-dimensional connection (infinite chains). For $Y < 2$ the Zachariasen–Warren rules would no longer permit glass formation.

Yet Stevels and Trapp [40] synthesized glasses with $Y < 2$, i.e., an SiO_2 content $< 50\%$ (e.g., (mol%) 15 Na_2O, 15 K_2O, 15 CaO, 15 BaO, 40 SiO_2). The properties of such glasses can no longer be determined primarily by a 3-dimensional network, but rather by the modifier ions ("invert glasses"). Stevels held the high degree of disorder responsible for glass formation from such melts; in Dietzel's terminology: There are so many competitors for the oxygen ions in an "invert glass" melt that none can prevail, so that no crystalline compound can be precipitated.

Experience with optical glasses agrees with such concepts. Extreme optical properties often call for glasses at the limit of the glass-formation region. The resulting tendency toward crystallization can be suppressed by increasing the number of network-modifiers and thus the degree of disorder. Unfortunately, demands for definite optical properties interfere with the full exploitation of this approach.

Huggins [41] developed what he called a "structon" theory. Deviating some-what from the network theory, he hoped to explain property discontinuities by a scheme of certain atomic groupings.

Tilton [42] coined the so-called "vitron theory". He postulated the arrangement of SiO_4 tetrahedra in five-rings, forming a pentagonal dodecahedron supposedly consisting of 20 tetrahedra. Growth of such entities called "vitrons" by additional five-rings, which was to explain their combination over regions of low ordering, is not possible without geometrical distortion. But these regions would have to be unrealistically large.

Neither Huggins' structon nor Tilton's vitron theory have achieved decisive importance. They deserve mention, however, among attempts to understand glass structures on the basis of entities more similar to molecular ones. Many other reputable researchers (for example J. E. Stanworth, W. E. S. Turner, E. Kordes, and N. J. Kreidl) were attracted by the network theory and soon contributed to its further development and our deeper understanding of glass structures.

4.6 Lebedev's Crystallite Theory

The crystallite theory of Lebedev [43] originated in parallel with the development of the network hypothesis. It was supported by X-ray investigations, in

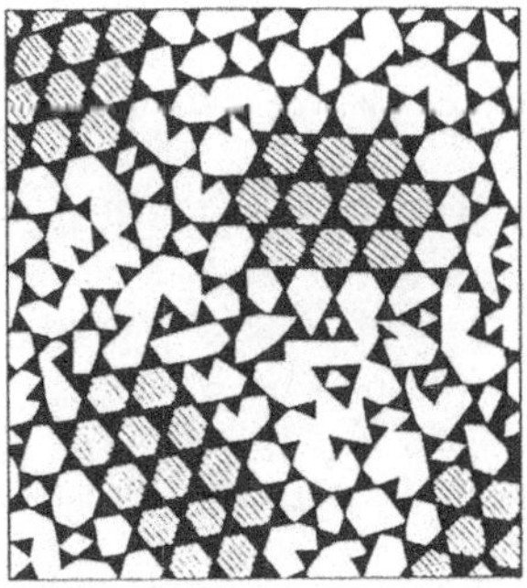

Fig. 4.8. Schematic of Lebedev's crystallite theory. Dark areas represent the crystallites surrounded and glued by a less ordered SiO_2 glass (Valenkov and Porai-Koshits [45]

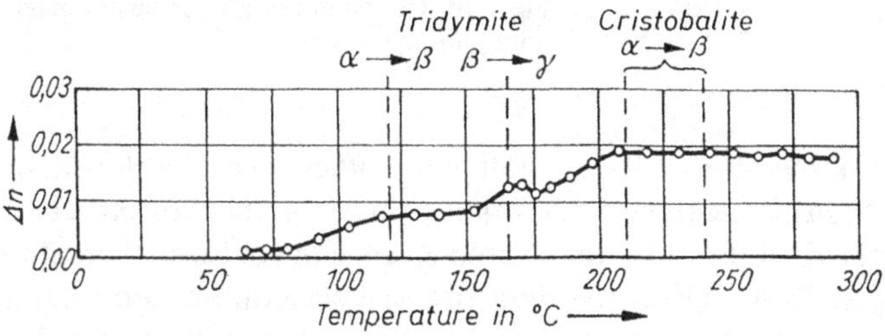

Fig. 4.9. Change in refractive index with temperature in a binary sodium silicate glass (23% Na_2O) according to Tudorovskaya [46]

this case those of Randall et al. [44] and Valenkov and Porai-Koshits [45]. At the start, the crystallite theory was called the opposite extreme of the network hypothesis. Whereas the statistical distribution of the large cations (modifiers) in a disordered network is a characteristic of the network theory, the crystallite theory is characterized by the significantly high order of these cations, i.e., the assumption of crystallites.

According to Lebedev [43], the structure of glasses can be regarded as an accumulation of microcrystalline entities, for example, in silicate glasses those consisting of SiO_2 and various silicates. Even pure SiO_2 glass was supposed to be structured in this manner (see Fig. 4.8). In compounded glasses the microcrystallites were to be either chemical compounds or solid solutions which must coincide with the phase diagram corresponding to the composition of the particular glass.

These microcrystallites were not to be visible optically or in the UV, but had to be conceived as strongly deformed structures which corresponded only approximately to the normal lattice. It was for that reason that they were to be distinguished from microcrystals by the term "crystallite [43]." The crystallite theory offers an explanation for discontinuities in the property curves of glass systems. These discontinuities were unexplained by the network hypothesis. Thus, Tudorovskaya [46], for example, attributed the discontinuous change in the refractive index of a sodium silicate with increasing temperature to $\alpha \rightarrow \beta$ transformations of the SiO_2 crystallites contained in these glasses (see Fig. 4.9).

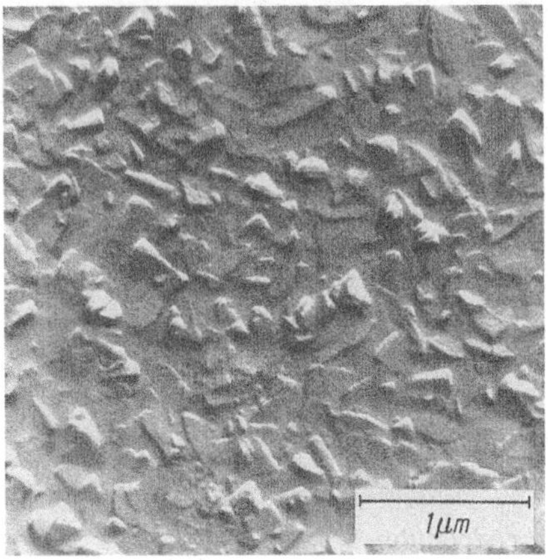

Fig. 4.10. Crystallites in a clear binary magnesium phosphate glass

Like the network hypothesis, the crystallite hypothesis was developed and modified by many glass researchers, especially those of the Soviet school. Valenkov and Porai-Koshits [45], for example, concluded from theoretical calculations based on X-ray diffraction data that the structure of a crystallite was most regular in its center, practically equal to that of the crystal. Order does decline toward the periphery of the crystallite, resulting in the formation of a strongly amorphous interlayer connecting all crystallites. The diameter of the crystallite was usually given as 0.8 to 1.5 nm (8 to 15 Å). Such dimensions correspond to only a few unit cells, so that one finds oneself on the borderline regarding validity of the lattice concept.

In the end, Lebedev [47] himself modified his original interpretation of the crystallite theory by deemphasizing the expression "crystallite." In his opinion, disordered as well as ordered regions in the form of defined chemical compounds are present in a glass. Lebedev did not feel that unsurmountable differences remained between the network theory and his concept of "microheterogeneity." Before Lebedev's modification, however, Warren had stressed the profound differences of the crystallite theory from his idea of a statistical distribution of network-modifiers.

Today one has to state objectively that there are experimental examples even for the first and most extreme formulations of the crystallite theory (see Fig. 4.10). It is true, however, that in such cases (consider also the glass-ceramics, which are to be treated later) much larger "crystallite" dimensions are observed. Also, glasses whose structure corresponds to the crystallite theory are clearly extreme or boundary cases. The majority of glasses appear to assume a middle position between network and crystallite theory. During development of the crystallite concepts, it may have been unfortunate that many experiments were conducted on SiO_2 glass prepared from ground rock crystal. Considering the extremely high temperatures required to melt the modifications of crystalline SiO_2, the lattice may not have collapsed completely, especially in the center of the grains, when this older process was used. Experimentally demonstrated

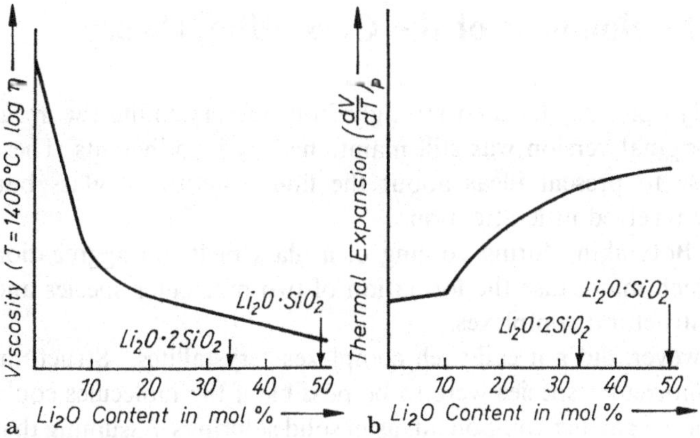

Fig. 4.11. Discontinuous property changes depending on composition in a binary lithium silicate glass. **a** Schematic for viscosity (Dietzel [32]; **b** schematic for thermal expansion (Mackenzie [48]

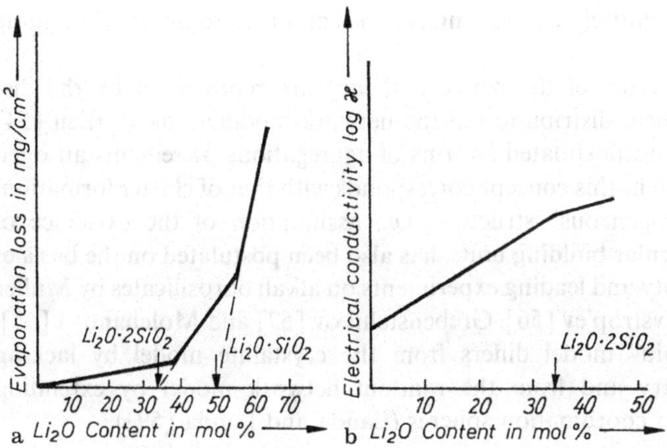

Fig. 4.12. Discontinuous property changes depending on composition in a binary lithium silicate glass. **a** Schematic for evaporation loss (Turner [49]; **b** schematic for electrical conductivity (Kusnetsov [50]

"crystallites" thus may not have been ordered regions formed from a homogeneous melt, but residual structures. Silica glass produced according to modern processes from the gas phase (i.e., via hydrolysis or plasma techniques from pure $SiCl_4$) are evidently much more homogeneous.

The crystallite theory was used to explain discontinuities in the relations of properties to composition (see Figs. 4.11 and 4.12). The observed discontinuities, maxima, or minima were assumed to relate to the appearance of a new crystallite species. This conclusion seemed plausible since these discontinuities are frequently located at compositions corresponding with defined chemical compounds.

4.7 Further Development of the Crystallite Theory

Botvinkin's [51–53] opinions deviated strongly from the crystallite theory at a time when the original version was still maintained by its adherents. These opinions were close to present ideas about the fine structure of glass but, unfortunately, they received little attention.

According to Botvinkin, during cooling of a glass melt ion aggregation processes occur which may cause the formation of two molecular species and two differentiated structural complexes.

Botvinkin, however, did not call such complexes "crystallites." Structural complexes of two molecular species were to be possible if the molecules could enter the "lattice" of the partner component, as in solid solutions. Assuming that building units in a glass correspond to defined stable crystallized compounds, Botvinkin explained the similarity of some glass and crystal property curves, as well as the coincidence of discontinuities in glass properties with the location of stable compounds in the phase diagram (see Fig. 4.12).

Today Botvinkin's ideas must be considered a decisive step toward assimilating the originally entirely contrasting theories and toward an understanding of most glasses.

Similarly, adherents of the network theory are represented by the discussions of a statistical distribution of the network-modifiers by Hartleif [54] and Dietzel [32], who postulated swarms of aggregations. It remains an open question to what extent this concept corresponds with that of cluster formation.

A "microheterogeneous" structure, i.e., assumption of the existence of defined stable molecular building units, has also been postulated on the basis of electrical conductivity and leading experiments on alkali borosilicates by Müller and Markin [55], Evstrop'ev [56], Grebenstchikov [57] and Molchanova [58]. A recent "crystalloid" model differs from the crystallite model by lacking translation symmetry and from the random network model by extending beyond the first two coordination spheres (Landa and Landa [59]).

4.8 Kinetic Theory

Since glass formation has been found in materials of any compositional, bonding, or structural type, it has become more useful to consider how rapidly a liquid or vapor of any type must be cooled to avoid a detectable (e.g., 10^{-6}) volume fraction of crystallization as a characteristic of glass formation (Uhlmann [60–62]).

Assuming that nucleation frequency (I) (s^{-1}) and growth rate (u) ($cm \cdot s^{-1}$) are known as functions of temperature, the time needed to obtain a volume fraction of 10^{-6} can be plotted as a function of temperature from Eq. (4.1):

$$10^{-6} = (\pi/3)\, I u^3 t^4 \qquad (4.1)$$

which is easily understood if it is considered that It is the frequency of nucleation over time t and u^3t^3 is the growth in three dimensions in time t. The plot thus obtained, called a T-T-T (time-temperature-transformation) curve – or, from its shape "nose curve" – is illustrated in Fig. 4.13. The shortest time needed for growth at any temperature is called the nose time t_n, the corresponding temperature of fastest crystallization is called the nose temperature T_n.

The nose results from the fact that both I and u have more or less sharp, somewhat different, maxima as a function of cooling below the melting temperature. Both I and u increase first with decreasing temperature because the thermodynamic driving force increases; but on further cooling this influence is overcome by the kinetic hindrance due to the increased viscosity.

From the T-T-T ("nose") curve one is able to calculate the rate of cooling needed at each temperature to avoid a fraction of crystallized volume above 10^{-6}. A rough estimate for the critical rate is given by Eq. (4.2):

$$\mathrm{d}T/\mathrm{d}t \approx \Delta T_\mathrm{n}/t_\mathrm{n} \tag{4.2}$$

where $\Delta T_\mathrm{n} = T_1 - T_\mathrm{n}$, the distance from the liquidus (T_1) to the nose (T_n) temperature.

In practice, nucleation is mostly promoted by heterogeneities which have a strong effect on the critical cooling rate (Onorato et al. [63], (Table 4.4)).

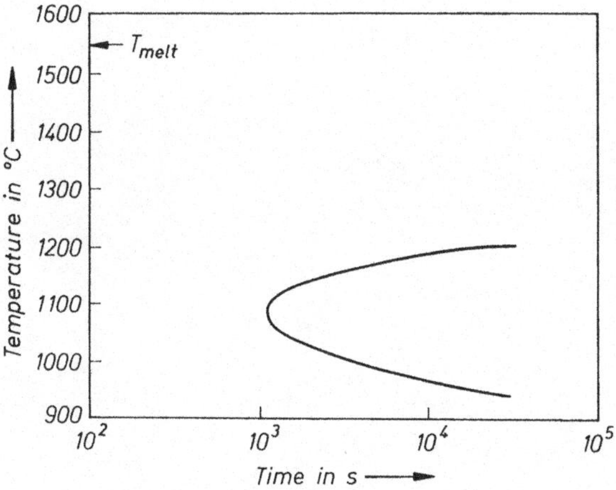

Fig. 4.13. "T-T-T" ("nose curve"). Plot of temperature T vs. time τ required to obtain 10^{-6} of the volume crystallized. The curve shifts very little for lower volume fractions, but depends strongly on composition

Table 4.4. Critical cooling rates of some materials[a]

Material	Homogeneous Nucleation	Heterogeneous nucleation contact angle (deg)		
		100	60	40
SiO_2 glass	9×10^{-6}	10^{-5}	8×10^{-3}	2×10^{-1}
GeO_2 glass	3×10^{-3}	3×10^{-3}	1	20
$Na_2O\cdot2SiO_2$ glass	6×10^{-3}	8×10^{-3}	10	$3\times10^{+2}$
Typical metal	9×10^{8}	9×10^{9}	10^{10}	5×10^{10}

[a] After Onorato and Uhlmann [63]

dT/dt may have to be 5–10 times faster if heterogeneities are present (see table). The effect of heterogeneities is largest if the contact angle glass heterogeneity is small ($\ll 80°$); it is negligible if it exceeds $120°$.

From a practical viewpoint, the influence of viscosity on nucleation and growth kinetics indicates that, for glass formation, (a) high viscosity is advantageous at T_1 and (b) even if the viscosity is not very high at T_1, a steep increase below T_1 is advantageous. Also, compositions close to eutectics and complex compositions hindering arrangement and growth of nuclei are often advantageous. Generally, of course, low-viscosity liquids will require very rapid cooling, for example, spat cooling ($10^6\ K \cdot s^{-1}$) or vapor deposition ($> 10^6\ K \cdot s^{-1}$). Metal glasses, once believed unobtainable, are examples of new industrial products and uses (VanDerSande et al. [64], Haasen [65]). A recent extended theory of disordered networks (Gupta and Cooper [66, 67] interestingly relates general aspects of topology with glass formation. Interpretations of the structure of simple non-crystalline solids were competently "revisited" by a team of leading specialists in the field. (Wright et al. [68, 69]).

5 Methodology in Glass Research

The number and multiplicity of methods serving glass research today are constantly increasing and enlarging. This is no doubt the consequence of the higher complexity and lower regularity of constituent species and sites. Unlike the case of crystal chemistry, it is not easy to arrive at a unique structural model using Fourier analysis. As a rule one depends on the interaction of a multiplicity of indications, through various measuring techniques, to obtain indirect proofs for the most probable structure. The number of publications of such investigations poses a problem in itself for adequate presentation such as has been achieved by, for example, Scholze [70]. For this reason the present volume will have to concentrate on selected essential topics.

On the other hand, there is increasing evidence for the influence of the structure of the melt from which the glass is produced on the resulting glass structure. Not all structural elements of the crystal are destroyed right above the liquidus (Trostel [71]). It has therefore become necessary to investigate the structure of liquids, for instance, by (a) segregation of aggregations or microphases at various temperatures (Hummel et al. [72, 73], Johnson et al. [74]) or (b) use of "color indicators" to assess transmission or reflection changes from the UV to the IR from the liquid to the solid state. There are indications that such methods will provide answers regarding the formation of complexes and aggregations in the liquid.

Weyl [75] has introduced concepts for comparing types of glass structure with the different models of liquids proposed by Bernal [76], Frenkel [77], and Stewart [78]. It will not be possible, would indeed be presumptive, to discuss and assess to a proper extent in this book even the most important methods used to elucidate glass structure such as: X-ray diffraction; small-angle X-ray diffraction (Hoffman et al. [79], Porai-Koshits et al. [80, 81], Zarzycki et al. [82], Becherer et al. [83], Patel et al. [84], Urnes [85], Ruland [86]); UV, visible, IR spectroscopy (Scholze [87], Florinskaya [88], Bates [89]); Mössbauer spectroscopy (Kurkjian [90]); nuclear magnetic resonance (Hatton et al. [91], Silver et al. [92], Bray et al. [93–106], Leventhal et al. [107], Landsberger et al. [108], Kriz et al. [109–112], Park et al. [113], Kim et al. [114, 115], Hendrickson et al. [116], Müller-Warmuth et al. [117–120], Bishop et al. [121, 122], Kramer et al. [123], Jellison et al. [124, 125], Kline et al. [126], Rhee [127–129], Schulz et al. [130], Weaver [131], Holzmann et al. [132], Mosel et al. [133], Lechus [134], Milberg et al. [135], Otto et al. [136], Baugher et al. [137], Panek et al. [138], Van Vleck [139], Lamb [140], Ramsey [141], Krogh-Moe [142, 143], Greenblatt et al. [144], etc.); electron microscopy; measurement of other optical, mechanical, magnetic or electrical properties; scattering measurement (Maurer [145], Goldstein [146, 147], Hammel [148]).

A more recent, extensive, and excellent text embracing all spectroscopic methods is by Wong and Angell [18].

Two methods will be described here: nuclear magnetic resonance and electron microscopy. Nuclear magnetic resonance, applied particularly by Bray and collaborators, provides definite information on the smallest regions, particularly the coordination and bond type of individual constituents. Electron microscopy has supplied new insight into large structural units and their influence on various properties.

5.1 Structure of Liquids and Melts

Since structural research on glass will increasingly stress the origin of the formation of microstructure in order to better understand the fine structure of glasses, as well as to better control micro processes, it appears necessary to briefly treat some models developed for liquids.

Weyl [75], in particular, attempted to relate three of the most important models, those of Bernal [76], Frenkel [77], and Stewart [78], to the probable structure of glass melts.

Bernal's model attributes the disorder of liquids to a very small concentration of particles with coordination numbers deviating slightly from the normal, e.g., from 6 to 5 or 7. Bernal describes the liquid by the coordination number of the particles and a disorder function. All rules pertaining to crystals are valid, if viewed statistically, for this model liquid which resembles an ideal crystal. The disorder of a Bernal liquid contains neither crystalline areas nor cavities capable of accepting particles. According to Bernal the liquid is "flawless," i.e., does not contain typical defects. Because of considerable netting it must have high viscosity at the melting temperature, which favors freezing to a glass. As an example, a cristobalite melt comes close to this model.

Frenkel's model is an opposite extreme of Bernal's. It is assumed that numerous bonds spontaneously open on an atomic scale and close on other locations, while other bonds are broken. In this manner the entire liquid contains a system of cavities and fractures. The high fluidity follows from the ease with which such a melt forms fissures. That is why the model is also termed "fissured." Crystals melting to such liquids cannot be overheated, nor their melts undercooled. This is why glass formation cannot be anticipated for them. An example for this model is an NaCl melt.

Stewart's model admits molecular aggregations with a certain degree of order, not, however, identical with ordered crystalline regions or crystallites. To differentiate these aggregations from crystalline regions, Stewart called them "cybotactic regions." This means that a crystal need not melt to a clear isotropic liquid, but may form a mixture of "two phases," one of which may be orientable, causing the entire system to be birefringent. As a rule, with increasing temperature the "ordered" regions decrease in size, the liquid becoming optically clear and isotropic. Thus Stewart's model somewhat resembles crystalline liquids.

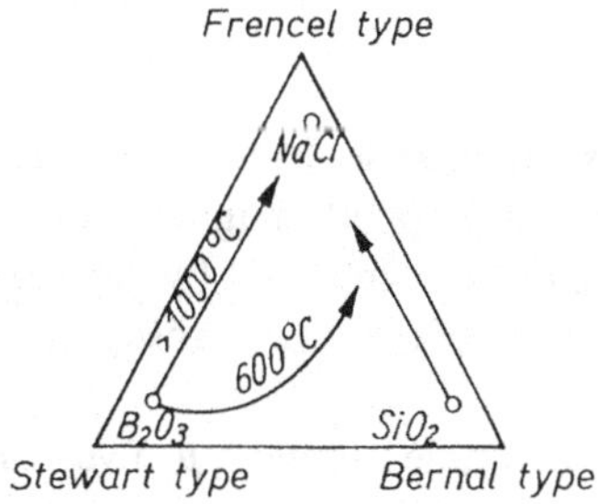

Fig. 5.1. Classification of melts according to various types of liquid structures (after Weyl [75])

As shown in Fig. 5.1, Weyl in a way classified glass melts according to these three types of liquids. As a rule, glass-forming melts tend to resemble the Bernal type. However, basic phosphate glasses were shown experimentally to contain orientable chain elements. If the glasses are stretched to fibers, the chain elements orient in the direction of tension and cause, in fact, birefringence. B_2O_3 glass behaves in a similar manner; apparently the planar triangles are somewhat oriented in the direction of tensions.

The arrows in Fig. 5.1 indicate which type of liquid an SiO_2 or B_2O_3 melt is shifted toward when an alkali is added. An increase in temperature also shifts, on principle, all glass structures toward the Frenkel type. Pressure, in contrast, often causes a shift toward the Bernal type, i.e., toward improved glass formation [149, 150].

5.2 The Nuclear Magnetic Resonance Method as Applied to Glass Research (P. J. Bray)

5.2.1 Introduction

Nuclear magnetic resonance (NMR) spectra have been observed in glasses since 1951 [91]. Much of the subsequent work [92–104, 107–110, 113–116] has focused on boron–oxygen bonding and configurations in borate glasses, but the studies have also involved hydrogen [117, 118], lithium [92, 97, 99, 100, 102, 116, 122, 123], beryllium [92, 99, 100], oxygen [124], fluorine [96, 97, 126], sodium [92, 100, 126, 127], aluminum [92, 119, 121, 130], silicon [91, 105, 117, 131–133], phosphorus [96, 99, 100], scandium [99], vanadium [99–102, 104, 108], selenium [134], cadmium [99], tin [104], tellurium [100–102, 104], cesium [99, 100, 128, 135, 136], thallium [100–102, 104, 136–138] and lead [95–97, 99, 104, 107].

Some examples of this work will be reviewed after presentation of the basic NMR theory.

5.2.2 Basic NMR Theory

NMR studies are possible because of energy levels which arise from the interaction of nuclear magnetic dipole moments with an applied magnetic field. Transitions between these Zeeman energy levels can be stimulated by electromagnetic radiation of the proper frequency; monitoring of the power absorption yields the absorption line shape.

The atoms of most nuclei do possess a magnetic dipole moment, $\boldsymbol{\mu}$, a vector quantity that can be expressed as

$$\boldsymbol{\mu} = g\mu_0 \boldsymbol{I} \tag{5.1}$$

where $\boldsymbol{I}$ is the dimensionless spin vector of magnitude $[I(I+1)]^{1/2}$; the scalar I (called the spin of the nucleus) is an integer or half integer, μ_0 is the Bohr nuclear magneton, and g (a pure number) is the nuclear g factor. Application of a magnetic field produces the Zeeman interaction and a set of energy levels given by

$$E = -\boldsymbol{\mu}\cdot\boldsymbol{H} = -g\mu_0 \boldsymbol{I}\cdot\boldsymbol{H} = -g\mu_0 Hm \tag{5.2}$$

where m is the magnetic quantum number having $2I+1$ possible values ranging from $-I$ to $+I$ in integer steps. In the absence of other interactions, the $2I+1$ energy levels are equally spaced, as shown in Fig. 5.2 a for the case $I = 3/2$.

Transitions between adjacent energy levels in Fig. 5.2 a can be produced by electromagnetic radiation whose frequency ν_0 satisfies the Bohr resonance condition

$$h\nu_0 = \Delta E = g\mu_0 H \tag{5.3}$$

where ΔE is the difference in energy between adjacent levels. There is only one frequency in this case, since the Zeeman levels are equally spaced. If the magnetic field H or radio frequency ν is swept slowly through the resonance condition of Eq. (5.3), power is absorbed and the resonance is detected.

Interactions other than the Zeeman interaction can split the Zeeman energy levels and produce structure or broadening of the absorption line shape. Several of those interactions (dipolar, chemical shift, quadrupolar) will be analyzed in the following sections, with presentations of examples from NMR studies of glasses.

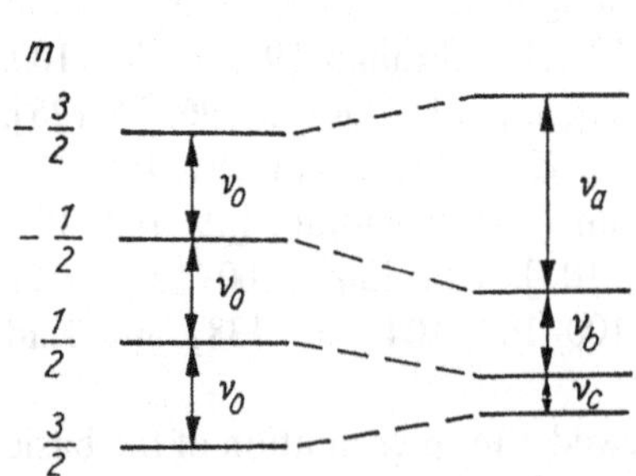

Fig. 5.2. Energy levels arising from the interaction of the nuclear magnetic dipole moment with a magnetic field. **a** No electrical quadrupole interaction; **b** small quadrupole interaction

5.2.3 Dipolar Interaction

A reasoning nucleus in a solid will, in general, experience a set of local magnetic fields arising from other nuclei in the material. If the ith nucleus were to experience such a field from only one other nucleus labeled the jth nucleus, the dipole-dipole interaction energy would be

$$\frac{\boldsymbol{\mu}_i \cdot \boldsymbol{\mu}_j}{r_{ij}3} - \frac{3(\boldsymbol{\mu}_i \cdot \boldsymbol{r}_{ij})(\boldsymbol{\mu}_j \cdot \boldsymbol{r}_{ij})}{r_{ij}5} \tag{5.4}$$

where $\boldsymbol{\mu}_j$ and $\boldsymbol{\mu}_j$ are the magnetic moments of the ith and jth nuclei, and $\boldsymbol{r}_{ij}$ is the position vector of the jth nucleus relative to the ith nucleus. When more than one other nucleus produces a local field at the ith nucleus, the appropriate sum of interactions must be computed. Each nucleus of type i will, in general, experience a different total local field. Since all nuclei of type i contribute to the NMR spectrum, the resulting distribution of local fields will split the Zeeman energy levels and broaden the absorption curve or give it structure. It is the dipolar broadening that converts the Δ-function response of Eq. (5.3) into a smooth bell-shaped or structured curve. Figure 5.3 displays the ^{11}B resonance for four-coordinated borons in a glass of composition $Na_2O \cdot 2B_2O_3$. The curve is relatively narrow but dipolar broadening does produce a peak-to-peak width of about 6 kHz in this recording of the first derivative of the absorption curve.

It is possible in principle to calculate the line shape $g(\nu - \nu_0)$ of the dipolar-broadened resonance. However, this is too difficult in practice and is generally replaced by calculations of the "moments" of the line shape. The so-called "second moment" of the line is most commonly employed; it is defined as

$$M_2 \equiv \int_{-\infty}^{\infty} (\nu - \nu_0)^2 g(\nu - \nu_0) \mathrm{d}\nu \bigg/ \int_{-\infty}^{\infty} g(\nu - \nu_0) \mathrm{d}\nu \tag{5.5}$$

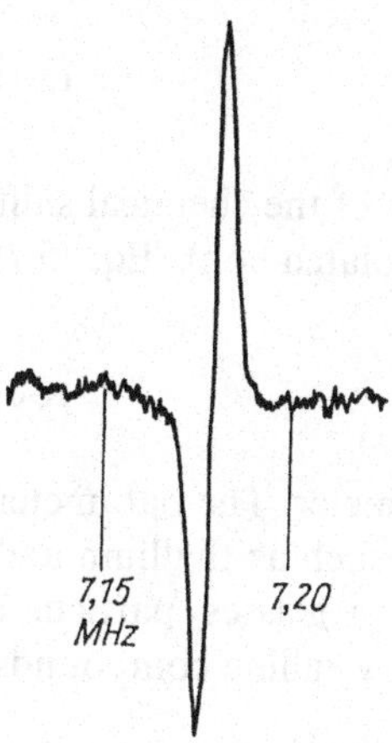

Fig. 5.3. NMR spectrum of ^{11}B in BO_4 units of a glass of composition $Na_2O \cdot 2B_2O_3$ (first derivative of the absorption curve)

and has been calculated by van Vleck [139]. For a glass, in which all orientations of the vector $\boldsymbol{r}_{ij}$ are equally probable, the angular dependences must be averaged over all angles and the result is

$$M_2 = K_1 \sum_j r_{ij}^{-6} + K_2 \sum_k r_{ik}^{-6} \tag{5.6}$$

The first term involves interactions among the resonating nuclei; the second term involves interactions of the resonating nuclei with one type of dissimilar nucleus. K_1 and K_2 depend on the properties of the resonating and dissimilar nuclei, respectively. If there is more than one dissimilar nucleus, additional terms of the second type must be added. Values of M_2 computed from Eq. (5.6) on the basis of particular structural models for the glass can be compared with values of M_2 obtained from Eq. (5.6) by employing the line shape $g(\nu - \nu_0)$ determined from the actual NMR spectra.

Müller–Warmuth and coworkers [133] have measured the dipolar line width of ^{23}Na- and ^{29}Si-NMR spectra for sodium silicate glasses containing between 8 and 33 mol% Na_2O. They found that the second moments obtained from Eq. (5.6) on the assumption that the sodium ions are uniformly distributed in the glasses are in substantial disagreement with the experimental results. The large experimental M_2 for sodium shows that most of the sodium ions are grouped in pairs or clusters; the separation of the ions is considerably less than that which would occur if the distribution were uniform.

5.2.4 Chemical Shift

The magnetic field $\boldsymbol{H}$ at a nucleus during an NMR experiment will generally differ in both magnitude and direction from the applied field $\boldsymbol{H}_{\text{app}}$. This effect is in addition to, and apart from, the dipolar broadening interaction discussed above, and originates [140, 141] in diamagnetic and second-order paramagnetic changes of the atomic electron distribution, caused by the applied field. The components of $\boldsymbol{H}$ are related to the components of $\boldsymbol{H}_{\text{app}}$ by

$$H_i = \sum_{j=1}^{3} (\delta_{ij} - \sigma_{ij})(H_{\text{app}})_j \quad i = 1, 2, 3 \tag{5.7}$$

where δ_{ij} is the delta function and the σ_{ij} are components of the chemical shift tensor. For the case of spherical symmetry (e.g., an isolated ion), Eq. (5.7) becomes

$$\boldsymbol{H} = (1 - \sigma)\boldsymbol{H}_{\text{app}} \tag{5.8}$$

where the single scalar σ is called the chemical shift parameter. The parameter σ is relatively large (e.g., of order 10^{-3}) for heavy atoms such as thallium and lead, and this isotropic shift has been used to identify, in glasses, particular metal-oxygen bonding configurations that occur in the crystalline compounds of the systems [95, 107].

If spherical symmetry is lacking, often there is still present an axis of 3-fold or higher symmetry (e.g., a single chemical bond, or one axis of a tetrahedral bonding arrangement). In that case, the magnitude H of the field at the nucleus can be expressed as

$$H = H_{\text{app}}[1 - \sigma_{\perp})^2 \sin^2\theta + (1 - \sigma_{\parallel})^2 \cos^2\theta]^{1/2} \tag{5.9}$$

where θ is the angle between $\boldsymbol{H}_{\text{app}}$ and the symmetry axis, and $\sigma_{\perp}$ and $\sigma_{\parallel}$ are the values of σ in Eq. (5.8) when $\boldsymbol{H}_{\text{app}}$ is perpendicular ($\sigma_{\perp}$) or parallel ($\sigma_{\parallel}$) to the symmetry axis.

The value of $\boldsymbol{H}_{\text{app}}$ required to produce $\boldsymbol{H}$ for the resonance condition (Eq. (5.3)) can be obtained from Eq. (5.9) if $\sigma_{\perp}$, $\sigma_{\parallel}$, and θ are known. But all values of the orientation angle θ are equally probable in a glass, so the observed NMR response will encompass all values of $\boldsymbol{H}_{\text{app}}$ at which resonance will occur as θ is varied from 0 to π. The resulting powder pattern is displayed at the top of Fig. 5.4. (The pattern is given as a function of the variable spectrometer frequency for a fixed value of $\boldsymbol{H}_{\text{app}}$, rather than as a function of $\boldsymbol{H}_{\text{app}}$ for a fixed frequency ν_0. Equation (5.3) provides the conversion between frequency and magnetic field.) If dipolar broadening is invoked to smooth the curve, and the first derivative is taken for direct comparison with the recorder tracings obtained from the spectrometer, one obtains the line shape given by the dotted line in the bottom half of Fig. 5.4.

An actual NMR spectrum for this case of an axially symmetric chemical shift is shown in Fig. 5.5. Here the splitting between the "shoulder" n the left of the trace and the "well" at the right of the response is about 15 G at a spectrometer frequency of 16 MHz. This splitting varies [100–102, 104] as a function of composition, revealing the presence of more than one type of tellurium–oxygen bonding configuration in these glasses. A second example,

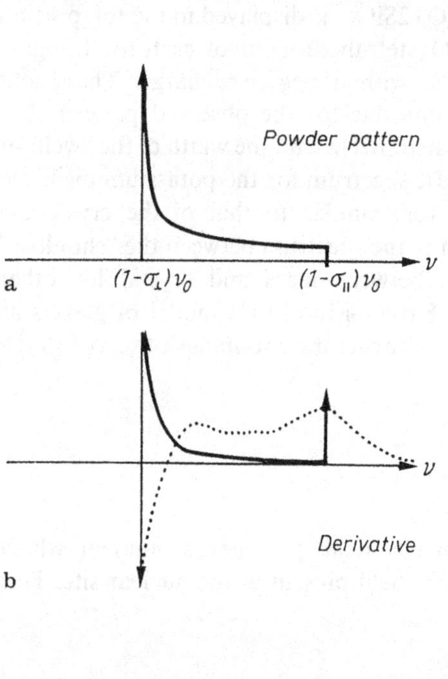

Fig. 5.4. Powder spectrum. **a** Powder pattern exhibiting an anisotropic axially symmetric chemical shift; **b** derivative exhibiting dipolar broadening (dashed line)

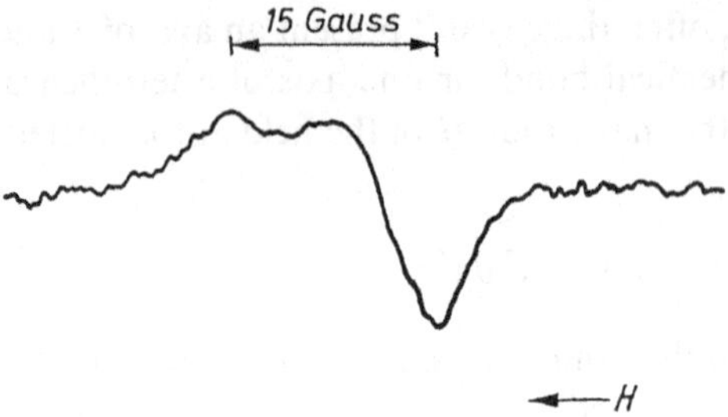

Fig. 5.5. ^{125}Te-NMR spectrum at 16 MHz in a glass of composition 70 mol% TeO_2, 30 mol% V_2O_5

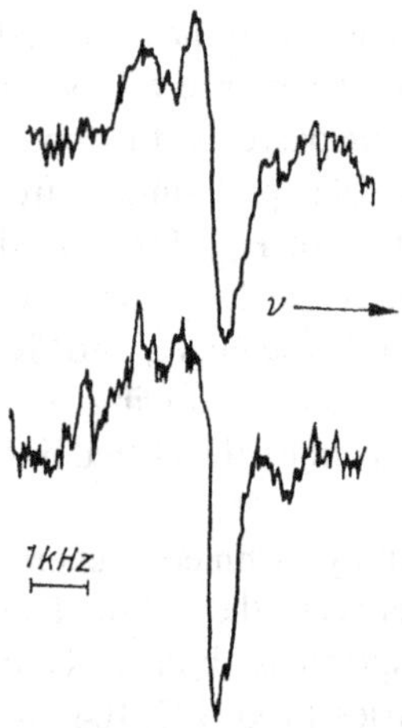

Fig. 5.6. ^{29}Si-NMR spectrum at 10 MHz for polycrystalline (top) and vitreous (bottom) material of composition $K_2O \cdot 2SiO_2$

observed for ^{29}Si in polycrystalline $K_2O \cdot 2SiO_2$, is displayed in the top portion of Fig. 5.6. In this material, each SiO_4 tetrahedron involves three bridging oxygens, and one nonbridging oxygen with a negative charge. The 3-fold symmetry about the latter axis is responsible for the observed pattern. The splitting in Fig. 5.6 is only about 1.4 G at 10 MHz, and the width of the "well" at half-depth is less than 0.33 G. The NMR spectrum for the potassium disilicate glass (bottom portion of Fig. 5.6) is very similar to that of the crystalline material, though there is some filling in of the spectrum between the "shoulder" and the "well." Similar relationships between glass and crystal for other K_2O–SiO_2 compositions support the Krogh–Moe [143] model of glasses as composed of random arrangements of the structural groupings observed in the crystalline compounds.

5.2.5 Quadrupole Interaction

A nucleus of spin $I \geqslant 1$ possesses an electrical quadrupole moment which interacts with the gradient of any electric field present at the nuclear site. This interaction shifts the Zeeman energy levels as depicted in Fig. 5.2b. For cases in which the quadrupole interaction is sufficiently small with respect to the Zeeman interaction, the quadrupole interaction can be treated as a perturbation, yielding an NMR transition frequency given by

$$\nu = \nu_0 + \nu_Q\left(m - \frac{1}{2}\right)A + \frac{\nu_Q^2}{\nu_0}C + \frac{\nu_Q^3}{\nu_0^2}\left(m - \frac{1}{2}\right)E. \tag{5.10}$$

Equation (5.10) is calculated to the third order using perturbation theory. Here ν_0 is the resonance frequency when only the Zeeman interaction occurs (Eq. (5.3)), and

$$\nu_Q = 3Q_{cc}/2I(2I-1)$$
$$Q_{cc} = e^2qQ/h$$
$$\eta = (V_{xx} - V_{yy})/V_{zz}$$

where eQ is the electrical quadrupole moment of the nucleus, η is the asymmetry parameter, and the V_{ij} are the components of the electric field gradient (EFG) tensor in its principal axis system, defined so that $|V_{zz}| \geqslant |V_{yy}| \geqslant |V_{xx}|$ and $eq \equiv V_{zz}$. The terms A, C, and E in Eq. (5.10) are complicated functions of m, I, $\cos\theta$, $\cos 2\phi$, and η. The quantity m denotes the transition from the energy level labeled by the magnetic quantum number m to the level labeled by $m-1$, and θ and ϕ are the Euler angles of the magnetic field $\boldsymbol{H}_{\text{app}}$ with respect to the principal axis system of the EFG tensor.

When the sample is polycrystalline or vitreous, all of the angles θ and ϕ are equally probable. Therefore, the resonance condition (Eq. (5.10)) must be averaged over all possible values of θ and ϕ. The resulting pattern is called a powder pattern. Figure 5.7 depicts this pattern (dashed lines) and the pattern with dipolar broadening (solid line) for a small quadrupole interaction ($A_1 = Q_{cc}/4$). An actual recorder trace of the first derivative of the absorption curve for this case is presented in Fig. 5.8 where the ^{11}B resonance in crystalline BPO_4 has been detected. The so-called "central" transition ($m = \frac{1}{2} \leftrightarrow m = -\frac{1}{2}$) is unaffected in this case of a very small Q_{cc}, because it is influenced only in second and higher orders (see Eq. (5.10)). The "wings" in the resonance arise from the structure at $\pm A_1$ in Fig. 5.7. The coupling constant Q_{cc} is only 50.4 kHz in this

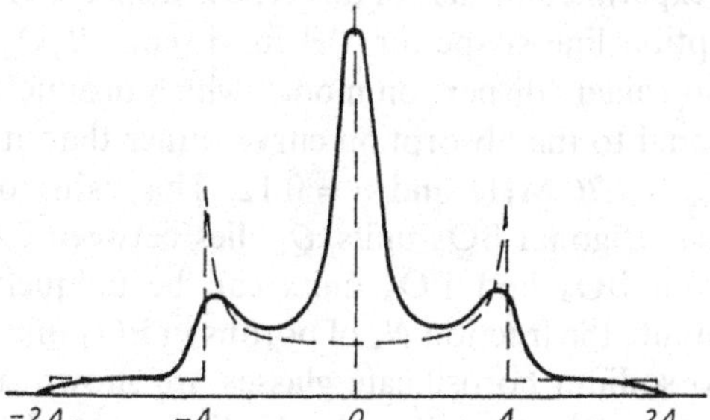

Fig. 5.7. Predicted resonance line shape (solid curve) for a nucleus with spin $I = 3/2$ (^{11}B) with a small quadrupole interaction in a polycrystalline powder or a glass

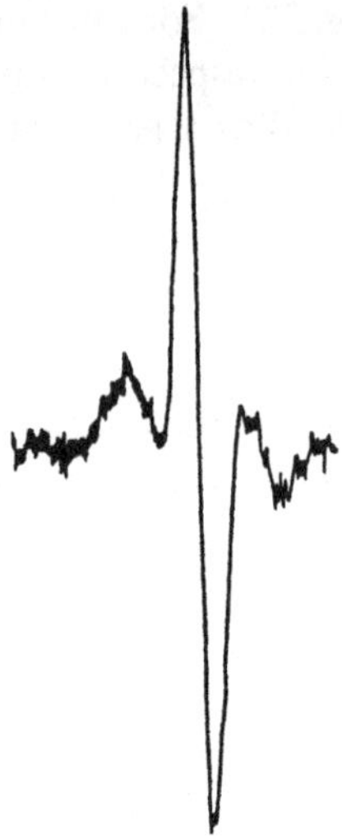

Fig. 5.8. NMR spectrum of ^{11}B polycrystalline boron phosphate (BPO_4). $\nu_0 = 7.177$ MHz

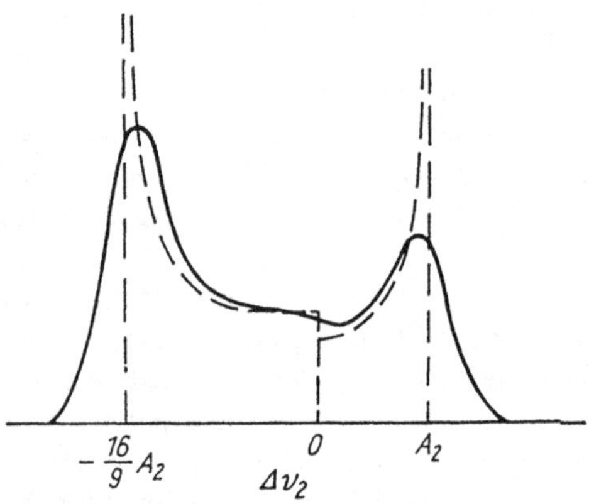

Fig. 5.9. NMR resonance line (solid curve) for a nucleus with a large quadrupole interaction Q_{cc} and $\eta = 0$ in a glass or a polycrystalline powder

case, whereas the Larmor frequency ν_0 is 7.177 MHz. Boron atoms in BO_4 configurations characteristically give rise to the narrow central line with $Q_{cc} \leqslant 850$ kHz; the 'wings" in Fig. 5.8 can be detected in the glasses in some cases, but with difficulty.

A much larger value of Q_{cc} produces the powder pattern displayed in Fig. 5.9 when η is small or zero. An experimental case of this type is displayed in Fig. 5.10 which presents the absorption line shape for ^{11}B in vitreous B_2O_3. (The spectrometer was run in the so-called "dispersion mode" which produces a response that is directly proportional to the absorption curve rather than its derivative.) For vitreous B_2O_3, $Q_{cc} = 2.76$ MHz and $\eta = 0.12$. This value of Q_{cc} is characteristic for ^{11}B in planar trigonal BO_3 units; Q_{cc} lies between 2.4 and 2.8 MHz, so the responses from BO_4 and BO_3 units can be uniquely identified and used [94, 120] to compute the fraction N_4 of borons in BO_4 units in a glass. The ^{11}B spectra for two sodium borosilicate glasses are shown in Fig. 5.11, where the lines from BO_4 and BO_3 units are clearly distinguished.

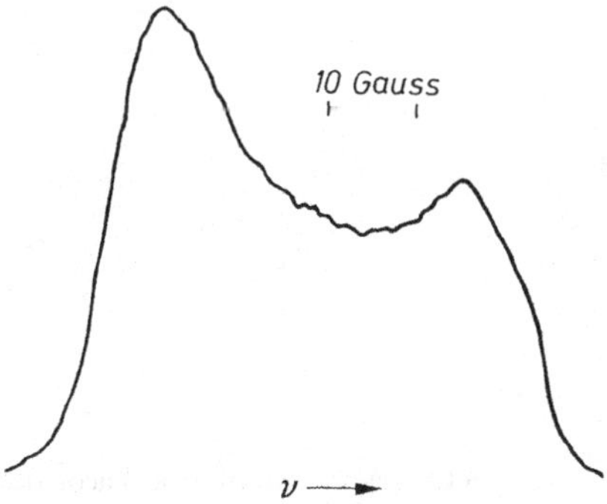

Fig. 5.10. ^{11}B spectrum for vitreous B_2O_3 at 16 MHz (dispersion mode)

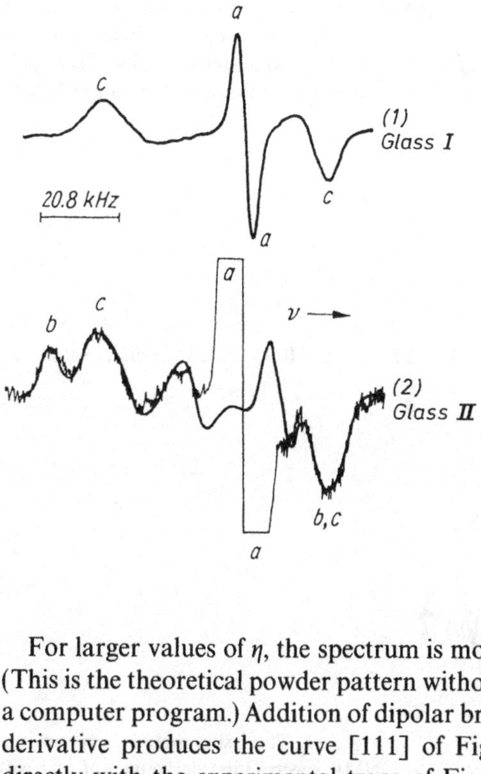

Fig. 5.11. ^{11}B NMR spectrum (first derivative) at 16 MHz for two sodium borate glasses (I. 0. 11 Na_2O, B_2O_3, SiO_2; II. 1.3 Na_2O, B_2O_3, SiO_2). The peak *a-a* arises from B atoms in BO_4 units in each glass, the peaks *c* from symmetrical BO_3 units, the peaks *b* in glass II from asymmetrical BO_3 units. The heavy solid glass for glass II is a computer simulation for the spectrum from both types of BO_3 units

For larger values of η, the spectrum is more complex, as shown in Fig. 5.12a. (This is the theoretical powder pattern without dipolar broadening simulated by a computer program.) Addition of dipolar broadening and generation of the first derivative produces the curve [111] of Fig. 5.12(b), which can be compared directly with the experimental trace of Fig. 5.12(c) for polycrystalline calcium metaborate. This material contains the chains $(BO_2^-)_n$ depicted in Fig. 5.13; here each BO_3 unit has one nonbridging and two bridging oxygens. The 3-fold symmetry about the axis perpendicular to the BO_3 plane is destroyed, and $\eta = 0.54$. These asymmetric units can also be detected quantitatively in glasses [99, 100, 102, 104, 110, 113] as can be seen from the spectrum for glass No. 11 in Fig. 11.

Another case of a large Q_{cc} and large η is depicted in Fig. 5.14, which presents the ^{17}O-NMR spectrum (first dervative) for oxygen in vitreous B_2O_3. The simulated spectrum (smooth line) in Fig. 5.14 involves only one oxygen site

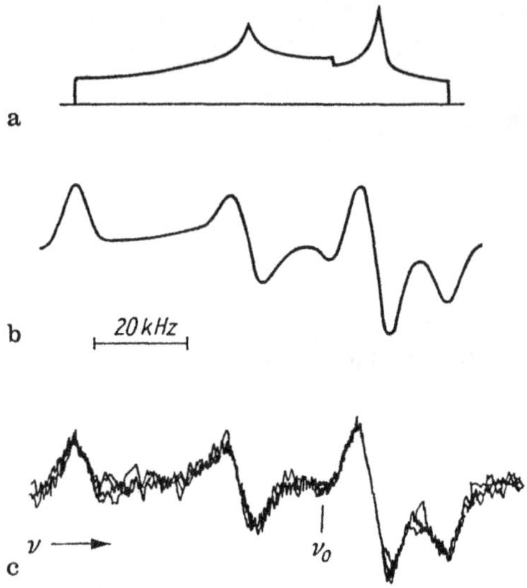

Fig. 5.12. NMR spectrum. **a** Theoretical powder pattern for the central transition ($m = \frac{1}{2} \rightarrow m = -\frac{1}{2}$) with $I = 3/2$. $\nu_0 = 16$ MHz, $Q_{cc} = 2.56$ MHz, and $\eta = 0.54$, **b** First derivative of (a) after dipolar broadening; **c** Superposition of four experimental traces for ^{11}B in polycrystalline calcium metaborate at 16 MHz

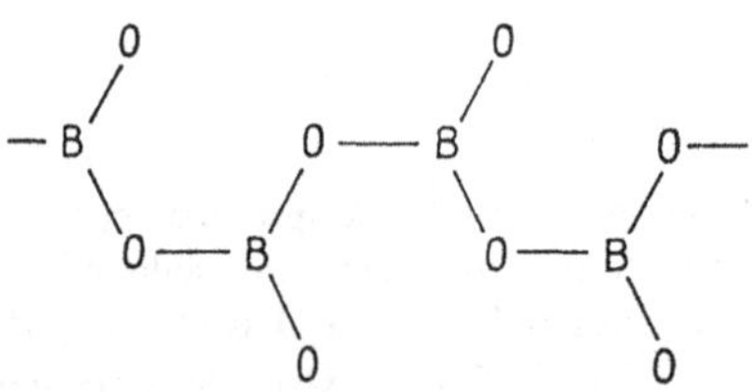

Fig. 5.13. Part of BO_2^- chain in calcium metaborate

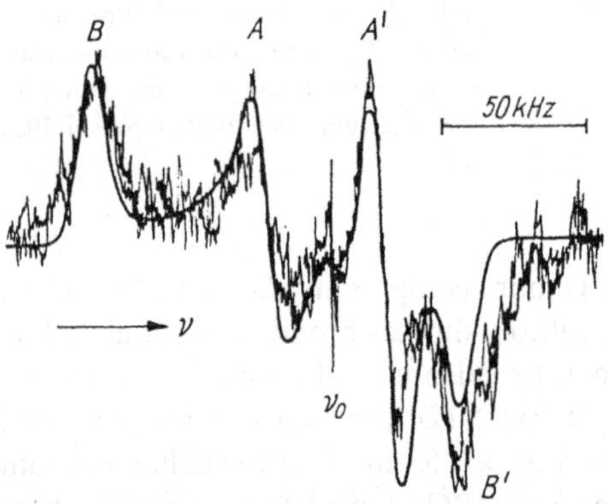

Fig. 5.14. The superposed experimental NMR spectra (first derivative) of ^{12}O with inscribed one- site simulated curve

with $Q_{cc} = 4.69$ MHz, $\eta = 0.58$ but no distributions in either quantity. There is a good fit of the features A, A′ and one concludes that this portion of the spectrum arises from the oxygen O(R) in the boroxol rings (Fig. 5.15) which Krogh-Moe [143] concluded were the major structural arrangement in this glass. The rings are stable and relatively rigid, so that no large variations of angles and bond lengths should be present to produce distributions in Q_{cc} and η. But there is clearly disagreement in the features B, B′ between the experimental

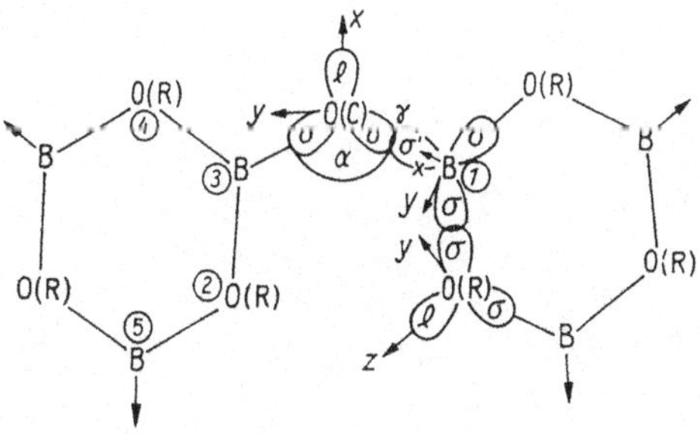

Fig. 5.15. Boroxol ring model for B_2O_3 glass. O(R) ring oxygens; O(C) connecting oxygens

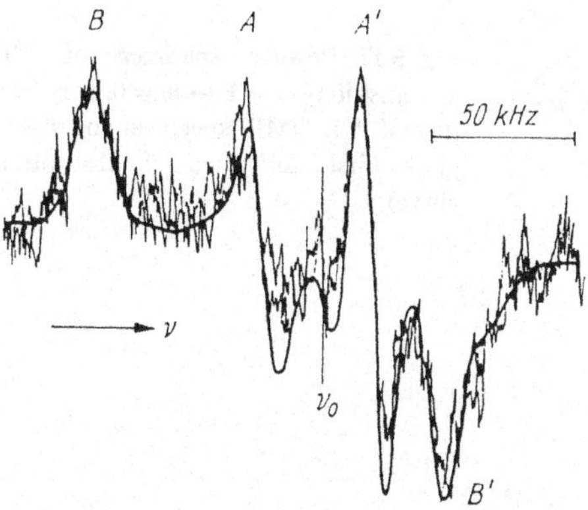

Fig. 5.16. Two-site fit with distributions (smooth curve) of the superposed experimental curves for ^{17}O (first derivative of Fig. 5.14)

results and the one-site computer simulation. There must, however, be oxygens O(C) that connect the boroxol rings, and the B–O–B bond angles for those oxygens must vary over an appreciable range. Correspondingly, a second oxygen site was involved with $Q_{cc} = 5.75$ MHz, $\eta = 0.4$, and a Gaussian distribution of width $\sigma_\eta = 0.2$ in η. The excellent agreement of this two-site fit with experiment is shown in Fig. 5.16.

The powder pattern for the observable transitions ($m = 0 \leftrightarrow m = -1$ and $m = 1 \leftrightarrow m = 0$) for ^{10}B is shown in Fig. 5.17a. This nucleus of spin $I = 3$ is about 30 times more sensitive for determinations of Q_{cc} and η than is ^{11}B. Figure 5.17b displays the experimental spectrum (first derivative) for ^{10}B in vitreous B_2O_3, displaying the features predicted by the powder pattern (Fig. 5.17a). The ^{10}B spectra have been obtained for a set of glasses in the binary system Na_2O–B_2O_3; that portion of the spectrum labeled A in Fig. 5.17 is shown for these glasses in Fig. 5.18. Agreement between the experimental results (solid lines) and computer simulations (open circles) is excellent. The simulations are based on a model put forward by Krogh-Moe [142], and assume that only boroxol rings, tetraborate groups, and diborate groups are present in the glasses. It should be stressed that only one parameter is adjusted for fitting all of the glasses; it simply determines the relative weighting of the spectra for the three types of structural unit involved in this model for the glasses. It is clear that the ^{10}B-NMR spectra can be used to identify and

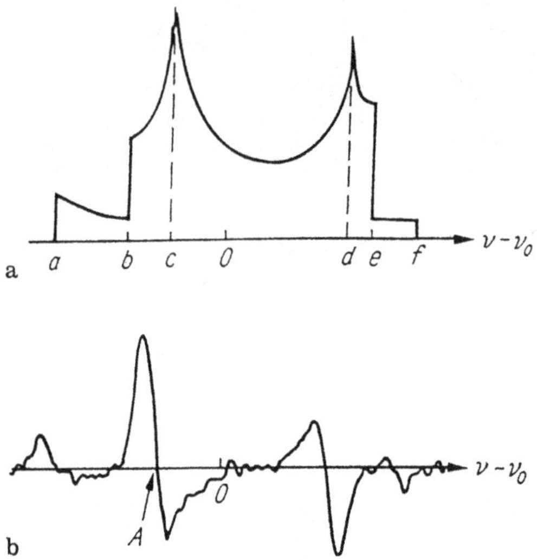

Fig. 5.17. Powder specimen of ^{10}B **a** Transitions $m = 1 \leftrightarrow m = 0$; **b** experimental ^{10}B-NMR spectrum for B_2O_3 glass (first derivative of absorption curve)

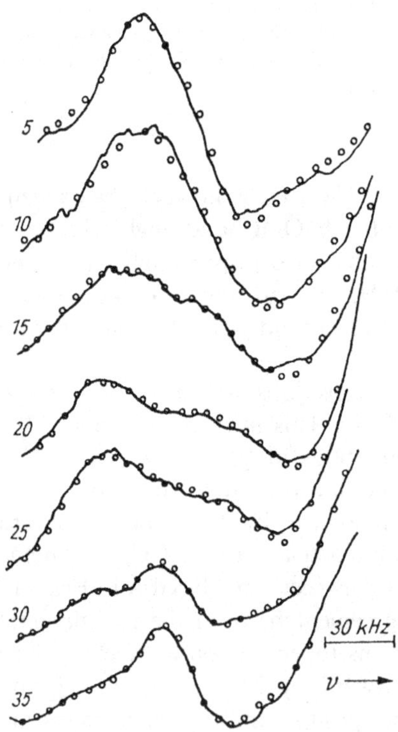

Fig. 5.18. Experimental ^{10}B-NMR spectra on an expanded scale displaying the main feature A of Fig. 5.17b for B atoms on BO_3 units in such Na_2O–B_2O_3 glasses (solid lines). Computer-simulated spectra are indicated by open circles

determine quantitatively the presence in the glasses of large structural units that occur in the crystalline compounds of the borate systems.

5.3 Electron Microscopy

5.3.1 Introduction

The impulse given to glass research by the appearance of electron microscopy is comparable to the one given to science and medicine in general by the corrected Abbe microscope. Today the electron microscope is no longer a specialist's special tool: Every solid state chemist should master its use as well as that of the microscope.

This section will not replace the pertinent literature (e.g., Vogel [151], Reimer [152], Picht [153], Glavert [154, 155], Skatulla et al. [156, 157], Müller [158], Wyckoff [159], Borries [160]), but will hopefully provide the necessary basis for the utilization of electron microscopy in glass research, at least to help the understanding and critical evaluation of its results.

Because of their emphasis on smallest and extended regions, respectively, nuclear magnetic resonance and electron optical investigations may be considered complementary methods in glass research.

5.3.2 Relations Between Light and Electron Microscopy

It is known that the distance of two points that can just be resolved by a light microscope is given by:

$$d = \frac{\lambda}{n \sin \omega}$$

where d = distance of points that can just be resolved, λ = wavelength of light from about 400 to 700 nm (4000 to 7000 Å), n = refractive index of the immersion liquid, and ω = half aperture of the objective's opening angle. $n \sin \omega$ is also called the numerical aperture. In general, resolution is related rather to one-half the wavelength used so that more correctly

$$d = \frac{\lambda}{2n \sin \omega}$$

The wavelength of the electrons used in electron microscopy is about 100000 times smaller than that of visible light. Consequently, much smaller objects can be imaged. The resolution of a contemporary electron microscope is about 0.2–0.3 nm (2–3 Å). Its function is analogous to that of a light microscope: The light source is replaced by an electron source (Fig. 5.19) and the glass lens system is replaced by a system of magnetic or electric electron lenses. One distinguishes magnetic or electrostatic instruments.

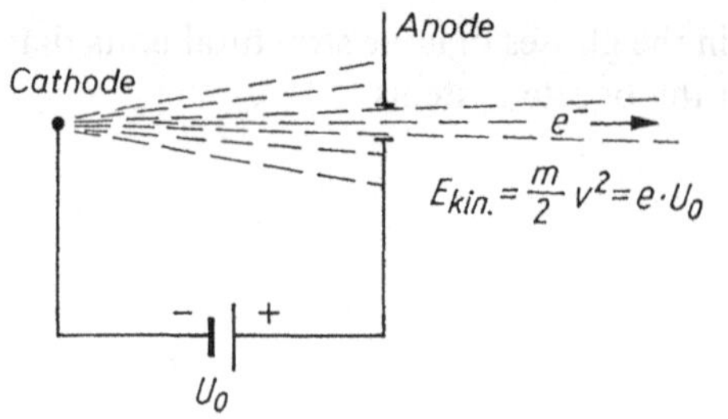

Fig. 5.19. Light source in electron microscopy; thermal (glow) electron emission. The kinetic energy (velocity) of electrons behind the anode depends on the acceleration voltage U_0

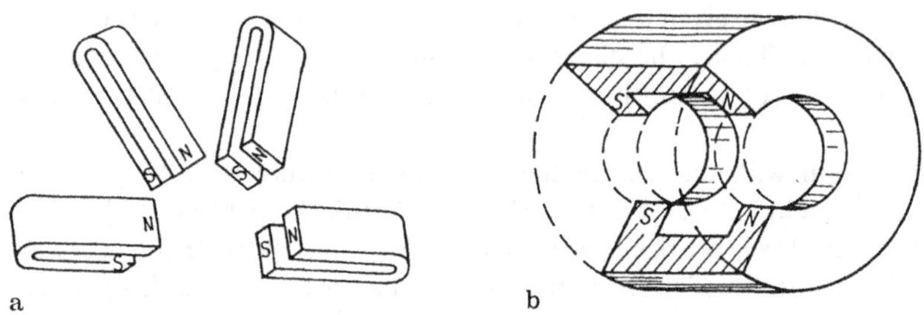

Fig. 5.20. Schematic of magnetic electron lenses. **a** Ring-shaped arrangement of horseshoe magnets; **b** the magnetic focus as a sequence of an infinity of horseshoe magnets represents a permanent-magnet lens

The electron source is shown in the schematic of Fig. 5.19. From a thin cathodic tungsten (wolfram) wire heated in a vacuum of 1.33×10^{-2} to 1.33×10^{-3} Pa, electrons emerge and are directed toward a hole anode by means of the applied electrical field V. The strength of this field primarily influences the speed, and thereby the penetration power of the electrons into the sample under investigation. The usual voltage of the beam is 50–100 kV; however, high-voltage instruments are known to work with more than 1 MV. Advantages and disadvantages of instruments when used in glass research will be discussed later. The function of lenses is assumed either by ring magnets (Fig. 5.20) (compared to a light microscope or an electrical field (Fig. 5.21).

5.3.3 Imaging and Preparation of Samples

On principle, a sample may be investigated in the electron microscope by bringing it directly into the beam – just like in the light microscope. However, since electrons do not penetrate as much – the depth is a function of the voltage U_0 (Fig. 5.19) – one is forced to prepare extremely thin samples, e.g., powders placed on thin structureless carrier films. In the case of glass powders only, the wedge-shaped edges are passed by the beam; the rest appears black.

The contrast in the image is based on "scattering absorption," i.e., elastic and inelastic scattering of electrons at the atoms of the sample. The scattered electrons do not reach the imaging screen of the photographic plate because of

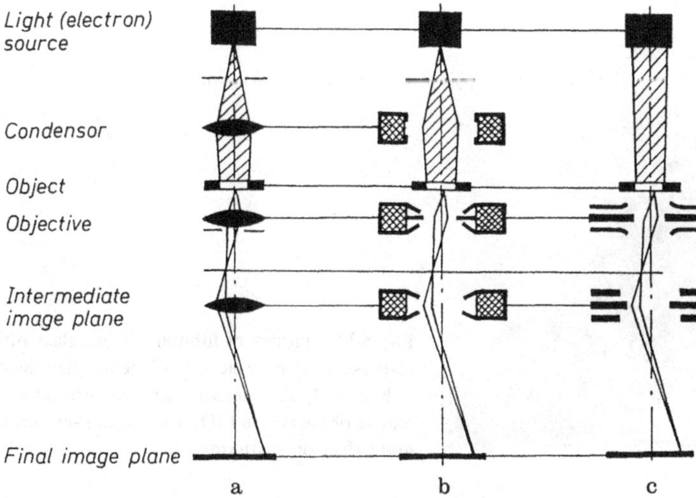

Fig. 5.21. Ray path in a light microscope and in electron microscopes. **a** Light microscope; **b** magnetic electron microscope; **c** electrostatic electron microscope

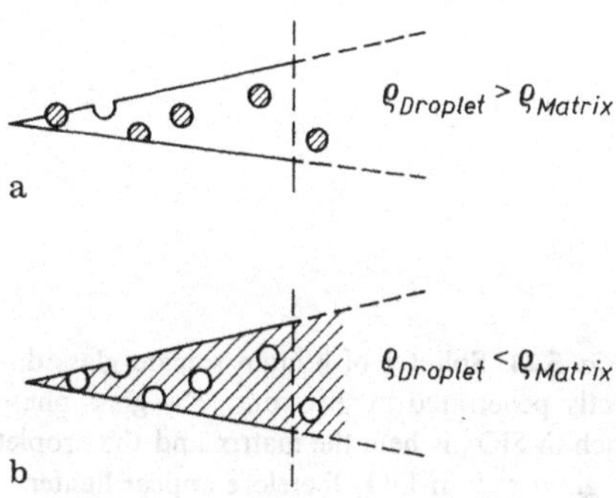

Fig. 5.22. Schematic of the direct penetration by electrons of a wedge shaped splinter of glass. **a** The droplet shaped immiscibility regions have a larger density than the glass matrix. The droplets appear as dark regions in the electron micrograph; **b** the matrix has the larger density therefore the droplets appear as lighter regions

the use of a small aperture. Therefore samples of a large mass density (g/cm^2) will appear darker.

5.3.3.1 Direct Penetration of the Sample by Electrons

The observation of a glass sample under direct penetration presents the advantage that the full resolution of the electron microscope is used. Only penetration of the borders of glass fragments can be considered since microtome slices of glass can not generally be made. The direct penetration of thin glass sheets (produced by glass blowing) is also meaningless because the structure of a glass quenched to a thin sheet never corresponds to that of a bulk sample. Two cases of direct penetration of a glass fragment are shown schematically in Fig. 5.22: a) the droplet phase is denser than the matrix and b) vice-versa. The denser phase is in principle darker on the electron optical image as is experimentally confirmed in Figs. 5.23 and 5.24.

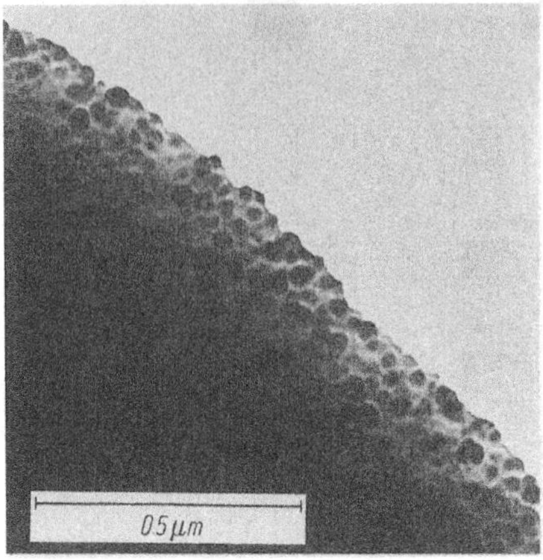

Fig. 5.23. Splinter of lithium silicate glass directly penetrated by electrons. The droplet phase rich in SiO_2 and denser than the surrounding matrix phase rich in LiO_2 appears darker due to more electron scattering

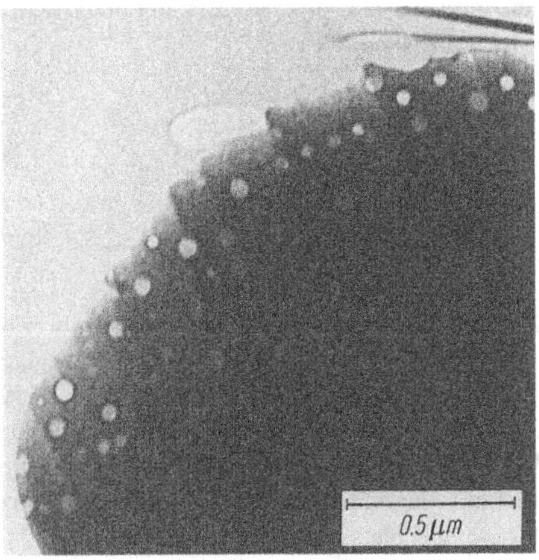

Fig. 5.24. Splinter of lithium silicate glass directly penetrated by electrons. The glass phase rich in SiO_2 is here the matrix and the droplet regions rich in LiO_2 therefore appear lighter

The voltage of the electron beam plays an important role in the direct penetration of glass. Only very thin samples can be observed with low voltages. While a very good contrast is obtained; the risk of sample modification always exists since low voltages often lead to a high absorption of electrons. The sample is thereby over heated, (melting may occur and easily reduced ions can be reduced, for example lead ions to metallic lead. Thus structure effects may be suggested which have nothing to do with the real glass structure are observed (see Fig. 5.25). Thicker glass samples can effectively be observed under direct penetration when high voltages are used but contrast is lost. An optimal voltage of the electron beam for the observation of glass under direct penetration exists between 100 and 200 kV and is dependent on the type of glass. Several electron microscopic methods should be used since interactions between the electron beam and the sample can never be excluded. Structure elements smaller than 50 Å can be observed under direct penetration provided that sufficient

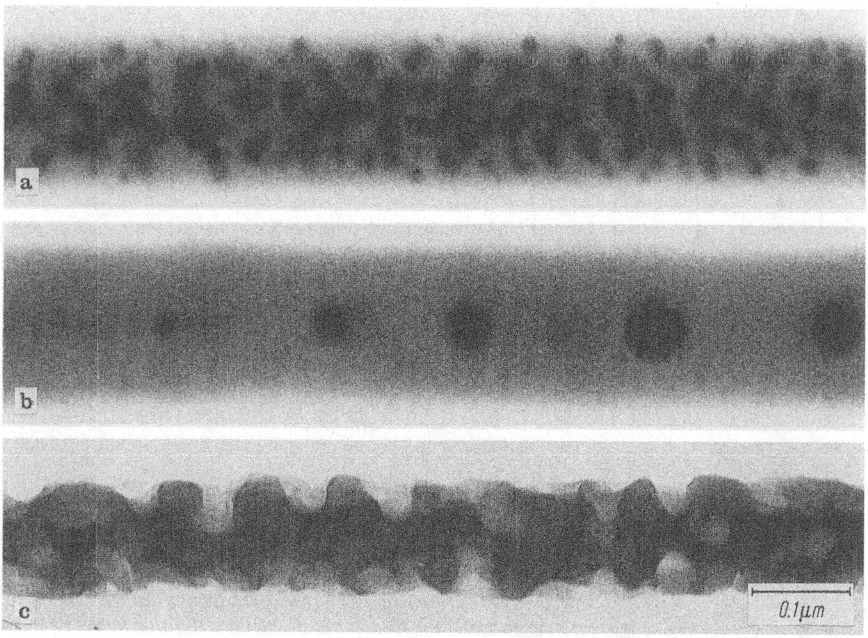

Fig. 5.25. Lead glass filament directly penetrated by electrons (100 kV) (Kuhn and Schimmel [161]). **a** After brief irradiation, lead ions are reduced to metal and small spherulets are precipitated in the glass; **b** on continued irradiation, the small lead spherulets combine into large spheres; **c** on further irradiation, the metallic lead evaporates completely, leaving a residual skeleton

differences in dimension exist. On principle, all observations under direct penetration may be suspect because of beam-glass interaction. Therefore indirect (replica) preparations are preferable (Skatulla and Horn [151, 156, 157]) as will be discussed more extensively in the following sections.

5.3.3.2 Carbon Replica Method after Bradley

To avoid wrong conclusions from the relief of the glass surface in any replica technique, a fresh glass surface should be used, produced, if at all possible, in a high vacuum. Fracture obtained by heating or a sharp impact causes a high speed of fracture which may eliminate structural detail.

Therefore, it is advisable to break the sample in a bending mode after scoring. Under such circumstances the fracture velocity is small and the fracture follows "grain boundaries," producing a good relief. Figure 5.26 shows the significant steps in the replica process (Bradley [162–164]). If the glass consists of a matrix containing droplet-shaped regions of immiscibility, perhaps of μm dimension, the surface created in fracture (Fig. 5.26 a) may look like that in Fig. 5.26 b. The droplets either remained in the surface or broke out, leaving behind a hole. Some also might have been sheared off. Vertical deposition of carbon produces a thin, temperature-resistant, stable (chemically, physically,

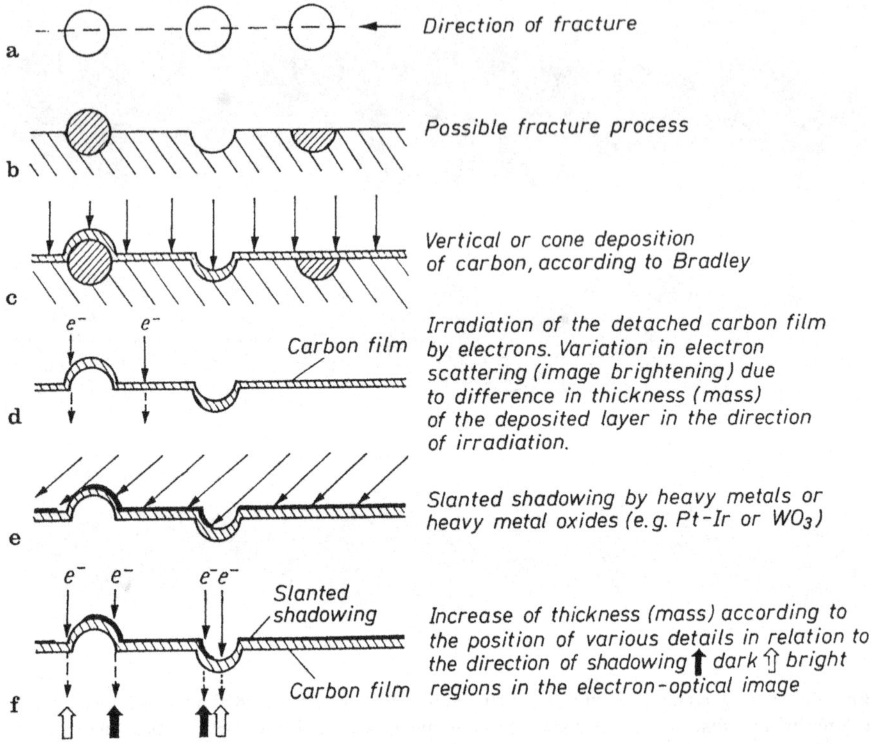

Fig. 5.26. Carbon replica preparation according to Bradley [162–164]. Example of a one-step replica process (partial steps a–f)

and against electrons) amorphous replica film (Fig. 5.26 c). Submersion in H_2O or diluted HF causes lifting and floating of the film which, after drying, is prepared for electron microscopy. Figure 5.26 d shows how different the path, and therefore the scattering of electrons, is at the different features of the sample. This causes contrast. Since the basic scattering power of low atomic number carbon is relatively small, a slanted beam of a heavy material, e.g., WO_3, is superimposed on the released carbon film. This produces shadowing (Fig. 5.26 e) which provides a good contrast in the electron beam image (Fig. 5.26f).

In a later development of the replica process Pt, Ir, or both replaced WO_3. Advantages were that the grain dimensions were smaller and that shadowing can be done while the carbon film remained attached. The WO_3 films did not sufficiently resist the removal of the carbon film by chemical agents. To dispose of remaining doubts about the "reality" of structural features – they might still be artifacts inherent to the films – Skatulla et al. [151, 156] further modified the replica process.

5.3.3.3 Further Development of the Experimental Technique of Bradley's Carbon Replica [165]

It has been shown that abrupt fractures are inappropriate for electron microscopic observation of fine structural inhomogeneities with the carbon replica technique. Only fractures which are slowly progressing can be used. Figure 5.27 shows an instrument which produces a fresh fracture surface in a high vacuum of 10^{-6} Torr or under atmospheric pressure. At times the preparation in a high vacuum is absolutely necessary, since the contact of the fresh fracture surface with air leads, in specific glasses, to undefined modifications. Once a fresh fracture surface is obtained, vacuum metallizing of the sample (that is the creation of a Pt–Ir–C mixed layer) follows. An evaporator electrode arrangement shown in Fig. 5.28, represents a decisive step forward in technology. The carbon electrode contains a welded Pt/Ir bead which acts as a point-shaped source of metal for deposition and guarantees excellent tracing. An angle of deposition between 30 and 60 degrees depending on the profile of the surface has been shown to be satisfactory. Two operations used to be necessary for the preparation of a replica film (perpendicular carbon vacuum deposition and slanted shadowing with metals or WO_3). The aforementioned technique therefore represents an important advance and produces images of significantly improved quality. Carbon particles deposited on a glass surface using the slanted vacuum deposition method at an angle of 30 to 60 degrees with the aforementioned Pt–Ir–C electrode, still possess a high mobility so that an integrated carbon film is created in the shadow areas. However the Pt and Ir particles mixed with C particles only deposit on the side of the sample facing the vapor phase. The Pt–Ir–C mixed layer has a surprisingly fine-grained structure and consequently shows a very small tendency for recrystallisation and grain dilatation under electron penetration (effects possibly due to thermal).

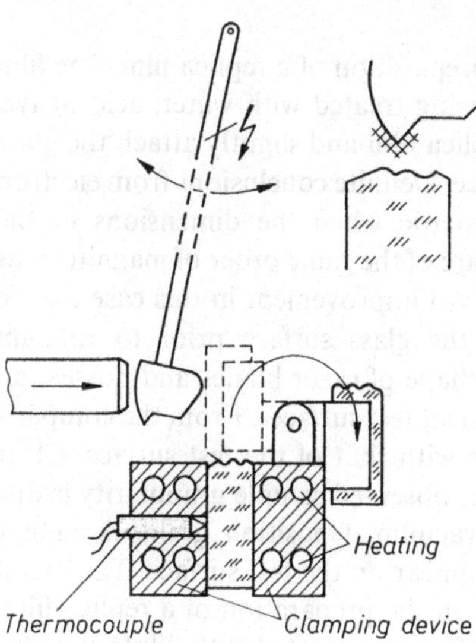

Fig. 5.27. Schematic of the apparatus for the new standard preparation method of glass samples. Modified carbon replica method with production of an MoO_3 crystal test surface according to Skatulla and Horn [156]

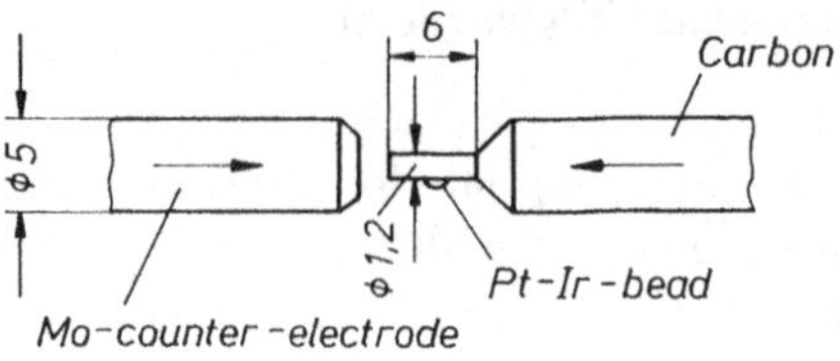

Fig. 5.28. Electrode arrangement for the production of a Pt–Ir–O mixed layer

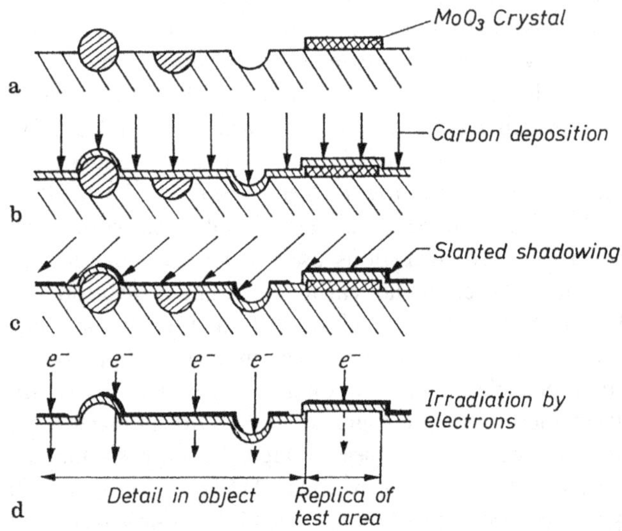

Fig. 5.29. Modified carbon replica method with production of an MoO_3 crystal test plane according to Skatulla and Horn [156]. (Partial steps a–d)

Figure 5.29 schematically shows the preparation of a replica film. The film is removed from the glass surface, while being treated with water, acid or lye. The treatment agents diffuse under the replica film and slightly attach the glass thus removing the film from the glass surface. Definite conclusions from electron microscopic observations become problematic when the dimensions of the photographed structural inhomogeneities are of the same order of magnitude as the grain of the vacuum deposition agent. An improvement in this case can be obtained by placing MoO_3 crystals on the glass surface prior to vacuum deposition [154]. These crystals have the shape of razor blades and possess an ideally smooth surface thus providing excellent test surfaces. From the comparison of the granularity of the glass sample with that of the test surface. Clear conclusions can be drawn as to whether the observed sample granularity is due to structural inhomogeneities or to the vacuum deposition grain. A similar granularity would, in the latter case, also appear on the test surface. The use of MoO_3 crystals requires an additional step in the preparation of a replica film. Once the film is removed from the glass surface, it is treated with dilute soda lye causing dissolution of residual MoO_3 crystals forming sodium molybdates. The addition of test surfaces enhances the resolution of structural inhomogeneities to ca. 50 Å, compared to ca. 120 to 150 Å, in the past.

In air MoO_3 crystals are placed on the fresh fracture glass surface at normal pressure by exposing the surface to MoO_3 fumes. In a high vacuum, the nose of the handle in the fracturing device should be covered with MoO_3 crystals before the high vacuum is applied (see Fig. 5.27 top right). The glass sample to be fractured faces the nose covered with MoO_3 crystals. A high voltage spark which jumps from the handle nose to the clamping jaw is produced. The transfer of a sufficient quantity of MoO_3 crystals onto the fresh fracture surface is thus accomplished. In this condition it is subjected to vacuum deposition as described above.

The replica film now also contains the replica of an ideally smooth MoO_3 crystal face which cannot be confused. A granularity not associated with the glass structure should also appear on this test surface. Thus a proper replica technique permits a much more reliable interpretation than the transmission technique.The resolution limit for features in the glass is determined by the granularity of the deposits. Thus the resolution associated with the electron lens cannot be exploited fully. Glass features smaller than the grain dimension of the deposits cannot be observed. The glass structure might also appear homogeneous when the inhomogeneities are smoothly cut during the generation of a fresh fracture surface and therefore a good relief cannot be obtained.

Figure 5.30 shows droplet-shaped immiscibility regions corresponding to the schematic of Fig. 5.29. Figure 5.31 shows a case where only the insertion of an MoO_3 test face allowed us to ascertain the very small immiscibility regions in the glass.

The electric spark generated during the preparation of a fresh fracture surface covered with MoO_3 crystals may jump from the handle nose to the surface of a glass with a certain electric conductivity (as for instance) with

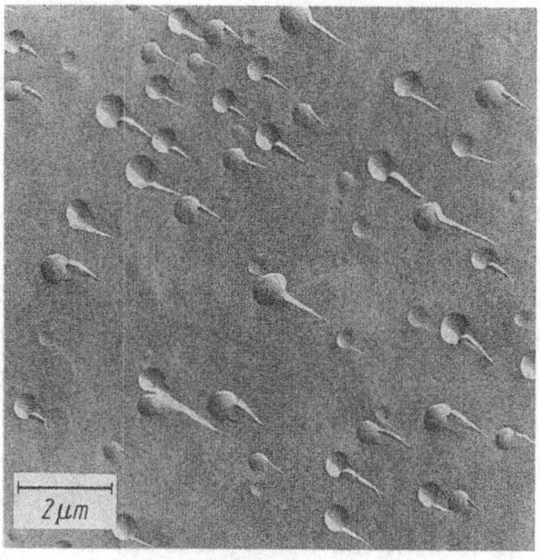

Fig. 5.30. Droplet-shaped immiscibility regions in a glass fracture plane (replica process). The droplets remaining protruding in the surface, as well as those broken out and leaving a cavity, usually exhibit "fracture flags." The dark shadow lies to the left or right of the droplet. A pair of the droplets is cracked right through the middle

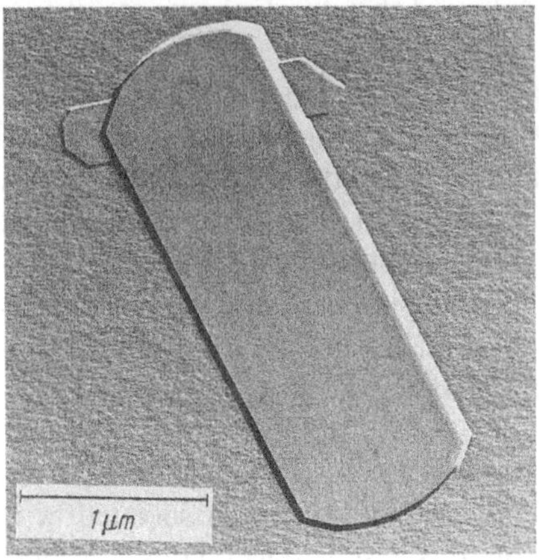

Fig. 5.31. Fracture plane of a glass exhibiting only a weak tendency toward immiscibility (replica process). In the center of the picture, two MoO_3 test planes. A comparison of the granularity of the sample must be attributed to the glass structure (very small immiscibility regions)

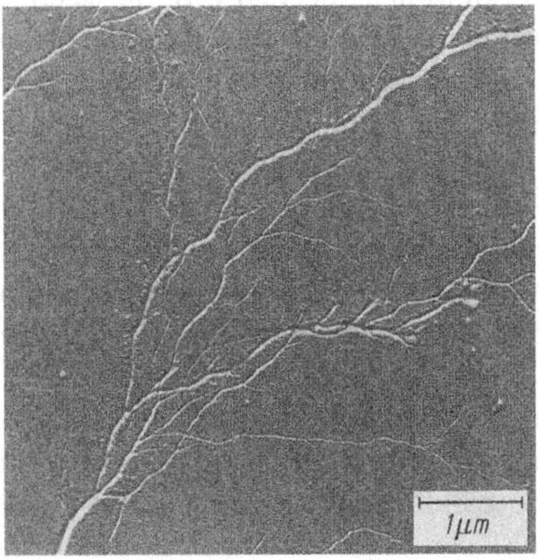

Fig. 5.32. Surface of a tellurite glass hit by sparks during the introduction of MoO_3 crystals. This leads to ditch-like depressions due to the partial evaporation of the glass

tellurite or chalcogenide glasses. This can lead to a totally false interpretation of the electron micrograph. Figure 5.32 shows the traces left by such a spark on the fresh fracture surface of a tellurite glass. After partial evaporation and condensation of the evaporated material, the glass surface was uniformly covered with micro-crystallites which have nothing to do with the structure of the glass (see Figs. 5.33 and 5.34). A modified fracturing device (shown in Fig. 5.35) offers a reliable preparation technique for electrically conductive glasses [165]. The MoO_3 crystals are deposited onto the glass surface with an additional Pt electrode previously covered with these crystals. The spark will in this case jump from the Pt electrode to the handle. The same glass as in Figs. 5.33 and 5.34 was prepared using this modified fracturing device. The result is shown in Fig. 5.36. The small granularity observed on the sample can be reliably interpreted as structural inhomogeneities after comparing it with the MoO_3 test surface.

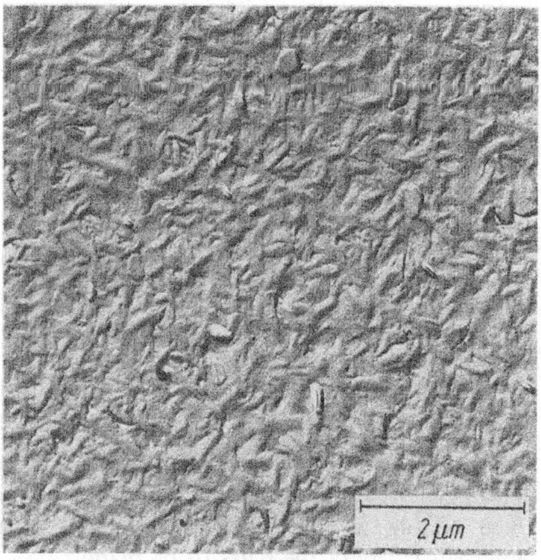

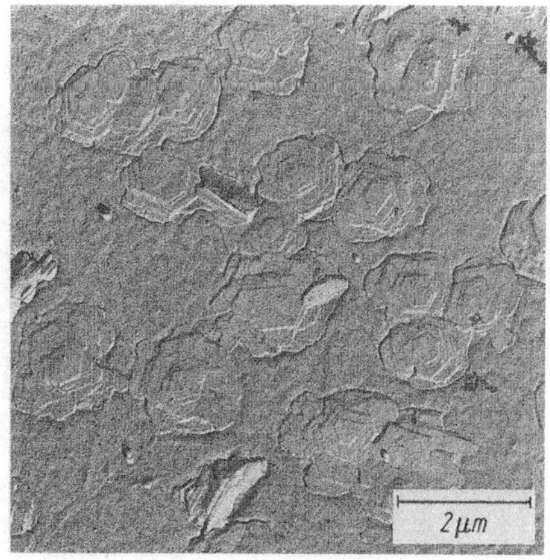

Figs. 5.33. and 5.34. Crystal phases which were created by vapour condensation on the surface of a tellurite glass. They have absolutely no connection with the microstructure of the glass

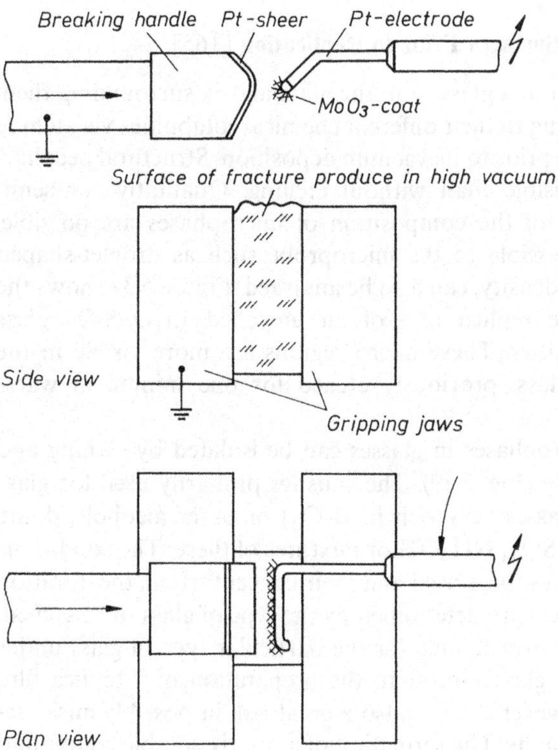

Fig. 5.35. Schematic of the apparatus for the preparation method of glasses with a lower electrical resistance

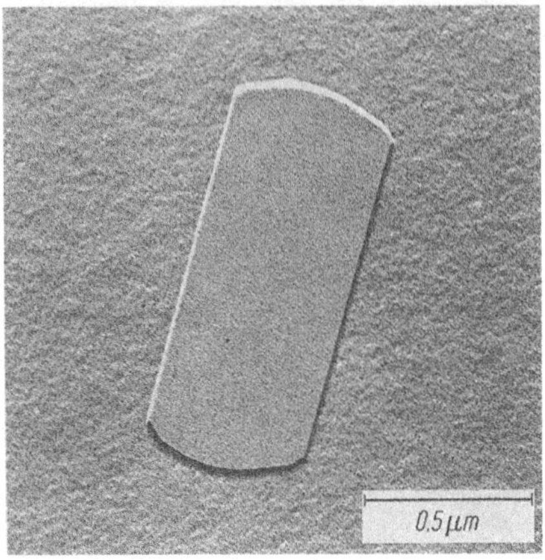

Fig. 5.36. Micrograph of a tellurite glass with an MoO_3 crystal test plane. Comparison of the granularity of the test surface and fracture surface enable a reliable interpretation of the microstructure of the glass to be made

Many optical and technical glasses have an electric resistance of 10^{14} to 10^{17} Ωcm. In this case the older preparation technique is sufficient. However glasses with a resistance of 10^8 to 10^{12} Ωcm (still smaller values are possible) should only be prepared using the modified fracturing device.

5.3.3.4 Treatment of Glass Surfaces Prior to Replication [165]

The inhomogeneities present in a glass or in the glass matrix surrounding them can be differentiated according to their different chemical solubilities via etching of the fresh fractured surface prior to its vacuum deposition. Structural peculiarities can be made more visible than without etching. Qualitative or semi-quantitative interpretations of the composition of microphases are possible. Regions which are not accessible to the microprobe such as droplet-shaped regions with a high packing density, can also be analyzed. Figure 5.37 shows the electron micrograph of the replica film of an unetched Li_2O–SiO_2 glass containing microheterogeneities. These micro regions are more visible in the micrograph of the same glass, previously etched for one minute in water (see Fig. 5.38).

In many instances microphases in glasses can be isolated by etching and therefore analyzed separately (Fig. 5.39). The caustics primarily used for glass are water, methanol (for glasses very rich in B_2O_3) or other alcohols, dilute solutions of HF, HNO_3, H_2SO_4, NH_4OH or mixtures of these. The conditions for etching such as the caustics to be used and their concentration, the duration and temperature of the process are determined by the type of glass to be etched. Optimal conditions must be investigated for the particular type of glass under observation. The etching of glasses prior to the preparation of a replica film offers great advantages. However there is also a great risk in possibly misinterpreting the electron micrographs. The formation of a poorly soluble compound from caustics and glass is possible and cannot be ignored. Figure 5.40 shows an electron micrograph of a PbO–B_2O_3–SiO_2 glass previously etched for one

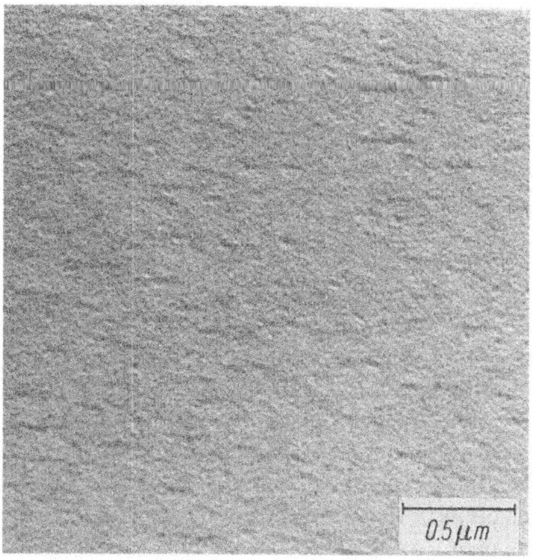

Fig. 5.37. Fracture plane of a lithium silicate glass (Pt–Ir–C replica technique). Evidence of microphase separation can only be recognized with difficulty or by an experienced observer

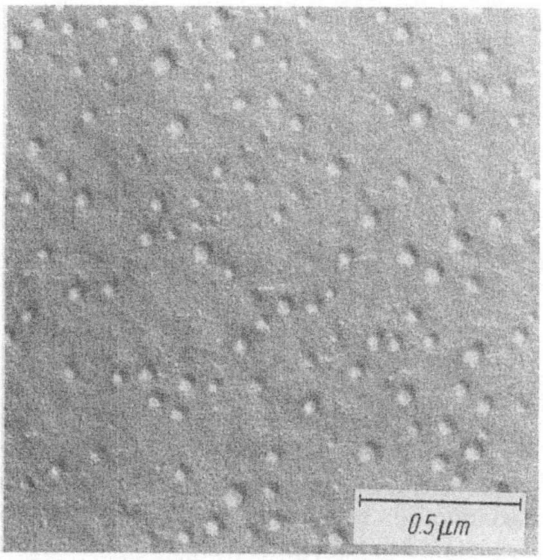

Fig. 5.38. Micrograph similar to Fig. 5.37. The glass surface was etched for 30 sec. in water prior to the Pt–Ir–C vacuum deposition. The droplet shaped immiscibility regions are more clearly visible after the chemical treatment of the surface

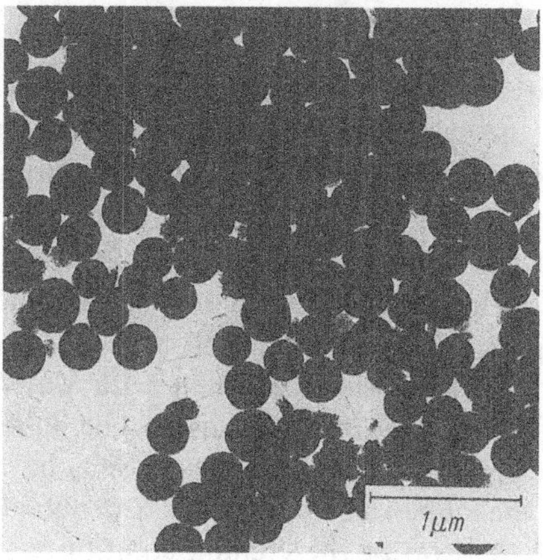

Fig. 5.39. Droplet regions rich in SiO_2 isolated by chemical treatment of a barium borosilicate glass and directly penetrated by electrons

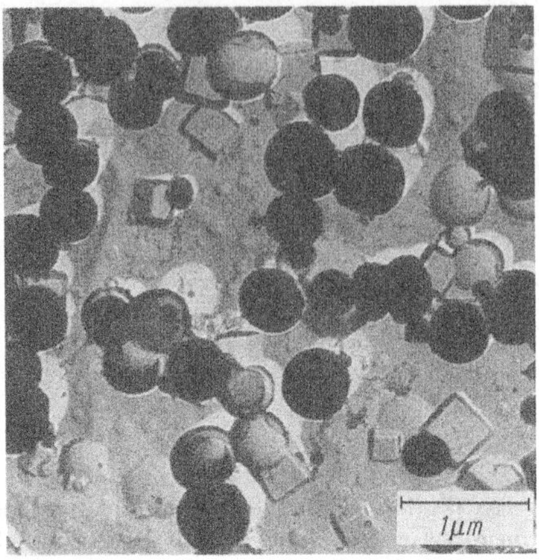

Fig. 5.40. Electron micrograph of a $PbO–B_2O_3–SiO_2$ glass previously etched in a 1N solution of HNO_3 (for 1 minute). The micrograph shows cubic crystals in the vicinity of droplet-shaped immiscibility regions (clear droplets are in the glass, the dark droplets were already isolated and have adhered to the replica film). The cube shaped crystals are due to an error in the sample preparation and have no relation to the glass structure

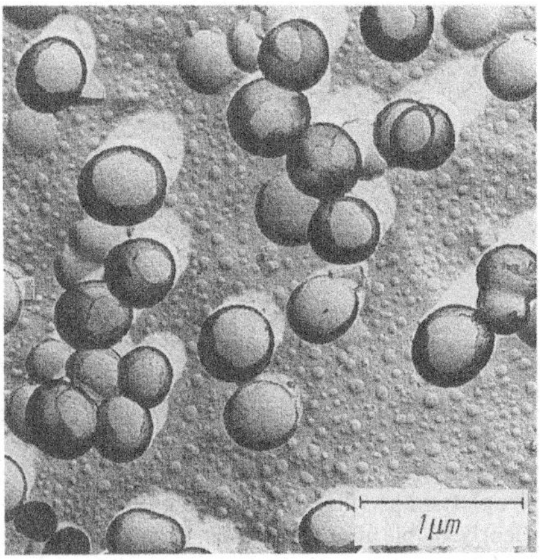

Fig. 5.41. Electron micrograph of the same glass as in Fig. 5.40. The sample was cleaned with chemical etching in water, alcohol and acetone, and cellite film. It only shows the real structured inhomogeneities

minute in a 1N solution of HNO_3. Cube-shaped crystals are observed in the vicinity of the droplet-shaped immiscibility regions. This micrograph is a perfect example of the possible errors involved in the preparation of an etched glass sample. The cube-shaped crystals are crystals of $Pb(NO_3)_2$, formed during etching of the sample with HNO_3 which crystallized on the glass during the drying process. The dark spheroids are droplet-shaped enclosures of SiO_2 which were fully isolated by the etching process. They adhered to the replica film and were directly penetrated by electrons. Figure 5.41 is a micrograph of a similarly etched glass surface which was cleaned by successive treatment with water, alcohol and acetone. An additional Cellit purification step is used. The cube-shaped $Pb(NO_3)_2$ crystals along with the dark isolated enclosures of SiO_2 have disappeared. Additional evidence that the cube-shaped crystals are due to a secondary effect is shown in Fig. 5.42. A drop of the etching solution was

Fig. 5.42. Etching solution removed from the surface of the $PbO–B_2O_3–SiO_2$ glass shown in Fig. 5.40 and dried on a substrate. The same cubic Pb $(NO_3)_2$ crystals as in Fig. 5.40 are observed

removed from the surface of the $PbO–B_2O_3–SiO_2$ glass and left to dry on a substrate. The result shows the same cube-shaped crystals as in Fig. 5.40. Well-defined shades behind the cubic crystallites demonstrate the resolution of the described vacuum deposition and replica technique. The ionic etching technique (G. Schimmel [166] and H. Bach [167]) must be mentioned as complementary to the aforementioned chemical etching technique. Under specific conditions, it is a good alternative. It should however be mentioned that there is a large risk of interaction between the glass surface and the ionic beam which would lead to sample modification via the formation of apparent structures.

5.3.4 The Scanning Electron Microscope and Electron Microprobe

Further developments exploiting the interaction of an electron beam and a solid include the scanning electron microscope and the electron microprobe. They present specific advantages for glass research and often give increased information.

If a high-energy (some kV) electron beam hits a solid sample the following effects beyond those described above are observed (Fig. 5.43): secondary electrons, backscattered electrons, Auger electrons, X-rays, and cathode luminescence.

A more detailed analysis of an electron beam on a sample "transmitting" electrons is presented in Fig. 5.44a. Secondary electrons (SE), which possess much lower energy (<50 eV), are detached (D_1) from a relatively thin surface layer. Also, backscattered electrons (BE) appear (D_2) in a wide energy range. The critical range (L_1) defines the excitation range for characteristic X radiation (D_3). The radiation (D_4) occurs in range L_2, which is determined by the range of

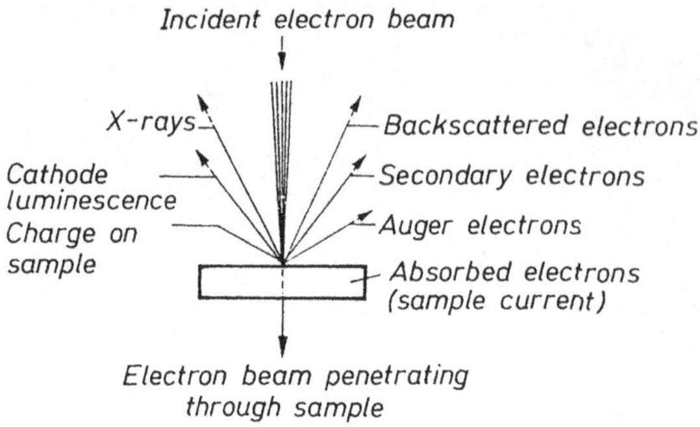

Fig. 5.43. Secondary signal types during irradiation of a sample with high-energy electrons

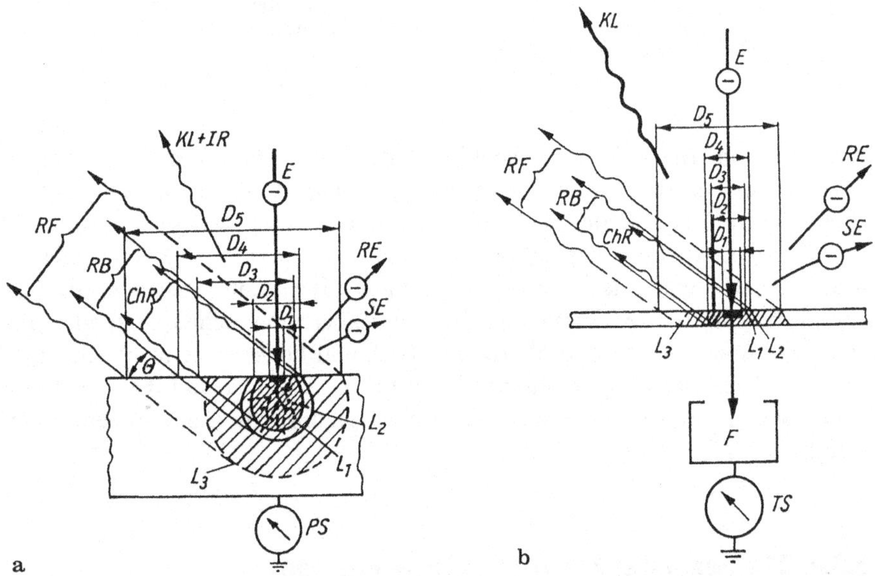

Fig. 5.44. Measured signals caused by the interaction of electron beam and sample during electron beam-X-ray analysis, and their range of spreading (Tögel [168]). **a** Thick, **b** thin samples. E electron beam, RE backscattered electrons, SE secondary electrons, L_1 critical range of electrons (excitation range of the characteristic X-ray spectrum measured in each case), L_2 range of electrons, L_3 region of secondary excitation of the X-ray spectrum (fluorescence excitation), KL thermal infrared radiation, PS sample current, TS transmission current, *F* Faraday beaker, θ angle of X rays vs. spectrometer, Emission ranges and bundle widths measured: D_3, ChR characteristic X-ray spectrum, excited primarily, D_3, RF characteristic X-ray spectrum, excited secondarily, D_4, RB X-ray radiation due to retarding of particles, D_1 secondary electrons, D_2 backscattered electrons

the electrons. In range L_3 a secondary excitation by X-rays occurs, i.e., fluorescence (D_5). In the case of some materials, cathode luminescence and thermal IR radiation can be found. Electrons absorbed in the sample produce a probe current. For thin samples, the effects are the same but are cut in a certain depth. Figure 5.44a and b are, of course, only sections through the conditions in a sample irradiated by electrons. In actuality, the secondary signals must be visualized as in a hemispherical distribution.

The principle of the scanning electron microscope involves the reception of backscattered (BE) or secondary (SE) electrons by a detector and their conversion to light, which is registered in a photomultiplier. Line-by-line registration of the light impulses of a plane sample yields an electron optical scanning micrograph, analogous to a TV image. Classical replica electron microscopy (Figs. 5.26 and 5.29) demands a good relief. If structural inhomogeneities are sheared off, the result fakes homogeneity. Scanning electron microscopy does not require this relief. On the contrary, it requires a plane surface. Often the sample face is polished and coated with a carbon, gold, copper or aluminum layer to avoid undesired electrostatic charges. The scanning electron microscope thus offers the advantage that a certain depth, i.e. a volume of bulk glass, is probed. Therefore one need not rely on the indirect evidence of a surface profile which depends on so many factors. This represents a significant technological advance. Moreover a much better three dimensional image is usually obtained. The contrast observed in this type of image either results from backscattered electrons (material contrast) or from the geometric structure of the surface (topographical contrast). However there are a range of disadvantages associated with scanning electron microscopy and they should not be overlooked:

- The interaction between electron radiation and the sample leads to electron absorption which will generate an incomparably higher thermal charge than that with the classical replica technique. The risk of sample modification is thus inherent for certain glasses. A melting of the glass surface might be observed.
- Resolution is inferior and only entities greater than 50 Å can be analyzed.
- Non-conductive samples must be vacuum coated with carbon or gold in order to avoid electrical charging during analysis (it can seriously affect analysis). A carbon vacuum deposition provides a good material contrast while a gold vacuum deposition will primarily provide a good topographical contrast.

The principle of the electron microprobe involves the qualitative, or even quantitative, analysis by X-ray spectrometry of the characteristic X-ray radiation of elements excited when a focused electron beam hits the surface of a solid. Two types of spectral analysis are possible.

Wavelength Dispersion Analysis

The reflection of X-ray radiation from a crystal with the corresponding interplanar distance of the crystal follows the Bragg relationship:

$$n \cdot \lambda = 2d \cdot \sin\theta$$

where n is the diffraction order, λ is the wavelength, d is the interplanar distance of the crystal and θ is the reflection angle.

Energy Dispersion Analysis

The whole spectrum is allowed to reach an appropriate semiconductor detector, in which every single X-ray count produces a number of electron-hole pairs corresponding to its energy. The spectrum can therefore be analyzed by energy dispersion. The entire spectrum can be quickly observed on the screen with the help of a multi-channel analyzer (Figs. 5.45 and 5.46). It has become possible to determine quantitative concentrations of 10^{-13} to 10^{-14} g of elements with a resolution of about 1 μm. If the sample is followed along a line setting the detector to a certain wave length of a certain element (e.g., $Si_{K\alpha}$ radiation), the resulting profile of the X-ray intensity provides the distribution of the concentration of this element along that line through the sample. It is also possible to do this line-by-line over the entire sample, as in the scanning electron microscope. In Fig. 5.47 this is done for barium ($Ba_{L\alpha}$ radiation). Some modern instruments combine scanning electron microscopy and electron microprobe. In this way it is possible to superimpose the X-ray intensity curve (i.e., the distribution of an element) along an interesting section on a scanning electron image, exemplified in Fig. 5.48 for Ca.

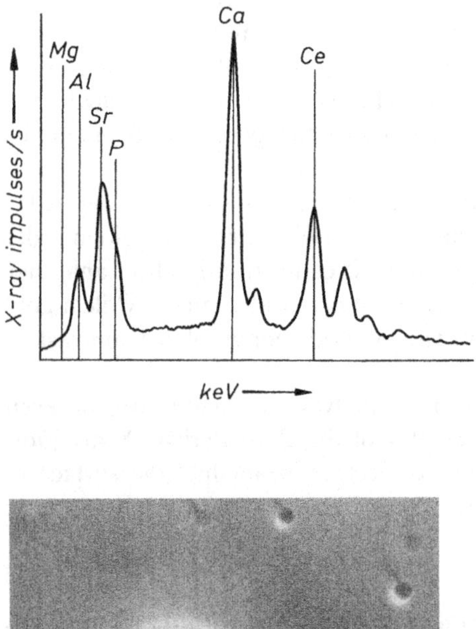

Fig. 5.45. X-ray spectrum of an optical glass analyzed by energy dispersion. The intensity of the emitting X ray (measured in impulses/s) is represented as a function of the quantum energy (in keV). The height of the peak gives information about the concentration of the corresponding element in the sample

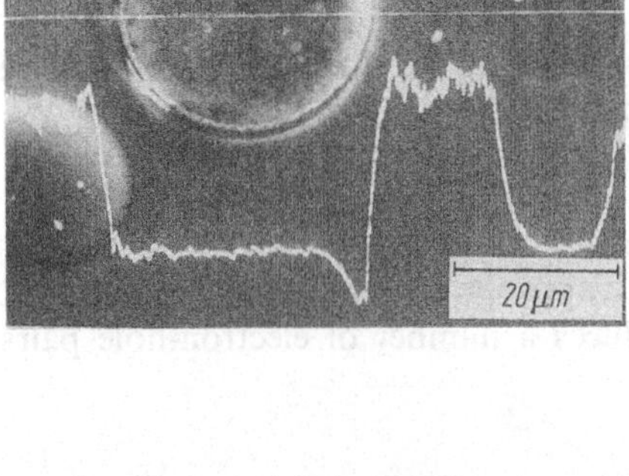

Fig. 5.46. X-ray intensity profile of a $CaO–B_2O_3–SiO_2$ glass measured by wavelength dispersion. The electron beam was run along the white horizontal line on the sample. The X-ray intensity curve for the $Ca_{K\alpha}$ radiation has been registered in the scanning electron micrograph (in impulses per second). Information about the distribution of the Ca^{2+} ions in the structure inhomogeneities observed in the scanning electron micrograph can be obtained this way

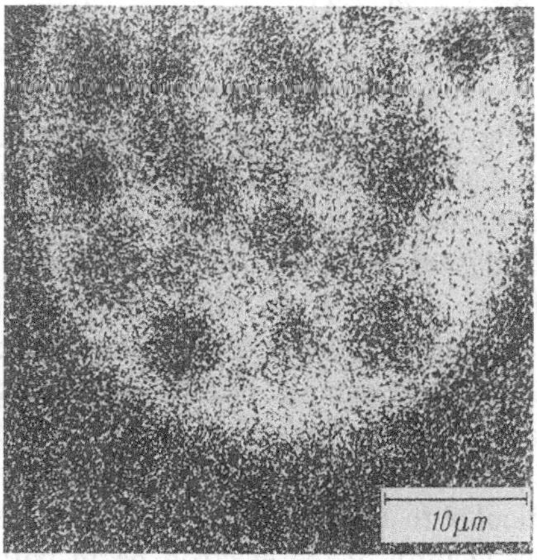

Fig. 5.47. $Ba_{L\alpha}$ X-ray scanning micrograph of a $BaO–B_2O_3–SiO_2$ glass. Ba is enriched in the primary large-droplet phase

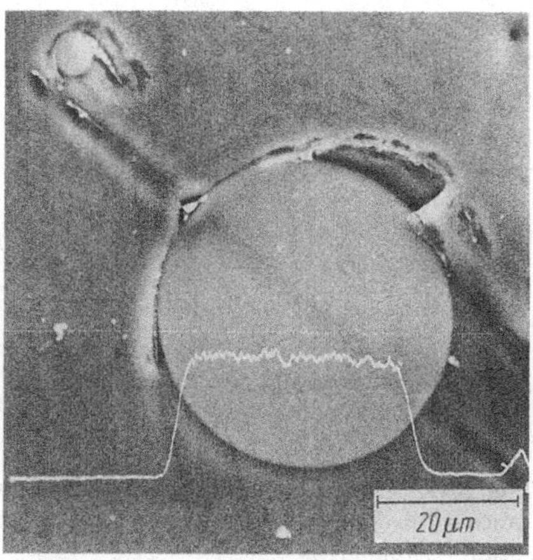

Fig. 5.48. Electron scanning micrograph of a $CaO–B_2O_3$ glass with inscribed X-ray intensity curve for $Ca_{K\alpha}$ radiation. The microprobe electron beam has been led along the white horizontal line across the sample. The intensity curve lying below shows unequivocally that the droplet phase is rich in Ca

Classical replica electron microscopy demands a good relief. If structural inhomogeneities are sheared off, the result fakes homogeneity. Scanning electron microscopy does not require this relief; on the contrary, it requires a plane surface. Often the sample face is polished and subsequently coated with a C, Au, Cu, or Al layer to avoid disturbing electrostatic charges.

The scanning electron microscope thus offers the advantage that a certain depth, i.e., a volume of bulk glass, is probed so that one need not rely on the indirect evidence of a surface profile which depends on so many factors. Add to this the possibility of ascertaining concentration and distribution of elements, at present down to atomic number 3 (Li). Of course, sensitivity decreases with atomic number.

The electron beam microanalyzer, used point-by-point, works down to 1 to 2 μm^3. This, of course, is a much smaller resolution than that offered by *classical transmission* electron microscopy (0.2–0.5 nm (2–5 Å)) or replica electron microscopy (2–4 nm (20–40 Å)). Moreover, the microprobe does not allow one to recognize the volume of excitation if one comes close to the limit of 2 μm. Problems and uncertainties arise by using the microprobe when the object to be analyzed (e.g. a droplet-shaped microphase) is smaller than 3 to 5 μm and only constitutes part of the volume under excitation. Special care is necessary for microprobe analyses of relatively unstable glasses. Microprobe analyses usually involve scanning speeds of the electron beam of 2 to 50 $\mu m\,min^{-1}$. Melting of the glass along the path of the electron beam often occurs causing distinctive modifications of structure and concentration in the sample. Glasses containing sodium ions are especially sensitive as can be seen in Figs. 5.49 and 5.50. The electron beam left a deep trace over the glass probe (Fig. 5.49). Figure 5.50 is a micrograph of a glass section perpendicular to the trace and clearly shows structural modifications. The glass in the vicinity of the electron beam path

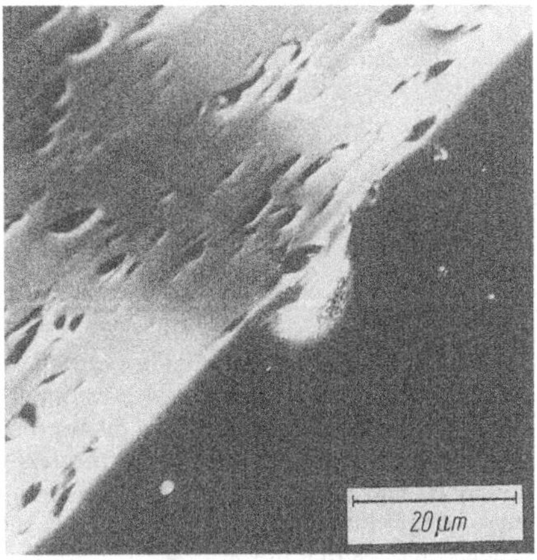

Fig. 5.49. Scanning electron micrograph of the trace left by an electron beam with the electron beam microanalysis

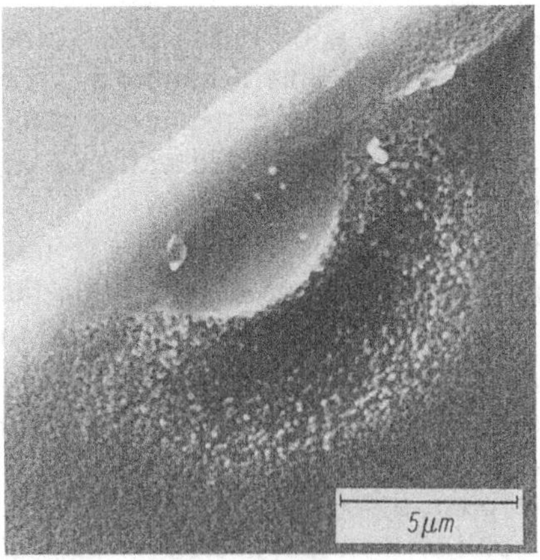

Fig. 5.50. Enlarged section of Fig. 5.49. Fracture surface perpendicular to the trace. The micrograph clearly shows structure modifications of the glass caused by thermal effects (electron absorption) in the immediate vicinity of the trace

quickly cooled again after melting and is relatively homogeneous. It is surrounded by a ring of strong phase separation. Only glass in greater depths remained unaffected by the electron beam. These observations must be taken into special consideration for all microprobe analyses of glasses. The effects can change considerably depending on the glass type. All this shows is that there is no single optimal method. Rather, the three methods described here complement each other; they are all needed in contemporary glass research. For more detail consult the specialized literature, particularly Bethge [169, 170], Picht et al. [153], Vogel et al. [165], Skatulla et al. [151, 156, 157], Goddhew [155], Beeston et al. [154], Reimer [152], Brümmer [171], Bach [172], Völksch et al. [173], Zarzycki [174].

6 Microphase Separation

6.1 Early History

The phenomenon of phase separation in glasses, which since about 1950 has become an essential topic of modern glass research, was recognized as early as 70 years ago. Schott gives the following description [2] "My melting experiments revealed that SiO_2 and B_2O_3 were compatible, while P_2O_5 exhibits hostile behavior. While $SiO_2 + B_2O_3$ as a glass-forming double acid accept bases in any proportions, the addition of P_2O_5 almost always causes milky opalescence, apparently because phosphates are as insoluble in the melt as oil in water." It now seems appropriate to summarize 20 years of world wide research into phase separation in glasses. One must at first realize that for quite some time two structure theories were providing guidance in glass research: the network theory of Zachariasen and Warren [29, 30] and the crystallite theory of Lebedev [43]. More or less extensive adjustments later occurred.

The attitude toward separation phenomena after 1930 was, at first, somewhat affected by adherence to the crystallite or network theory (see the section on experimental evidence). Landmarks on the way to later thorough studies of phase separation in glasses, were, on the side of adherents of the network theory, the work of Hartleif [54], Dietzel [32], and Warren et al. [175]; on the side of crystallite theory, that of Markin [176], Evstrop'ev (quoted by Markin [176]), Muller (quoted by Markin [176]), J.W. Grebenstchikov (quoted by Markin [176]) and, particularly, Botvinkin. [51–53]. The followers of the network theory distanced themselves from the statistical distribution of network-modifiers in the basic glass network and instead discussed swarm-like aggregates of network-modifiers. At the same time the followers of the crystallite theory Qabandoned the favored high-order in crystallite regions (at least beyond 9Å) while accepting swarm-like molecular aggregates. These modifications of the two structure theories brought them much closer to each other. After the reports of the first experimental results of phase separation, ardent defenders of the network theory justifiably asked: aren't the swarm-like aggregates in glass purely accidental? A conventional statistical distribution of network-modifiers in a basic glass network also admits aggregates or clusters.

Laves [31] investigated the distribution of a guest component in a mixed lattice. This is, to some extent, analogous with the investigation of network-modifiers in a basic glass network. Figure 4.5 shows Laves' schematic representation of his results with the distribution of 50% black and 50% white squares. He reached the conclusion that the distribution of a guest component in a mixed lattice lies on the limit of a statistical distribution, that is between the states represented in Fig. 4.5a and 4.5c. The ionic aggregates in glass (discussed from 1940 to 1950) were therefore to be understood as accidental statistical fluctuations of composition. Starting in the fifties, however, experimental investigations of the fine structure of glasses were initiated with new methods for structure research, including electron microscopic methods. These investigations extended and led to convincing observations of the fine structure of glasses.

6.2 Electron Microscopy Evidence for Immiscibility Phenomena in Glasses

The proof for the existence of microinhomogeneities in glasses is inseparably associated with the development of electron microscopy as a new structural tool. However, in glass research electron microscopy was first used only with extreme caution.

Slayter [177] and Prebus and Michener [178] first investigated glasses by transmission electron microscopy. Selyubsky [179] and Oberlies [180] first used the replica method in their electron optical investigations of glass.

These authors concluded more or less unequivocally that, on the one hand, glasses may not always be as homogeneous as postulated by the random network theory but, on the other hand, also not as ordered as postulated by the crystallite theory. In addition, they suggested that the inhomogeneities were not caused by insufficient homogenization of the melt only, but by structural processes. These suggestions immediately provoked a wave of disbelief and polemics. Unbiased objections pointed out that significant modifications of the structure may occur due to the direct penetration of glasses by electrons and the effects observed could only be secondary effects. This viewpoint, as we will see later, has considerable significance. On the other hand, the inhomogeneities observed with the indirect replica technique were described as secondary effects, caused by the interaction between moist air and the glass, that is by corrosion. At any rate this is an opinion that deserves to be seriously examined, too.

Today their conclusions should be considered as generally vindicated, even though the evidence was not 100 percent secure and some of the criticism (effect of electron beams in transmission, secondary effects in the replica structure, corrosion effects, etc.) has some foundation. The general protest that followed their statements caused these authors to either abandon this work or, at the very least, to no longer be convinced of the interpretations. Their referenced works remain solitary efforts. In the meantime, Vogel and collaborators [181–183] had started systematic studies of BeF_2 glasses which, according to Goldschmidt [28] represent "weakened models" of SiO_2 glasses. The term "model structure" was assigned to one of two substances which crystallize in the same lattice and whose ions have the same or similar radii but different charges. Model and original merely differ by their lattice forces.

The bonding forces in BeF_2 are only one quarter of those of the original SiO_2 because of the difference in charge between the ions for oxygen and fluorine on one hand and for silicon and beryllium on the other hand. One can expect that processes can be observed more easily during their formation because of this relative weakness in the bonding force of BeF_2 glasses. During the initial investigations of binary alkali BeF_2 glasses with the usual physicochemical methods, an opalescent turbidity appeared in unexpected composition ranges despite very clean experimental conditions. It was soon observed that, even under the most careful experimental conditions, opalescence appeared in unexpected compositional areas. Originally interpreted as defects, the phenomenon was recognized as due to droplet-shaped immiscibility regions as soon as electron microscopy was mobilized after light microscopy and X-ray diffraction had failed. Even in "model" glasses showing no Tyndall effect (turbidity), the electron microscope revealed droplets as large as 30 to 60 nm (300–600 Å). It was concluded that the unmixing was basically due to structural processes of a general character: it was merely easier to verify in the fluoride "model" glasses.

Figure 5.30 in the preceding chapter shows a replica of a turbid beryllium fluoride glass, with droplets of the immiscible phase either protruding, removing leaving a hole, or sheared off.

From that time on, droplet-shaped immiscibility regions were clearly identified in various special glasses (Vogel et al. [151, 181–192]) and – after development of the improved techniques (Skatulla et al. [151, 156]) described in Chapter 5 – also in more conventional (silicate, borate, borosilicate, etc.) glass systems. As expected, much smaller droplets, which had been harder to detect, were found in the latter case. Figure 6.1 shows phase separation in an entirely clear (no Tyndall effect) binary lithium glass as evidenced on a wedge-shaped splinter by transmission electron microscopy. Figure 6.2 shows phase separation in an equally clear sodium borate glass. In both cases the droplets are again much smaller than those found in the "model" (fluoride) glass (Fig. 5.31).

The entire development was supplemented in a decisive way by refined methods of small-angle X-ray scattering (Hoffman et al. [79], Porai-Koshits

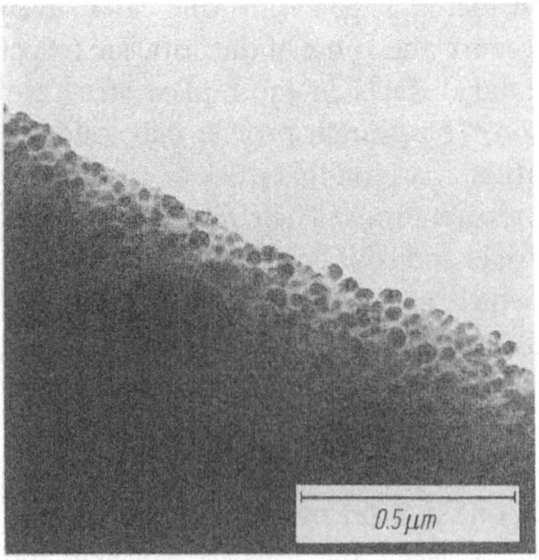

Fig. 6.1. Electron-optical micrograph of a Tyndall-free binary lithium silicate glass (28.6 mol% LiO_2, 71.4 mol% SiO_2) after direct irradiation of the wedgelike edge of a glass splinter. The LiO_2-rich denser droplet regions appear darker than the SiO_2-rich matrix because of stronger absorption of electrons

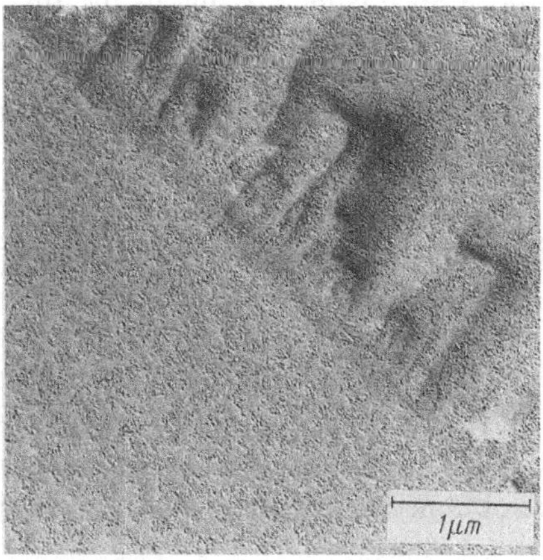

Fig. 6.2. Electron-optical micrograph of a Tyndall-free sodium borate glass (16 mol% Na_2O, 84 mol% B_2O_3) after replica preparation. A droplet-shaped Na_2O-rich borate glass phase is embedded in a B_2O_3-rich matrix

et al. [80, 81], Zarzycki et al. [82], Urnes [85], Ruland [86], Patel et al. [84], Becherer et al. [83], and Andreev et al. [193, 194]), and light scattering analysis. (Maurer [145], Goldstein [146, 147], and Hammel [148]). At present, all three methods represent the most important complementary complete processes for proving the existence of structural inhomogeneities in glass. In this review on microphase separation we must point out that these new concepts of glass structure, though strongly supported by experiments, at the beginning had difficulty gaining acceptance. Yet, later they became a major topic of glass research.

During the past 20 years, after initial resistance broke down in the face of such demonstrations, so many investigations of microphase separation have contributed to our present knowledge that it is hard to do them justice.

Selected references are Hammel et al. [195], Vogel et al. [196–209], Gerth et al. [210], Kühne et al. [211, 212], Mazurin et al. [213], Porai-Koshits et al. [214], Ohlberg et al. [215, 216], Moriya, [217, 218] Seward et al. [219], Shaw et al. [220], Srinivasan et al. [221], Tran [222, 223], Watanabe [224], Zarzycki et al. [225, 226], Zdanov et al. [227], Charles [228], Rindone [229, 230], Rötger et al. [231, 232], Hinz et al. [233], Hummel et al. [72, 73], Johnson et al. [74, 234, 235].

Up to almost 1960, it was believed that the observed immiscibility effects had their origin in statistic fluctuations or were due to exceptional conditions. It was only then that immiscibility was recognized as a basic phenomenon subject to rigorous theoretical treatment (see next section). In 1961, Kim addressed the Congress with the question: In what systems or under what conditions has no phase separation been observed? On the basis of what is known today one may answer by the following survey of phase separation in the most common systems.

(a) All combinations of the classical glass-formers (B_2O_3, SiO_2, P_2O_5) tend more or less toward phase separation, even though each conforms to the Zachariasen–Warren rule. Dietzel's [14] field strength concepts give an explanation (see Chapter 4). Although quenched B_2O_3–SiO_2 melts appear, to the human eye, to be clear, this appearance is merely due to the small dimensions of the separated region. Immiscibility, as predicted by Skatulla et al. [184] and Vogel [186, 188] was confirmed by experiment in 1960 (Bondarev et al. [236]). Immiscibility for the 80 $B_2O_3 \cdot 20$ SiO_2 and 40 $B_2O_3 \cdot 60$ SiO_2 glasses is shown in Figs. 7.2b and 7.2c.

(b) Binaries of SiO_2 and B_2O_3 with elements of the first column (Li, Na, K) tend to unmix in certain areas and droplet-shaped regions of phase separation have been found.

(c) Binaries of SiO_2 and B_2O_3 with elements of the second column (Be, Mg, Ca, Sr, Zn, Cd) show a much greater tendency toward unmixing. The phase diagrams of the SiO_2 systems show stable thermodynamic immiscibility gaps in the liquid phase. Even if the addition of a third component closes this gap and glasses clear for the human eye are obtained, microphase separation can be shown to persist.

(d) Binary B_2O_3 and SiO_2 systems with elements of the third group (Al, Ga, Tl, Y, La) also show immiscibility gaps in the phase diagram. After addition of a third component to close the gap, microphase separation is still observed, just as with second column elements. The special case of aluminosilicate glasses deserves emphasis. They had been considered homogeneous for a long time, but it has been determined that, because of the high melting temperature, necessitating a fast cooling rate on casting, the immiscibility droplets are merely very small. McDowell et al. [237] first unequivocally demonstrated phase separation in Al_2O_3–SiO_2 glasses, an important finding considering the technological importance of Al_2O_3 in glasses.

(e) Binary B_2O_3 or SiO_2 systems with elements of the fourth column tend to exhibit drastic microphase separation.

(f) Borate and silicate glasses containing fifth and sixth column elements (Nb, Ta, Mo, W) also show a pronounced tendency toward unmixing. The control (suppression) of this tendency plays an important role in the manufacture of new optical glasses containing Ti, Zr, Th, Nb, Ta, and W.

(g) Germanate glasses behave similarly to silicate glasses (Fig. 6.3). (Topping et al. [238])

(h) Tellurite systems, now applied in optical lenses, also show immiscibility, (Fig. 6.4). (Rindone et al. [230]).

(i) The introduction of column 7 elements (F, Cl, Br) and of $(SO_4)^{2-}$ ions as a rule strongly increases existing immiscibility tendencies in simple borate or silicate glass (Figs. 6.5, 6.6). The glass represented in Fig. 6.5 is melted without addition of fluorine whereas the one in Fig. 6.6 contains 3 mol% fluorine. The huge influence on phase separation is clearly demonstrated (Vogel [151], Vogel et al. [196]).

(j) One must also mention that corresponding combinations of P_2O_5 with a network modifier in significantly smaller quantities also tend to immiscibility (Vogel [151, 187], Rindone et al. [230], Murthy [239]). This can be expected from Dietzel's rules (see Chapter 4) based on field strength. This field requires more experimental studies. Conditions change rapidly when we consider combinations of P_2O_5 with either one or both B_2O_3 and SiO_2.

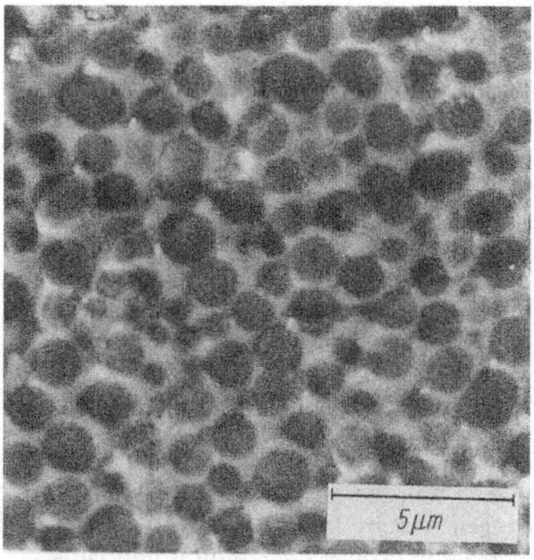

Fig. 6.3. Electron micrograph of a chalcogenide glass (67.5 mol% Se, 22.5 mol% Ge, 10 mol% Pb). Droplet-shaped immiscibility regions also in this type of glass

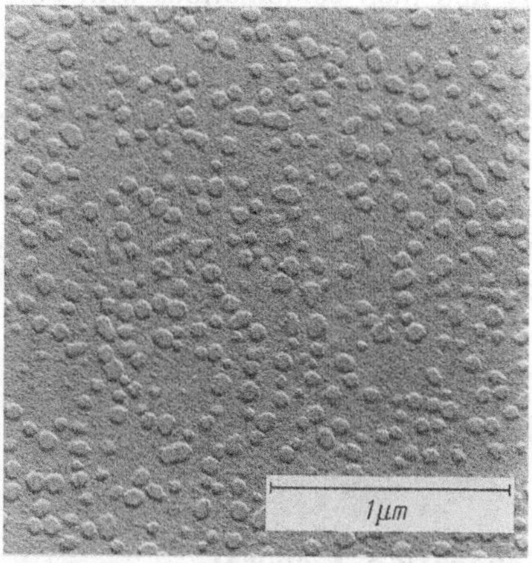

Fig. 6.4. Electron-optical micrograph after replica preparation of a telluride glass (37.3 mol% TeO_2, 42 mol% B_2O_3, 20.8 mol% K_2O) which had been treated with methanol. Unequivocal proof for droplet-shaped immiscibility regions also in this type of glass

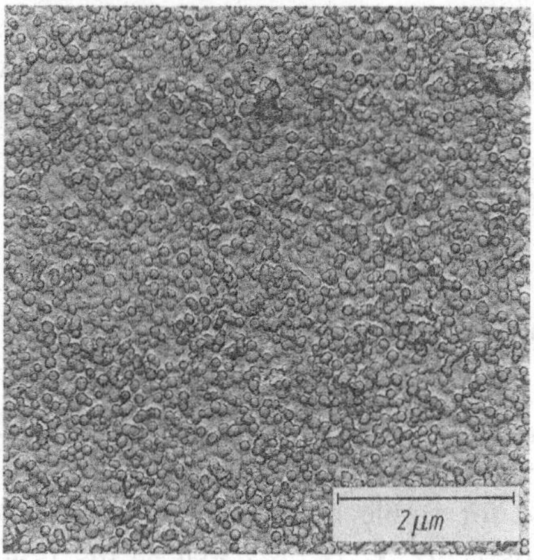

Fig. 6.5. Electron-optical micrograph of a binary sodium silicate glass after replica preparation (18.0 mol% Na_2O, 82.0 mol% SiO_2)

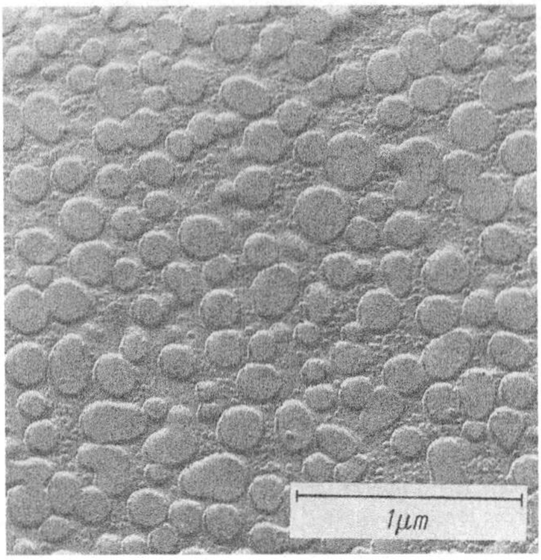

Fig. 6.6. Electron-optical micrograph of a sodium silicate glass after replica preparation (as in Fig. 6.5), but after replacing 3 mol% of the oxygen ions by fluorine ions. An unequivocal increase in droplet dimension is noticeable

(k) Non-oxidic glasses based on S, Se, Te and their combinations with other constituents, particularly As, Ge, Pb, Te, P, and I, are the subject of numerous studies, particularly because of their transparency in the IR [240–249] and their electronic properties. In most of these systems, regions of immiscibility have been observed (see, for instance, Fig. 6.3).

From all these qualitative observations it may be concluded that the absence of microphase separation is the exception. It is this conclusion which greatly stimulated the theoretical studies of phase separation, reported in the next section.

6.3 Theoretical Treatment

6.3.1 Thermodynamics of Phase Separation. General (Kortüm [250])

The phenomenological theory of classical thermodynamics explores given states and their change with composition, temperature, and pressure. It does not ask for their origin. It limits itself to equilibrium conditions of reversible changes of state which are not functions of time. Real processes, however, proceed irreversibly; they are non-equilibrium states for which classical thermodynamics has validity only as an approximation.

6.3.1.1 Conditions of Equilibrium and Stability

Under what external conditions is a system in equilibrium? Experimentally, equilibrium is attained when properties no longer change with time: The system has arrived at the lowest free energy. Analytically, this is expressed in equilibrium conditions of the first order, valid for stable or metastable states. At equilibrium the thermodynamic functions U, H, F, G – as well as the total entropy in a closed system – have an extreme value for given values of the characteristic variables. In order to distinguish stable from metastable equilibrium states, equilibrium conditions of the second order must be considered.

In a one-component system, the thermal and hydrostatic condition of stability must be fulfilled for the existence of a pure homogeneous phase, i.e., at a given temperature and a given pressure it must not decompose in two coexistent phases.

For the stability of mixed phases one more condition must be added, namely: The mixed phase must not decompose at constant temperature and constant pressure into two coexistent phases of different composition, i.e., spontaneous unmixing must not occur.

6.3.1.2. Derivation of Stability Conditions for a Binary Mixed Phase

The symbols used are $\bar{G}$ mean free enthalpy, a activity, f activity coefficient $= a/x$, $a/1-x.x$ concentration of second component, μ chemical potential $+ RT$ ln a, μ standard potential of pure component, and $\mu - \mu^0$ rest potential $= RT \ln a$ Indices 1 and 2 refer to components 1 and 2. The mean free enthalpy of a binary mixed phase is defined by (at constant temperature and pressure)

$$\bar{G} = (1-x)\mu_1 + x\mu_2 \tag{6.1}$$

Separating into standard and rest potentials:

$$\bar{G} = (1-x)\mu_1^0 + x\mu_2^0 + (1-x)RT\ln a_1 + RT\ln a_2 \tag{6.2}$$

For $x = 0$, $\bar{G} = \mu_1^0$; for $x = 1$, $\bar{G} = \mu_2^0$, i.e., the curve for $\bar{G}$ as a function of x (Fig. 6.7) terminates at the chemical potentials of the pure components. The dashed line connecting μ_1^0 and μ_2^0 represents the contributions of the standard potentials to $\bar{G}$, i.e., that of the first two terms of Eq. (6.2). Since $1 > a > 0$, the contribution of the last two terms is always negative, i.e., the $\bar{G}$ curve must lie under that straight line. The tangent in each point is

$$\left(\frac{\partial \bar{G}}{\partial x}\right)_{p,T} = \mu_2 - \mu_1 = \mu_2^0 - \mu_1^0 + RT\ln\frac{a_2}{a_1}$$

$$= \mu_2^0 - \mu_1^0 + RT\ln\frac{x}{1-x} + RT\ln\frac{f_2}{f_1} \tag{6.3}$$

Since limits of $\ln f_2$ and $\ln f_1$ are ∞ and 0 for $x \to 0$, and $x \to 1$, respectively, the $\bar{G}$ curve must hit the end values μ_1^0 and μ_2^0 with a tangent of infinity (Fig. 6.7).

Now let us explore the center part of the curve for two cases, I and II (upper and lower curves, respectively, in Fig. 6.7).

A composition O will separate spontaneously into two phases Q and R only if the energy of the average of Q and R (point P) is lower than that of O. This is true in case II, but not in case I, where the energy at P is higher than at O. This condition implies, as can be seen in Fig. 6.7, that the $\bar{G}$ curve must be

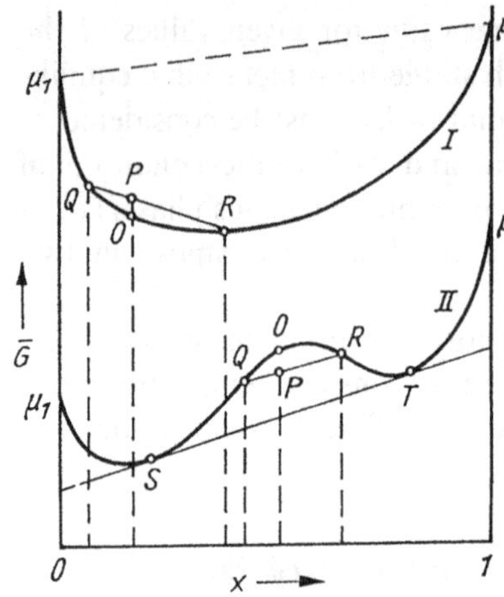

Fig. 6.7. Free enthalpy G of a binary phase as a function of composition x

concave for the region of separation, i.e.

$$\left(\frac{\partial^2 \bar{G}}{\partial x^2}\right)_{p,T} < 0 \quad \begin{array}{l}\text{(case II)}\\ \text{between Q and R}\end{array} \tag{6.4}$$

For the system to be stable (or metastable), and no spontaneous separation to take place, the $\bar{G}$ curve must be convex, i.e.,

$$\left(\frac{\partial^2 \bar{G}}{\partial x^2}\right)_{p,T} > 0 \quad \text{resp.} \left(\frac{\partial^2 F}{\partial x^2}\right)_{v,T} > 0 \quad \begin{array}{l}\text{(case I)}\\ \text{(everywhere)}\end{array} \tag{6.5}$$

The mixture in case II is unstable in the concave part of curve G and decomposes into two coexistent phases, whose composition is given by the contact points S and T of the common tangent. According to the general equilibrium conditions for phase transitions at a given pressure and temperature, the following applies for the coexisting phases: $\mu_1' = \mu_1''$ and $\mu_2' = \mu_2''$.

Introducing G, the following equations are obtained,

$$(G + (1 - x)(\partial G/\partial x))'_{p,T} = (G + (1 - x)(\partial G/\partial x))''_{p,T} \tag{6.6}$$

$$(G - x(\partial G/\partial x))'_{p,T} = (G - x(\partial G/\partial x))''_{p,T}$$

$$(\partial G/\partial x)'_{p,T} = (\partial G/\partial x)''_{p,T} \tag{6.7}$$

Equation (6.7) signifies an equal slope for the tangents at points S and T and Eq. (6.6) equal intercepts. Both points S and T with a common tangent thus correspond to the coexisting phases. The transition between convex and concave, thus between stable or metastable and unstable happens at the points of inflection, Q and R, of curve II, where $(\partial^2 G/\partial x^2) = 0$. The homogeneous phase can therefore not exist between Q and R. Although the homogeneous phase could exist in the region between S and Q or between R and T, since $(\partial^2 G/\partial x^2) > 0$ in these regions; it is however metastable with regard to the decomposition into phases S and T because a tangent drawn here does not completely lie under the curve. The stability condition

$$(\partial^2 F/\partial x^2)_v(\partial^2 F/\partial v^2)_x = (\partial^2 F/\partial x \partial v)^2 > 0 \tag{6.8}$$

still needs to be satisfied in order to differentiate between the stable and metastable states of a binary mixed phase. A generalization of Gibbs' findings for the conditions of phase stability in one or two component systems can be formulated in the following manner: Stable equilibrium occurs when the partial second derivatives of the functions state with respect to one of their characteristic variables (V, S, n) is positive.

In addition, sufficient stability conditions must be considered for two independent components. When dealing with a system of more than two components, necessary and sufficient stability conditions are frequently considered in their abbreviated form:

$$[\Delta^2 G(x_1, x_2 \ldots x_{n-1})]_{p,T} - [\Delta^2 G(p, T)]_x > 0$$

resp. (6.9)

$$[\Delta^2 F(V_1 x_1, x_2 \ldots x_{n-1})]_T - [\Delta^2 F(T)]_{v,x} > 0$$

6.3.1.3. Characterization of the Regions of Immiscibility in Binary and Ternary Systems.

Immiscibility phenomena are usually represented as solubility curves in an isobaric temperature (T)-composition (x) diagram, plotting the composition of both stable coexisting phases as functions of temperature (Fig. 6.8) In this way one obtains either open gaps miscibility gaps with an upper or lower critical temperature, or closed miscibility gaps with two critical temperatures.

The two phases coexisting at a given temperature having compositions x^{I} and x^{II} – corresponding to the tangent section S-T in Fig. 6.7 – are connected by the conode S-T in Fig. 6.8. At the critical temperature the conode degenerates to a single point (K_0), i.e., the two phases become identical when the binary mixture is homogeneous.

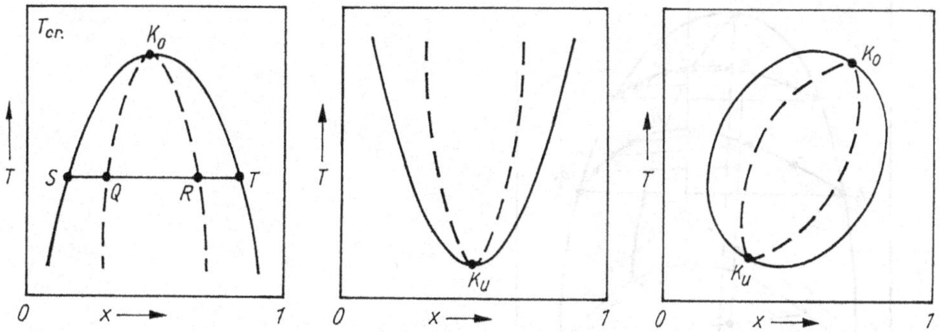

Fig. 6.8. Miscibility gaps in binary systems

In a ternary system, a maximum of three phases may appear at given pressure and temperature according to the phase rule, $P + F = n + 2$, Where only two phases occur, one more degree of freedom is available, which means that there is not just one pair, but a continuous series of such pairs with compositions given by the binodal curves.

In binary systems, determination of the solubility curve as a function of temperature suffices to characterize the immiscibility gap. The compositions of the coexisting phases are determined unequivocally by the terminals of the conodes in the T-x plane.

In ternary systems the compositions of coexisting phases can not be read from the T-x sections through the 3-dimensional gap. One must mobilize isothermal sections which are often termed conodal or binodal curves (Fig. 6.9). The figure presents the 3-dimensional T-x relation of a ternary system with miscibility gap (Vogel [251]). From several isotherms of such a system, the temperature dependence of solubility of ternary mixtures can be obtained. The resulting solubility plane (binodal plane) fK_1K_4 extends in space in the shape of a semidome. On it are placed the ternary phase pairs saturated in each other. The critical curve K_1K_4 is contained in this solubility plane. It is determined by the critical points of the isothermal sections and yields the temperature dependence of the compositions of critical mixtures.

The composition of a mixture p on heating can be followed easily at the hand of the 3-dimensional diagram. At temperature t_1, p is split into the A-rich phase f_1 and the B-rich phase φ_1. At temperatures t_2 and t_3, the compositions and concentrations of the phases are given by conodes $f_2\varphi_2$ and $f_3\varphi_3$ which cross the vertical line above p. At f_3 the A-rich phase f_3 equals p, i.e., the B-rich

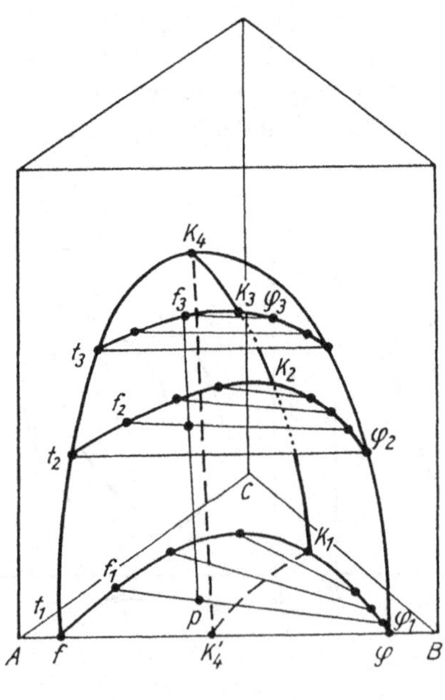

Fig. 6.9. Miscibility gap with upper critical point in a ternary system

phase vanishes: The transition to the homogenous region has been reached. K_4 is the critical point for this ternary system (i.e., $\partial_3\bar{G}/\partial x^3 = O$).

6.3.2 Thermodynamics of Immiscibility in Glasses

Depending on the composition and the resulting bonding conditions, undercooling will result in most glasses in aggregation processes which comprise all transitions between homogeneous mixtures and completely separated phases. The theoretical treatment of phase separation in glasses is based chiefly on the thermodynamics of mixed phases and the application of mechanisms of nucleation and growth of crystals.

The thermodynamic premise for phase separation is a positive free mixing enthalpy ΔG_m of the participating components. After the Gibbs-Helmholz relation, a positive ΔG_m can be obtained just as well from a positive ΔH_m as from a negative ΔS_m. This means that immiscibility can be caused by the relations of bonding as well as by those of ordering. In the first case the immiscibility gap will have an upper critical point, in the second a lower critical point. In glasses, the first case is observed in almost every system. Nevertheless, Brückner et al. [252], has identified the second case in cabal glass melts. The Gibbs stability criteria [250] for binary mixtures were first applied to vitreous systems by Cahn et al. [253]. As has been shown above, the following conditions obtain for the stability of a binary mixture against composition fluctuations at constant temperature and pressure (volume):
For stability as well as metastability:

$$\left(\frac{\partial^2 G}{\partial x^2}\right)_{p,T} > 0, \quad \left(\frac{\partial^2 F}{\partial x^2}\right)_{v,T} > 0$$

For the boundary of stability, the spinodal:

$$\left(\frac{\partial^2 G}{\partial x^2}\right)_{p,T} = 0, \quad \left(\frac{\partial^2 F}{\partial x^2}\right)_{v,T} = 0$$

For instability:

$$\left(\frac{\partial^2 G}{\partial x^2}\right)_{p,T} < 0, \quad \left(\frac{\partial^2 F}{\partial x^2}\right)_{v,T} < 0$$

For the critical point:

$$\left(\frac{\partial^3 G}{\partial x^3}\right)_{p,T} = 0, \quad \left(\frac{\partial^3 F}{\partial x^3}\right)_{v,T} = 0$$

With these conditions in mind, consider Fig. 6.10. Figure 6.10 a is a typical schematic of a phase diagram with a miscibility gap in the liquid state. Figures 6.10 b, c, and d represent the free-energy curves corresponding to temperatures T_0, T_1, T_2 in Fig. 6.10 a. Figure 6.10 a also shows the liquidus and solidus for a third, crystalline component. The construction of tangents gives the

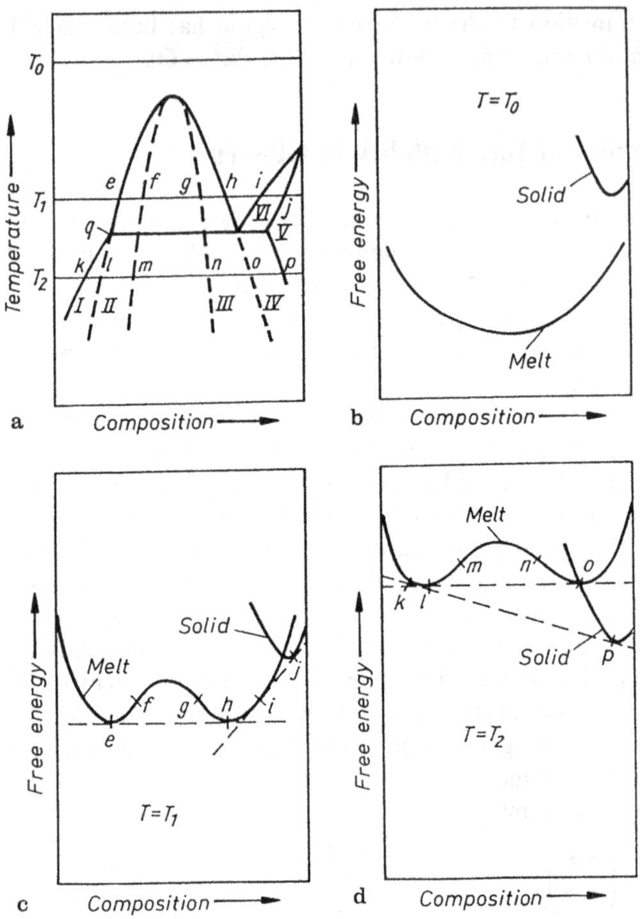

Fig. 6.10. Miscibility gap in a binary system and crystallization of a third component with corresponding curves (a–d) for the change in free-energy composition

compositions of coexisting phases. A horizontal tangent gives the composition of the coexisting phases in the gap and tangents touching the liquidus and solidus curves determine the composition of the liquid and solid phases.

At the temperature T_0 (Fig. 6.10b), the liquid has the lowest free energy over the entire compositional range. The free energy has a simple minimum and the homogeneous phase is stable for all concentrations. At T_1 (Fig. 6.10c), the liquidus curve has two minima, i.e., between compositions e and h the melt decomposes into two liquid phases. On the right side of the diagram the solid phase has the lowest energy; therefore a liquid phase of composition i is in equilibrium with a solid phase of composition j.

At T_3, (Fig. 6.10d), the solid phase of composition o is in equilibrium with a homogeneous liquid phase of composition k. The phase separation into compositions l and o is now metastable (i.e., is observed only if crystallization is avoided and a glass is formed). Composition q, termed the immiscibility boundary, becomes important when glass formation is observed. To the right of q the

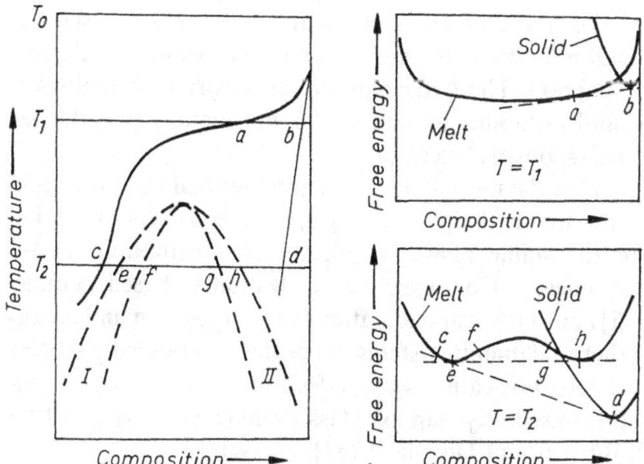

a b

Fig. 6.11. Metastable subliquidus miscibility gap with corresponding curves (a and b) for the change of free energy with composition

melt contains two phases so that, on cooling without crystallization, no homogeneous glass can be obtained, except if one cools extremely rapidly from above the critical temperature. Below the liquidus, six metastable regions exist for a homogeneous phase. Cooling into regions I, IV, V, or VI cannot lead to immiscibility. In regions II and III, phase separation may occur. Since they are metastable, nucleation, overcoming an energy barrier, is necessary.

The spontaneous unmixing in an instability region (spinodal immiscibility) where infinitesimal fluctuations lower free energy will be discussed later.

Whether a stable (above liquidus) or metastable (below liquidus) immiscibility occurs depends on the relation of the free-energy curves of liquid and solid phases. Consider the free-energy curves in Fig. 6.11 and the derived phase diagram (in which the entire immiscibility region is below the liquidus!)

Since $\partial F/\partial T = -S$, and since S is larger for crystal than liquid, the F curve of the liquid on cooling rises faster than that of the crystal. Now, if at the liquidus temperature the F curve of the crystal is below that of the liquid, an immiscibility appearing only at that temperature will be metastable. If this temperature is above T_g for the glass, and crystallization has been avoided, the resulting glass may develop metastable immiscibility. If this temperature is below T_g of a glass, the immiscibility will not appear in spite of the thermodynamic driving force, for kinetic reasons: The viscosity is too high for the required rearrangement to occur.

6.3.3 Kinetics of Immiscibility in Glasses

Two mechanics of immiscibility in glasses can be discussed on a theoretical basis: (1) a nucleation and growth mechanism for the microphases and (2) the spinodal mechanism.

Figure 6.12 shows the free enthalpies of mixing of a binary system in dependence on composition for various temperatures and (below) the derived gap in the *T-x* (phase) diagram. The latter diagram also shows the regions of stability, metastability, and instability as well as the morphology (• = droplet structure, ∞ = connective (sponge) structure).

As pointed out above, in the metastable range infinitesimal fluctuations of concentration increase the free enthalpy, making spontaneous separation impossible. Separation to the stable phases requires finite fluctuations at the expense of a nucleation energy. Homogeneous nucleation is based (Volmer [254], Becker et al. [255]) on statistical fluctuations in composition and energy which characterize all thermodynamic systems consisting of coupled particles. Nonstationary homogeneous nucleation was, indeed, proved to occur in the metastable region of the miscibility gap of glass-forming systems (Ohlberg [256], Hammel [256], Burnett and Douglas [257], Neilson [259]).

The spinodal mechanism has been postulated for the instable region. According to Cahn [260, 261] and Cahn and Hilliard [262], the nucleation energy drops to zero at the spinodal. Statistical fluctuations will grow without any barrier as they lower free energy. The process is one of diffusion, which can be described by differential equations such as that by Cahn:

$$\frac{\partial c}{\partial t} = D\Delta c - 2\alpha M\Delta^2 c + \frac{\partial D}{\partial c}(\nabla c)^2 - 2\alpha\frac{\partial M}{\partial c}[\nabla c\nabla(\Delta c)]$$

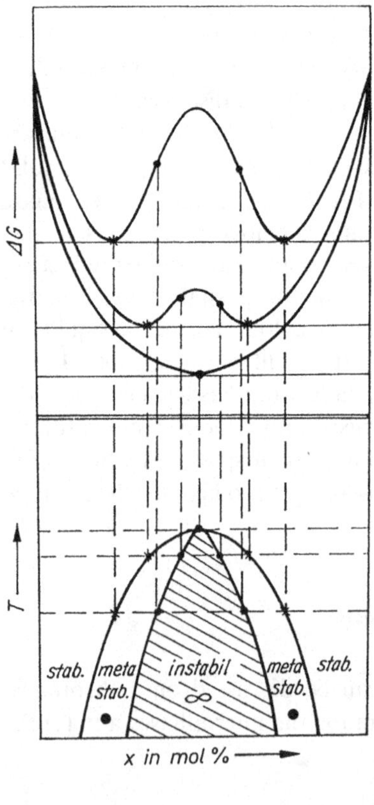

Fig. 6.12. Free enthalpy of mixing *G* of a binary glass system on dependence of composition for various temperatures, and the miscibility gap derived from them

where $D = \partial^2 f/\partial c^2 M$ = diffusion coefficient, M = mobility of particles, f = free energy, c = composition, t = time, and α = a coefficient depending on temperature and composition, which can be calculated for the case of regular solutions.

The first two terms describe early stages. The solution of the thus-simplified equation through Fourier series gives hints as to how the concentration changes:

$$c - c_0 = e^{R_{mt}} \sum_{\beta m} (A(\beta)\cos(\beta r) + B(\beta)\sin(\beta r))$$

where R is the so-called factor of reinforcement and β the wave number. It follows that a definite distance between the centers of fluctuation is attained, which, at constant temperature, does not change during the early stage, the so-called wavelength λ:

$$\lambda = 4\pi \left[\alpha V \Big/ \left(\frac{\partial^2 S}{\partial c^2} \right) \right]^{1/2} (\Delta' T_s)^{-1/2}$$

where S = molar entropy and T_s = spinodal temperature. λ increases as the temperature falls relative to the spinodal. Cahn concluded from the mathematical solution that a connective structure would form.

The higher, nonlinear terms of the diffusion equation reveal the further growth of these fluctuations: (a) gradual sharpening of the diffuse phase boundaries and slow growth of the centers of fluctuation and (b) continuous increase in the compositional differences between the phases. This may amount to the formation of separate particles.

In summary, the two mechanisms may be characterized as follows:

(1) In the metastable region, nucleation and growth of a droplet phase occur. Different compositions of matrix and droplets (Fig. 6.13 a) are defined from the beginning.

(2) In the instable (spinodal) region dimensions of connective phases are defined from the beginning and there is a continuous increase in the differences of composition and the volume of the phases (Fig. 6.13b).

Several authors interpreted the observation of connective structures as a proof for the spinodal mechanism. (Cahn and Charles [253], Ohlberg [256]). However, an exact proof requires the experimentally difficult investigation of the early stage. This was clearly recognized when Haller [263] showed that, during growth of the droplet phase, connective structures may form if nucleation density is appropriate (Fig. 6.14). This process was also demonstrated by electron microscopy for barium and sodium borosilicate glasses (Fig. 6.15). (Vogel et al. [198, 199, 209], Vogel [200, 201]). In this connection, Ostwald's [265] experimental and theoretical demonstration of a functional change in microphases of oil-water emulsions in the large region of 22:78 to 78:22 vol% deserves mention.

Filipovich [266] and Andreev et al. [267] also made valuable contributions to the theory of immiscibility in glasses. They used the same differential equation as Cahn but interpreted the solution more broadly, permitting the formation of separated particles.

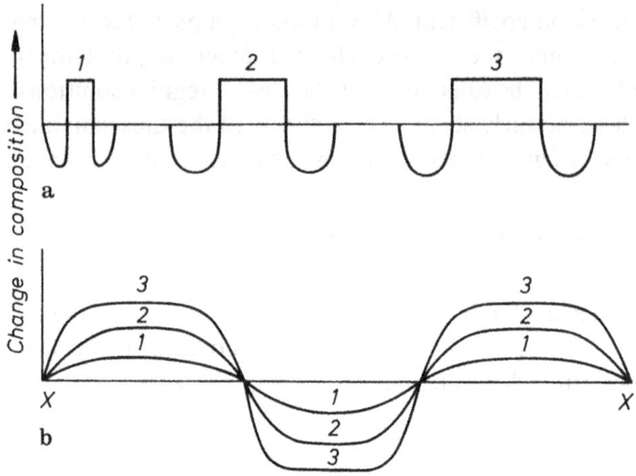

Fig. 6.13. Comparison of nucleation growth (**a**) and spinodal mechanisms (**b**). In (**a**) a particle of constant composition grows (in time steps 1, 2, 3) in diameter along the x axis. The surrounding matrix is depleted as it supplies components. In (**b**) a particle of constant diameter ("wavelength Λ") grows (in time steps 1, 2, 3), changing composition until completion

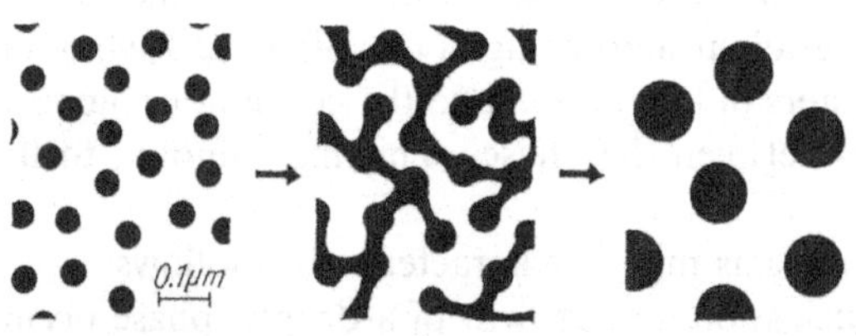

Fig. 6.14. Schematic of growth of droplet-shaped microstructures in a glass matrix (Tashiro [264], Haller [263])

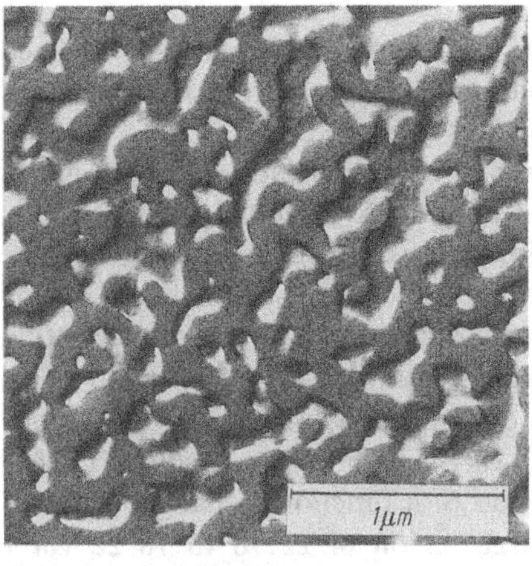

Fig. 6.15. Electron-optical micrograph of a sodium borosilicate glass after replica preparation. The picture shows one typical connective structure. To emphasize the phase boundaries, the sample was treated with water before replication

For various reasons, theories of immiscibility in glasses at present remain limited in their applicability to observed phenomena. On principle, the calculations undertaken so far are applicable only to one- and two-component systems, and even there with limitations. They do not, for example, take sufficient care of the changes in the matrix. Besides, the spinodal represents a sharp boundary only if not more than one of all possible variables of the system changes when other parameters change. This is not the case under real conditions. More detailed conclusions are possible on the basis of the thermodynamics of irreversible processes. Thus, Kluge and Wiedemann [268] were able to explain theoretically, as an interface problem between two phases of a ternary system, the formation of envelopes around the droplets found by electron optics (Vogel et al. [197]). So far, an unequivocal proof of the spinodal mechanism in phase-separated glasses has not been achieved. But the investigations of the early stages of phase separation did lead to the discovery of the supercritical fluctuations near and above the critical point (Zarzycki and Naudin [269]). They also may form near the spinodal (Porai-Koshits [269]). They grow with decreasing temperature and can be identified by X-ray (and neutron) small-angle scattering (SAXS).

In his critical review of recent studies of inhomogeneities in the glass structure, Porai-Koshits [270] concludes that a revision of the theory of spinodal separation on the basis of the theoretical results of Cook [271] and Morral and Cahn [272] is one of the most relevant tasks within the field of immiscibility. Furthermore, the mechanism of formation, as well as the relaxation of the various forms of inhomogeneous structures and their mutual transformations. e.g., that of statistic fluctuations into supercritical fluctuations, and of the latter into immiscibility structures require investigation.

Nor has it become clear whether the composition in the final stage is, in fact, identical with those of the binodal curve, i.e., the terminals of the conodes. Because of the high viscosities characteristic for systems freezing to glasses and the resulting large time dependence, conditions usually fall short of equilibria. As a consequence, multiple phase separation may occur, making a theoretical treatment much more difficult.

6.4 Experimental Evidence

6.4.1 Functional Change of Microphases

The systematic study of phase-separation phenomena in simple binary glass systems has indicated that the maximum of immiscibility usually corresponds to a composition between two defined stable chemical compounds. Various authors (Porai-Koshits and Averyanov [214], Johnson and Hummel [234, 235], Vogel and Ehrt [209], Seward and Uhlmann [273]) found subliquidus immiscibility domes between two stable compounds in the phase diagram of their systems, as illustrated schematically in Fig. 6.16. The domes can be asymmetrical, a feature of specific influence on multiple phase separation (see next section).

Investigations of series of compositions in simple binary systems under equal cooling conditions have shown that, near the composition corresponding

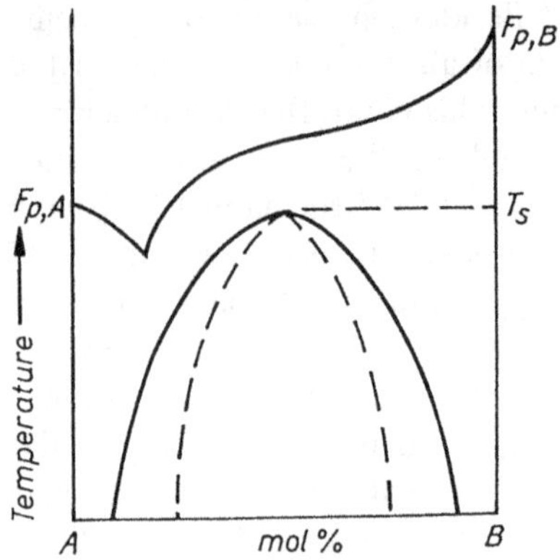

Fig. 6.16. Schematic of a subliquidus immiscibility region in a binary melting diagram between compounds A and B. F_{pA} melting point of compound A, F_{pB} melting point of compound B, T_s critical point

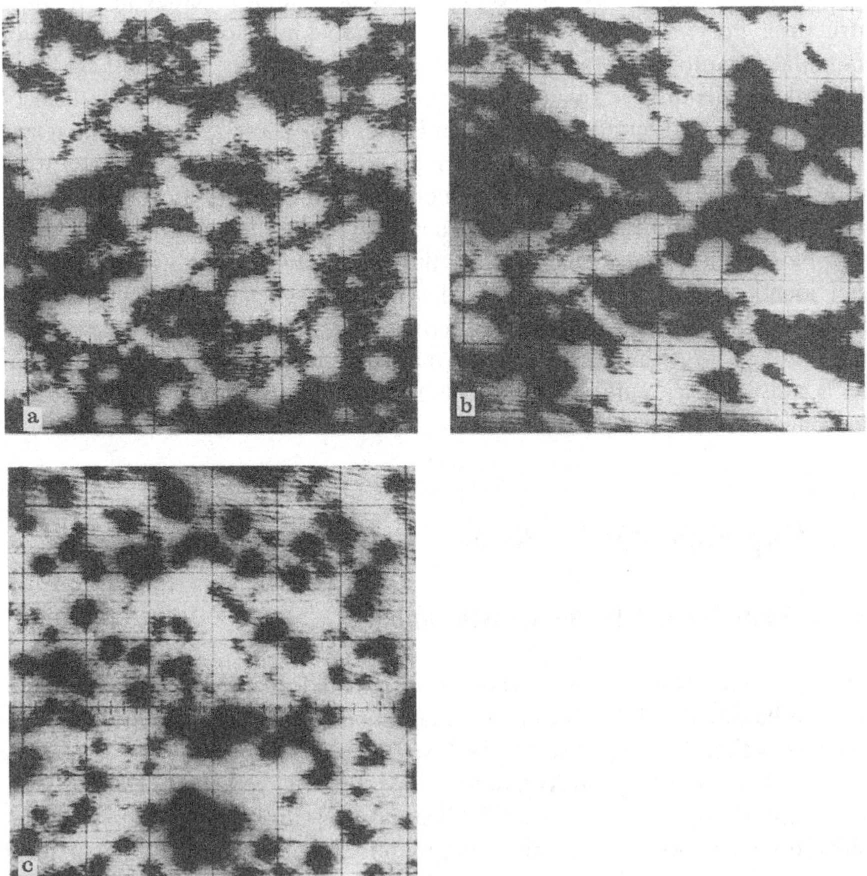

Fig. 6.17. $Ba_{L\alpha}$ X-ray scanning micrographs of a barium borosilicate glass whose composition lies to the left **(a)** to the right **(c)**, or (b) within maximum immiscibility. A change in function of the microphases can be recognized, i.e., the droplet phase takes on the matrix function. At maximum immiscibility the typical connective structure appears **(b)**

to the maximum of the immiscibility dome, a functional change of the microphases occurs. The microphase which appeared as a droplet phase takes over the matrix function, while the matrix phase becomes the droplet phase. Transition and penetration structures are observed in a narrow range near the maximum, i.e., neither phase is droplet-shaped and both phases are connected and interpenetrate (Fig. 6.17). This range of interpenetration need not necessarily be due to a spinodal mechanism.

The phenomenon of a functional change of the microphases is enormously important for the control of glass properties such as chemical resistance, electrical conductivity, etc. It is by no means irrelevant whether the mobile ion is concentrated in the matrix or the droplet phase.

6.4.2 Multiple Phase Separation [151]

In most cases, electron microscopy demonstrates droplet-shaped immiscibility regions embedded in a homogeneous matrix, the result of a simple primary separation process.

In 1958, for the first time, a two-step phase separation process was discussed for the case of Vycor-type[1] glasses and, subsequently (1958/1959) demonstrated experimentally (Skatulla et al. [184], Vogel, [185, 186], Kühne et al. [211]). According to these findings, these glasses consist of three microphases: two droplet phases of vastly different dimensions embedded in the base glass matrix. Later, many exemplary cases of secondary-phase separation and three-phase glasses were confirmed (Vogel [208], Porai-Koshits and Averyanov [214], Schönborn [274], Vogel and Byhan [275]). These examples showed that the processes involved in the formation of structures, even on cooling simple melts, were much more complicated than had been assumed because of the decisive role of kinetic factors.

Thus, for the first time as many as up to eight different microphases could be identified by electron microscopy after stepwise phase separation in the system $BaO–B_2O_3–SiO_2$ (Vogel et al. [197], Gerth and Rehfeld [210]). Figure 6.18 shows six experimentally demonstrated microphases.

Figure 6.19 is a schematic of a stepwise phase-separation process leading to more than two microphases. Through primary phase separation, two phases of different viscosity are formed which, during further cooling, behave almost like two separate systems with their own separation processes. A concentration adjustment in the total system corresponding to the conodes will hardly be possible, i.e., it will be limited to the separated primary phases.

This will easily lead to non-equilibria below equilibrium conditions, or entirely local near-equilibria within the microphases. Depending on the temperature and viscosity conditions of the cooling melt, this kinetically driven step mechanism of phase separation may continue in a limited way. Multiple

[1] Corning Glass Works, Corning, NY.

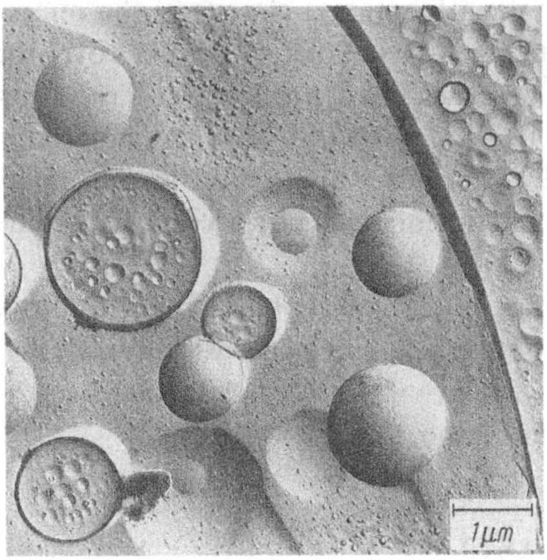

Fig. 6.18. Electron-optical micrograph of a barium borosilicate glass after replica preparation. Because of stepwise unmixing, six microphases of different composition were formed

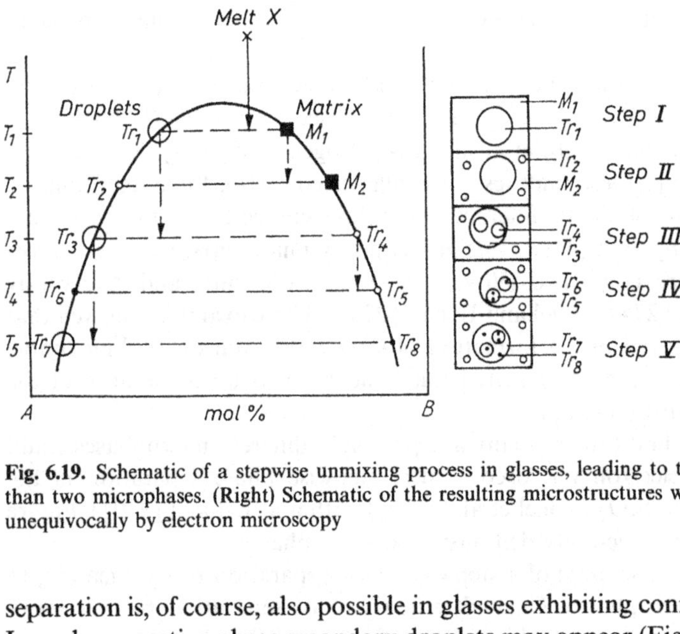

Fig. 6.19. Schematic of a stepwise unmixing process in glasses, leading to the formation of more than two microphases. (Right) Schematic of the resulting microstructures which could be proved unequivocally by electron microscopy

separation is, of course, also possible in glasses exhibiting connective structures. In each connective phase secondary droplets may appear (Fig. 6.20). At present experimental results seem sufficiently understandable. In the future more quantitative explanations will have to be provided.

6.4.3 Shells around Microphases (Vogel [151], Vogel et al. [197])

Multiple phase separation is the consequence of impeded or incomplete diffusion processes during cooling. These processes may lead to the formation of shells (halos) around microphase droplets, as has been demonstrated experimentally (Fig. 6.21). As the temperature dropped, some ion diffusion could be

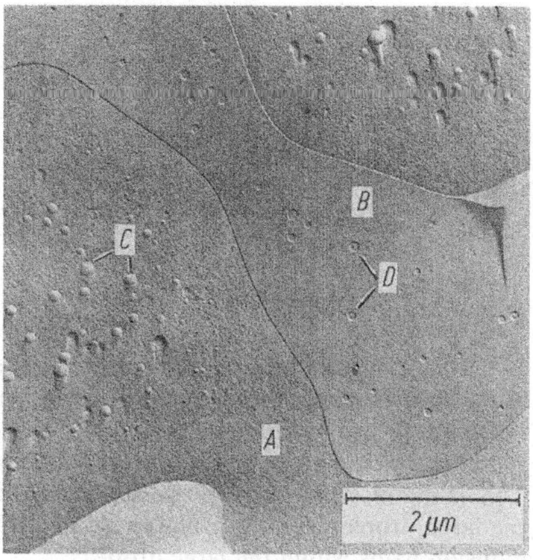

Fig. 6.20. Electron-optical micrograph of a barium borosilicate glass after replica preparation. The composition is that of maximum immiscibility. The primary phases *A* and *B* penetrate each other (connective structure). Both primary phases show secondary unmixing. In phase *A*, secondary phase *C* and in phase *B*, secondary phase *D* were precipitated in droplet shapes

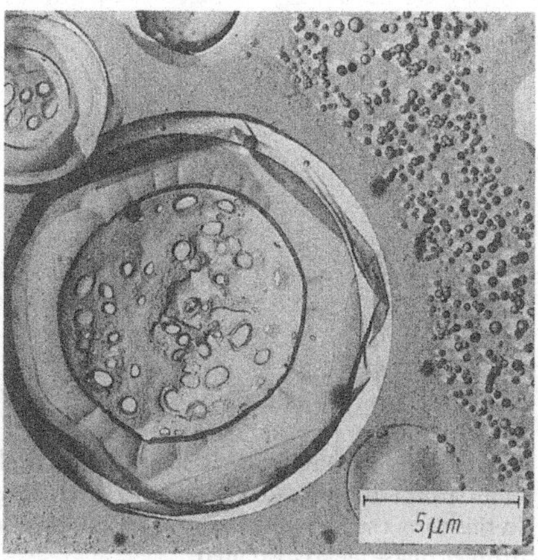

Fig. 6.21. Electron-optical micrograph of a barium borosilicate glass after HNO_3 etching and replica preparation. The glass consists of four glass phases. In the matrix phase, diffusion halos around the large-droplet phase can be recognized. The large droplets are surrounded by an SiO_2 shell

maintained in the matrix phase. But, since the large droplets had already been frozen solid, the necessary exchange of concentration was possible only within the area of halos. The components diffusing in the direction of the droplets were able to precipitate only in the form of a shell, in this case SiO_2. Farther from the large droplets, the corresponding share of SiO_2 has formed small droplets in a secondary separation process. The reverse case also occurs: The diffusion zone exists within the droplet. Now the composition of the melt is on the other side of the dome, and diffusion is easier in the droplet phase (Fig. 6.22). The diffusion process within the large droplet B in the direction toward the matrix generates a diffusion halo within the large droplet. At the same time, a halo also can be identified around the secondary phase C. This is understandable, if one recognizes that the matrix phase A and the secondary phase C are – if not quite identical – rather similar (Fig. 6.19, M_2, Tr_4).

The barium borosilicate glass of Fig. 6.21 had been etched weakly with HNO_3. The SiO_2 shell was not attacked, but the partial solution of the barium

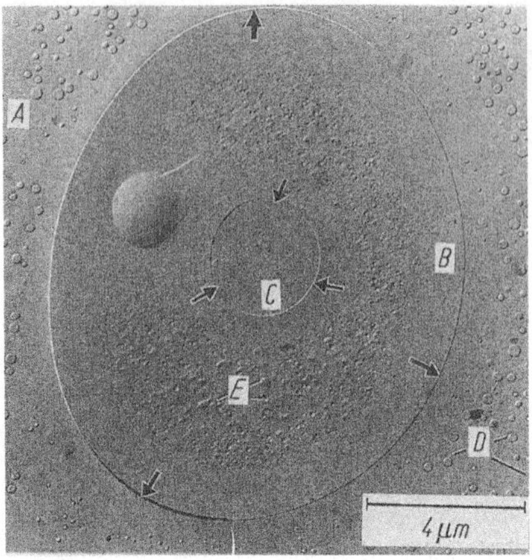

Fig. 6.22. Electron-optical micrograph of a multiphase barium borosilicate glass after replica preparation. In this case, the droplet phase is that in which changes in function are still easy. This is manifest by the stepwise separation of the large droplets into four phases and by the diffusion halos within the large droplets

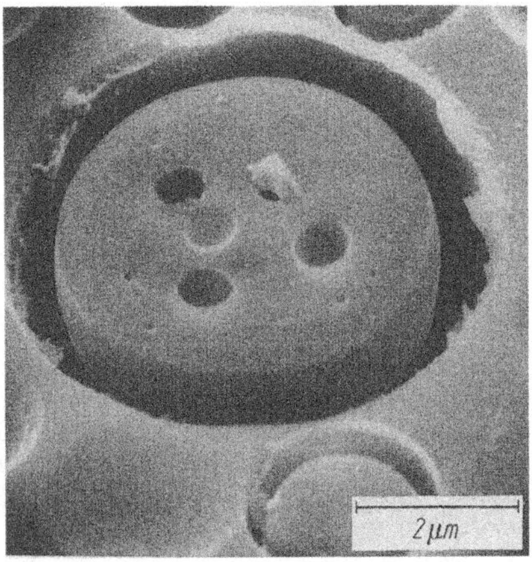

Fig. 6.23. Electron backscattering micrograph of a barium borosilicate glass after HF etching. In this case, the SiO_2 shell was dissolved and the droplet phase nearly isolated

borate-rich droplets caused the support of the SiO_2 shell to be partly impaired, which caused some creasing. The glass of Fig. 6.23 was the same as in Fig. 6.21, but had been etched with HF. Now the SiO_2 shell was almost entirely dissolved and the droplet nearly isolated.

Shells around droplet-shaped microphases are not exceptions. Once the formation of shells had been recognized, analogous phenomena could be found unequivocally when older electron optical pictures were reinspected. It is quite understandable that shell formation will strongly affect the chemical properties of a glass.

6.4.4 Droplet Agglomeration after Secondary Phase Separation

A fundamental phenomenon was discovered by Vogel, Höland, Horn and Völksch [276] working on bioglass ceramics (to be described in a later chapter), specifically on phase separation in glasses of the system $CaO–P_2O_5–SiO_2$. All glasses investigated were on the line joining the $CaO \cdot SiO_2–3CaO \cdot P_2O_5$ pseudo system in the $CaO–P_2O_5–SiO_2$ system (see ([277] in Fig. 6.24). This system reveals a low marginal miscibility, that is the calcium metasilicate phase has a low solubility in the calcium orthophosphate phase, and the latter in the calcium metaphosphate phase.

When melts of these pseudo-binary systems are cooled quickly glass formation occurs: In an experiment described here, the composition of the melt was richer in $3CaO \cdot P_2O_5$. A precipitation of large SiO_2-rich droplets in a P_2O_5-rich matrix glass occurs in a primary immiscibility process. As the system is further cooled, secondary immiscibility occurs in both phases, that is P_2O_5-rich microdroplets form in SiO_2-rich large droplets and SiO_2-rich microdroplets form in a P_2O_5-rich matrix glass phase (Figs. 6.25 and 6.26). The blackberry-type aggregation of the microdroplets is interesting in both cases and can be especially well recognized in Fig. 6.25. We are dealing in both cases with the removal of the marginal miscibility of $CaO \cdot SiO_2$-rich and of $3CaO \cdot P_2O_5$-rich phases in the other component. The aggregation of the microdroplets formed in the secondary immiscibility step is the result of a further approximation by the microdroplets' composition to one of the chemically stable compounds $CaO \cdot SiO_2$ and $3CaO \cdot P_2O_5$ (see the general diagram of multiple phase separation in Fig. 6.18).

The schematic Fig. 6.27 is, together with the electron micrographs of Figs. 6.25 and 6.26, illustrates the result of extensive investigations and identifications. The existence, in the majority of glasses, of defined molecular building

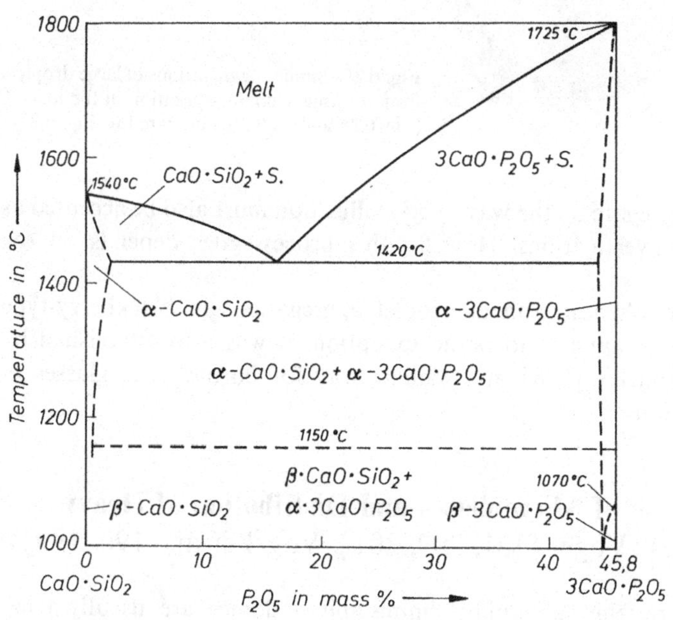

Fig. 6.24. Pseudo-binary system $CaO \cdot SiO_2–3CaO \cdot P_2O_5$ [277]

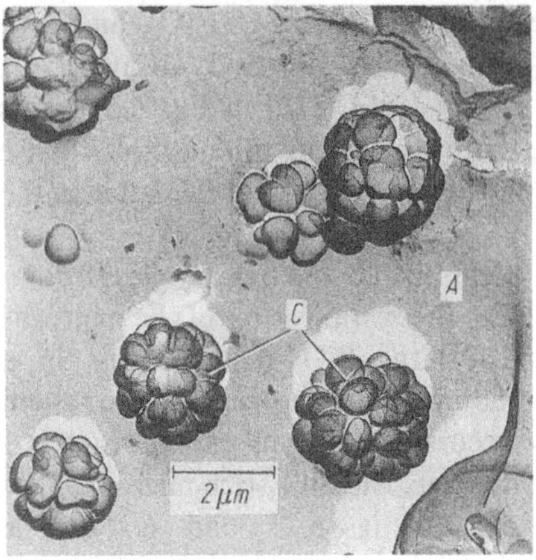

Fig. 6.25. Electron micrograph (Pt–Ir–C replica method) after a weak surface etching for 30 s in 2% citric acid. Blackberry-type aggregation of droplets in the matrix glass phase after a secondary immiscibility process in this phase

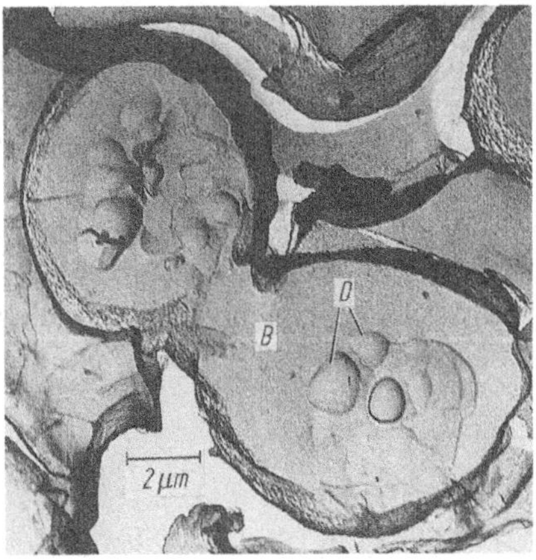

Fig. 6.26. Similar aggregations of large droplets from a primary phase separation in the matrix. (Fig. 6.25 and 6.26 are compared to Fig. 6.27)

elements which aggregate on the way to crystallization must also be accepted as a result of these investigations. How far this process goes depends on the undercooling.

The described phenomenon of droplet aggregations in blackberry-type formations in glasses appears to be no exception. It was also established by Kashieva and Dimitriev [278] in a similar way for borotellurite glasses of different compositions.

6.4.5 Composition of Microphases and Distribution of Heavy Metal Ions (Vogel [151, 200, 201], Vogel et al. [196, 199])

As indicated before, the subliquidus immiscibility domes are usually lying between two compounds in the phase diagram. In the case of slow cooling, i.e., if

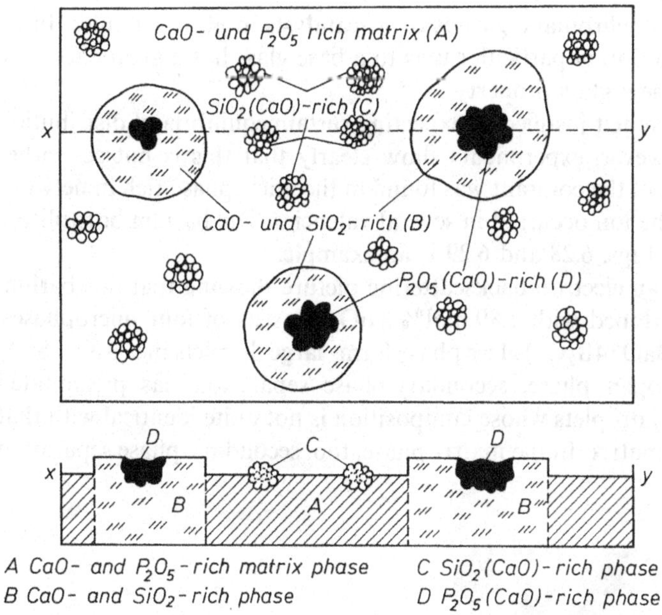

Fig. 6.27. Schematic of the phase separation behavior of the base glass from the pseudo-binary system $CaO \cdot SiO_2$–$3CaO \cdot P_2O_5$

equilibria are fully established, the two resulting phases should, according to the conodes at lowest temperatures, closely approach the composition of these compounds. On rapid cooling, the microphases would differ somewhat more in composition from that of the compounds. At any rate, the compositions of the microphases somewhat approach those of stable chemical compounds, depending on the degree of undercooling. The isolation of phases, or separation by high-temperature centrifuge, followed by analysis, proves this postulate unequivocally.

Figure 5.48 represented a scanning electron microscopy picture of a Ca-containing borate glass superimposed by an X-ray intensity curve for the $Ca_{K\alpha}$ radiation. The intensity curve proves without any doubt that the Ca concentration in the matrix has a constant and very low value, whereas it jumps to a high constant value in the droplet phase.

Equilibrium clearly has been reached to a large extent in this glass. Most certainly, the droplet-shaped areas cannot be interpreted as incidental fluctuations. If they were, the compositions would not be so constant. Everything points to a tendency toward microphase separation for glasses containing two building elements, one phase of each being rich in one of these elements.

Once some measure of knowledge had been attained on the basic immiscibility phenomena, further, more subtle, problems could be addressed such as the question of how 3d elements (e.g., Co, Ni, Cu, and other color indicators) would be distributed among the microphases of the base glass. This problem is related to the enormous effects which small additions may have in filter, laser, photosensitive, photochromatic glasses or as catalysts in glass ceramics. In all cases the small addition of particular ions to a base glass has a great effect, the development of a new glass property.

One might have felt justified in expecting certain equilibria of distribution for such ions. However, experiments show clearly that this is not so; rather almost 100 percent of the colorant was found in the microphase rich in network modifiers, even if the ion occurred in several valencies. The barium borosilicate glass illustrated in Figs. 6.28 and 6.29 is an example.

Figure 6.28 is an electron backscattering picture showing that this barium borosilicate glass, doped with 1.89 mol% NiO, consists of four microphases. A barium borate ($BaO \cdot 4B_2O_3$) glass phase forms large droplets in a B_2O_3–SiO_2 matrix. In the droplet phase, secondary-phase separation has precipitated smaller B_2O_3–SiO_2 droplets whose composition is not quite identical with that of the B_2O_3–SiO_2 matrix. In the matrix phase, too, secondary-phase separation

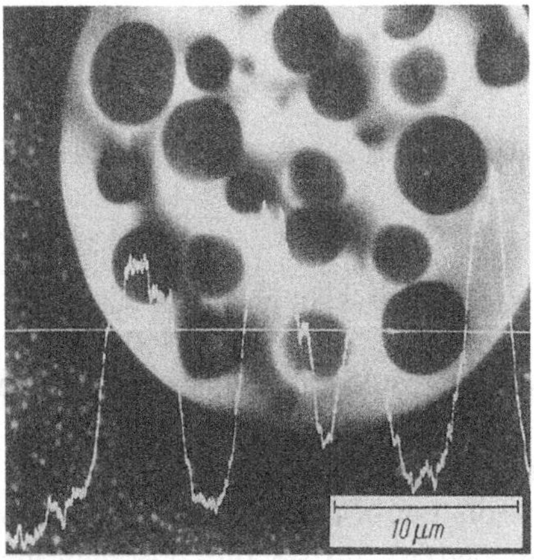

Fig. 6.28. Electron backscattering micrograph of a barium borosilicate glass with 1.89 mol% NiO doping with insection of a $Ba_{L\alpha}$ X-ray intensity curve. The electron beam was led along the white horizontal line across the sample. The X-ray curve shows that the barium ions are etched in the large-droplet phase, which appears white

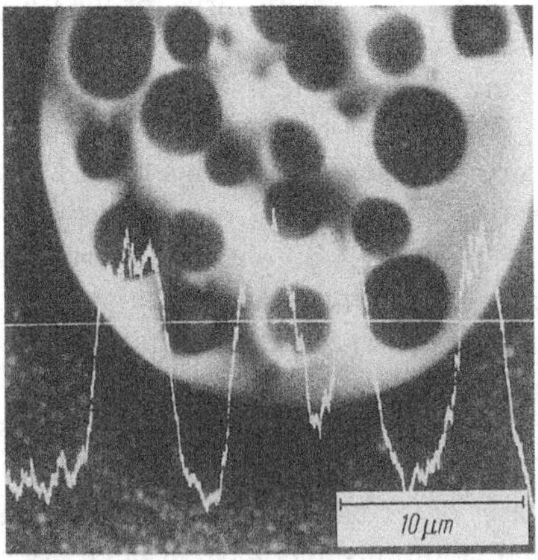

Fig. 6.29. Electron backscattering micrograph of a barium borosilicate glass (1.89 mol% NiO doping) with inserted $Ni_{K\alpha}$ X-ray intensity curve. The electron beam was led along the white horizontal line across the sample. The X-ray intensity curve shows that the nickel ions are enriched in the barium borate-rich large-droplet phase, which appears white

has precipitated a second, very small, droplet phase. Since these droplets are too small for microprobe analysis, they will be ignored in what follows.

In Fig. 6.28, the $Ba_{L\alpha}$ X-ray intensity curve is superimposed on the picture. As in the foregoing cases, the Ba intensity curve is very low in the matrix, jumps to a high value in the droplet phase, but is again low within the smaller droplets. It is, however, a little higher than in the matrix, in agreement with the step mechanism schematically illustrated in Fig. 6.19.

In Fig. 6.29 the $Ni_{K\alpha}$ X-ray intensity curve is superposed on the picture. The curve follows the Ba intensity profile closely. Neither the matrix nor the smaller droplets contain Ni.

In another barium borosilicate glass doped with 4 mol% CoO the matrix was rich in barium borate, with the B_2O_3–SiO_2 microphase appearing in droplet form. Acid treatment allowed separation of the droplets from the dark-blue glass. They were entirely colorless. The sensitive Co analysis was negative.

The experiment with CoO was repeated for comparisons in which the barium borate glass phase was either matrix or droplet phase. The droplets were large enough for light-microscopy investigations. In the barium borate glass phase enrichment in colbalt occurs in both cases. In the first case, colorless droplets always occurred in a dark-blue matrix; in the second-case dark-blue droplets were embedded in a colorless matrix.

Similar results were obtained with Ni and Cu.

6.4.6 General Conclusions on Immiscibility Behavior and Microstructure [209]

The following conclusions can be drawn concerning the probable structure of glasses based on the extensive experimental results on phase separation phenomena.

(a) One-component glass melts (SiO_2, P_2O_5, B_2O_3, GeO_2, BeF_2, etc), i.e., glasses consisting entirely of one network-former, usually solidify to a homogeneous glass. Their structure conforms well with the network theory.
(b) Glass melts whose compositions correspond with a defined stable chemical compound, i.e., that contain only one kind of building element, also solidify to a homogeneous glass.
(c) Glass melts whose compositions lie between two defined stable compounds tend more or less toward phase separation. Essential criteria are field strength difference of participating cations, interface tension of the generated microregions of different composition, and, last but not least, volume requirements of different structural units.
(d) Depending on the volume relation of two primary microphases, one of them may have the droplet or matrix function. Near maximum immiscibility, penetration structures are observed if the undercooling is the same.
(e) The chemical composition of the microphases tends to approach that of defined stable compounds, the degree depending on the rate of cooling and the stability of the structural elements, i.e., the chemistry of the system.

(f) Certain additives, such as 3d elements, are not distributed statistically among the microphases. Depending on the rate of cooling, the additive is present nearly 100 percent in the microphase offering the best coordination possibilities, i.e., highly charged cations are always concentrated in the phase richer in anions (e.g., non-bridging oxygen). In a homogeneous base glass the distribution is statistical.

6.4.7 Control of Phase Separation

The control of phase separation represents a wide field of applied glass research, of unquestionable importance for the development and production of glasses.

Control is possible, for instance – based on the findings reported in the preceding sections – by choosing the composition near to (homogeneity) or far from (phase separation) that of a chemical compound: or by going from a modifier of high to modifiers of low field strength (decreased phase separation), for instance, in the series Li_2O–SiO_2, Na_2O–SiO_2, K_2O–SiO_2, Rb_2O–SiO_2, Cs_2O–SiO_2.

Combining several binary systems which have a strong tendency to phase-separate also decreases this tendency and may even close the gap completely.

Small amounts of certain additives may strongly increase phase separation by strongly affecting the structure or interface tension. They may induce a decrease in viscosity or increase in ion mobility, facilitating separation kinetics. Fluorine, in particular, has a remarkable effect when replacing oxygen, which amounts to breaking a network bridge (compare Figs. 6.5 and 6.6 [151, 196]). For the same reason, the introduction of OH (Kreidl and Maklad [279], Maklad and Kreidl [280], Boulos and Kreidl [281]) increases phase separation in simple glasses. Rindone and Ryder [230] established the enhancement of phase separation by small additions of noble metals (e.g., Pt) to simple phosphate and silicate glasses (see also Murthy [239]), an important feature in the development of photosensitive and similar glasses.

Even a small CaO content, introduced, for instance, from a dissolving tank block, may lead to an enormous increase in phase separation, leading to defects in the product. Without an understanding of the phase-separation mechanism, this negative influence of a major constituent of conventional glasses would be incompressible.

Figure 6.30 shows the 10-fold increase (from 0.1 μm) in droplet size in a phosphate opal glass after introduction of 3 mass% CaO. On the positive side, this permits control of opacity.

One of the most important methods of control of phase separation is heat treatment of the melt or the solidified glass. It makes, of course, quite a difference whether the same glass is heat-treated at the same temperature after cooling from the melt or after heating from room temperature, because completely different processes, resulting in different structures, are often induced.

One of the most important utilizations of the generation of definite structures by controlled heat treatment, with subsequent leaching of one of the

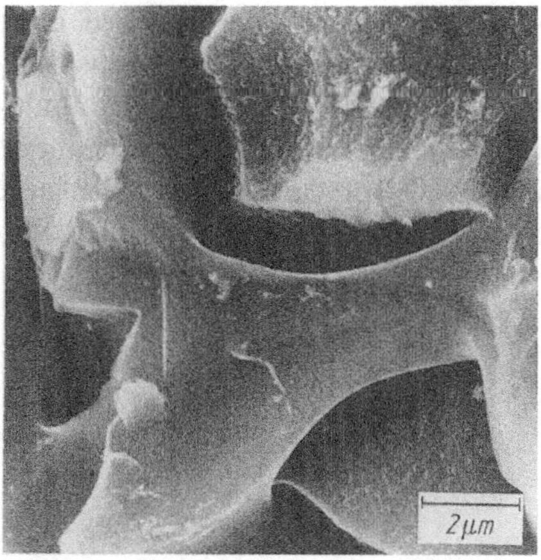

Fig. 6.30. Electron-optical micrograph (replica preparation) of a borosilicate glass opacified by a phosphate (3 mass% Ca). Because of the added Ca, the droplet dimension has increased 10-fold

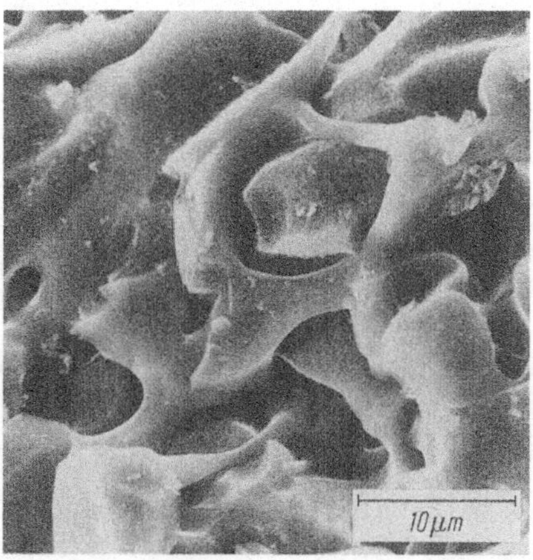

Fig. 6.31. Electron backscattering micrograph of a sodium borosilicate glass leached by acid treatment with connective structure. The residue is almost pure, highly porous SiO_2 glass

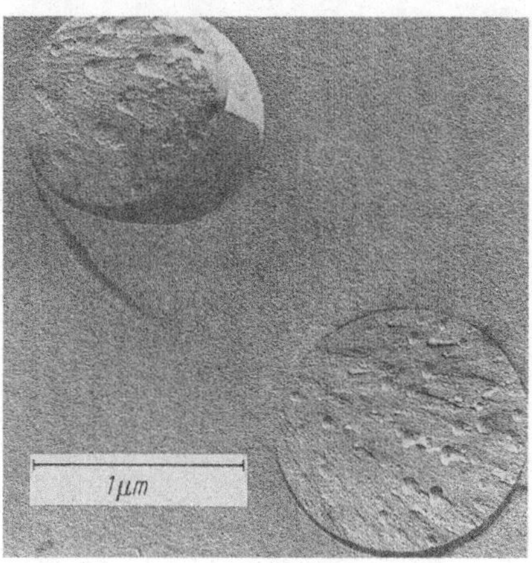

Fig. 6.32. Electron backscattering micrograph of a sodium borosilicate glass leached by acid treatment. Because of secondary unmixing, very small. SiO_2-rich regions were precipitated in the sodium borate-rich phase. Although, close to the pore walls after leaching, none of the very small. SiO_2-rich droplet-phase regions can be ascertained (see also the diffusion halos of Figs. 6.20 and 6.21), a large accumulation of many small SiO_2-rich droplet regions can be recognized within the large pores

generated microphases, is the Vycor process, which will be described more extensively in the next chapter. Figure 6.31 depicts the leached porous SiO_2 glass obtained in this process. Under favorable conditions of cooling, a multiphase Vycor glass can be obtained. Acid treatment reveals that the channels shown in Fig. 6.31 contain small secondary SiO_2 precipitations in the large pores (Fig. 6.32). The pore size of a Vycor-type glass can also be modified with a different temperature treatment.

7 Structure and Properties of Colorless Glasses

An exhaustive treatment of the relations between structure and properties of all glasses known today is not possible within the scope of "Glass Chemistry." This chapter represents just the basis for deeper studies required for special fields of interest. The reader is referred to the abundant literature, and to the texts of Scholze [70], Rawson [282], Volf [283], Rothenberg [284], Doremus [285], etc. More recently, the structure of inorganic glass-forming systems was reviewed extensively by Kreidl and Angell [286], who give an ample bibliography.

7.1 Silica Glass

Pure SiO_2 glass can be considered the prototype of glass. The first hand-made glasses were, however, not pure SiO_2 glasses but combinations of SiO_2 with K_2O, Na_2O, CaO and other oxides (network modifiers). The reason lies without any doubt in the extremely high melting temperature of silica. The production of a pure and bubble-free SiO_2 glass is still not possible by the usual methods, even at a temperature of 2500 °C. The value of log η for the viscosity η in dPa·s) is almost 8 even at 2000 °C according to Brückner [287]. We already described the structure of pure SiO_2 glass in the presentation of the network theory of Zachariasen and Warren (see Fig. 4.1). Accordingly, the formation of a disordered three dimensional SiO_4 tetrahedron network is characteristic for a SiO_2 glass and thus explains the very high viscosity of this glass. The six-member SiO_4 tetrahedron ring predominates over higher and lower ring formations.

Results of glass structure investigations in the 20s and 30s were naturally relatively imprecise and also ambiguous. Originally Warren's X-ray investigations supported the network theory whereas investigations by Randall, Rocksby, Cooper and Porai-Koshits supported the crystallite theory. The following is of particular significance regarding these contradictory interpretation of X-ray investigations.

Lebedew and his followers have carried out many investigations of pure SiO_2 glass (at that time "quartz glass"). SiO_2 glass was produced for the first time by Gaudin in 1839 [288] and most likely also by Lebedew (1921) [43] by melting quartz granules. During the process, the ordered SiO_2 tetrahedron frame of the rock crystal collapses on the surfaces of the granules but not on the inside even at the extremely high melting temperature of quartz. The X-ray evidence of structural inhomogeneities thus referred to "melt relics" and not to crystallites originating from an homogeneous melt.

Considerable advancements in X-ray technique after 1960, such as the analysis of X-ray low angle scattering led to definite evidence of structural inhomogenities in SiO_2 glasses (Porai-Koshits [289], Zarzycki [269], Uhlmann and Kreidl [290]). They were at times described as "density fluctuations" which could be due to a small concentration of OH in SiO_2 glasses. The structural inhomogeneities or density fluctuations clearly identified by X-ray low angle scattering have absolutely nothing to do with the appearances of microphase separation [181–185] discovered by electron microscopy in most glasses (except in pure SiO_2 glass) in 1956–57. Phase separation is hardly possible in pure one-component glasses.

H. Rawson [291] impressively represented (Fig. 7.1) the influence of different production methods and different trace impurities on the local density fluctuations (optically detectable granulate structure) of optical silica glasses. Glasses a) and b) were obtained from rock crystal and glass c) by the gas phase method which will be described later. The methods of production of more or less pure SiO_2 glasses have decisively progressed in the last 50 years, especially in the last 10 years. They necessitate the use of completely different starting materials. Different structure property relationships are obtained and SiO_2 glass has found different technical applications.

The term silica (or vitreous silica) glass should be used since in most cases the starting material now is rarely quartz sand or rock crystal. The following processes are used today to produce silica glass:

- quartz or transparent silica glass is produced by heating coal or graphite sticks (electrical heat conductors) embedded in quartz sand with a composition of at least 99.5% SiO_2 (by weight) and at most 0.2% Al_2O_3 (by weight) or 0.02% Fe_2O_3 (by weight). Complete melting does not occur at 1800 °C thus a transparent silica glass with numerous enclosed bubbles is obtained. After removal of the coal heat conductor, the melt can be further shaped into bars or other devices [292]. The "fulgurit" which occurs in nature can be compared

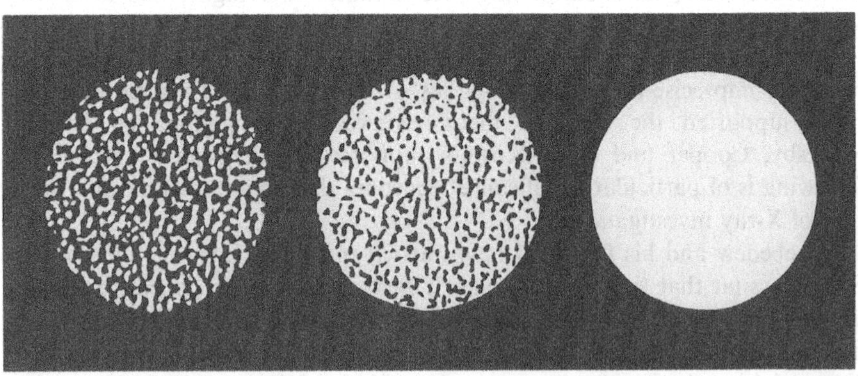

a b c

Fig. 7.1. Homogeneity of silica glasses produced differently (Rawson [297]). **a** and **b** are produced from rock crystal, **c** is a synthetic silica glass. A completely different granularity is observed with the quartz or transparent silica glass. "Fulgurit" occurs in deserts due to lightning striking sand and thus provoking it to melt or sinter together.

- transparent silica glass or "quartz glass" is obtained by burning pure rock crystal in an oxyhydrogen flame or electrically [292]. It can be further modified for various applications such as optical lenses, prisms, bars but also as tubes, cups etc. This glass has a very high UV transmittance (see Fig. 7.2) and can therefore be used for special applications in the optical industry. The two silicate glasses (quartz and quartz glass) are characterised by the following common properties: an extremely low thermal expansion $\alpha = 5.8 \cdot 10^{-7}$ per degree, a temperature resistance of up to 1400 °C, low electrical conductivity also at high temperatures, and an excellent chemical resistance to acids and water.
- the production of silica duped-glass light transmitting fibers 10 years ago revolutionised the field of modern communications. Totally new methods of production of silicate glass or silica glass were developed. R. Clasen has written an excellent review [294]. These new processes are the following:
  * the PCVD (plasma-activated chemical vapor-deposition) method
  * the MCVD (modified chemical vapor deposition) method
  * the OVD (outside vapor deposition) method and
  * the VAD (vapor axial deposition) method

By all four methods Silica glasses are produced via $SiCl_4$ in the vapor phase. The processes are shown in Fig. 7.3 (see also [294]). It is essential in all four methods that the starting material $SiCl_4$ can be obtained extremely pure (by distillation). The absorption effect of 3d trace impurities and of OH groups must be minimized and light losses due to scattering on structural inhomogeneities must be

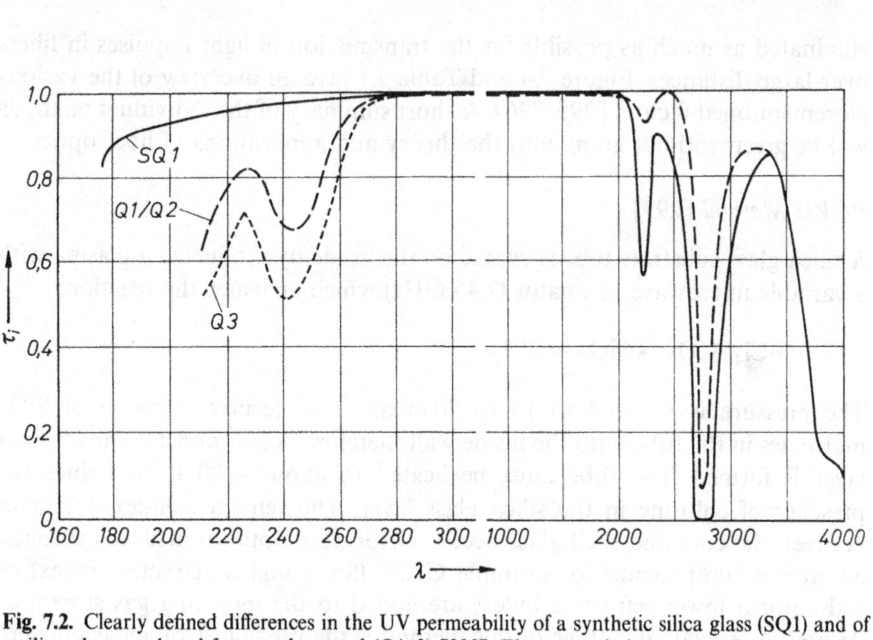

Fig. 7.2. Clearly defined differences in the UV permeability of a synthetic silica glass (SQ1) and of a silica glass produced from rock crystal (Q1–3) [293]. The transmission curves also confirm the different granularity observed by Rawson for the three silica glass types (see Fig. 7.1)

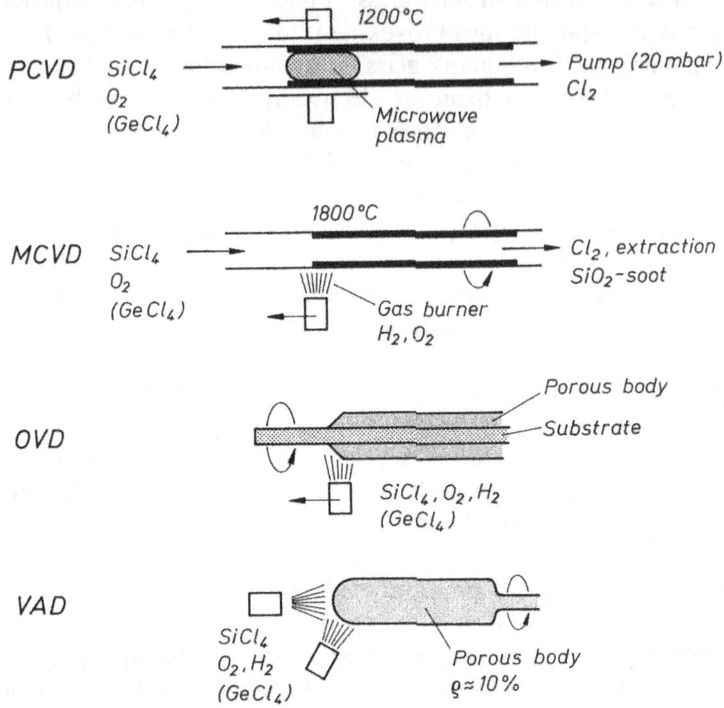

Fig. 7.3. Schematic of the four most important processes for the production of silica glass for light-transmitting fibres (Clasen [294])

eliminated as much as possible for the transmission of light impulses in fibers over large distances. Figure 7.4 and Table 7.1 give an overview of the various aforementioned factors [295, 296]. A short summary of the individual methods will be given without going into the theory and applications of fiber optics.

PCVD Method [297]

A silica-glass substrate tube is coated on the inside by producing a plasma with a variable microwave resonator (2.45 GHz) which activates the reaction:

$$SiCl_4 + O_2 \rightarrow SiO_2 + 2Cl_2$$

The pressure is reduced to 13 to 20 mbar. A molecular diffusion of SiO_2 molecules in the tube onto the inside wall therefore occurs and a compact glass layer is formed. The tube must be heated to about 1200 °C to reduce the presence of chlorine in the silica glass layer. The refractive index difference between the core and shell glass necessary for fibre optics is achieved through additional components: for example $GeCl_4$ (for a higher refractive index) or C_2F_6 (for a lower refractive index) are added to the incoming gas stream as desired. A special advantage of this method is the possibility of achieving very high (about 3.5 mass%) fluorine content [298].

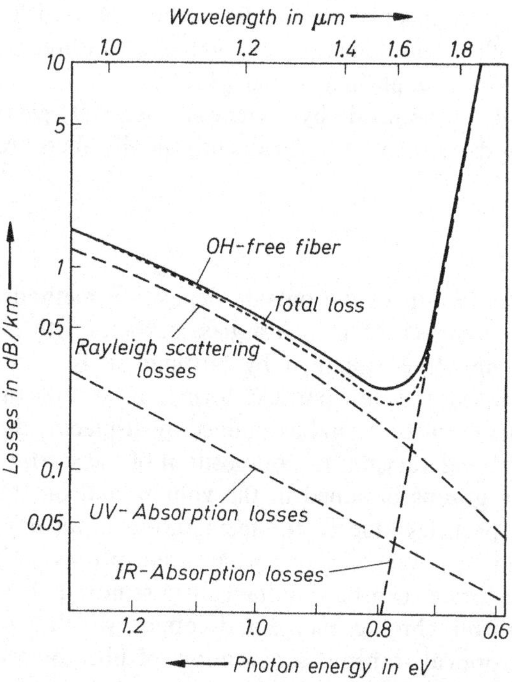

Fig. 7.4. Optical attenuation loss (fitted curve) and mechanism for light-transmitting fibres [295, 296]. 1 Decibel (dB), q 10 $\lg(I_1/I_2)$, I_1 non-attenuated wave, I_2 attenuated wave

Table 7.1. Optical absorption losses due to side group elements [296]

Element	Loss [dB/km·ppm] at	
	800 nm	1300 nm
V	1050	500
Cr	610	720
Mn	20	110
Fe	50	140
Co	20	440
Ni	30	160
Cu	10	

MCVD Method [299]

This method resembles the previous one except that the reaction is thermally activated. The substrate tube must be heated strongly, since activation can only take place from the outside of the tube (to avoid contamination). Since the reaction runs at atmospheric pressure, fine, submicroscopic silica glass particles (soot) are created by homogeneous nucleation. They can only be separated by thermophoretic forces, which requires a high temperature gradient on the inside wall of the tube. The yield therefore amounts to only 50% of the starting material used. In order to obtain a compact silica glass layer, the soot particles must be sintered with the same [300]. This demands very high temperatures up to 1800 °C, which can be attained with a plasma burner [301].

The silica glass tubes coated on the inside by extremely pure SiO_2 glass using the PCVD or MCVD method are collapsed subsequently, finally fibers are pulled out from this precursor.

OVD Method [302]

The OVD method belongs to the group of the outside deposition methods. A porous body is produced from submicroscopic silica glass particles with an average diameter of about 100 nm. It is obtained by burning $SiCl_4$ in an oxyhydrogen flame and by separating the soot particles formed onto a thorn. A body with the desired component structure can be obtained by displacing the burner, rotating the substrate rod and varying the composition of the starting gases. The unavoidable hydroxy groups obtained in the volume and on the surface of the small silica glass particles (due to the incineration in an oxyhydrogen flame) can be removed by a very effective purification process. All impurities which form volatile chlorides can be simultaneously removed. This also applies to the 3d elements iron, chrome, nickel and copper, which lead to undesired absorption in the optical transmission window of fiber optics. After the purification process, the rough body is sintered to compact silica glass. It collapses like the PCVD and MCVD forms and is finally extracted to form fiber optics.

VAD Method [303]

The VAD method resembles the OVD method except that an axial production of submicroscopic silica glass particles occurs on the frontal surface of a rough body in addition to the radial production. This is obtained by mounting a second burner in the axis of rotation or at an angle smaller than 90 °C. The rough body is pulled out in proportion with the growth rate of the burner. A substrate rod is no longer required as in the OVD method since a complete rod-shaped product is obtained. The collapsing step also becomes obsolete.

- Two additional methods for the production of silica glasses need not be mentioned at this point: the Vycor and the sol-gel method. Their chemical mechanism will be explained later. Glasses of different composition and also silica glasses can be obtained by the sol-gel method, for use in the development of wave conductors [294].

7.2 Alkali Silicate Glasses

Binary alkali silicate glasses have relatively little technological significance; primarily, they are valuable models for the treatment of basic problems essential for the discussion of technologically significant glasses.

It is only in the form of their aqueous solutions ("water glasses") that they have found technical application (Wels [304], Iler [305]). Binary Na_2O–SiO_2 or K_2O–SiO_2 melts with about 50 mol% alkali are quantitatively soluble when boiled in water and find various applications in the form of a syrupy mass.

Regarding the exact definition of the glass forming area in binary alkali silicate melt systems, Rawson [282] has already pointed out that this is problematic, and depends on the cooling rate of the melts. Statements about the higher glass building limit, that is the alkali concentration at which an alkali silicate melt solidifies into a glass without any traces of crystals, considerably vary with different authors. The reason for this variation is the various melt volumes (5 cm^3, 100 cm^3, 1000 cm^3 or more) or more precisely the cooling rate, which is limited by the volume. An international agreement for a standard cooling rate of the melts is necessary in order to compare properties which were determined by different authors.

The glass-forming region in the binary systems M_2O–SiO_2 (M = alkali) is generally limited to about 50 mol% M_2O (Moore and Carey [306], Imaoka and Yamazaki [307]). The boundary moves toward higher M_2O contents in the series Li–Na–K–Rb–Cs. Unfortunately, most of the numerous literature data do not sufficiently specify sample sizes, cooling rates, and cooling conditions, which critically affect those boundaries. It is high time that recognition is given to this fact, either by international standardization or, at least, a precise description of experimental procedures (Rawson [282]).

Dietzel [14] and Dietzel and Wickert [308] measured crystal growth in the binary system Na_2O–SiO_2 and plotted $1/u$ vs. composition, defining $G = 1/u$ as the "glassiness" of the composition. In line with general experience in other systems and theoretical considerations, maxima of "glassiness" G were found near the eutectics between compounds $Na_2O \cdot SiO_2$ and $Na_2O \cdot 2SiO_2$ and $Na_2O \cdot 2SiO_2$, and SiO_2 at 25 and 38 mol% Na_2O (Fig. 7.5). An observation from glass development is therefore confirmed. It is known that a good glassy solidification is always obtained when a multitude of chemical building units is present in the melt, and hinders both ordering and crystallization processes. (Sometimes called the "confusion" factor). The existence of the aforementioned invert glasses should be explained. The simplest examples for such mutual influences exist in binary eutectic melts. With this example, the problem of determining a glassiness number G based on the determination of the crystal growth rate, also becomes clear. A sodium silicate melt which contains about 18 mol% Na_2O and is produced in a 1 L crucible solidifies after founding as a completely clear glass. The same melt produced in a 30 L crucible, solidifies as an opaque glass, due to the much lower cooling rate of such a large glass volume. It was not opaque due to the formation of crystals but rather due to the formation of droplet shaped glassy immiscibility regions. True, a has been obtained but it cannot be characterized with the glassiness number G. Since the concept of glassiness is associated with clarity and transparency it would be misleading to ascribe a higher glassiness value to the absence of crystals. It was already pointed out, in a previous section, that the simplest binary alkali silicate glasses tend to immiscibility, and exhibit a maximum for immiscibility in the

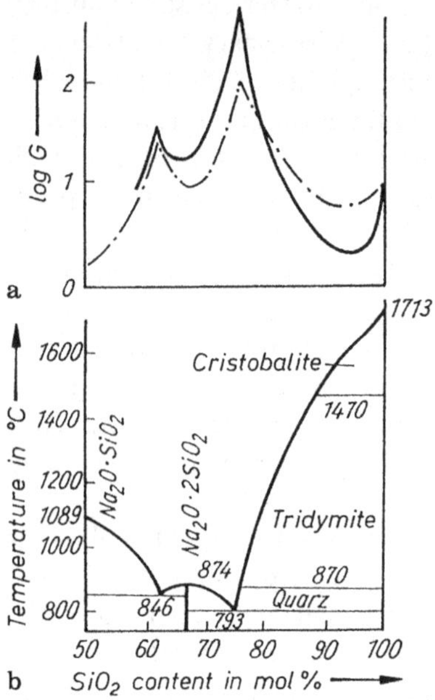

Fig. 7.5. Binary system Na_2O–SiO_2. **a** "Glassiness"; (G) in the system Na_2O–SiO_2 (Dietzel [14]. Dietzel and Wickert [308] - - - experimental; –.–.–.–. calculated; **b** melting system Na_2O–SiO_2 (Kracek [309])

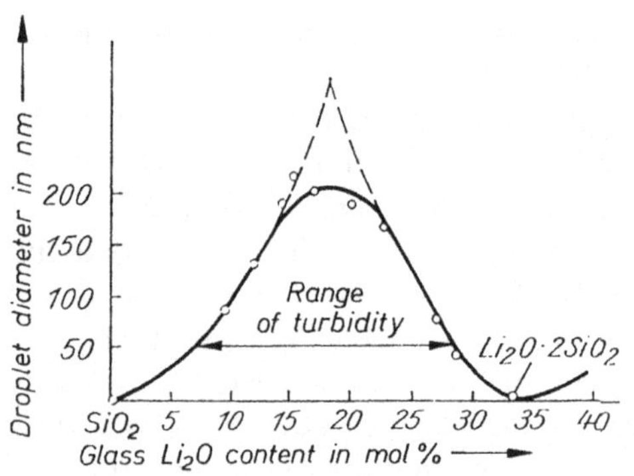

Fig. 7.6. Dependence of droplet size on composition in the system Li_2O–SiO_2 (Vogel [151])

concentration region between 0 and 33 mol% of alkali (see Fig. 7.6). The immiscibility tendency decreases from the Li_2O–SiO_2 system to the Na_2O–SiO_2 or the K_2O–SiO_2 system but it is still unambiguously detectable.

Many properties of alkali silicate glasses can be understood on the basis of the principles governing silicate structures introduced in Chapters 3 and 4. Every Na_2O added to SiO_2 provides an extra oxygen to the network, eliminating one bridging oxygen (–O–) and creating two non-bridging oxygens (O)

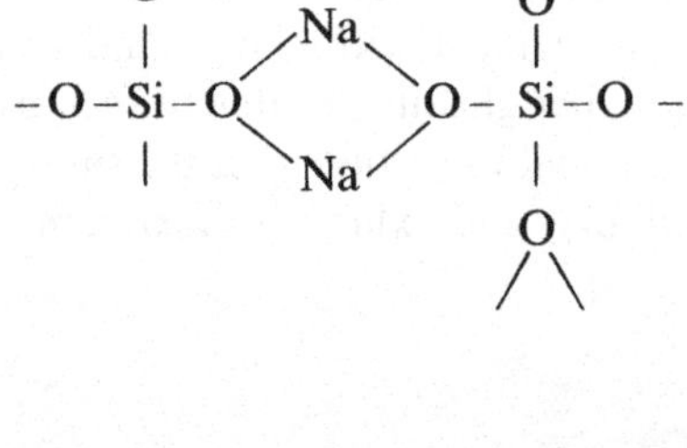

The anionic SiO_4 tetrahedra groupings became more mobile with every bridge broken, and the Na cations become increasingly more mobile through the interstices of the network. The presence of non-bridging oxygen can be demonstrated directly by many techniques, e.g., spin resonance (Chapter 5), or photoelectron spectroscopy (Brückner et al. [310]). According to the random network theory the non-bridging oxygens and alkali ions should be distributed statistically, but numerous experiments indicated ordering or clustering (see for example, Greaves [311–314]), reminiscent of the building elements in compounds and microphases in immiscibility regions.

Most recently, for instance, X-ray data suggested that $M_2O \cdot SiO_2$ (Yasui et al. [315]), as well as $M_2O \cdot 2SiO_2$ glasses (Imaoka et al. [316]) resemble the chain and sheet structures of the corresponding crystalline solids, with just some changes in bond angles. In the case of mixed-alkali glasses, the definitely coordinated locations available for the alkali ions accordingly will hinder transport in certain directions for the larger ion. Frequently later workers ignored that in his classical papers Warren had suggested two possible interpretations of the X-ray analyses of alkali silicate glasses: (a) random insertion of alkali ions, and (b) arrangements resembling those of crystalline alkali silicates. Increasingly evidence for the second interpretation (at least at low temperatures) has replaced earlier acceptance of the first one. A recent example is NMR (Farnan et al. [317], Bulkermann et al. [318] n-diffraction (Wright et al. [69]) evidence for some ordering in potassium silicate glasses.

Density and refractive-index curves of binary alkali silicate glasses are shown in Figs. 7.7 and 7.8. The density curves (see Fig. 7.7) show the awaited dependency in the region of lower alkali concentrations. They lie on top of each other in the order Li–Na–K, in accordance with increasing atomic weight. However the Na and K curves nearly coincide. This is due to the relatively small ion density of potassium with respect to the one of sodium. Almost all elements of the argon series (element 19 to 36) show unusually low ion densities. Above 30 mol%, the Na and K curves intersect. According to Dietzel et al. [308, 319, 320], this occurrence can be explained by the different use of cavities by the alkalis during the insertion into a SiO_2 network. According to these authors' calculations (calculated for a glass with 33 mol% alkali) Li ions occupy not only up to 100% of the available cavities, but theoretically even up to 127% after breaking

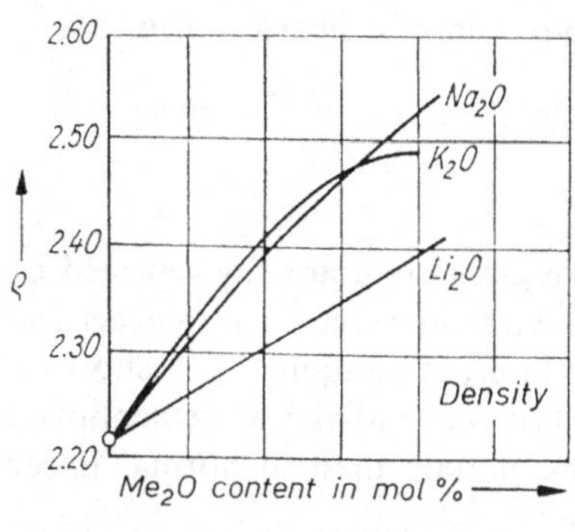

Fig. 7.7. Graphic representation of the change in density in binary alkali silicate glasses (Dietzel and Sheybany [319])

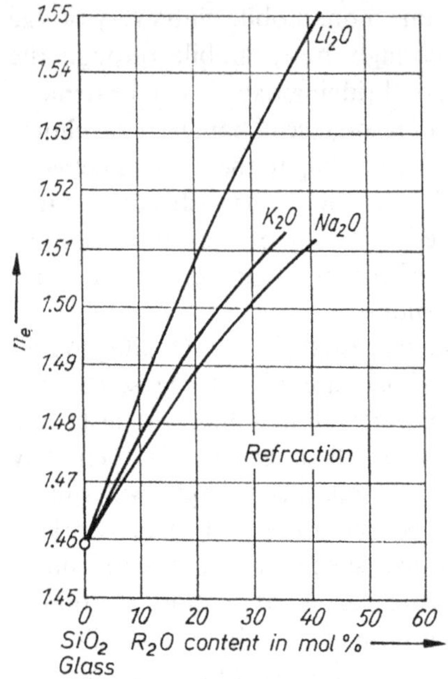

Fig. 7.8. Graphic representation of the change in refractive index in binary alkali silicate glasses (Dietzel and Sheybany [319])

of the oxygen bridges. This implies that a contraction of the network must have taken place. The oxygen introduced merely causes a volume increase due to the breaking of bridges and its size. According to the calculations, Na ions should occupy up to 70% and K ions up to 27% of the available cavities. The larger density of sodium silicate glass with respect to potassium silicate glass is a result of the lower demand for new volume by the insertion of lighter Na ions relative to K ions in a SiO_2 glass.

These considerations could also explain why the refractive index somewhat surprisingly decreases in the series Li_2O–Na_2O–K_2O in spite of increasing polarizability and number of electrons. The refractive index is, however, determined by the electron density of the system. Also, no doubt the introduction of alkali increasingly expanding the network in this series affects the polarizability of oxygen. Both effects contribute to the relatively larger refraction of Na and Li. This consideration, involving the role of the polarizability of oxygen, is of practical importance for the development of optical glasses. It suggests the possibility of obtaining high refraction without using the heaviest elements.

7.2.1 The Mixed-Alkali Effect

If the sodium ions in a binary sodium silicate glass are gradually replaced by potassium ions, all properties dependent on the transport mechanism (electrical conductivity, self-diffusion, dielectric loss or internal damping . . .) show an abnormal change. Non-linear behavior is observed and the aforementioned properties are lower by several orders of magnitude than in normal linear changes of properties. This phenomenon is called the "mixed-alkali effect". According to Scholze [70] the mixed-alkali effect is first observed at a total alkali content of about 10 mol%. It also decreases with an increase in temperature. The mixed-alkali effect was revealed for the first time during Otto Schott's solving of the glass thermometer problem in 1884 (see pp. 20, 21). Today the purely empirical treatment and elimination of the effect by the development of thermometer glass (already over 100 years ago by Otto Schott) can no longer be a satisfactory structural interpretation. Therefore during the last 10 to 15 years numerous glass researchers have tried hard to find scientific interpretations. This work is partly based on further experimental investigations by means of modern measurement techniques but also on only purely theoretical considerations [321, 70, 322]. A review of these very different and at times opposing views does not appear useful here. It has been experimentally observed that the "mixed-alkali effect" appears to be especially strong when the two alkali ions present in the glass have very similar field strengths F, that is ΔF is very low as in the combination Na/K. The mixed-alkali effect diminishes at larger ΔF values as in the combinations Na/Rb or Na/Cs. The effect has also been observed experimentally when an alkali ion and an alkaline earth ion or two alkaline earth ions are combined in a glass. The largest effect for these combinations is also observed at a low ΔF value and the smallest effect at a large ΔF value for the two network modifier ions. One should more correctly speak of a mixed-oxide effect on the basis of these experimentally determined and extended results [33, 70, 322–324].

As was stated declared by Dietzel [323] in 1983 even today no theoretical explanation appears entirely satisfactory, and that further experimental investigations are necessary. At this point, the following statement might be made: the mixed-alkali effect is surely due to ionic aggregation and immiscibility processes, during the temperature decrease of a melt.

In alkali silicate glasses, a statistical distribution on the SiO_2 network takes place until about 10 mol%. Starting at 8 to 10 mol% of alkali oxide, immiscibility is also observed and can be more or less experimentally demonstrated. This means that an alkali rich droplet glass phase (or just aggregates without phase borders) have settled in a SiO_2-rich matrix glass phase. A drastic change in the viscosity conditions occurs at about 10 mol% of alkali oxide. Viscosity values higher than expected for a statistical distribution of alkali ions have been determined. (see Section 13.6 and especially Fig. 13.2).

The immiscibility behavior of the alkali silicate glasses is in agreement with Scholze's statement that the mixed alkali effect can only be measured with a total alkali content of at least 10 mol%.

An enrichment of the aggregation or immiscibility regions with two alkali ions of about the same field strength (that is ΔF is almost 0) will also lead to the formation of two different stable molecular building units of the droplet glass, for example sodium and potassium. In the case of the combination of network modifier ions of very different field strengths (that is $\Delta F > 0$) as is the case for Na/Cs or Na/Ca, according to today's knowledge of the immiscibility behavior of glasses, only one stable molecular building unit exists, that is the network modifier ion with the highest field strength.

Also in the case of nucleation and crystallization in immiscibility regions, it is primarily a crystal phase of the cation with the highest field strength that is formed, first, and eventually but not simultaneously other cations of lower field strengths.

Let us return to the aforementioned importance of applying the mixed alkali effect to solve the glass thermometer problem (see Otto Schott, Section 1.3). Investigations by Weber, Schott and Wiebe [325–327] also preceded it. Old thermometers were manufactured from glasses which contained both sodium and potassium oxides. They exhibited an ice-point depression of -0.3 to $-0.6\,°C$ after use at higher temperatures. After long storages at room temperature, the readings returned to 0. This was described as the so-called continuous rise ("Säkulärer Anstieg").

On the basis of what was presented in this chapter, today's opinion on the ice-point depression is that it is due to the simultaneous formation of two stable molecular silicate building units with sodium and potassium in the aggregation regions after more or less advanced immiscibility processes. When the thermometers are used at higher temperatures, a three-dimensional structural equilibrium state for the building units will occur according to the temperature. As the thermometer is cooled, the lower energy levels, at lower temperatures, are not achieved due to the two existing stable molecular building units of sodium and potassium. This is also supported by the higher viscosity of the glass which is due to the forward immiscibility tendency.

The hindrance of molecular building elements in glass formation is already known. Sterical hindrance for organic melts with giant molecules prevents crystallization and leads to glass formation (see Section 2.1).

The so-called invert glasses are introduced in the next chapter. Melt compositions with much less than 50% network-former oxides, which according to the Zachariasen–Warren structure hypothesis can absolutely not solidify as glasses, will still form a glass when cooled. Obviously crystallization or an ordering process is prevented by the reciprocal hindrance of several molecular building units.

The electrical resistivity of the $Na_2O \cdot K_2O \cdot 4SiO_2$ glass ($\log \sigma = 12.6$) is more than two orders larger than that of either $Na_2O \cdot 2SiO_2$ ($\log \sigma = 8.1$) or $K_2O \cdot 2SiO_2$ ($\log \sigma = 9.1$) glass (Lengyel and Boksay [328], Doremus [285]). In the system Li_2O–K_2O–SiO_2 it is about five orders of magnitude larger. Recently (Moynihan et al. [329]), this behavior was discussed in terms of the weak electrolyte model (postulating participation of only a small concentration of mobile ions in transport), assigning the foreign ion one of several inhibiting effects. But neither this effect nor the weak electrolyte theory itself appears firmly established for these glasses (Jain et al. [330]).

A new intensive loss peak in internal friction measurements appears, beginning with small additions of a second alkali (Rötger [331]), which has been associated with electrically inactive elastic dipoles in which the diffusion coefficient of the slower moving ion is involved (Day [324]). The effect of the difference in masses postulated by Hendrickson and Bray [332] can at best be a small contributory one (Jain et al. [330]).

Ingram [333a, b] discusses the mixed-alkali effect in terms of the paired-interstitialcy(*) mechanism (Day [324]) under the justifiable assumption that unlike pairs are preferred over like pairs, and that the mixed pairs (e.g., $LiK^{*})^{2+}$ are effectively immobilized. But the "peaks" of the mixed-alkali effect are not always at the same concentration of the added alkali (e.g., in internal friction) so the complete interpretation remains an open question, perhaps to be further elucidated by studies of crystalline β-Al_2O_3 (Ingram [333a,b]).

Dietzel [33a,b] still believes that the effect of difference in field strength can be mobilized as a fundamental variable in the phenomenon.

The role of incipient phase separation should be considered for many compositional ranges. The effect on other properties has been summarized by Day [324] and is much smaller for properties not involving transport.

The mixed-alkali effect has significant practical implications. In thermometer glasses, for example, the presence of a second alkali has to be avoided to prevent instability due to volume relaxation. On the other hand, low di-electric loss glasses are obtained by incorporating two alkalis (e.g., Li, K in Corning 7070: 71 SiO_2, 26 B_2O_3, 0.5 Li_2O, 1.0 K_2O or Jena 3079 III, 30 PbO, 52.7 B_2O_3, 5.0 Al_2O_3, 4 Li_2O, 8 K_2O.

7.3 Alkaline Earth and Alkali-Alkaline Earth Silicate Glasses

Binary alkaline earth silicate glasses have hardly any practical significance. Except for the system BaO–SiO_2, all alkaline earth silicate melts exhibit open (above liquidus) immiscibility gaps, and, even in the system BaO–SiO_2, the S shape of the liquidus suggests subliquidus immiscibility, which is verified experimentally (Fig. 7.9). The gaps extend from about 2 to 40 mol% for MgO, 2 to 30 mol% for CaO, 2 to 20 mol% for SrO, and 2 to 16 mol% (below the

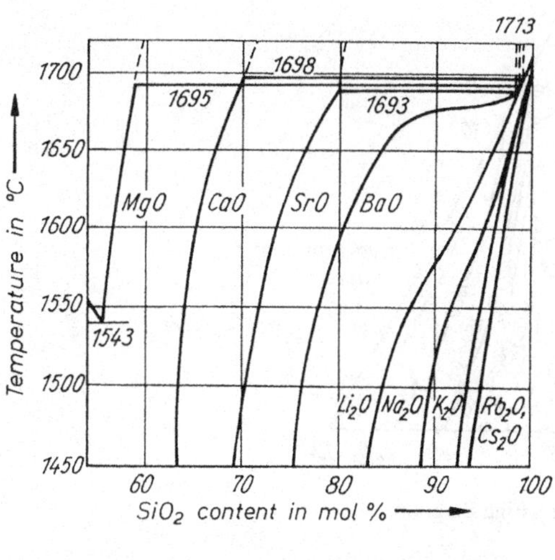

Fig. 7.9. Melting temperatures in binary silicate systems on the SiO_2-rich side. Open miscibility gaps are present, increasing in the series SrO, CaO, MgO [Kracek]

liquidus) for BaO. In a way, the system $BaO–SiO_2$ represents a transition to the alkali silicate systems, also shown in Fig. 7.9, where the S shape becomes less pronounced with decreasing field strength (Ba–Li–Na–K) until it is no longer recognizable for Rb_2O and Cs_2O.

As in the alkali silicate glasses, phase separation typically produces two phases, one almost pure SiO_2, the other rich in the oxide of the cation. The droplets are usually so large as to promote turbidity, making the preparation of clear glasses impossible.

The combination of two binary systems suppresses – as had been noted before – the extent of phase separation. At least, the competition for oxygen coordination causes the formation of microphase droplets so small as to become invisible. At the same time, the addition of CaO significantly increases the chemical resistance of sodium silicate glasses.

One simple example of the greatest technological importance is the system $Na_2O–CaO–SiO_2$, from which container, window, and optical crown glasses are derived (phase diagram Fig. 7.10). Glasses based on this system are called soda-lime glasses. A typical average composition is (in wt%): 72 SiO_2, 14 Na_2O, 11 CaO, 1 MgO, 2 Al_2O_3. In practice, a small amount (0.1 wt%) of sulfate is added to remove bubbles introduced by carbonates used as raw materials.

Even in the perfectly clear ternary ($Na_2O–CaO–SiO_2$) glasses, small droplet phases have been found (Oberlies [180], Ohlberg and Hammel [216], Ohlberg and Parsons [334]). Immiscibility, that is a separation into an alkali rich and a SiO_2-rich phase, will facilitate leaching of the alkali rich phase by

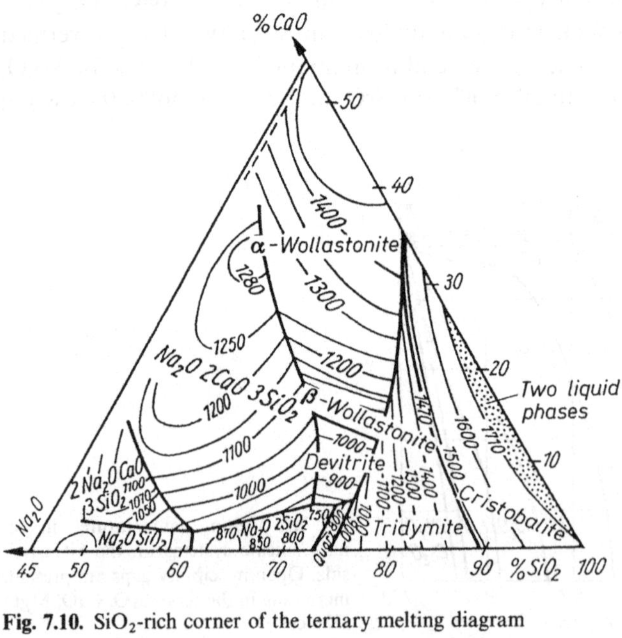

Fig. 7.10. SiO_2-rich corner of the ternary melting diagram

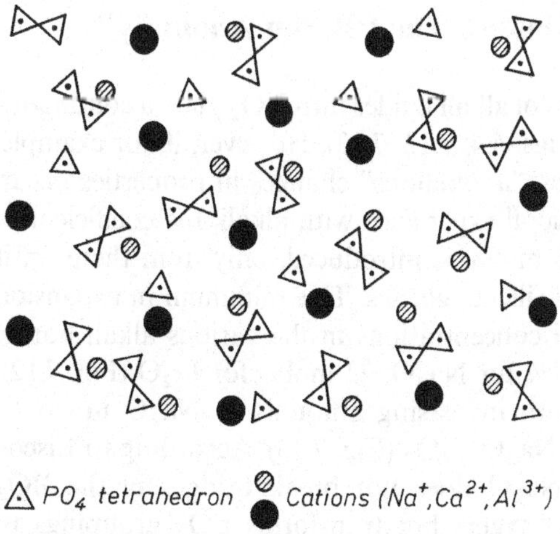

Fig. 7.11. Schematic of an invert glass structure (Trapp and Stevels [40])

water or air. The so-called "stain (sensitivity)" in optical crown glass can be traced back to that as well. A better hydrolytic resistance had already been reached without knowledge of the imiscibility phenomena in fully clear glasses, by "stabilizing" the glass via addition of 2 to 3 mass % Al_2O_3. A simple combination of three immiscible binary systems, namely the Na_2O–SiO_2, CaO–SiO_2 and Al_2O_3–SiO_2 systems, lead to a further repression of phase separation thus enhancing the hydrolytic resistance of the glass [237].

Alkali-alkaline earth glasses containing less than 50 mol% SiO_2 ("invert glasses") have been obtained by Trappand Stevels [40] by combining the binary systems of SiO_2 and Na_2O, K_2O, CaO, BaO. This combination decreases immiscibility and crystallization tendencies, in part due to decreasing symmetry, to the point that clear glasses are obtained in the absence of a 3-dimensional network (Fig. 7.11).

7.4 Borate and Borosilicate Glasses

The production of borosilicate glasses plays an important and diversified role all over the world: optical glasses, laboratory (heat- and corrosion-resistant) glasses, solder glasses, etc. Borate glasses have been studied extensively in their structural relation to borosilicate glasses as well as because of their interesting structures and properties, especially the so-called boron oxide anomaly. This name is given the phenomenon of maxima and minima of properties in the binary systems.

The structure of pure B_2O_3 has been considered a network of BO_3 triangles (Biscoe and Warren [335], Svanson et al. [336], Bray [337], Bray and O'Keefe [106], Mozzi and Warren [338]), organized in the 3-member boroxol ring (Fig. 7.16). In various other models, the planar structure has been modified (e.g., Richter et al. [339] Amini et al. [340], Bell et al. [341]).

7.4.1 Binary Alkali Borate Glasses. The "Boron Anomaly"

If one introduces increasing amounts of alkali oxides into SiO_2 glass a continuous change in properties is observed (Figs. 4.4, 7.12, 7.13). However, if, for example, Na_2O is introduced into B_2O_3 glass, "anomalous" changes in properties occur. For instance, in contrast to the general experience with alkali, the coefficient of expansion decreases until about 16 mol% is introduced; only from there on it increases as it always does in alkali silicate glasses. This minimum in expansion occurs at somewhat different alkali concentrations in the various alkali borate systems: 14 mol% for K_2O, 16 mol% for Na_2O, 22 mol% for Li_2O (Fig. 7.12).

Similarly, density increases with increasing amounts of Na_2O up to 16 mol%, more so than in the binary Na_2O–SiO_2 (Fig. 7.13). According to Biscoe and Warren [335], up to 16%, Na_2O does not break bridges in the BO_3 network in forming non-bridging oxygen, but transforms BO_3 groupings to BO_4 tetrahedra (Fig. 7.14). One Na_2O causes the formation of two BO_4 tetrahedra, which now are participating in a 3-dimensional network, thus strengthening the structure. This behavior first suggested that property maxima and minima might be connected with a maximum of BO_4 tetrahedra formed at 16 mol% Na_2O. The occurence of the boric acid anomaly can be summarized in the following way:

$$[BO_3] \rightarrow [BO_4] \rightarrow [BO_3]$$

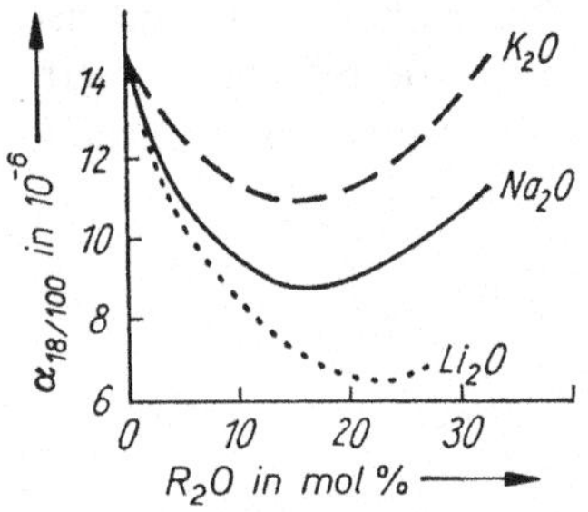

Fig. 7.12. Change of the coefficient of linear thermal expansion α as a function of composition in binary lithium, sodium, and potassium borate glasses

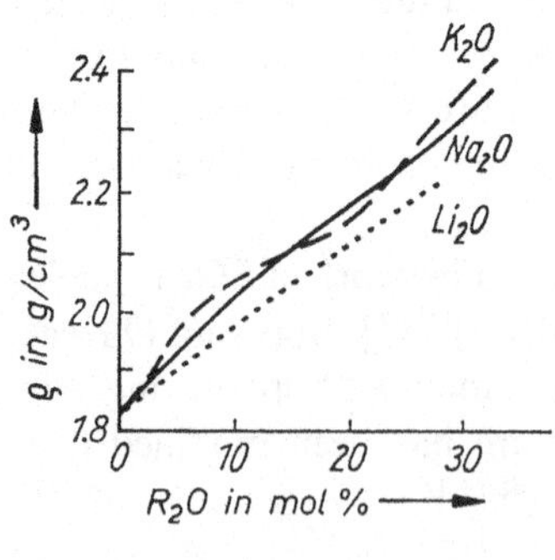

Fig. 7.13. Change of density as a function of composition in binary lithium, sodium, and potassium borate glasses

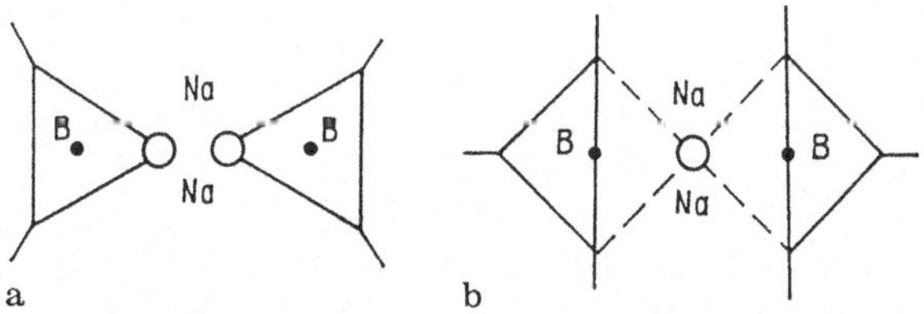

Fig. 7.14. Schematic of the possible ways for structural incorporation of Na_2O onto a B_2O_3 glass. **a** Breaking bridges and formation of nonbridging oxygens; **b** formation of BO_4 tetrahedron and strengthening of the network

Boron's endeavour to optimally coordinate itself is a phenomenon known to the chemist for a long time. It should however be emphasized that the conventional analysis by titration is not applicable to boric acid due to the very low dissociation of orthoboric acid. A multivalent alcohol, such as glycerol, mannite, glucose or fructose, must be added to the solution. A medium to strong monobasic acid is formed via the complex formation of boron which now can be titrated with sodium hydroxide. The endeavour of boron to maximally screen itself by 4-coordination is also satisfied via this complex formation.

The bridging oxygen atoms take over the role of glycerol's OH groups (see previous equation).

7.4.1.1 Temperature Dependence of the Boric Acid Anomaly

According to Dietzel [14, 32], the boric acid anomaly should not exist in glass at high temperatures. The association of $[BO_3]$ planar triangles with oxygen atoms into $[BO_4]$ groups should first occur at lower temperatures. This assertion can be verified by at least two observations: the viscosity behavior of alkali borate glasses at different temperatures and their immiscibility behavior (which will be treated later).

A discontinuous density modification of a B_2O_3 glass by introducing increasing amounts of Na_2O must logically result in an abnormal viscosity modification of the glass. Shartsis, Capps and Spinner [342] have clearly proved this effect (see Fig. 7.15). While the viscosity curve of binary alkali borate glasses continuously drops with an increase in the alkali content at a temperature of 1000 °C, a increasingly distinct point of inflection appears at about 16 mol% alkali at temperatures of 900 °C, 800 °C, 700 °C and 600 °C. The $[BO_4]$-tetrahedron formation therefore first occurs below 1000 °C. At present, the structure of alkali borate glasses is best described according to X-ray (Svanson et al. [336]), nuclear magnetic resonance (Bray and his school [92–98, 106, 107, 112, 113, 129, 144]), and Raman spectroscopy (Konijnendijk [343], Beekenkamp [344]) as the combination of BO_3 triangles and BO_4 tetrahedra in groupings as illustrated in Fig. 7.16. In pure B_2O_3, the chief structural element is the boroxol ring connecting three BO_3 groups through three oxygens upto 20% Na_2O, tetraborate groups, having the definite BO_4: BO_3 ratios shown in the figure, form while boroxol groups decay; only a few loose BO_3 triangles and

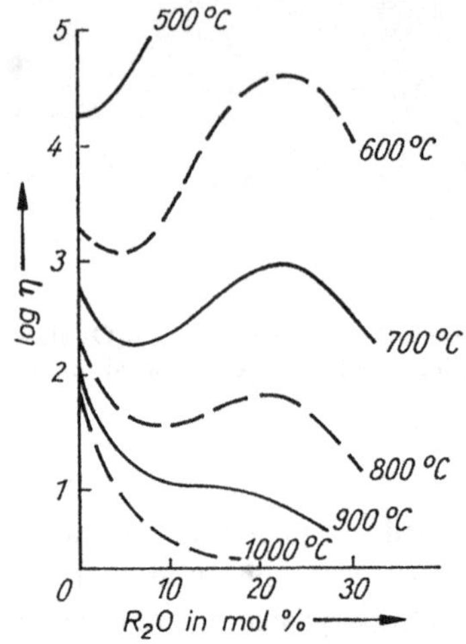

Fig. 7.15. Viscosity of binary alkali borate melts at various temperatures (Shartsis et al. [342])

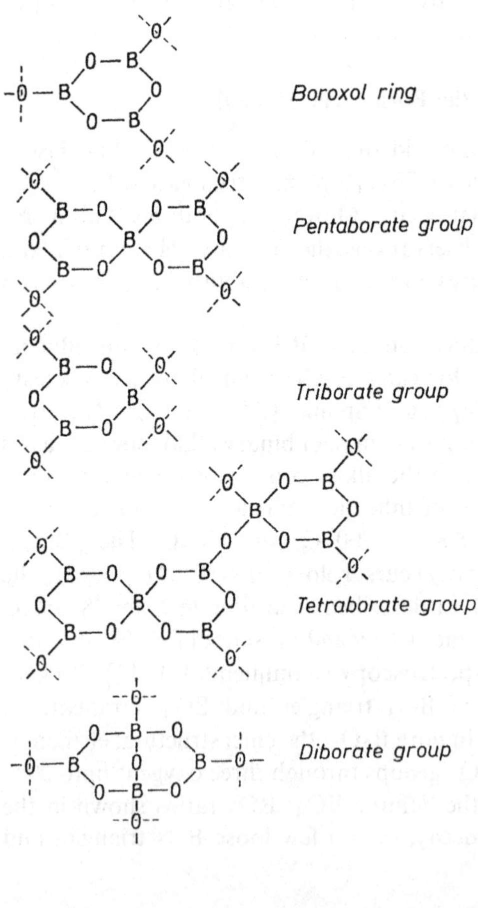

Fig. 7.16. Possible structural elements (as verified experimentally in various water-free crystalline alkali borates) of borate glass

BO_4 tetrahedra may be present. Between 20 and 35% Na_2O, diborate and ditriborate groups form first-pairing BO_4 tetrahedra. From 40 to 50% Na_2O, diborate, ortho-, pyro-, and meta ring borate groups occur, and nonbridging oxygens appear in larger numbers, as in silicate glasses.

Thus no maximum of BO_4 tetrahedra exists at 16% Na_2O and early interpretations of the "boron oxide anomaly" had to be discarded. More recent interpretations will be summarized in the succeeding section. Studies dramatize the large effect of species (Cs, Rb, K, Na, Li) and content in alkali borate glasses (Kamitsos [345]). Pure nuclear quadrupole resonance (NQR) (no magnetic field) providing one order of magnitude higher accuracy seems to establish 85% of boroxol rings and 15% chains or triangles like vitreous in B_2O_3 like in crystalline B_2O_3. (Bray and Gravina [346], Bray et al. [347]).

Typically boroxol groups might e.g. be found having vanished at $R = Li_2O{:}B_2O_3 \approx 0.25$. Up to $R = 0.5$ diborate groups may prevail. Towards $R \approx 1$ each additional Li_2O is found destroying one diborate group forming 2.5 meta and 1.5 loose BO_4, this structure essentially seeming achieved at $R = 1$. Above $R = 1$ these groups appear to change into 0.6 pyro and 0.4 ortho, i.e., BO_4 groups have vanished and BO_3 acquire increasing non-bridging oxygen.

Also, previously unknown sites in Li_2O–B_2O_3 could be identified. With the addition of larger alkali ions (e.g. Rb) less BO_4 and more non-bridging oxygen is shown to form (Raman spectra). (Soppe et al. [348], Chryssikos et al. [349]). Generally, at high temperatures BO_4 grouping decreases.

7.4.1.2 Tendency Toward Immiscibility

A tendency toward immiscibility and the presence of droplet-shaped immiscibility regions of extremely small dimensions in binary sodium borate glasses was demonstrated by electron microscopy as early as 1958 (Vogel [181], Skatulla et al. [184, 186]). The gap had a maximum around 16 mol% Na_2O. With an increase of the Na_2O content beyond 16 mol%, a decrease in immisibility occurs. Only one phase, homogeneous glass is observed in the electron microscope at a content of about 26 mol% of Na_2O (see Figs. 7.17 to 7.19). A relationship to the "boron anomaly" was considered and may be founded in the presence of entirely different groupings such as boroxol and tetraborate arrangements. The small droplet-shaped regions of Fig. 7.18 consist of an alkali-rich borate phase which certainly must approach the composition of the tetraborate. Figures 7.20 and 7.21 show the dependence of the microstructure of the glass at maximum immiscibility (16 mol% Na_2O) on cooling rate. The quenched glass of Fig. 7.20 shows no immiscibility regions, or only regions of dimensions bordering on the resolving power of the electron microscope. In the glass in Fig. 7.21, which was cooled at a conventional rate, the immiscibility regions are clearly recognizable and show a tendency to clustering. The sample shown in Fig. 7.22 had been treated at 500 °C for 2.5 h. It showed no visible opalescence, but the aggregation of droplet had progressed much farther. It

Fig. 7.17. Clear binary sodium borate glass (10 mol% Na_2O, 90 mol% B_2O_3). The glass contains small, sodium-rich immiscibility regions in a B_2O_3-rich matrix phase (electron optical micrograph after replica preparation)

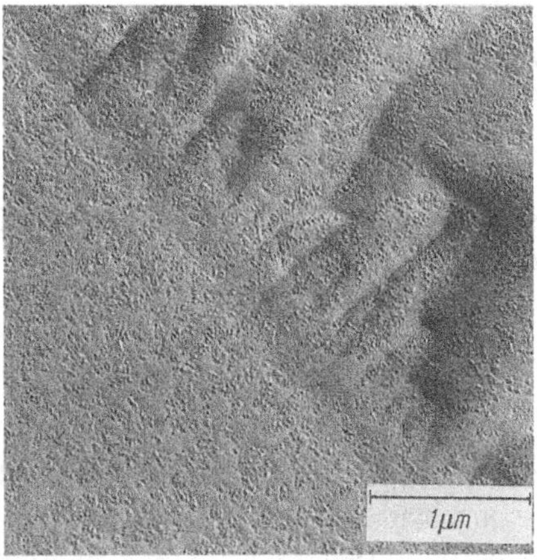

Fig. 7.18. Clear sodium borate glass (16 mol% Na_2O, 84 mol% B_2O_3). Maximum of unmixing into droplet-shaped sodium borate glass phase and B_2O_3 glass phase

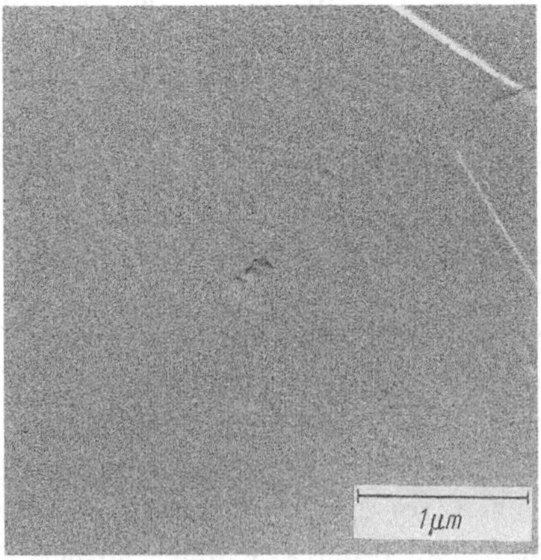

Fig. 7.19. Clear binary sodium silicate glass (28 mol% Na_2O, 72 mol% B_2O_3). Comparison of molybdenum trioxide crystal test plane (upper right) shows that this glass contains only one single phase. (Electron-optical micrograph after replica preparation.)

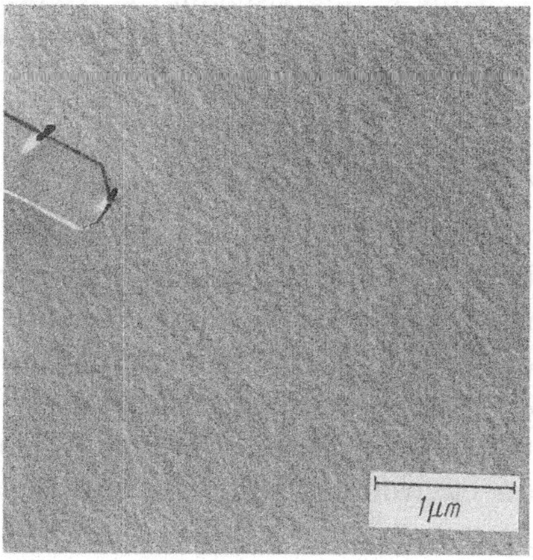

Fig. 7.20. Clear sodium borate glass (16 mol% Na_2O, 84 mol% B_2O_3). This glass was quenched radically after melting. Although immiscibility regions are present (as shown by comparison with the molybdenum trioxide crystal test plane), their dimensions are very small

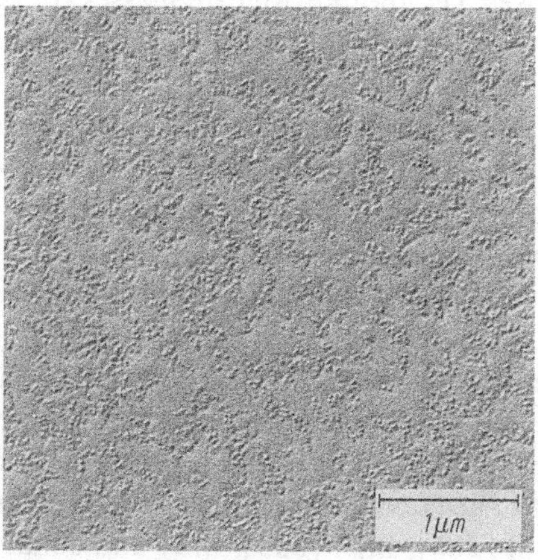

Fig. 7.21. Clear sodium borate glass (16 mol% Na_2O, 84 mol% B_2O_3). This glass was cooled normally after melting. The droplet-shaped immiscibility regions already tend to aggregate

Fig. 7.22. Clear sodium borate glass (16 mol% Na_2O, 84 mol% B_2O_3). This glass was heat-treated for $2\frac{1}{2}$ h at 500 °C after normal cooling from the melt. Now the aggregation of the droplets is clearly recognizable. However, they do not combine into larger droplets. The glass is perfectly clear and shows no visible turbidity

could be assumed that the organization of the borate groupings is temperature-dependent, a factor that can also be associated with the temperature dependence of the "boron anomaly."

7.4.1.3 Present State of Interpretations of the "Boron Anomaly"

The minimum in expansion and the maximum in phase separation near 16 mol% Na_2O in a borate glass used to be explained by the assumed change in coordination (from 3 to 4) of boron up to 16 mol%, involving a structural tightening, and the assumed appearance of "normal" disruption of the network and non-bridging oxygen above 16 mol%. Similarly, the immiscibility behavior appeared to be explained plausibly by the need of sodium ions to associate primarily with the BO_4 groups to promote electron neutrality. The structural complexes so formed certainly would have a different demand for spatial arrangement than the BO_3 groupings, thus inducing phase separation. The expansion minimum and immiscibility maximum are clearly significant for the properties of important commercial glasses, i.e., the borosilicate glasses (types Pyrex, Vycor, shock-proof household glasses, and optical materials), to be treated in later sections.

However, the reliable investigations of Krogh-Moe and collaborators have shown that the change in the coordination of B from 3 to 4 continues beyond 16 mol% Na_2O to 30 mol%. It is only above 30 mol% that the classical appearance of nonbridging oxygen at BO_3 groupings becomes significant (Fig. 7.23).

The understanding of these property maxima and minima is of practical importance, but the complexity of the phenomenon and the corresponding variation of the sharpness and position of the "anomaly" with property involved, alkali species, and temperature must defy a simple explanation. Particularly, the influence of the changes in the borate groupings and the amount and kind of alkali ions have different, and at times contrasting, effects (Uhlmann and Shaw [350]). Griscom [351b], Kreidl [286], Bray [352], Bray and collaborators [106, 112, 113, 129, 144]. Svanson et al. [336], Beekenkamp [344], and Konijnendijk [343] agree that, in the pure B_2O_3 glass, interconnected boroxol

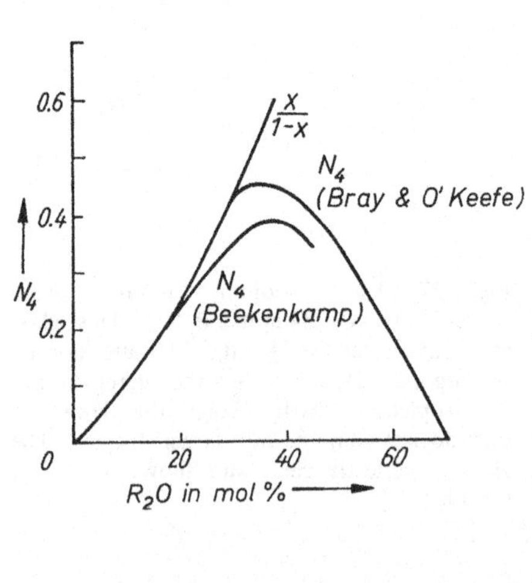

Fig. 7.23. Representation of the portion of B ions in 4-coordination as a function of composition in binary alkali borate glasses (after Bray and O'Keefe [106] and Beekenkamp [344])

groups are the chief structural element. In alkali borate glasses, pentaborate and triborate groups (primarily in the tetraborate connection), as well as the diborate group are considered the chief structural elements (see Fig. 7.16).

As far as the coefficient of expansion is concerned, we believe that for smaller Na_2O concentrations the formation of relatively compact penta- and triborate groupings with B in 4-coordination will cause a decrease in expansion. The tetraborate group, however, clearly has a looser structure (Fig. 7.16), in spite of the increased number of 4-coordinate B per structural unit. A looser structure usually will cause the coefficient of expansion to increase. Also, and significantly so, the alkali ion itself with its asymmetrical surrounding may contribute to an increase in expansion.

As to the immiscibility behavior of binary alkali borate glasses, it should be recalled that usually a melt containing two kinds of structural elements with differing spatial requirements will tend to phase-separate. This is certainly the case for boroxol rings as one, and penta- (or tri-) borate groups as the other structural element. This would make understandable the tendency toward phase separation increasing toward 16 mol% alkali. In the region of higher alkali concentration, the tendency decreases and can no longer be found at 24 to 30 mol% Na_2O. Apparently this is caused by the decrease in boroxol groupings, and perhaps more so by the appearance of additional groupings like tri- and diborate, since in most systems the simultaneous occurrence of several structural elements decreases immiscibility.

7.4.2 Borosilicate Glasses

7.4.2.1 The Ternary System Na_2O–B_2O_3–SiO_2

The technically most important glasses of this type are the sodium borosilicate glasses. Most recently, the liquidus surface for the alkali borosilicate systems was explored by Rocket et al. [353]. As in borate glasses, the addition of alkali causes the formation of 4-coordinated BO_4 groupings, but in competition with nonbridging oxygen formation at SiO_4 tetrahedra. Raman spectroscopy (Konijnendijk et al. [354]) permits the identification of the following units: SiO_4 with no, one, or two non-bridging O; boroxol rings consisting of BO_3, triangles: metaborate rings; and six-rings with one or two BO_4 groups. Additional detailed information on the concentration of different groupings containing BO_4 units has been obtained by nuclear spin resonance (Bray [352], Yun et al. [355], Xiao [356], Yun et al. [357]). At high SiO_2, the structure contains the $(BSi_4O_{10})^{-1}$ units found in the mineral reedmergnerite: If Na:B exceeds 1, non-bridging oxygen is appearing at SiO_4 tetrahedra. With this wealth of different groupings, it is not surprising that a tendency toward immiscibility is found in wide areas. Additional evidence is available from photoelectron spectroscopy (Smets et al. [358]).

The distribution of alkali between silicate and borate groups remains a matter of unresolved discrepancy. B NMR studies (Zhong et al. [359]) e.g.

postulate for the $m\ Li_2O.\ B_2O_3$–$kSiO_2$ system that up to some alkali level (m_0) at which all alkali goes to boron groups the alkali going to silicon groups (m) is:

$$m_{si} = (R - R_0)K(1 + K) \quad \text{for all } m \geqslant m_0$$

$$m_{si} = 0 \quad \text{for all } m \leqslant m_0$$

where m_0 increases with SiO_2 content from 0.270 (for $k = 0.5$) to 0.408 (for $k = 4$).

In contrast, most recent MAS NMR results seem in agreement to allotted proportional sharing of alkali between silicate and borate groups (Martin et al. [360, 361]).

When considering the influenc^{max}e of composition on the glasses of the ternary system Na_2O–B_2O_3–SiO_2, it is therefore necessary to recognize that all three constituent binary systems (Na_2O–B_2O_3, Na_2O–SiO_2, B_2O_3–SiO_2) tend toward immiscibility and that their microheterogeneous structure has been demonstrated by electron optics. In all three cases, the microphases found in clear glasses were small (Müller and Vogel [362]).

The ternary system Na_2O–B_2O_3–SiO_2 exhibits an immiscibility dome with an upper critical temperature of about 760 °C. This very interesting system was investigated by numerous authors, Molchanova [58], Vogel [185], Kühne [211], Kühne and Vogel [184], Kühne and Skatulla [212], and Porai-Koshits et al. [363]. The system contains a domain in which heat treatment produces a visible phase separation (Fig. 7.24). The first exact delineation of the domain was disclosed by Molchanova [58]. We have introduced into this figure the straight line representing increasing SiO_2 content at constant Na_2O:B_2O_3 ratio = 16:84. The line corresponds to the maximum of immiscibility in the

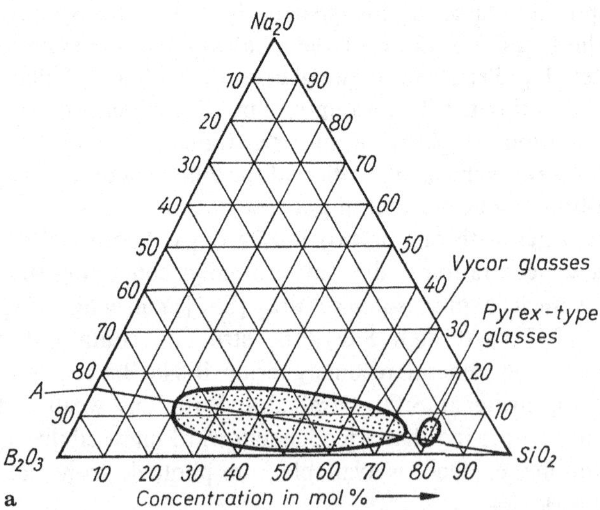

a

Fig. 7.24. Ternary Na_2O–B_2O_3–SiO_2 glass system with an immiscibility region (Molchanova [58] and location of Vycor- and Pyrex-type glasses on this diagram binary, and the "anomaly" concentration. It is seen that the ternary immiscibility region hugs this line. Each departure from the 16:84 region leads not only to decreasing immiscibility but also to increasing coefficients of expansion.

The glasses in the left part of the region shaded in the figure are clear when cast and cooled at a conventional rate, but show droplet regions of 20 to 30 nm (200–300 Å) in the electron microscope. They grow more than 100-fold after heat treatment (Fig. 7.25). The glass is turbid. In this region of lower SiO_2 content, the droplets are an SiO_2-rich phase embedded in an alkali-rich borate matrix. With increasing SiO_2 content, the melting temperature increases and the phase separation within the borate matrix is suppressed. The minima of expansion along the 16% line indicate that the structural processes underlying the "boron anomaly" must have been completed. At SiO_2 contents above 50 mol%, the matrix is silica-rich and the droplets are the alkali-rich borate phase. In the center of the immiscibility region, the phase functions change through the spinodal area and the microstructure is interconnected (Fig. 7.26). Frequently,

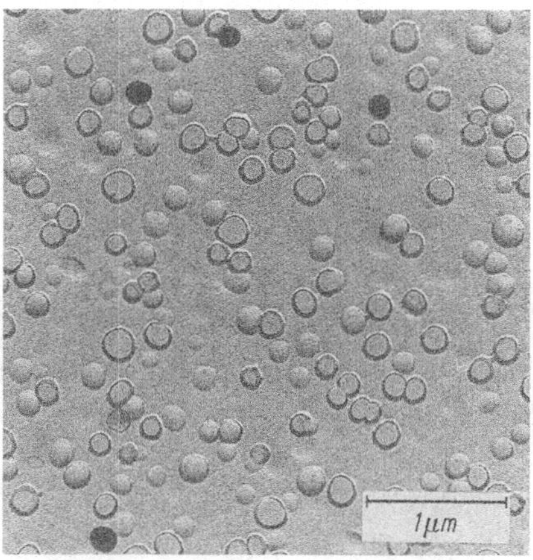

Fig. 7.25. Sodium borate glass situated in the left portion of the immiscibility region marked in Fig. 7.24, after reheating. The droplet dimension has been increased 100-fold compared to that in rapidly cooled glass. (Electron-optical micrograph after replica preparation.)

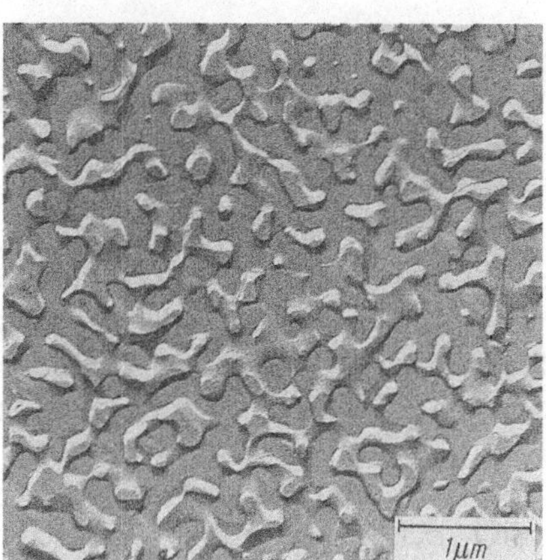

Fig. 7.26. Sodium borosilicate glass from the center of the immiscibility region marked in Fig. 7.24. The connective structure indicates the change in function of the microphases. (Electron-optical micrograph after replica preparation.)

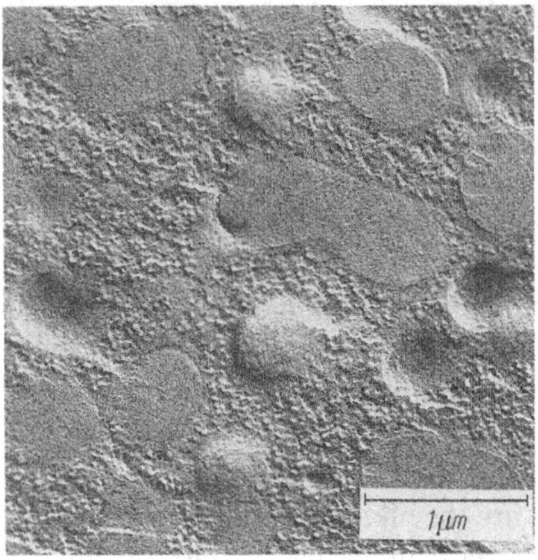

Fig. 7.27. Sodium borosilicate glass from the sodium borate-rich side of the immiscibility region. The micrograph shows multiple phase separation: (1) primary precipitation of large SiO_2-rich regions in an alkali borate-rich matrix phase. (2) In a secondary process (according to Fig. 6.18), a second, SiO_2-rich smaller-droplet phase is precipitated in the matrix. (Electron-optical micrograph after replica preparation.)

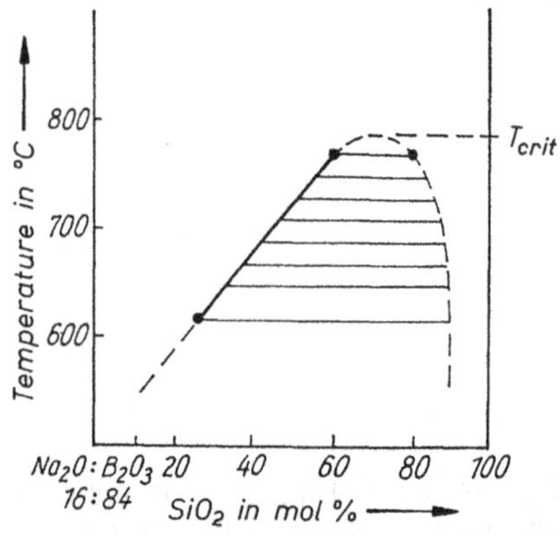

Fig. 7.28. Asymmetrical immiscibility dome. Schematic of the tendency toward phase separation in the system Na_2O–B_2O_3–SiO_2, alongside the anomaly line

multiple separation is observed, based on the step process described earlier, and illustrated in Fig. 7.27 for a glass whose composition lies on the left side of the immiscibility region (Fig. 7.28).

Figure 7.29 shows a glass lying on the right (SiO_2-rich) side of the immiscibility dome. The immiscibility trend in the SiO_2-rich matrix is relatively small, as made evident in Fig. 7.28. The right side of the immiscibility dome is quite steep, i.e., the solubility of the alkali borate phase is small even at higher temperature.

This is in contrast to the left borate-rich side of the dome where the curve is not steep, indicating a strong increase of borate solubility with increasing temperature. This, in turn, creates favorable conditions for the stepwise multiple unmixing event in Figs. 7.27 and 7.29 (Vogel et al. [200, 364]). The glass of Fig. 7.27 lies on the alkali borate-rich side of the dome. In the primary separation process, large SiO_2-rich droplet regions had formed in the sodium borate-rich matrix. On further cooling, a very small SiO_2-rich droplet phase was precipitated in a secondary separation process, as the solubility of SiO_2 dropped

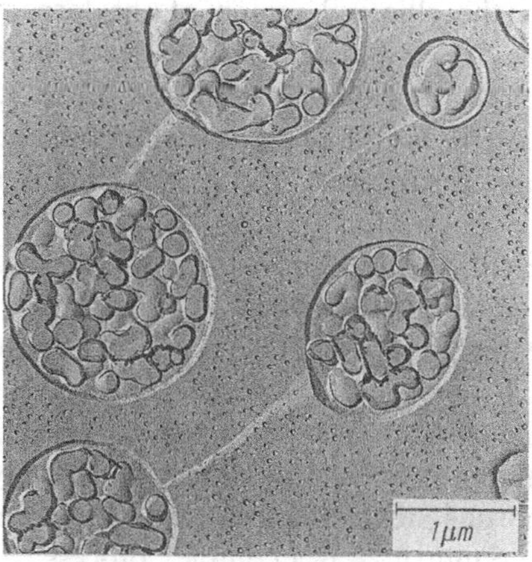

Fig. 7.29. Electron-optical micrograph (replica preparation) of a sodium borosilicate glass of the Vycor®-type composition, located on the SiO_2-rich side of the immiscibility dome. Large droplets of primary phase separation representing a sodium-rich borate glass phase can be recognized, which during cooling have been subjected to a secondary phase separation of SiO_2-rich droplets. In the SiO_2-rich matrix, there also occurred a secondary phase separation of very small sodium borate-rich droplets. A four-phase glass results

significantly. The result was a three-phase glass. This interpretation corrects earlier interpretations ([184, 185]).

In Fig. 7.29 the glass lies on the right SiO_2-rich side of the dome. The large droplets are the alkali borate-rich phase. Since it is in this phase that the high solubility of SiO_2 at high temperature is much reduced on cooling, a noticeable precipitation of SiO_2 droplets occurred within the large droplets. In the SiO_2 matrix, however, the small solubility and the small change of solubility on cooling illustrated by the steep fall of the dome on that side led only to the precipitation of very small droplets of alkali borate. The picture clearly shows the difference in the secondary-phase separation in the two primary phases. The resulting microstructure shows four phases: (1) a primary SiO_2-rich matrix with (2) secondary, small, borate-rich droplets; (3) primary, large, borate-rich droplets with (4) secondary, SiO_2-rich droplets. When the first systematic phase separation studies were carried out on sodium borosilicate glasses [184, 185] by the end of the fifties, it was believed that the very small droplet phase in Fig. 7.27 was an alkali borate rich phase in agreement with the removal of an alkali rich borate glass phase in sodium borate glasses (see Fig. 7.16 to 7.18 or 7.19 to 7.21). This assumption must be corrected in view of the experimentally proven observations carried out with a modern investigation technique in the last 20 years. The disappearance of a phase boundary between the alkali borate phase and the B_2O_3 glass phase in glasses of the ternary system Na_2O–B_2O_3–SiO_2 obviously follows the SiO_2 absorption by the binary Na_2O–B_2O_3 glass. Higher melt temperatures result and consequently also higher quenching rates in the cooling process of the melt.

7.4.2.2 Vycor-Type Glasses

In many glass systems the microphases formed by phase separation differ in chemical resistance. In extreme cases one of the phases can be extracted completely. The Vycor process (Hood et al. [365], Nordberg [366, 367], Elmer et al.

[368–370]) utilizes this phenomenon in the elliptical region of the ternary Na_2O–B_2O_3–SiO_2 described in the preceding section (Fig. 7.24). A nearly pure SiO_2 glass is obtained by melting a glass in the compositional field 55 to 75 mass% SiO_2, 20 to 35% B_2O_3, and 5 to 10% Na_2O, phase-separating it, and leaching the sodium borate-rich connective microphase. The process includes five steps:

(1) The sodium borosilicate glass is melted and fashioned into objects (e.g., beakers, dishes, plates, tubing) in a conventional manner.
(2) The clear glass objects are heat-treated at about 500 to 600 °C. Phase separation and opalescence result.
(3) The opalescent glass objects are treated with a mineral acid, e.g., 3N H_2SO_4 at about 90 °C. The soluble sodium borate phase is almost completely extracted and the object now has a porous SiO_2 structure. This extraction, typically proceeding at about 0.8 to 1.5 $mm \cdot h^{-1}$, is a complex process.
(4) After washing, residual water is removed in a cautious drying operation. The resulting glass is almost pure SiO_2, opaque, with a mean pore diameter (2 to 5 nm) (20–50 Å) and a void volume of about 25 to 40%. The specific surface is very large and can be controlled between 100 and 300 $m^3 \cdot g^{-1}$ by the heat treatment of step 2. In this intermediate step the porous ("thirsty") Vycor glass finds various applications, such as in chromatography, as a carrier of high-temperature catalysts, as a membrane for gas diffusion, as a filter in biological engineering, for disposing of nuclear waste, etc.
(5) The porous object is heated to about 1100 °C and sinters to a perfectly clear, nearly pure SiO_2 object, undergoing about 30% volume contraction.

A typical commercial Vycor-type glass contains: 95 to 98 mass % SiO_2, 0.3 to 0.6 mass% Na_2O, 2.5 to 3.5 mass% B_2O_3. Characteristic properties of Vycor-type glasses are:

Linear expansivity $\alpha = 7$ to $8 \times 10^{-7} \cdot K^{-1}$
Maximum thermal shock about 1100 °C
Thermal resistance 800 to 900 °C
Chemical resistance (acid, base and water resistance) = class 1. (German standard)
Density $\rho = 2.18\ g \cdot cm^{-3}$
Softening point 1500 to 1530 °C

In contrast to the process of obtaining SiO_2 glass objects from a quartz melt requiring 2000 °C, the Vycor process requires only conventional melting and shaping temperatures. Many objects could not be fashioned from quartz melts. For these reasons the Vycor process has found many applications, particularly for the production of laboratory ware and of tubing for halogen lamps.

The successful operation of the process requires much more restriction in composition than the original patents would suggest. Of course, connected microstructures (Kühne [211, 212], Vogel et al. [200, 202, 205]) must be obtained and secondary phase separation may affect the leaching process. Secondary immiscibility occurences in both primary phases are possible in base glasses via droplet as well as connected structure formations. Consequently an acid-extraction process (see Figs. 6.30 and 6.31) can proceed in an extremely complex way and is still not completely understood.

If Na_2O is replaced by K_2O or Li_2O similar reactions are observed, but the regions of immiscibility and connectivity are shifted (Taylor et al. [371]).

Al_2O_3 is often added to the composition to suppress crystallization and to control phase separation. Therefore, the finished product frequently contains a few percent Al_2O_3.

7.4.2.3 Pyrex-Type Glasses

The Pyrex-type glasses also lie near the 16% Na_2O ("anomaly") line of minimum expansion in the ternary Na_2O–B_2O_3–SiO_2. However, they are outside the range of visible phase separation at higher SiO_2 contents (80%) (see Table 1.5 and Fig. 7.24). However, even though transparent laboratory glass of fairly low expansivity is produced from Pyrex® glasses, a tendency toward immiscibility (Kalsing [372]) on a small scale remains present. This may affect chemical resistivity. However, small additions of Al_2O_3 control both phase separation and resistivity. In these functions an equalizing effect on the surface tensions of the microphases plays an important role. Fluorine also acts in this manner. Commercial Pyrex® glasses represent a compromise between minimum expansivity and maximum chemical resistance. Yet some characteristics of the Na_2O–B_2O_3 binary remain (Skatulla et al. [184], Vogel [192]).

The tendency toward immiscibility in borosilicate laboratory ware becomes manifest in the daily routine control of chemical resistivity in production.

Routine Control of Rasotherm Glass according to DIN 12111

Thick-wall ($\geqslant$4 mm) objects show a larger alkali loss than thin-wall ($\leqslant$2 mm) objects (Figs. 7.30 and 7.31). Moreover, the thick-wall glass loses more alkali after annealing (54 γ compared to 27 γ). A second heat treatment (1 h at 550 °C) further reduces resistivity (Fig. 7.31). Thin tubing does not show any of these effects. The temperature region of immiscibility is passed much more quickly than in thick-wall objects. The control data also reflect day-by-day process variations.

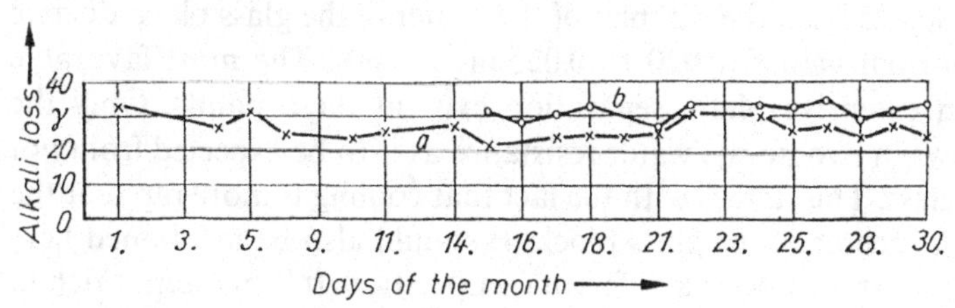

Fig. 7.30. Daily control of the alkali loss of an apparatus glass: **a** thin (< 2 mm) wall: **b** thick (≥ 4 mm) wall

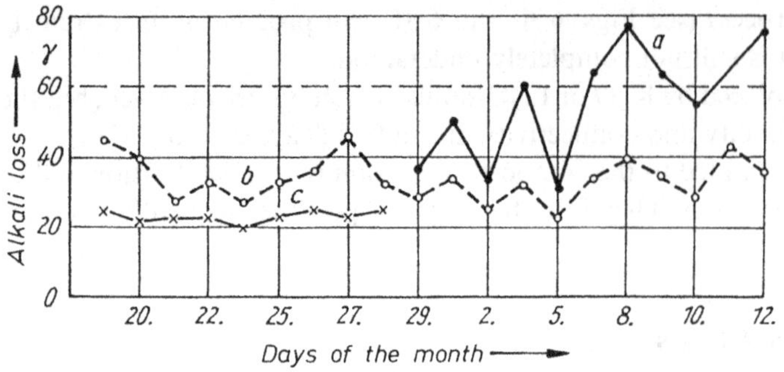

Fig. 7.31. Daily control of the alkali loss of an apparatus glass: **a** thick (≥ 4 mm) wall, additional annealing (550 °C, 1 h); **b** thick (≥ 4 mm) wall, lehr annealing; **c** thin (< 2 mm) wall, lehr annealing

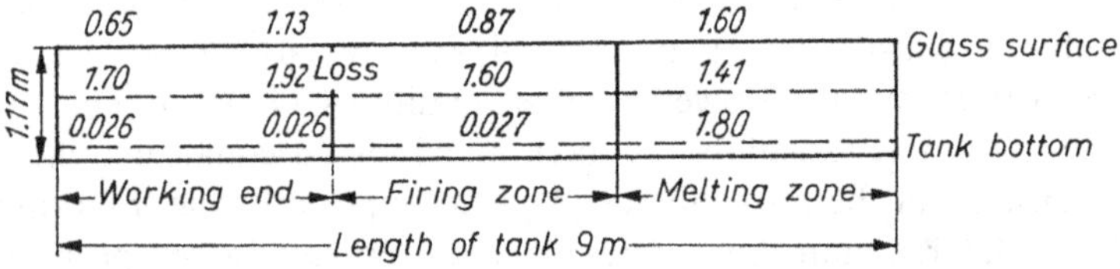

Fig. 7.32. Alkali loss of glass samples from a cooled tank. Section along the center of the cooled tank. The numbers represent the alkali loss in Na_2O (mg) established according to the German Standard (DIN 12 111). At the same time, the numbers indicate where the samples were taken. Normal value: 0.020 to 0.025 Na_2O

Investigations on Glass from a tank

The demonstration of water resistance convincingly proves that the aforementioned cases are in fact immiscibility processes. Relatively small modifications arise in production, inducing phase separation which is when the tank is accentuated. The slow temperature drop should strongly promote the process in a glass tending to immiscibility. As the glass block is broken, glass samples are taken out of the center of the tank at various places along its length (see Fig. 7.32). The values for alkali emission at the locations of the glass samples are given in the figure. The alkali losses of the annealed bath glass increase by almost two orders of magnitude. The glass hydrolytic grade has changed from 1 to 5. The alkali losses from the samples of the center of the glass block deviate the most from normal values (0.020 to 0.025 mg Na_2O). The most favorable conditions for an extensive phase separation exist at these points. Only the bottom glass shows in two points water resistance as is to be expected from not phase separated glass. This agrees with the fact that cooling is more rapid at the bottom than at the center of the glass block. It should also be mentioned here that glass at the bottom possesses a different composition. It is in essence rich in Al_2O_3. Al_2O_3 is known to improve the water resistance of glass. It also lowers the tendency to immiscibility as mentioned in the preceding section.

Heat Treatment Experiments with Glass Tubing

– Temperature Dependence

The observations made until now were obtained from glass samples taken during the course of production. The following experiments actually show that via additional heat treatment, immiscibility can be obtained with glass tubing, which has the least tendency to show phase separation due to the special temperature conditions during production. The duration of the additional treatment was kept constant at 2 hours for all temperatures. The temperature dependence of the immiscibility process is shown in Fig. 7.33. A very rapid increase of the leaching values occurs above 580 °C. After two-hour treatment at 630 °C, the water resistance of the glass approaches the transition from grade 2 to 3. In the region (550 °C), the glass tubing does not show any unusual deviations from normal values (0.020 to 0.025 mg Na_2O).

– Time Dependence

Heat treatment at 600 °C was performed for various lengths of time (up to 8 h) and the results are shown in Fig. 7.34. At first the leaching values increase

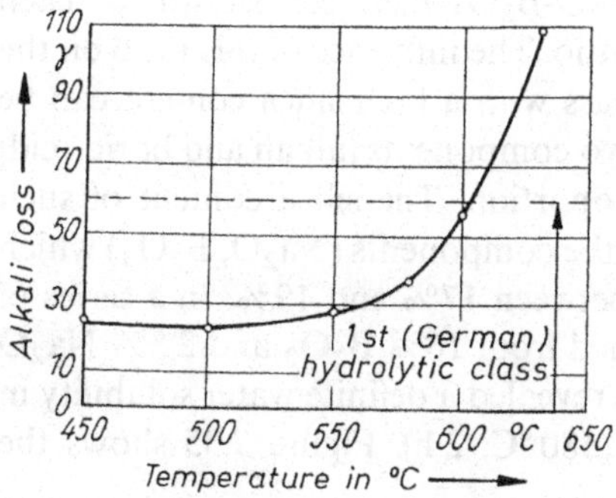

Fig. 7.33. Dependence of water resistance on temperature of heat treatment. Heat-treatment time 2 h; samples: tubing 15 to 18 mm in diameter, 1 to 2 mm wall thickness

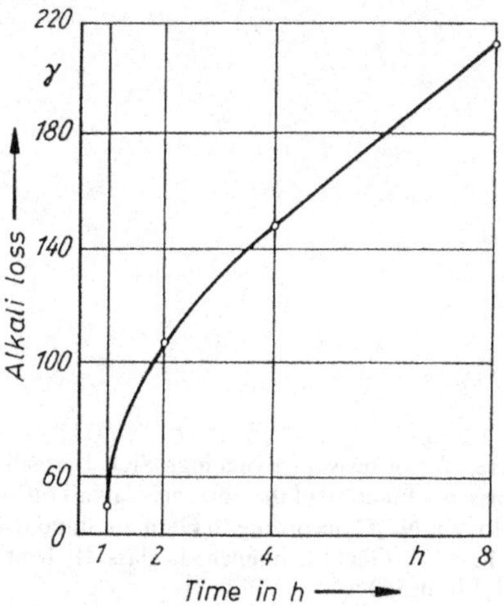

Fig. 7.34. Dependence of water resistance on the time of heat treatment. Temperature: 600 °C; samples: tubing 15 to 18 mm in diameter, 1 to 2 mm wall thickness

proportionally fast. After longer times the slope of the curve decreases. The curve approximates a parabola which lies symmetrically to the time axis.

One could question the fact that the values for 600 °C for 2 hours do not coincide when the curves in Figs. 7.33 and 7.34 are compared. For each experimental curve, a standard pipe traction was used, however these two glass samples were taken from the tank quite different times. Therefore the difference in the measured values can be explained by deviations in the glass composition. The characteristics of both curves is however, not affected by the differences between the glass samples.

Immiscibility and Composition

Phase separation in technical glasses strongly depends on composition. The three components in *Vycor* glass for instance, silica, boric acid and alkali are so well balanced that after proper annealing of the glass, an approximately complete leaching capacity is obtained. At least one more component comes into play with technical (conventional) glass: Alumina. The tendency towards immiscibility must be reduced to a minimum by a clever choice of the components. Here the boric acid-alkali ratio plays an important role. The coefficient of expansion in a two-substance system Na_2O-B_2O_3 has been shown to reach a minimum at a defined alkali-boric acid ratio. The influence of this ratio on the immiscibility of a technical (apparatus) glass with a high silica content can be demonstrated by keeping the sum of the two components (alkali and boric acid) constant while changing their relative proportion. The silica content of such glasses is approximately 80%. The sum of the components (Na_2O, B_2O_3) which are more readily soluble in water varies between 17% and 19%. In a series of eight experimental melts composition varied from 16% B_2O_3 and 2.5% Na_2O to 12.5% B_2O_3 and 6% Na_2O. Each glass revealed a definite water solubility in the quenched and in the annealed state (580 °C, 2 h). Figure 7.35 shows the average alkali loss from two equal series.

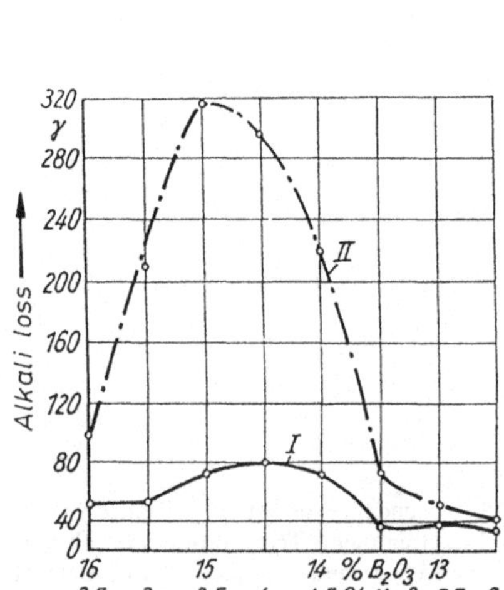

Fig. 7.35. Water resistance of a high-SiO_2 borosilicate glass as a function of the boric acid: alkali ratio. Alkali loss in Na_2O according to German Standard (DIN 12 111). Glass I: quenched: glass II: heat-treated 2 h at 580 °C

A clear maximum is observed demonstrating that one has to avoid this range of composition in the production of technical glasses. The phenomena described here are not exclusive to Rasotherm glass for instance. A comparison of the water resistance of Rasotherm and Pyrex glass shows similar behavior for both glasses under similar conditions

	Pyrex (Solidex) (Alkali loss in mg Na_2O)	Rasotherm
Normal cooling	30	22
Additional treatment 2 h at 630 °C	138	109

Electron Microscopic Investigations of Glasses of the Pyrex Type

Results given in the preceeding sections on the hydrolytic properties as well as the structure of glasses of the Pyrex type can also be confirmed by electron microscopy. Figure 7.36 shows the micrograph of a glass of the Pyrex type with normal properties. Practically no immiscibility regions are recognizable or they are too small to be detected by electron microscopy. Figure 7.37 shows a fraction of a glass of Pyrex type taken out of tank after shutting it off (from the center, Fig. 7.32). Immiscibility regions of various dimensions can be clearly seen. This phase separation was responsible for the aforementioned low water resistency of the glass. Figure 7.38 illustrates the enormous practical importance of these results. They allow better control of the glass properties. Without it production can fail, as was the case for distillation flasks which were not sufficiently resistant against prolonged water attack. With permanent use, the round flasks became useless due to strong corrosion (see Fig. 7.38). Phase

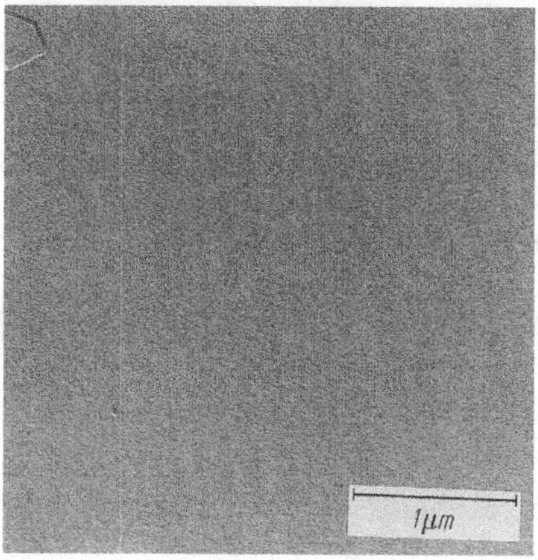

Fig. 7.36. Technical glass (Pyrex type) with normal properties (electron-optical micrograph). Comparison with molybdenum trioxide crystal test plane (top left) shows that, if any immiscibility regions exist, they would be below the level of resolution

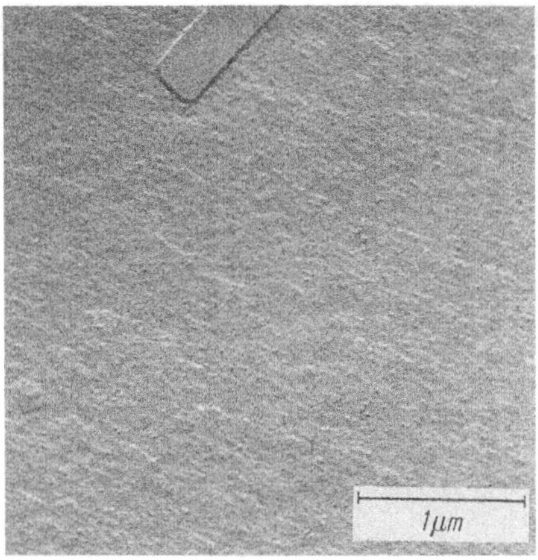

Fig. 7.37. Technical glass (Pyrex type) from the center of a cooled tank, with clearly worse hydrolytic properties. Comparison with molybdenum trioxide crystal test plane (top) clearly reveals phase separation in the form of the precipitation of a droplet-shaped glass phase. The precipitation of this sodium borate glass phase is responsible for the poor water resistance

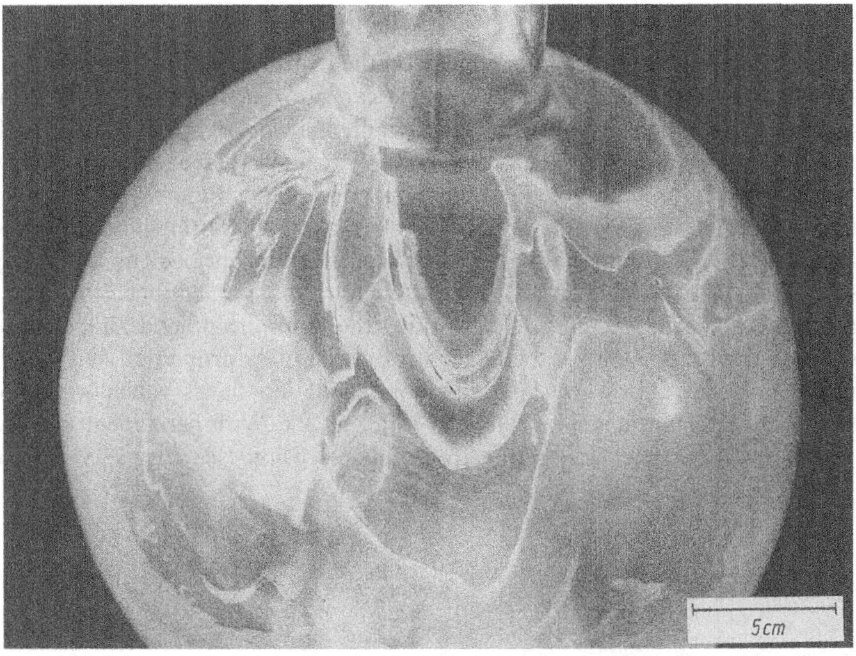

Fig. 7.38. Distillation flask of Pyrex-type glass. The flask was strongly corroded by continuously recirculating water, resulting in a serious disturbance of production by too advanced, insufficiently suppressed unmixing during melting of a Pyrex-type apparatus glass (magnification 1:3)

separation was not minimised during production thus facilitating the leaching of the water soluble borate glass phase.

Some Other Borate Glasses

There is an unlimited number of other borate glasses, in some cases leading to new applications. A few recent examples:

1. Barium borate: solvent for obtaining magnetic barium hexaferrite (Zagnazi et al. [373]).
2. Cadmium borate: photochromic (Dmitriev et al. [374]).

Aluminoborate glasses:
3. "Cabal" ($CaO–Al_2O_3–B_2O_3$) (also with Ca replaced by other cations) e.g. (4): known for a long time.
4. Sodium aluminoborate (doped): can be phase separated and leached to a porous structure like vycor (Elmer [375]).
5. Silver iodide borate: potential solid state battery
6. Lanthanum borate: basis for polynary high index low dispersion optical glass (see Fig. 1.13)

Borate glass structures were modeled successfully by Molecular Dynamics (Deppe [376]).

Aluminosilicate Glasses

This family of stable glasses is of particular importance for the production of various glass ceramics (Beall [377]) (see Chapter 10) as well as sol-gel processed coatings (see e.g. Prabakar et al. [378]).

In *alkali aluminosilicate glasses* the concentration of non-bridging oxygen decreases as Al_2O_3 is added to alkali silicate glasses since oxygen introduced by M_2O now can be used for the formation of AlO_4 tetrahedra. This effect saturates above $M_2O:Al_2O_3 \approx 1$ when Al^{3+} enters as a 6-coordinate network modifier. Structure and properties are strongly affected by small amounts of OH which introduces non-bridging oxygen type sites without the presence of stronger M–O bonding (Jewell et al. [379, 380]).

Among many complex systems based on $M_2O–Al_2O_3–SiO_2$ the addition of Y_2O_3 and ZrO_2 provides harder lower-expansion coatings (Makishima and Hara [381], Shelby et al. [382]). Complex variants are used in the form of radioactive beads for treatment of liver cancer. (Applewhite and Day [383]). Structures have been determined in detail by Kohli et al. [384].

Low SiO_2 *calcium aluminosilicate* compositions are developed as more stable nearly equally IR transmitting replacements of calcium aluminate glasses. (Higby et al. [385]).

Complex $MgO–Al_2O_3$ (cordierite) glasses are important precursors of glass ceramics used in electronic packaging (see Chapter 10).

7.5 Glasses of High Lead Content

7.5.1 Glass Formation in Lead-Containing Systems

High lead content characterizes the flint glasses – one of the two main families of optical glasses – and the so-called crystal glasses of fine table-ware, art, and jewelry.

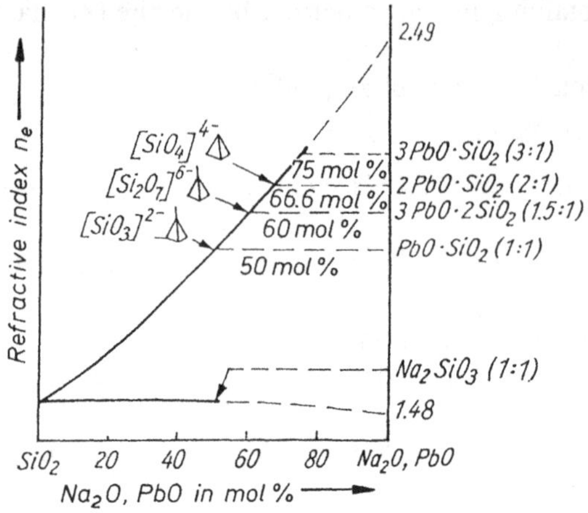

Fig. 7.39. Schematic of the dependence of light refraction on composition in the binary glass systems Na_2O–SiO_2 and PbO–SiO_2. The glass-forming region in the system Na_2O–SiO_2 extends to about 50 mol% Na_2O, that in the system PbO–SiO_2 from about 0 to 75 mol%

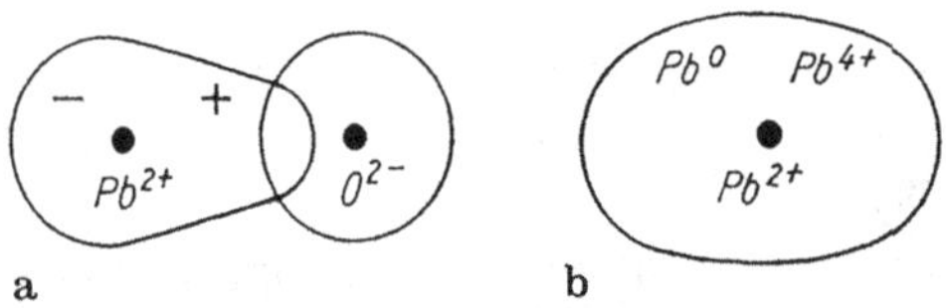

Fig. 7.40. Schematic of polarizability. **a** Pb^{2+} ion by O^{2-} ions; **b** creation of a Pb^{2+} ion with dipole character

If silicate glasses are described on the basis of the Zachariasen–Warren network theory, the existence of high-lead silicate glasses requires special consideration. If Na_2O oxide is added to SiO_2 in increasing amounts, Si–O–Si bridges are disrupted and the viscosity drops. When 50% Na_2O is introduced, one-half of the bridges are disrupted and the mobility of the building units becomes so high that the melt crystallizes easily and glass formation no longer occurs (Fig. 7.39).

In the binary system PbO–SiO_2, however, glasses are formed easily up to 70 mol% PbO, in small samples even to 75 mol% (92 mass%) PbO. According to the network theory, all bridges would be broken at 66% network-modifier concentration, and no glass could be formed. Plausible explanations from an atomistic viewpoint were offered by Fajans and Kreidl [386] and Stanworth [387].

The Pb^{2+} ion possesses two single 6s electrons in its outermost shell. (A similar situation characterizes the Tl^+ and Bi^{3+} ions, resulting in similar behavior, to be discussed later.) These two single 6s electrons are responsible for the high polarizability of Pb^{2+}, which can be polarized even by the oxygen anion, as represented schematically in Fig. 7.40. This polarization distorts the electron shell to such an extent that one may attribute dipole character to the lead ion. This also may be expressed by saying that one side of the ion functions more like Pb^0, the other like Pb^{4+}. This concept also helps to explain the behavior of surfaces of high-lead-content glasses. On the surface, Pb^{2+} is exposed to oxygen only on the inside, which gives the surface the metal-like

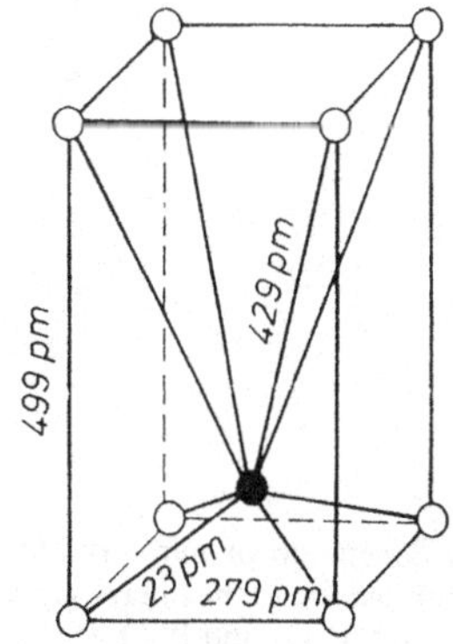

Fig. 7.41. Tetragonal cell of crystalline PbO (Fajans and Kreidl [386]). (Atomic distances in Å)

Table 7.2. Function of the Pb^{2+} and Pb^{4+} ions in a glass network

Ion	Ion radius r (in Å)	Field strength $F[Z/a^2]$	Function in the glass network
Pb^{2+}	1.32	0.27	Network-modifier
Pb^{4+}	0.84	0.8	Intermediate oxide (= Glass formation)

character affecting its behavior. More striking are the consequences of this deformation of Pb^{2+} in the structure of the tetragonal cell of crystalline PbO (Fig. 7.41).

As the first four oxygen ions approach a Pb^{2+} ion in the course of initial crystallization, polarization and "dipole" formation occur, as shown in Fig. 7.40. Consequently, the next four approaching oxygen ions will be placed at a much larger distance on the other side of the Pb^{2+}. The distances, as determined by X-ray analysis, are 0.23 nm (2.3 Å) for the first four and 0.43 mn (4.29 Å) (almost twice as large!) for the second four. At low concentrations of PbO, Pb^{2+} may be considered as acting like a Zachariasen–Warren network-modifier.

At high concentrations, however, the partial function, like Pb^{4+}, falls into the criteria for intermediates (smaller ionic radius, field strength 0.5 to 1) capable of participating in the "network" (Table 7.2). Nuclear magnetic resonance investigations (Bray [388]) confirm the 4-coordination of lead in high (> 50 mol%) PbO glasses.

In phosphate glasses, Na_2O can be introduced up to about 66%, which has been explained by the formation and entanglement of chains which have been shown to exist by chromatography. In lead phosphate glasses, however, less lead (only 66%) can be introduced than in silicate glasses (Fig. 7.42).

This may be explained by the strong counterpolarization of the higher charge of the P^{5+} cation (field strength 2.1, compared to 1.57 for Si^{4+}) (Fig. 7.43). The structure of such high-lead glasses was described by Mydlar et al. [389].

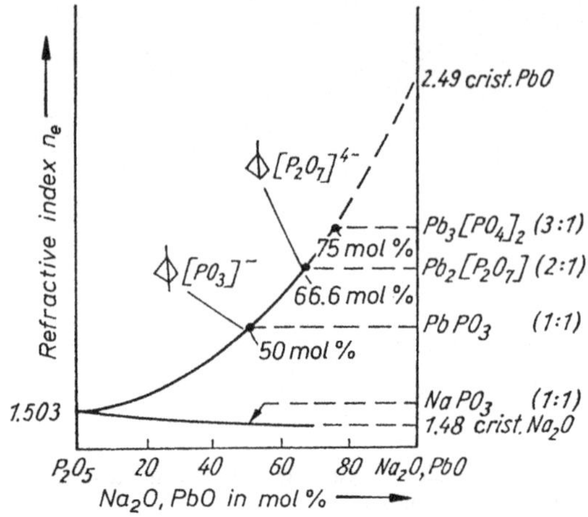

Fig. 7.42. Schematic of refractive index vs. composition in the binary glass systems Na_2O–P_2O_5 and PbO–P_2O_5. Glass formation extends between about 0 to 66 mol% Na_2O (PbO) in both systems

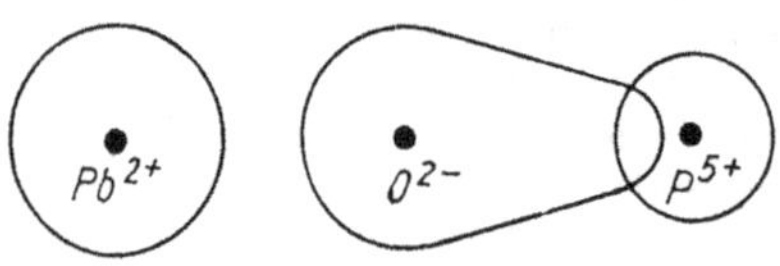

Fig. 7.43. Schematic of the counterpolarizing effect of the P^{3+} ion on the interaction between O^{2-} and Pb^{2+} ions (see Fig. 7.40 (a))

From a thermodynamic viewpoint, a melting point lower than the average bond strength between constituent elements may be considered favorable for glass formation. This postulate is based on the concept that a relatively low melting point is a consequence of a low ratio $\Delta E/\Delta S$, where ΔE is the difference between the energies and ΔS that between the entropies of melt and crystalline solid. High-lead silicates have, indeed, very low melting points.

7.5.2 Phase Separation in Glasses Containing Lead

Glassmelting practitioners are familiar with the appearance of samples taken from a melt of high-lead silicate glass: that of a reddish-yellow, milky mass. Only when cast at a temperature near T_g does the milky turbidity vanish and the glass become perfectly transparent. Suspicions of Berger [390] were justified by electron optical investigations (Vogel [187]) showing that even the clear transparent glass consists of two phases, one a high-lead matrix, the other a high-SiO_2 droplet phase (Fig. 7.44). This is interpreted as the consequence of a precipitation of SiO_2-rich droplets in the melt on lowering the temperature when –Si–O–Si bonds become rigid. These droplets float in the still quite fluid, high-lead phase like goulash in the gravy. At and below T_g, light scattering diminishes and to the human eye the glass appears homogeneous. Oberlies [391] was able to dissolve such droples from a microtome slice in HF, whereas treatment with HNO_3 produced isolated, SiO_2-rich droplets after dissolution of the high-PbO phase.

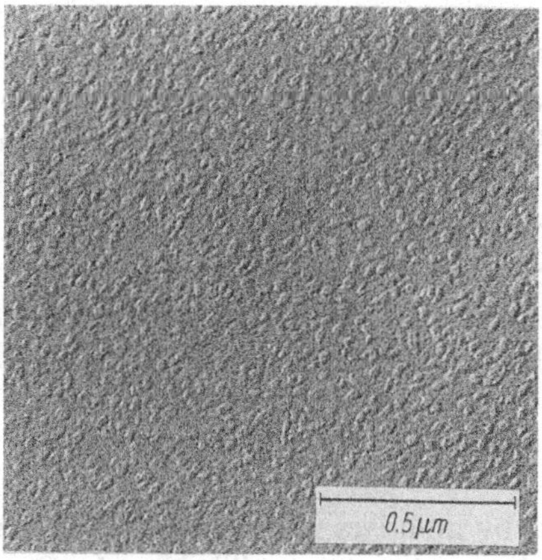

Fig. 7.44. Electron-optical micrograph of a binary high-PbO PbO–SiO_2 glass after replica preparation. A high-SiO_2 droplet phase is embedded in the PbO-rich SiO_2 phase

The tendency of high-lead silicate glasses toward immiscibility also becomes apparent in their crystallization behavior. Instead of lead silicates – which are precipitated after prolonged heat treatment according to the phase diagram – in practice crystallization as a defect produces cristobalite or tridymite, i.e., crystalline modifications of SiO_2.

The tendency toward immiscibility is suppressed in the ternary K_2O–PbO–SiO_2, clearly as a consequence of the virtual absence of immiscibility in the binary K_2O–SiO_2 due to the low field strength of K^+. No doubt this is the basis of the old useful "lead glass rule" of Zschimmer (quoted by Jebsen-Marwedel [20]), according to which a minimum content of mass% K_2O = (76 − mass% PbO)·0, 27 prevents devitrification. Zschimmer's "lead glass rule" indicates the minimum K_2O content necessary for a glass of a given PbO content (under 76 mol%) to avoid devitrification. In the light of today's knowledge, the "lead glass rule" simply illustrates the primary suppression of the immiscibility tendency of high-lead silicate glasses via the combination of two binary systems. In this case, the combination of the PbO–SiO_2 system with the K_2O–SiO_2 system reduces the tendency to immiscibility, since K_2O–SiO_2 glasses do not tend to immiscibility due to the very low field strength of the potassium ion.

To date, no immiscibility has been demonstrated in lead phosphate glasses. Apparently the competition in volume requirement is less for the typical chain structures of pure phosphate glasses. If, however, phosphate systems are combined with SiO_2, B_2O_3, or other glass-forming oxide systems, phase separation will occur.

7.5.3 Structurally Conditioned Coloration of High-Lead Silicate Glasses (Vogel et al. [392])

High-lead silicate glasses are usually greenish-yellow, mostly due to traces of iron, copper, or chromium ions. But it has been established that a change in the structure of the base glass has a decisive influence on the light transmission of lead glasses. Reference should be made to the preceding sections in which glass formation at high lead content was associated with the formation of PbO_4 tetrahedra which may involve an equilibrium $Pb^{2+} \leftrightarrows Pb^{4+}$. Earlier investigations have shown that the distribution of lead ions in the base glass network should not be taken from a statistical point of view. A SiO_2-rich lead silicate glass phase is deposited in a highly dispersed form in a base glass very rich in lead [187]. Thus oxidizing atmospheres, use of oxidizing nitrate, or introduction of lead as PbO_2 instead of PbO or PbO_4 may cause reversible brownish coloration due to finely dispersed PbO_2. Besides the equilibrium $Pb^{2+} \leftrightarrows Pb^{4+}$, a second equilibrium PbO_4-tetrahedra (network) $\leftrightarrows$ PbO_2 (dispersed molecule) is involved.

The brown coloration is suppressed by additives, decreasing the oxygen content of the melt. Arsenic is effective, since the equilibrium $As_2O_5 \leftrightarrows As_2O_3 + O_2$ at high temperature goes to the right, at low temperatures to the left (a phenomenon used in fining soda-lime glasses). This process deprives the lead of oxygen, suppressing the formation of Pb^{4+} and PbO_2. In spite of the analogous fining reactions of Sb_2O_3, this oxide on the contrary leads to stronger coloration, due to the formation of lead antimonate, as shown by Hahn-Weinheimer [393] in an investigation of Roman window glasses.

In Fig. 7.45, the light transmission of five 100 cm^3 binary lead silicate melts (60 mol% PbO) produced under equal conditions from different raw materials is compared: glass I: PbO, glass II: PbO_2, glass III: PbO + 0.2 mol% As_2O_3 glass IV: PbO_2 + 0.2 mol% As_2O_3, and glass V: PbO + 0.2 mol% Sb_2O_3. It can be seen that the color is intensified significantly if PbO_2 (glass II) replaces PbO (glass I). As little as 0.2 mol% As_2O_3 significantly increases transmission for both the PbO and the PbO_2 glass (compare glass I with glass III, II with IV).

Note that the introduction of As_2O_3 does improve even the glass melted from PbO, suggesting that even this glass contains Pb^{4+} or the bonding of PbO_2.

This interpretation is supported by the fact that γ irradiation discolors lower (< 50 mol%) lead silicate glasses yellow, higher lead silicate glasses reddish-brown. Under excessive radiation the glasses become black, indicating reduction to metallic lead.

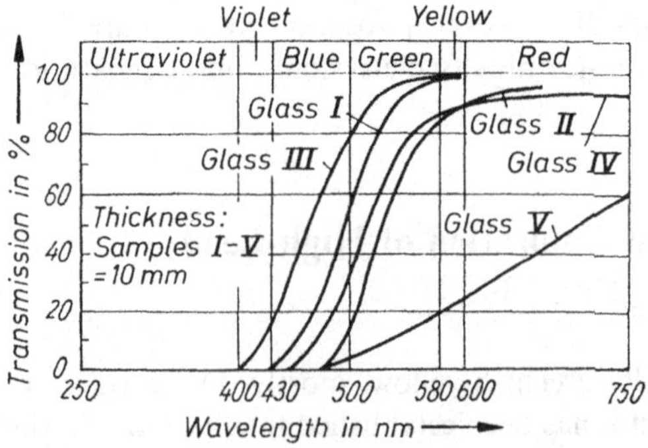

Fig. 7.45. Transmission of 5 binary 60 mol% PbO-40 mol% SiO_2 glasses from various raw materials

7.6 Phosphate Glasses

Pure phosphate glasses do not have the importance of silicate or borosilicate glasses, because of their poor chemical stability. Their application is limited to very special requirements, e.g., optical glasses, heat-absorbing glasses, HF-resistant glasses, dosimeter glasses used to measure high-energy radiation, etc. Combinations of P_2O_5 with other glass-formers have found more applications, since valuable specific effects of P_2O_5 may be retained at better chemical resistivity.

In this section only the typical properties of all phosphate glasses will be discussed briefly; the properties of special glasses will be treated in later sections referring to the respective uses.

7.6.1 Structure of Phosphate Glasses

In spite of a strong tendency toward sublimation, pure P_2O_5 may be molten under certain conditions and, after proper undercooling, obtained as a quite hygroscopic glass, which, of course, is of scientific interest only. X-ray analysis (Biscoe et al. [394]) has established the PO_4 tetrahedron as the principal building unit. Because of the pentavalency of phosphorus, only three out of four corners of a tetrahedron network – otherwise analogous to that of SiO_2 – are connected. The P–O–P angle was determined to be 140 °.

If basic oxides, like alkalis (M_2O), are dissolved in a pure P_2O_5 melt, good glass formation is observed (Table 7.3). However, when compared to analogous silicate glasses, these glasses exhibit peculiarities not expected from the basic network theory. Generally, properties like refraction, density, etc. change continuously in binary alkali phosphate glasses. However, much evidence exists for the formation of chain structures after introduction of increasing amounts of alkali. According to Van Wazer [400, 401], between 0 and 50%, chains are

Table 7.3. Glass-formation regions in binary phosphate systems M_2O ($M^{II}O$–)P_2O_5

System	Glass formation mol% M_2O ($M^{II}O$-)	References
Li_2O–P_2O_5	0 . . . 59.4	Kordes and Becker (306) [395]
Na_2O–P_2O_5	0 . . . 60.6	Kordes and Becker (306) [395]
K_2O–P_2O_5	0 . . . 47	Imaoka (320) [396]
BeO–P_2O_5	0 . . . 67.9	Kordes, Vogel, and Feterowsky (305) [397]
MgO–P_2O_5	0 . . . 60.0	Kordes, Vogel, and Feterowsky (305) [397]
CaO–P_2O_5	0 . . . 57.8	Kordes, Vogel, and Feterowsky (305) [397]
SrO–P_2O_5	0 . . . 56	Imaoka (320) [396]
BaO–P_2O_5	0 . . . 57.5	Kordes, Vogel, and Feterowsky (305) [397]
ZnO–P_2O_5	0 . . . 64.8	Kordes (307) [398]
CdO–P_2O_5	0 . . . 57	Kordes and Becker (306) [395]
PbO–P_2O_5	0 . . . 65	Kordes (308); Kordes, Vogel, and Feterowsky (305) [397]

connected to networks via oxygen, depending on basicity of X; above 50 mol% only chains and rings are formed and a network is no longer possible.

Thilo distinguishes polyphosphate glasses (M:O > 1), metaphosphate glasses (M:O = 1), and ultraphosphate glasses (M:O > 1). Polyphosphate glasses form open chain structures, metaphosphate glasses ring structures and ultraphosphate glasses networks.

For some time the results of paper chromatography demonstrating chain structure and chain length in high-alkali phosphate glasses [400–406] were questioned because of doubts about the integrity of the original structure after dissolution in a suitable solvent. It is true that in acidic phosphate glasses the structure is destroyed by the solvent and only isolated PO_4 units are found by paper chromatography. However, in basic ("polyphosphate") glasses (> 50% M_2O), chain and ring structures may be considered established by paper chromatography and supported by the evidence of Raman spectroscopy (Bobovich [407]), infrared spectroscopy (Müller [408]), and X-ray diffraction (Brady [409]). According to X-ray analysis, P–O–P angles in chain structures are $\approx 115°$.

Table 7.4, after Schulz et al. [47], gives a survey of chain elements identified by paper chromatography in aqueous solutions of alkali phosphate glasses. Statements about structures based on paper-chromatography investigations of glasses of the systems $BeO–P_2O_5$, $MgO–P_2O_5$, $ZnO–P_2O_5$ and also $Al_2O_3–P_2O_5$ are even more problematic. These glasses or combinations thereof possess very good chemical resistance. Thilo and Wieker [410] have shown that the structure of these glasses is dismantled to a more of less large extent with peptization. Paper-chromatography investigations can only give a partial account of the true facts. Between $M^{II}O:P_2O_5 = 1$ and 2, long polyphosphate chains and cyclic rings may be present (Schulz et al. [47], Kanazawa et al. [411]).

In BeO-, MgO-, ZnO-, P_2O_5 glasses, cations increasingly contribute to network formation: discontinuities ("anomalies") in properties were reported by Kordes et al. [395, 397–399] (see Fig. 7.46).

Table 7.4. Structural units of $Na_2O–P_2O_5$ glasses of various compositions in aqueous solutions (after Hinz [23])

Composition of glass Na:P	Quantity of solution in mass%-phosphate										
	Mono-	Di-	Tri-	Tetra-	Penta-	Hexa-	Hepta-	Okcta-	Nona-	Tri-me-ta-	High-poly-mer
5:3	0	22.9	47.2	20.4	6.1	3.4	0	0	0	0	0
6:4	0	5.6	28.0	29.0	17.4	9.4	5.8	4.7	0	0	0
7:5	0	5.7	19.0	24.4	17.3	13.1	9.7	?	?	0	10.9
8:6	0	2.6	10.3	15.6	14.2	12.0	10.1	8.3	6.5	0	20.5
8.8:6.8	0	2.5	7.4	12.3	13.0	10.6	9.9	8.0	7.9	0	28.8

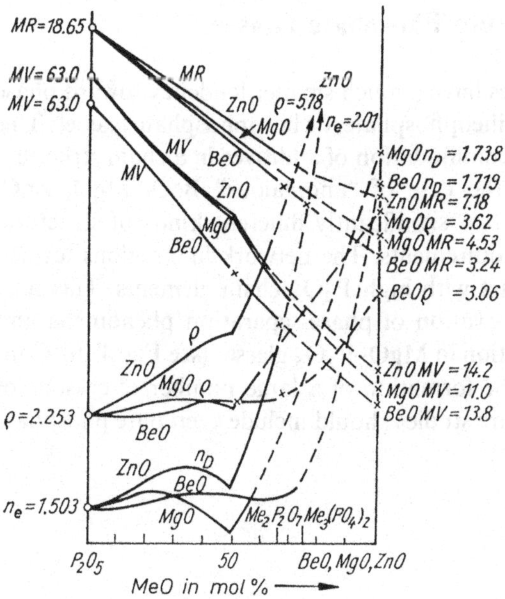

Fig. 7.46. Anomalous property vs. composition curves for binary Be-, Mg-, and Zn-phosphate glasses

In general, the weak structure of P_2O_5 caused by the presence of 25% non-bridging oxygen is, in contrast to that of SiO_2, fortified by $M_2^{III}O_3$, $M^{II}O$, and even M_2O.

An extreme case occurs in Al_2O_3–P_2O_5 glasses, where the combination of trivalent Al and pentavalent P in tetrahedral coordination allows for a structure resembling that of SiO_2:

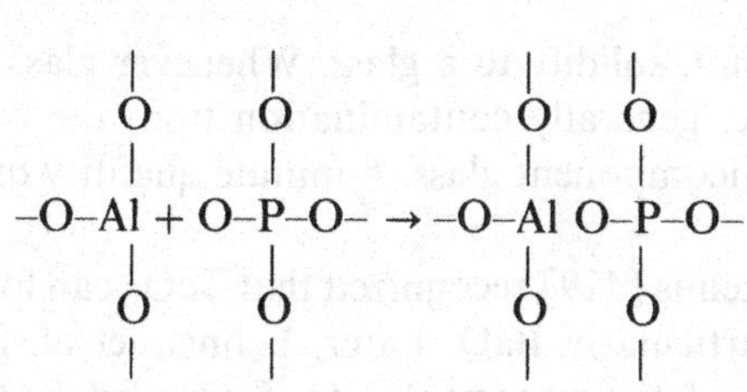

In fact, crystalline $AlPO_4$ (aluminum orthophosphate) is isostructural with crystalline SiO_2 (Hüttenlocher [412]), and X-ray diffraction patterns for aluminophosphate glasses resemble those of silicate glasses, with P–O, and Al–O distances of ca. 1.6 Å, like that of Si–O (Biscoe et al. [394], Wignall et al. [413]).

Phosphate glass systems and the wide range of fluorosphosphate glass systems have been used as hosts for Nd^{3+} as laser materials [414–418] because of advantageous compromises between a high gain and a low non-linear index (change of refraction with intensity), which in turn control failure and degradation.

7.6.2 Phase Separation in Pure Phosphate Glasses

Generally, pure phosphate glasses have a much smaller tendency toward phase separation than silicate, borate, silicophosphate, or borophosphate glasses. This may be attributed to the easy accommodation of additives in a chain structure.

This is no longer valid in the case of "anomalous" BeO, MgO, ZnO, Al_2O_3–P_2O_5 systems, in which two significantly different kinds of structural elements compete for volume requirements. The network-like regions formed near Be, Mg, Zn, and Al contrast with high P_2O_5 chain elements. This may underlie Vogel's [151, 187] observation of phase-separation phenomena and subsequent controlled crystallization in MgO–P_2O_5 glasses (see Fig. 4.10). Controlled crystallization involves the formation of a large number of crystals of equal size. This suggests that future studies should include even pure phosphate glasses.

7.7 Tellurite Glasses

Glasses containing TeO_2 as a main constituent possess extreme optical properties, e.g., $n_d = 2.1$ to 2.3, $\nu_d = 15$ to 18. Also, they exhibit good light transmission in the visible and the infrared regions, up to about 7 μm. For this reason, tellurite glasses are important for the design of scientific instruments and thus have become subject of thorough investigation.

7.7.1 Glass-Formation Range and Optical Properties of Tellurite Glasses

Melts of pure TeO_2 do not, solidify to a glass. Whenever glass formation was reported in the literature, generally contamination from the ceramic crucible actually produced a multicomponent glass. A minute quantity of Al_2O_3 suffices to effect glass formation.

As early as 1834, Berzelius [419] recognized that TeO_2 can form a glass with various metal oxides, particularly BaO. Later, Lehners et al. [420] observed glass formation in parts of the systems Na_2O–TeO_2 and K_2O–TeO_2. Since 1952, starting with work by Stanworth [421, 422], systematic investigations have been undertaken by various authors [423–429].

Special tellurite glasses have been studied intensely. In particular Imaoka [396] gave a relatively complete and exact account of the glass formation region of a great number of systems of TeO_2 with other metal oxides. Imaoka and Yamazaki (1968) [430] described the glass formation properties of ternary tellurite glass systems. In 1974, Vogel and coworkers [431] published the glass formation properties of binary tellurite systems as well as optical properties of these glasses. The glass formation limits stated by Imaoka were in essence confirmed and the insignificant deviations observed are obviously due to different melt volumes and their respectively different cooling rates. The binary melt combinations with the classic glass formers B_2O_3, P_2O_5, SiO_2 and GeO_2 as well as a large number of halogen and sulfate containing systems were also described (see Table 7.5 to 7.7). While the accommodation of F is not unusual, that of significant amounts of halides and sulfates is remarkable. Property data (density, n_d, ν_d) in the tables relate to the lower and upper limit of glass formation in each system. These limits differ considerably for the components shown. If halides or sulfates are melted with TeO_2, evaporation is more or less inevitable, so that ternaries of the volatile component, such as $MeCl_2$–Me_2O–TeO_2, result. The compositions listed in the tables represent analytical values. The optical data listed show that a new valuable family of optical glasses has been discovered.

Table 7.5. Binary tellurite glass systems. Density (ρ), refractive index (n_d), and Abbe number (ν_d) of glasses near the lower or upper boundary of the glass-forming region

System	Second Component Mol%	Properties		
		ρ in $g \cdot cm^{-3}$	n_d	ν_d
Li_2O–TeO_2	12.2 (Li_2O)	5.270	2.1080	18.4
	⋮	⋮	⋮	⋮
	34.9	4.511	1.9311	20.1
Na_2O–TeO_2	5.5 (Na_2O)	5.430	2.1483	17.0
	⋮	⋮	⋮	⋮
	37.8	4.143	1.8006	22.2
K_2O–TeO_2	6.5 (K_2O)	5.270	2.1224	17.2
	⋮	⋮	⋮	⋮
	19.5	4.607	1.9554	18.8
Rb_2O–TeO_2	5.6 (BeO)	5.357	2.1095	18.4
	⋮	⋮	⋮	⋮
	21.0	4.725	1.8929	20.1
BeO–TeO_2	16.5 (BeO)	5.357	2.1095	18.4
	⋮	⋮	⋮	⋮
	22.7	5.246	2.0749	19.1
MgO–TeO_2	13.5 (MgO)	5.347	2.1068	—
	⋮	⋮	⋮	⋮
	23.1	5.182	2.048	—
SrO–TeO_2	9.2 (SrO)	5.564	2.1283	17.8
	⋮	⋮	⋮	⋮
	13.1	5.514	2.1171	17.9

Table 7.5. (Continued)

System	Second Component Mol%	Properties		
		ρ in $g \cdot cm^{-3}$	n_d	ν_d
$BaO–TeO_2$	8.0 (BaO)	5.626	2.1556	17.6
	⋮	⋮	⋮	⋮
	23.1	5.533	2.0522	19.2
$ZnO–TeO_2$	17.4 (ZnO)	5.544	2.1245	18.4
	⋮	⋮	⋮	⋮
	37.2	5.466	2.0384	20.7
$Tl_2O–TeO_2$	13.0 (Tl_2O)	6.262	2.1902	15.9
	⋮	⋮	⋮	⋮
	38.4	7.203	2.2800	11.3
$PbO–TeO_2$	13.6 (PbO)	6.054	2.2018	16.5
	⋮	⋮	⋮	⋮
	21.8	6.303	2.2056	16.4
$La_2O_3–TeO_2$	4.0 (La_2O_3)	5.662	2.1596	17.6
	⋮	⋮	⋮	⋮
	9.9	5.707	2.1082	19.5
$Al_2O_3–TeO_2$	7.8 (Al_2O_3)	5.287	2.0817	19.2
	⋮	⋮	⋮	⋮
	16.8	4.850	1.9465	23.1
$TiO_2–TeO_2$	7.8 (TiO_2)	5.580	2.2028	16.0
	⋮	⋮	⋮	⋮
	18.9	5.242	2.2198	14.9
$WO_3–TeO_2$	11.5 (WO_3)	5.782	2.1924	16.7
	⋮	⋮	⋮	⋮
	30.8	5.953	2.1912	16.3
$Nb_2O_5–TeO_2$	3.6 (Nb_2O_5)	5.558	2.1924	16.7
	⋮	⋮	⋮	⋮
	34.4	5.159	2.2113	15.7
$Ta_2O_5–TeO_2$	3.8 (Ta_2O_5)	5.659	2.1903	17.0
	⋮	⋮	⋮	⋮
	33.4	6.048	2.1597	18.0
$ThO_2–TeO_2$	8.4 (ThO_2)	5.762	2.1653	17.7
	⋮	⋮	⋮	⋮
	17.1	5.981	2.1334	19.0

Table 7.6. Binary tellurite glass systems consisting of TeO_2 and the glass formers SiO_2, B_2O_3, P_2O_5 and GeO_2. Density (ρ), refractive index (n_d), and Abbe number (ν_d) of glasses near the lower or upper boundary of glass formation (clear glasses)

System	Second Component Mol%	Properties		
		ρ in $g \cdot cm^{-3}$	n_d	ν_d
GeO_2–TeO_2	10.2 (GeO_2)	5.437	2.1263	22.1
	⋮	⋮	⋮	⋮
	30.2	5.057	2.0112	18.4
P_2O_5–TeO_2	2.2 (P_2O_5)	5.433	2.1487	17.8
	⋮	⋮	⋮	⋮
	15.6	4.785	1.9930	21.9
B_2O_3–TeO_2	11.8 (B_2O_3)	5.205	2.0823	19.2
	⋮	⋮	⋮	⋮
	26.4	4.648	1.9644	21.7
SiO_2–TeO_2	Addition of SiO_2 in mol% to an Al_2O_3–TeO_2 base glass (1:5.2)			
	1.2	4.854	1.9589	22.2
	⋮	⋮	⋮	⋮
	17.9	4.484	1.8822	24.3

Table 7.7. Halide- and sulfate-containing tellurite glass systems. Area of glass formation. Density (ρ), refractive index (n_d), and Abbe number (ν_d)

System	Area of glass metal formation Halide or sulfate in Mol%	Metal Oxide in Mol%	Properties		
			ρ in $g \cdot cm^{-3}$	n_d	ν_d
LiF–TeO_2	—	—	—	—	—
$LiCl$–TeO_2	11.9 ($LiCl$)	4.3 (Li_2O)	5.197	2.1383	17.2
	⋮	⋮	⋮	⋮	⋮
	33.2	11.6	4.318	1.9816	19.2
$LiBr$–TeO_2	5.0 ($LiBr$)	6.5 (Li_2O)	5.363	2.1504	16.5
	⋮	⋮	⋮	⋮	⋮
	6.3	7.9	5.294	2.1373	16.8
Li_2SO_4–TeO_2	—	—	—	—	—
NaF–TeO_2	19.0 (NaF)	—	5.207	2.0578	18.2
	⋮	⋮	⋮	⋮	⋮
	32.0	—	4.901	1.9466	19.9

Table 7.7. (Continued)

System	Area of glass metal formation Halide or sulfate in Mol%	Metal Oxide in Mol%	Properties		
			ρ in $g \cdot cm^{-3}$	n_d	ν_d
$NaCl-TeO_2$	16.6 (NaCl)	0.8 (Na_2O)	5.001	2.0961	17.2
	⋮	⋮	⋮	⋮	⋮
	22.0	1.1	4.802	2.0591	17.3
$NaBr-TeO_2$	8.6 (BaBr)	4.6 (Na_2O)	5.196	2.1205	16.1
	⋮	⋮	⋮	⋮	⋮
	26.4	2.9	4.724	2.0349	16.3
$Na_2SO_4-TeO_2$	—	—	—	—	—
$KF-TeO_2$	16.1 (KF)	—	5.103	2.0516	18.2
	⋮	⋮	⋮	⋮	⋮
	42.0	—	4.241	1.7962	22.8
$KCl-TeO_2$	15.2 (KCl)	0.1 (K_2O)	4.956	2.0876	17.0
	⋮	⋮	⋮	⋮	⋮
	20.0	0.1	4.731	2.0498	17.3
$KBr-TeO_2$	10.8 (KBr)	2.7 (K_2O)	5.056	2.0962	16.5
	⋮	⋮	⋮	⋮	⋮
	23.8	2.5	4.594	2.0076	16.5
$K_2SO_4-TeO_2$	3.4 (K_2SO_4)	0.6 (K_2O)	5.283	2.1483	17.0
	⋮	⋮	⋮	⋮	⋮
	6.8	1.1	4.931	2.0793	17.5
$RbF-TeO_2$	12.2 (RbF)	2.5 (Rb_2O)	5.199	2.0018	18.3
	⋮	⋮	⋮	⋮	⋮
	23.9	1.4	4.952	1.8693	19.7
$RbCl-TeO_2$	21.9 (RbBr)	0.1 (Rb_2O)	4.777	2.0185	17.5
	⋮	⋮	⋮	⋮	⋮
	27.4	0.3	4.589	1.9646	18.1
$RbBr-TeO_2$	9.8 (RbBr)	2.1 (Rb_2O)	5.156	2.0988	16.5
	⋮	⋮	⋮	⋮	⋮
	18.0	0.5	5.000	2.0617	16.2
$Rb_2SO_4-TeO_2$	3.3 (Rb_2SO_4)	0.9 (Rb_2O)	5.292	2.1149	17.4
	⋮	⋮	⋮	⋮	⋮
	4.9	1.3	5.118	2.0812	17.7
BeF_2-TeO_2	—	—	—	—	—
$BeCl_2-TeO_2$	—	—	—	—	—
$BeBr_2-TeO_2$	—	—	—	—	—

Table 7.7. (Continued)

System	Area of glass metal formation Halide or sulfate in Mol%	Metal Oxide in Mol%	Properties		
			ρ in $g \cdot cm^{-3}$	n_d	ν_d
$BeSO_4$–TeO_2	3.8 ($BeSO_4$) … 17.2	3.7 (BeO) … 6.8	5.303 … 4.487	2.1007 … 1.8996	18.7 … 23.7
MgF_2–TeO_2	—	—	—	—	—
$MgCl_2$–TeO_2	5.0 ($MgCl_2$) … 15.7	1.2 (MgO) … 9.0	5.255 … 4.543	2.1227 … 1.9814	17.6 … 19.7
$MgBr_2$–TeO_2	3.0 ($MgBr_2$) … 8.4	3.4 (MgO) … 1.7	5.473 … 5.209	2.1468 … 2.1070	16.8 … 17.2
$MgSO_4$–TeO_2	6.7 ($MgSO_4$) … 43.5	4.1 (MgO) … 17.2	5.051 … 3.432	2.0245 … 1.6419	19.0 … 43.1
CaF_2–TeO_2	—	—	—	—	—
$CaCl_2$–TeO_2	6.6 ($CaCl_2$) … 15.5	0.8 (CaO) … 4.5	5.221 … 4.801	2.1402 … 2.0634	17.3 … 17.9
$CaBr_2$–TeO_2	3.2 ($CaBr_2$) … 4.8	2.9 (CaO) … 5.2	5.383 … 5.323	2.1652 … 2.1490	16.6 … 16.7
$CaSO_4$–TeO_2	—	—	—	—	—
SrF_2–TeO_2	—	—	—	—	—
$SrCl_2$–TeO_2	7.2 ($SrCl_2$) … 46.7	1.4. (SrO) … 9.1	5.271 … 3.966	2.1317 … 1.8326	17.3 … 26.5
$SrBr_2$–TeO_2	4.5 ($SrBr_2$) … 11.1	2.7 (SrO) … 10.1	5.420 … 5.155	2.1536 … 2.0627	16.5 … 17.1
$SrSO_4$–TeO_2	—	—	—	—	—

7.7.2 Structure of Tellurite Glasses

Brady [432, 433] published the first on the proposals of tellurite glasses. They are in essence based on the brookite-like structure determined by Ito and Sawada [434] for crystalline rhombic β-TeO_2 (tellurite). Tellurite glasses should accordingly consist of a network of strongly deformed [TeO_6] building units. This opinion, which was also shared by Jakhkind [435, 436], contradicts the network theory according to which the formation of a three dimensional disordered glass network should only be possible with coordination numbers of 3 or 4 for the network-former cations. As was later discovered, in particular on the basis of Zemann's work [437–439], the assumption of [TeO_6] polyhedra in tellurite glass in analogy to the structure of the crystalline β-TeO_2 (tellurite) was based on an incorrect structure determination of this compound. Zemann proved that the Te^{4+} ion is in coordination number 4 in crystalline rhombic β-TeO_2 and in coordination numbers 3, (3 + 1) and 4 in tellurite minerals.

Two extreme cases are known for the coordination of oxygen around the Te^{4+} cation in both forms of crystalline TeO_2, the rhombic or β-TeO_2 (tellurite) and the tetragonal or α-TeO_2 (paratellurite):

a) coordination of 3 in form of a trigonal pyramid. The Te–O distance is about 195 pm and the Te–O–Te bond angle at about 95°. The free electron pair lies in the tetrahedon gap not occupied by oxygen.
b) coordination of 4 in the form of a trigonal bipyramid (see Fig. 7.47) in which the corner of the polyhedron unoccupied by oxygen is filled by a free electron pair of tellurium. The average Te–O–Te distance is about 200 pm. Deviations from the "ideal pyramid" mainly arise from deviations of the Te–O–Te bond angle of up to 30° from 180° and of up to 20° in the equatorial plane from the ideal angle of 120°. Recent views on the structure of tellurite glasses [440, 441] derive from the para-tellurite structure (see Fig. 7.48) in which the [TeO_4] polyhedra are exclusively connected at the corners. This theoretically allows the formation of a three-dimensional network in glass, but does not seem to take place on quick cooling of a pure TeO_2 melt. The Te^{4+} ion's field strength (Z/a) is about 1.0. This suggests that TeO_2 is to be classified as

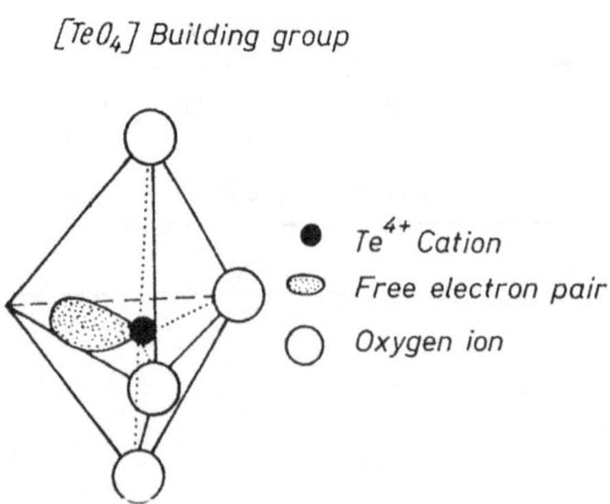

Fig. 7.47. Structural unit of crystalline α-TeO_2 (paratellurite) and β-TeO_2 (tellurite) with tellurium ions in 4-coordination

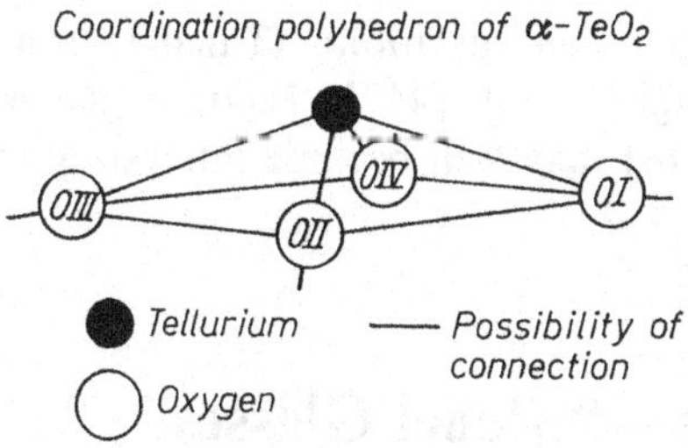

Fig. 7.48. Structural unit of tellurite glasses based on Fig. 7.47 with distances and bond angles.

bond angle:	distance:
OII-Te-OIV 90.9°	Te-OII 191 pm
OI-Te-OIII 162.6°	Te-OIV 191 pm
OII-Te-OIII 98.7°	Te-OIII 209 pm
OIII-Te-OIV 86.7°	Te-OI 209 pm

an intermediate oxide (like Al) according to Dietzel. i.e. not capable of glass formation on this own.

In linking combinations of TeO_2 with network modifying oxides the linking of $[TeO_4]$ polyhedra to chain structures is made responsible for the strong tendency to glass formation of these melts. These chain structures are analogous to the formation of chain-building units in crystalline di- or tritellurites (for example in denningite [(Mg, Ca, Zn) Te_2O_5] or in zinc tellurite [$Zn_2Te_3O_8$]).

The analogy with the structure of basic phosphate also fits glasses in some sense the formation of chain structures in multicomponent tellurite glasses. A similarly loose structure of entangled chain-building units in phosphate glasses was considered to be the cause of the relatively low tendency to immiscibility of pure phosphate glasses. The same observation can be made on pure tellurite glasses. Obviously such a loose structure of tellurite glasses also presupposes the possibility that a large portion of halogenides, and sulfates can be introduced without hindering glass formation. The incorporation of fluorine ions however, differs; oxygen ions by simply replaced by fluorine ions. As for the case of halogenides and sulfates, many questions remain unanswered and a wide field of future structural analyses remains open.

The immiscibility behavior of tellurite glasses changes drastically for combinations of TeO_2 with B_2O_3 and GeO_2. Such combinations show strong immiscibility (TeO_2–B_2O_3, TeO_2–GeO_2, see Fig. 6.4) due to the very different volume demands of the borate and germanate phases in comparison to those of the tellurite glass phase. In the borate and germanate phases, the three dimensional network structure elements are considerably denser than the chain building units in tellurite glass.

Strong immiscibility is, however, not observed in melt combinations of TeO_2 with P_2O_5 and also in phospho-tellurite glasses with additional oxides as network modifiers. Since phosphate and tellurite glasses contain primarily chain building units whose volume demands should be relatively similar, good miscibility ensues. The resulting glasses have a relatively homogeneous structure. Tellurite glasses also represent a possible compromise between the excellent host function of zirconium fluoride glasses with their low phonon energy (large cation, weak anion) for Erbium in up-conversion (for instance conversion of IR to visible laser light) and the better chemical and mechanical properties of oxide glasses. Acceptable non-radiative loss in oxide glasses requires high atomic weight and relatively low valency (oxygen concentration). (Tanabe et al. [442, 443], Hirao et al. [444–446], Yanagita et al. [447]. Tellurite glasses containing halides (Kozhukharov et al. [448] may well become interesting in this connection.

7.8 Beryllium Fluoride Glasses – "Model Glasses"

7.8.1 Theoretical Discussion of "Model Glasses"

Why "model glass" studies? A principal reason is the very high temperature ($\gg 1000\,°C$) required to melt conventional silicate glasses and the problems involved in accurate temperature control is that range.

It is well known that compounds such as molten resins, alkaloids, molten sugar, and molten phenolphthalein, with their giant molecules (molecular mass $\gg 5000$), when cooled yield glasses easily due to steric hindrance. From their behavior during melting and solidification, it is possible to draw, by analogy, conclusions for solidifying silicate melts which are also constructed from giant molecules in the form of infinite 3-dimensional disordered networks. This approach was taken by Tammann [11].

A second and more promising approach was proposed by Goldschmidt [28]. As expounded before, the constitution of a crystal lattice or any structure is, in his opinion, primarily dependent on the size ratio of the ions. In other words: Two compounds, constituted of different ions of well-comparable radii, will have the same lattice or structure, respectively.

It was in this connection that Goldschmidt [28] first coined the term "model properties." The characteristic of a "model" is the enlarged or reduced representation of a structure or a process to scale. In crystal chemistry, the term "model" is used if two substances crystallize in the same lattice and their ions have equal or very similar radii, but different valencies. Depending on whether the lattice forces of the model compound are larger or smaller than in the original compound, one speaks of a reinforced or a weakened model. The relationship "weakened model" to original exists between BeF_2 and SiO_2.

The radii of Be and Si, as well as of F and O ions, are similar. In both cases the valency differs by a factor of 2. For this reason both crystalline and vitreous BeF_2 may be regarded as doubly weakened models of the corresponding forms of SiO_2. The bonding forces between the ions, approximately calculated from the Coulomb relation, $P = Z_c Z_a e^2/a^2$, where Z_c = valency of cation, Z_a = valency of anion, a = distance cation-anion = $r_c + r_a$, e = elemental charge) are, for beryllium and fluorine in BeF_2, only one-fourth of those for silicon and oxygen in SiO_2.

Such model relations exist not only between the simple compounds and glasses BeF_2 and SiO_2, but also between complex compounds (see Table 7.8) and, conceivably, between such glasses.

Table 7.8. Model relations between radius ratios and simple and complex compounds

Model	Ion radiation (in Å) (Pauling)		Original
	Model	Original	
—	F^- 1.36	O^{2-} 1.40	—
BeF_2	Be^{2+} 0.31	Si^{4+} 0.41	SiO_2
LiF $LiBeF_3$ Li_2BeF_4	Li^+ 0.60	Mg^{2+} 0.65	MgO $MgSiO_3$ (Clinoenstatite) Mg_2SiO_4 (Forsterite)
NaF $NaBeF_3$ Na_2BeF_4 $NaLBeF_4$	Na^+ 0.95	Ca^{2+} 0.99	CaO $CaSiO_3$ (Pseudowollastonite) Ca_2SiO_4 (Olivine) $CaMgSiO_4$ (Monticellite)
MgF_2	Mg^{2+} 0.65	Ti^{4+} 0.65	TiO_2
CaF_2	Ca^{2+} 0.99	Th^{4+} 1.10	ThO_2

It was Goldschmidt [28] who discovered that BeF_2 melts solidify as glasses more easily than SiO_2 melts, and prepared and described them. His work is the basis for later studies of crystalline [449–459] and vitreous models of silicates. Brandenberger [460] had established the β-cristobalite structure of crystalline BeF_2. Warren and Hill [461] showed by X-ray diffraction that BeF_2 glass possesses a network structure analogous to that of SiO_2 glass. Heyne [462] described the preparation and, particularly, the IR and UV transmission of some BeF_2 glasses. Later, particularly Izumitani and Terai [463], Imaoka and Mizusawa [464], and Imaoka [465] investigated additional BeF_2 glass systems.

The weaker binding forces in a model glass cause a considerable decrease in property values such as those for the melting range, viscosity, chemical durability, and hardness. It may be expected that other phenomena related to structure can also be clarified on the basis of the concept of model glasses. It is known that the addition of fluorides facilitates the melting of silicates and borates. Evidently small additions of fluorides totally or partially substitute fluorine for oxygen in SiO_4 or BO_4 tetrahedra or BO_{3-} triangles. One might speak of a singly weakened model structure. Si–O and B–O bonds have been changed to Si–F and B–F bonds, reduced to one-half of the Si–O or B–O values, according to the approximate calculation given above. This causes melting and softening temperatures to drop with increasing fluorine content.

To complete the discussion of model glasses, Goldschmidt's [28] original opinion should be quoted, according to which perhaps also "strengthened"

Table 7.9. Original model bonding

Original	rModel	Ionic Bond	Relation of binding forces model: original $P_{Med}:P_{Org}$
	$BeF_2\,[BeF_4]^{2-}$	Be–F	1:4 weakened
	$SiF_4\,[SiF_6]^{2-}$	Si–F	1:2 weakened
	$BF_3\,[BF_4]^{1-}$	B–F	1:2 weakened
SiO_2		Si–O	
B_2O_3		B–O	
	BN	B–N	3:2 reinforced
	SiC	SiC	2:1 reinforced

model glasses should be possible, such as dinitrides ($M_3^{II}N_2$) or complex carbides (Table 7.9).

According to Goldschmidt, theoretically possible and practically available glasses could be classified as follows:

(1) Monovalent (anion) glasses: fluoride glasses.

(2) Divalent (anion) glasses: silicate, borate, phosphate, germanate, arsenate, sulfide, etc. glasses.

(3) Trivalent (anion) glasses: nitride glasses.

(4) Tetravalent (anion) glasses: carbide glasses.

"Strengthened model" (nitride, carbide) glasses should be expected to have valuable optical, chemical, and mechanical properties. They should be at least of scientific interest, and the required very high melting temperatures no longer present an insumountable problem.

That Goldschmidt's concept was not entirely esoteric might be demonstrated by the recent verification of the existence of amorphous carbides in steels, metal-glass films, and last but not least, Sialon glasses in which oxygen is partly replaced by nitrogen and Si by Al, as well as related oxygen-nitrogen glasses (see for example, the review by Löhmann [466]).

7.8.2 Ranges of Glass Formation in BeF_2 Model Systems. Properties of These Glasses (Vogel et al. [181–183, 204, 467])

Alkali-beryllium fluoride glasses represent, according to the preceding discussion, doubly weakened models of the respective alkaline earth melts. In contrast to silicate glasses, where one calculates about 50% homopolar bonding, in beryllium fluoride glasses homopolar bonding attains only about 21% (Mackenzie [35]). Consequently, their viscosity is much lower, and all transport processes (diffusion, crystallization, etc.) are much accelerated. This facilitates the study of typical behavior. Table 7.10 and Figs. 7.49, and 7.50 show the

Table 7.10. Glass forming areas in binary BeF_2 systems

System	MF or $M^{II}F_2$ content in mol%	
	Turbid area	Clear area
$LiF-BeF_2$	1 ... 36	36 ... 56.5
$NaF-BeF_2$	1 ... 27.4	27.4 ... 57.8
$KF-BeF_2$	1 ... 17.4	17.4 ... 51.3
$RbF-BeF_2$	1 ... 7.6	7.6 ... 33.9
MgF_2-BeF_2	... 29.5	Only after extremely rapid supercooling
CaF_2-BeF_2	... 27.5	
SrF_2-BeF_2	... 5.7	
BaF_2-BeF_2	No glass formation	
PbF_2-BeF_2	5 ... 10.1	0 ... 5
	Up to 30 mol% PbF_2 one phase is still vitreous, the other already completely crystallized	

ranges of glass formation in binary alkali and alkaline earth beryllium fluoride as well as in ternary alkali alkaline earth beryllium fluoride systems. In this connection, one should recall the ability of BeF_4 building units to form 3-dimensional networks (Fig. 4.2).

7.8.2.1 Density Plots

Plots of density and refraction for binary alkali-beryllium fluoride glasses (Fig. 7.45) show surprising features. One would expect that the introduction of cations of varying atomic weight would lead to increased density in the series Li^+, Na^+, K^+, Rb^+. In fact, however, the Na and K curves are nearly identical within experimental error, and extrapolation beyond 50 mol% would even lead to a crossover. But alkali silicate glasses show an analogous behavior offering an analogous explanation (Dietzel [320], Dietzel and Sheybany [319], Fig. 7.7). According to the calculations of these authors all Li^+ ions can be accommodated in the cavities of the SiO_4 tetrahedron network without demanding any additional volume. It is only the oxygen introduced with the Li^+ which causes an increase in volume as nonbridging oxygen is introduced. As a matter of fact, in a glass $Li_2O \cdot 3SiO_2$ there is room for 125% of the Li^+ ions introduced. This implies a shrinkage of the network. The Na^+ ions can be accommodated to 70%, the K^+ ions only to 27%, according to these calculations.

It is due to the difference in volume requirement and expansion, in combination with the relatively small ionic density of K as compared to Na (almost all elements of the argon series (19 to 35) have abnormally low ionic densities), that the Na and K plots coincide, even cross over. This interpretation also holds for alkali-beryllium fluoride glasses. The crossover occurs at about 38 mol% M_2O for silicate and at above 50 mol% MF for fluoride systems, evidently because in fluoride systems only one M per mole is introduced.

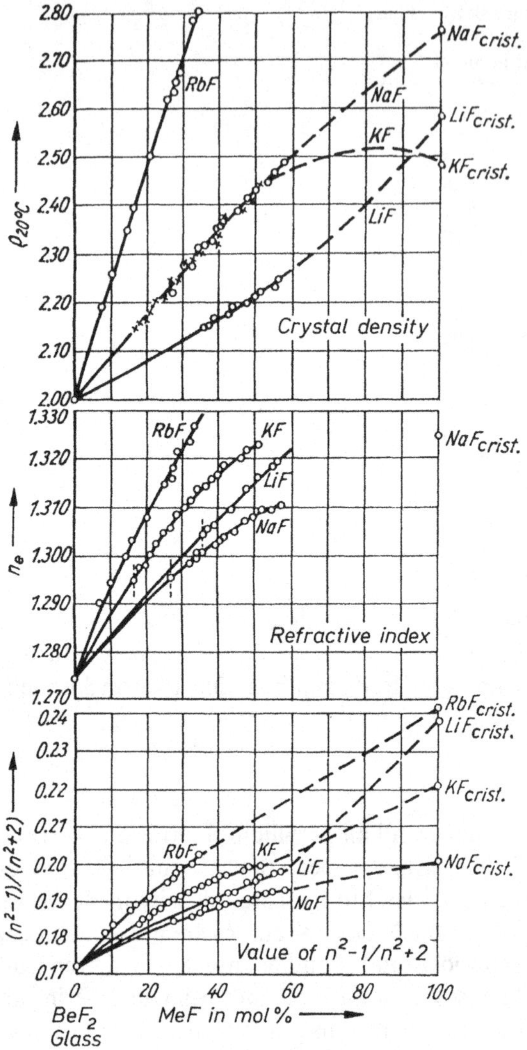

Fig. 7.49. Density, refractive index, and value of $n^2 - 1/(n^2 + 2)$ alkali beryllium fluoride glasses

7.8.2.2 Refractive Index Plots

The plots for the refractive index in the binary alkali-beryllium fluoride (Fig. 7.49) also do not show the expected relationship. The increase occurs in the order Na-Li-K-Rb. There is a discrepancy when the data are compared with those for density. The lithium glasses have the lowest density but a higher refraction than the sodium glasses. This, too, may be interpreted on the basis of Dietzel's concept about the insertion of alkali.

Refraction depends primarily on the total electron density. Density, however, depends on the mass per unit volume. If the insertion of cations causes a shrinkage of the network, the electron density may be increased even if the

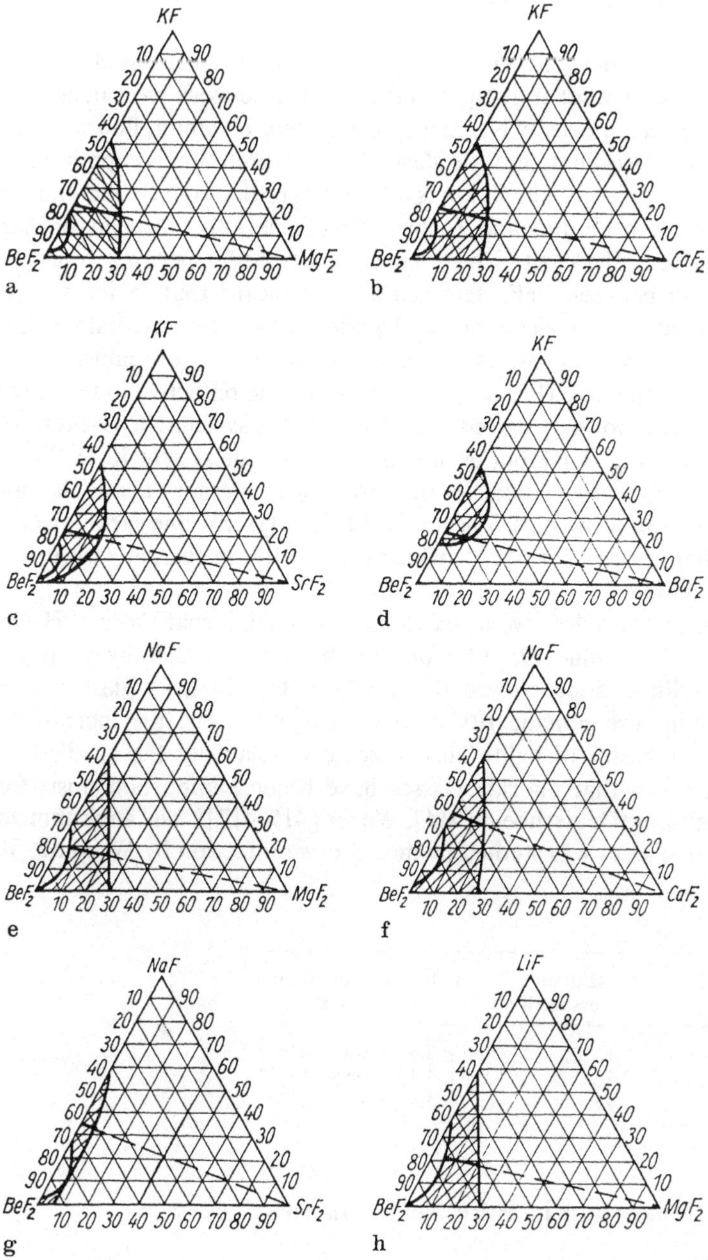

Fig. 7.50. Glass-forming areas (cross-hatched) in the ternary melting systems alkali-alkaline earth-BeF_2 (mol%)

cation itself has a smaller electron number than a cation causing no, or a smaller, shrinkage.

Introducing Li in BeF_2 glass causes the specific plus in electrons introduced by Na to be overcompensated by the shrinkage of the network so that, as a net result, the total electron density is larger in the lithium-beryllium fluoride glass than in the sodium-beryllium fluoride glass. This interpretation is supported by the plots for $n^2 - 1/n^2 + 2$. The Si ion is a little larger than the Be ion and the O ion is, according to Pauling and Zachariasen (Table 7.8), a little larger than the F ion. Therefore, the cavities between SiO_4 tetrahedra per mole must be larger than those between BeF_4 tetrahedra. This means that in the silicate system more Li ions can be accommodated without an increase in total volume, and also that the network contraction would become more noticeable.

This is indeed the case (Fig. 7.8). The curve for the refractive index in the system Li_2O–SiO_2 is not only placed above that for the system Na_2O–SiO_2 (cf. system NaF–BeF_2, Fig. 7.45), but even above that for the system K_2O–SiO_2.

According to extensive studies of the properties of ternary alkali-alkaline earth-beryllium fluoride systems (Vogel et al. [183, 204]), alkaline earth ions are network-modifiers, except for Mg^{2+}, which participates in the network in the form of MgF_4 units.

The extremely low refractive index values (eq. 33) (i.e., near those of H_2O!) and extremely high ν_D values (95–105) form the basis of a new family of optical glasses, the beryllium fluorocrowns (BeFK) (Fig. 1.8). In laser glasses, it is important to minimize nonlinearity of refractive index, i.e., the increase of refraction with intensity of light. This increase is relatively the smallest in low-index glasses, so that fluoride glasses have become important hosts for Nd^{3-} in laser glasses (Deutschbein [414], Weber [415–417]). The environment of Nd^{3-} in these glasses was studied by means of n diffraction by Wright et al. [468].

	Maximum cross section	Index	Non-linear index
BeF_2 glasses	4	< 1.4	0.3×10^{-13}
Fluorophosphate glasses	4.3	1.4	0.55×10^{-13}
Phosphate glasses	4.8	1.5	1.2×10^{-13}

7.8.3 Phase Separation in Pure BeF_2 Glasses

Turbidity, occasionally observed during systematic studies of BeF_2 glasses (Vogel and Gerth [181–183, 204, 467], Heine [462],) had first been attributed (Heine [462]) to small impurities of BeO, or devitrification (fine crystalline BeF_2). Extensive electron-optical investigations, however, showed that the phenomenon was due to microphase separation. Since the droplet-shaped immiscibility regions could be as large as 50 to 100 nm (500 to 1000 Å), the dependence of phase separation on various factors such as concentration of specific components, cooling rate, etc., was easy to follow. In the further course of the investigation it was found that even glasses free of Tyndall scattering contained immiscibility regions, now, however, only of 10–20 nm (100–200 Å) diameter.

Simultaneously a more elaborate preparation technique for the electron microscope also ensures the success of the transition to the investigation of original glasses (silicate, borate and borosilicate glasses among others). The dimensions of the immiscibility regions in these glasses are about one order of magnitude smaller than in BeF_2 model glasses. The observations with original glasses would have been seen as due to impurity effects had it not been for the knowledge acquired with the preceding model glass investigations, namely that the droplet regions are typical structure elements of glasses. This laid the foundation for world wide phase separation studies on glasses. These studies became one of the main fields in glass research and quite decisively influenced new developments in glass or glass-ceramics over the last 20 years.

Although the separated phases in the original oxide glasses were usually one order of magnitude smaller than those in the model fluoride glasses, the immiscibility gaps were almost exactly the same, e.g., MgO–SiO_2 like LiF–BeF_2, CaO–SiO_2 like NaF–BeF_2, SrO–SiO_2 like KF–BeF_2 (Fig. 7.9). The influence of the different field strengths of network-modifiers (alkali ions) could be demonstrated clearly for the first time (Fig. 7.51).

Today it is well established that the alkali fluoride-beryllium fluoride systems have asymmetrical immiscibility gaps extending from 0 to 50 mol% MF (= LiF, NaF, KF). The BeF_2 side of the gap is very steep. The change in function (matrix to droplet) has been ascertained. The old pictures already show the envelopes around the droplets which are only now understood. At that time they were attributed to faults in the technique.

One has to consider, though, that at that time the main thrust was to prove the reality of immiscibility phenomena, then generally doubted. Kreidl hence expressed this situation as follows: "Who talked about droplet-shaped micro-phases in glasses 20 years ago was burnt at the stake as a heretic: who now does not consider them, will be relegated to the bottom row."

7.8.4 Fluoride Glasses Free of Beryllium

Because of the toxicity of beryllium, much effort soon was spent on developing fluoride glasses free of beryllium (e.g., Heyne [462], Sun [469]). In early glasses of this type, AlF_3 was an important constituent. An example is 40 AlF_3, 24 PbF_2, 12 SrF_2, 24 MgF_2 (Sun [469]). It should be noted that neither AlF_3 nor NaF, KF, LiF fit the weakened model concept. AlF_3 contributes AlF_4 groups, allowing the novel combination $AlLiF_4$ to substitute for $2BeF_2$ (Krylova [470]). Excellent and extensive reviews of the development and study of halide glasses are by Baldwin et al. [471] and by Poulain [472].

More recently, systems based on ZrF_4 (HfF_4) and ThF_4 as major constituents have been investigated as candidate materials for infrared optics materials,

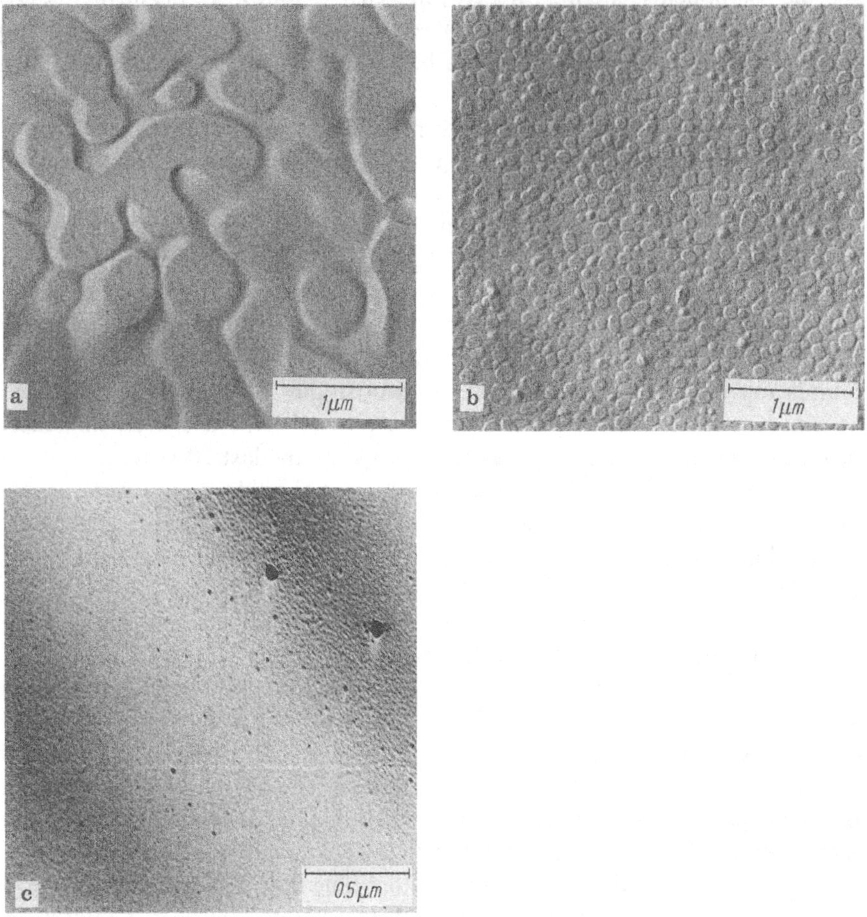

Fig. 7.51. Binary alkali-BeF_2 glasses with equal concentration, but different field strength of alkali (electron micrographs after replica preparation). **a** 15 mol% NaF; **b** 15 mol% KF; **c** 15 mol% RbF. It is clearly recognizable that, after equal thermal history, the dimensions of separated phases decrease greatly with decreasing field strength of alkali ions (droplet dimension decreases from 1000 nm (10 000 Å) for NaF to 25 nm (250 Å) for RbF). The NaF and KF glasses are turbid, the RbF glass is Tyndall-free

particulaly low-loss (1.10^{-3} $dB \cdot km^{-1}$) communication fibers. They are based on an accidental discovery during work on crystal growth in 1974 at the University of Rennes and reported by Poulain et al. [472]. Since 1980, more than 100 publications have followed this discovery (e.g., Lecocq et al. [473], Takahashi et al. [474], Drexhage et al. [475], Bendow et al. [476]). In 1983 alone, two major conferences were devoted to a large extent to these and related halide glasses [477, 478]. Further improvements for IR optics were achieved in rare earth fluoride glasses (Fonteneau et al. [479, 480] and Drexhage et al. [481]). They contain Zn, Th, Lu, Gd, La, and Bi. Lu increases glass formation but the cost is prohibitive; T_g is as high as 390 °C. Glass formation is extended most by small amounts of sodium, with T_g still remaining around 350 °C, but aging effects might be induced by Na. Abbe values are unusual. These glasses were considered useful for clad optical fibers (Drexhage et al. [481]).

Structures of $Zr(Hf)F_4$ glasses may be dominated by ZrF_x networks, but the larger Th is more likely randomly distributed in a packing of F^- ions, a model more reminiscent of ionic and metal glasses (Poulain [472]).

Fabrication of these glasses (Poulain [472], Maze [482]) involves melting and casting (a) of fluoride mixtures in glassy carbon, platinum, gold, or Pt–Au crucibles; or (b) of oxides with $NH_4F \cdot HF$, preferably in a controlled atmosphere within an SiO_2 envelope, fining to remove most of the OH, fiberizing from a rod in a brass mold near T_g, and cladding.

At this writing the main problems remaining before major utilization of such heavy metal fluoride fibers are: purity (to achieve theoretical low loss), aging, mechanical strength, and corrosion resistance. The potential usefulness of heavy metal fluoride glass in communication fibers has been assessed by Sigel and Tran [483, 484].

An up-to-date review on details is by Lucas [485]. Because of both contributing to an absorption edge at higher (by ≈ 1 μm) IR wavelength and to glass formation InF_3 has developed as a desirable component of complex (up to 7 component) fluoride glasses, replacing ZrF_4. (Poulain [486, 487]).

Since Rayleigh scattering decreases with the fourth power of wavelength the IR transmission of heavy cation fluoride glasses had promised losses in communication fibers as low as 0.001 db Km^{-1}. Yet by 1992 because of other than Rayleigh scattering mechanics, losses (e.g. concentration fluctuation (Izumitani et al. [488] inclusions etc.) remained near 0.5 db Km^{-1} comparable to those in otherwise preferable SiO_2 glass. And expectations are no better than 0.035. This situation limits application to sensors where short fiber length requires proper IR transmission without excessive demands on low loss. (Poulain [489].

Heavy cation fluoride glasses have become the leading candidate hosts for optically active (upconversion, amplifying) rare earth (especially erbium) doped glasses. (Hirao et al. [444–446] since their low phonon energy (heavy cation, weak anion) minimizes non-radiative losses.

Tellurium halide glasses are found to transmit to 12 μm where they have low losses. (Zhang et al. [490].

7.8.5 Fluorophosphate Glasses

The combination of fluoride and phosphate systems has further extended the families of optical glasses. Figure 1.8 locates the fluorophosphate crowns (FPK) and heavy crowns (FPSK) in the $n : \nu$ diagram.

In laser glasses, fluorophosphate glasses represent favorable compromises between the requirements of low non-linear refractive index (fluoride glasses) and high cross section (phosphate glasses).

Commercial fluorophosphate laser glasses are LG 812 (Schott), LHG 10 (Hoya). A typical composition is (mol%) 35 $Al(PO_3)_3$, 32 AlF_3, 12 NaF, 8 MgF_2, 27 CaF_2, 8 SrF_2, 8 BaF_2 (Stokovski et al. [491]).

Small amounts of phosphates stabilize extreme optical fluoride glasses against crystallization. Thus, the optical glass with the most advantageous anomalous partial dispersion FK54 contains (mass%) (cations as oxides) 48 F, 9.8 P_2O_5, 22.0 CaO, 4.0 Na_2O, 3.0 K_2O, 20.0 Al_2O_3, 3.9 MgO, 7.6 SrO, 2.5 La_2O_3 ($n_d = 1.437$, $\nu_d = 90.7$).

More recently, further progress allowed researchers to attain, and even surpass, the optical properties of crystalline CaF_2 lenses, minimizing phosphate additions (Vogel et al. [492–494]. The new glasses are based on near-eutectic compositions in the ternary AlF_3–CaF_2–SrF_2 with some MgF_2 additions.

7.9 Zirconium Fluoride Glasses

7.9.1 Glass Formation, Structure and Properties

Glass formation in binary or multiple component systems with ZrF_4 as a main component were discovered by Poulain and coworkers in 1975 [495, 496]. Whereas pure ZrF_4, whose structure was resolved by Lucas [497, 498] (see Fig. 7.52), is not capable of glass formation, glass formation is observed in combinations with other heavy metal fluorides, such as BaF_2, LaF_3, PbF_2 or ThF_4. The zirconium ion posesses the coordination number 8 in crystalline ZrF_4. ZrF_8 building units should be linked by the corner fluorine ions in a three dimensional network.

Melts of the binary system ZrF_4–BaF_2 (BaF_2 = 25–40 mol%) solidify as glasses with rapid cooling. Melts of the system ZrF_4–BaF_2–LaF_3 (see Fig. 7.53) experience even easier glass formation. In the zirconium fluoride glasses, the zirconium ion should possess the coordination number 6, 7, 8 and more rarely 9. This simultaneous existence of several coordination numbers enhances glass formation. This effect resembles the enhancement of "invert glasses" (see Section 7.3). The simultaneous presence of a larger number of network modifiers.

Fig. 7.52. Structure of crystalline ZrF_4 (Lucas [497, 498] Network of ZrF_8 polyhedra

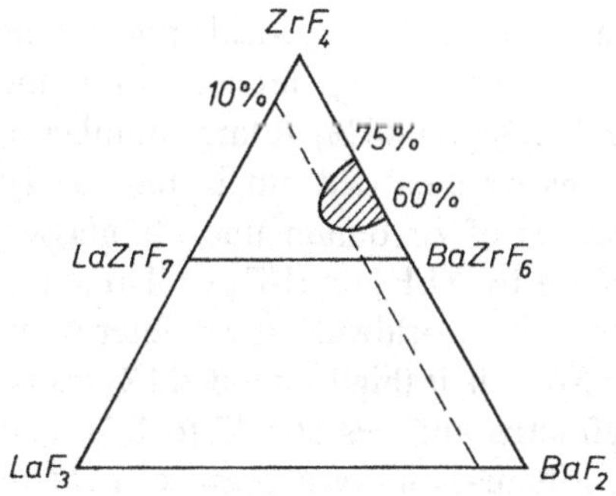

Fig. 7.53. System ZrF_4–BaF_2–LaF_3 with glass formation area (Lucas [497, 498])

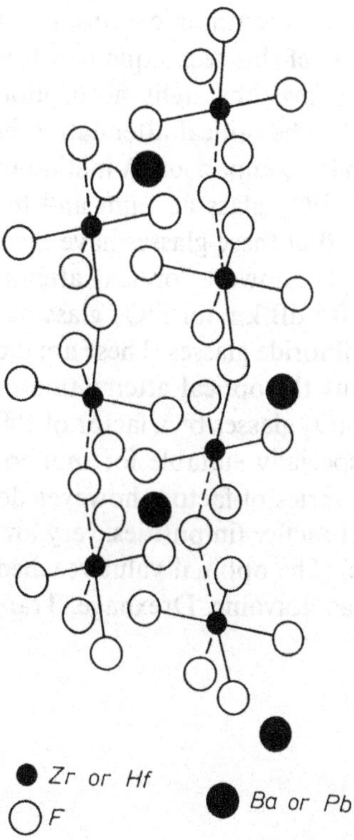

Fig. 7.54. Chain structure of Ba or Pb–Zr fluoride glasses (Almeida and Mackenzie). Two fluoride ions bridges are formed from one $[ZrF_6]$ unit to its neighbor. Similar glasses and structures should also be obtained with a complete substitution of the Zr ions by Hf ions. The represented structure corresponds to dizirconate or dihafnate glass

No unique opinion prevails concerning the structure of zirconium fluoride glasses. According to Lucas, an irregular three-dimensional network with inserted network modifiers is formed primarily with ZrF_7 or ZrF_8 polyhedra at the edges or corners. Almeida and Mackenzie [499] based their opinion on infrared and Raman investigations and believe that for example a Barium-zirconium fluoride glass should possess a chain structure (see Fig. 7.54), in which ZrF_6 octahedra are connected to chains by edges. At any rate, with this latter opinion the fact that no phase separation has yet been found in zirconium fluoride glasses is understandable, whereas for example for beryllium fluoride glasses which have a network structure, phase separation has clearly been observed. The most stable and more interesting glasses for practical applications come from the system ZrF_4–BaF_2–LaF_3 (see Fig. 7.53), but also from the composition 56 ZrF_4, 34 BaF_2, 6 LaF_3 and 4AlF_3 (in mol%). A large number of investigations and composition modifications were carried out in the last 10 years because of the special optical properties of zirconium fluoride glasses. ZrF_4 was at times partially or totally replaced by YbF_3 or HfF_4 or ThF_4 and ZnF_2, InF_3, BiF_3 among others followed. The worldwide great interest in zirconium fluoride glasses primarily results from their (high extended IR) transparency which reaches up to 8 μm with refractive indexes of 1.47 to 1.53 and more. In particular the very low "optical attenuation" of a beam of light in dB/km observed with these glasses raises considerable interest. Fibers of these glasses are especially suitable as wave guides in modern communications. Light impulses rather than electrical ones are at the basis of this technique which is now possible over long distances. Figure 7.55 shows the light absorption coefficient as a function of the wavelength in cm^{-1}. The optical attenuation in decibels per kilometer of a pure SiO_2 glass is shown for comparison. A minimum in light absorption at 10^{-6} cm^{-1} is observed for SiO_2 glass at 1 μm and for fluoride glasses at about 2.5 to 3.5 μm. This means that these glasses have their highest light transparency at these wavelengths. The lowest "optical attenuation" of a beam of light at these minima is about 0.5 dB/km for SiO_2 glass but 10^{-2} to 10^{-3} dB/km for the previously mentioned fluoride glasses. These are the theoretical minimal values of the attenuation. Thus the optical attenuation of fluoride glasses is theoretically lower than that of SiO_2 glasses by a factor of 100 to 1000 therefore making fluoride glass fibers especially suitable for modern comunications technology over long distances. A series of factors however do not allow these theoretical values to be reached in practice (impurities, very low water absorption of the glass fibers among others). The optimal value reached until now was 0.7 dB/km at 2.55 m. Poulain, Lucas, Ravaine, Drexhage, Tran,

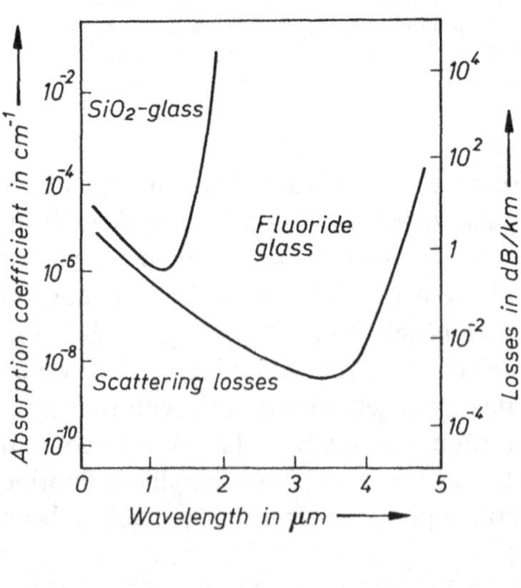

Fig. 7.55. Light permeability and optical damping dependent on the wavelengths in SiO_2 and fluoride glasses

Bray, Moynihan, Almeida, Mackensie, Miyashita and Manabe as well as a large number of other scientists [495–503] have done some pioneering work in this field.

7.10 Germanate Glasses

7.10.1 Glass Formation from GeO_2 and Germanate Melts

The vitreous solidification of GeO_2 melts was first reported by Dennis and Laubengayer [504]. It represents no surprise because of the close relationship of Ge and Si. Yet a comparison of silicate and germanate glasses yields a series of significant differences and peculiarities.

Glass formation in binary systems of GeO_2 with network-modifying as well as intermediate oxides occurs in relatively wide ranges (Table 7.11) (Imaoka [396], Riebling [505], and Murthy and Aguayo [506].

Under the conditions of Imaoka's experiments (1 to 3 g melts), visible immiscibility gaps were observed in the systems $CaO–GeO_2$ (0–15.5 mol% CaO), $SrO–GeO_2$ (0–14 mol% SrO), and $BaO–GeO_2$ (10–17.5 mol% BaO). This represents a certain analogy to silicate systems, although the lower field strength of Ge^{4+} ($F = 1.45$ compared to 1.57 of Si^{4+}) (Table 4.2) would lead one to expect a smaller tendency toward phase separation. To date, no systematic electron microscopy studies of phase separation have been made on germanate glasses, but even in germanate glasses containing alkali, zinc, lead, etc., cation microphase separation should be expected (Etrop'ev et al. [507]).

Relatively wide areas of glass formation are found in the ternaries $M_2O–Al_2O_3–GeO_2$ and $M_2O–Al_2O_3–GeO_2$, $M_2O–B_2–O_3–GeO_2$ $M_2O–B_2O_3–GeO_2$

Table 7.11. Boundaries of glass formation in binary GeO_2 glasses (Imaoka [396])

Melting system	Mol% of oxide added
$Li_2O–GeO_2$	0 . . 23.8
$Na_2O–GeO_2$	0 . . 38
$K_2O–GeO_2$	0 . . 59.5
$CaO–GeO_2$	15.5 . . . 59.5
$SrO–GeO_2$	14 . . 39
$BaO–GeO_2$	0 . . 10 and 17.5 . . . 29.6
$Tl_2O–GeO_2$	0 . . 47.5
$ZnO–GeO_2$	0 . . 48
$PbO–GeO_2$	0 . . 57
$Bi_2O_3–GeO_2$	0 . . . 34

(Murthy and Scroggie [508], Blinov [509, 510], Umes [511], Yoshimura et al. [512], Sakka [513]). Alkali-free borogermanate glasses also show phase separation (Caslavska et al. [514]).

7.10.2 Structure and Properties

Crystalline GeO_2 occurs in two modifications:

Hexagonal: high-quartz structure, coordination 4, density 4.228 $g \cdot cm^{-3}$, stability range 1033 to 1116 °C (melting temperature).
Tetragonal: rutile structure, coordination 6, density 6.739 $g \cdot cm^{-3}$, stability range 1033 °C to ambient temperature, significantly different from SiO_2.

The highly viscous GeO_2 melt solidifies to a clear glass of density $\rho = 3.64$ to 3.66 $g \cdot cm^{-3}$. This value, approximating that of the high-temperature modification, suggests 4-coordination of Ge. Zarzycki [515] confirmed this assumption by X-ray diffraction. Thus, the formation of a GeO_4 network is indicated.

When network modifiers (e.g., alkali M_2O) are introduced into GeO_2 melts, the resulting glasses show discontinuities, e.g., maxima. [516–518] The maxima in the curves for density (Fig. 7.56) and refraction (Fig. 7.57) are explained (Ivanov and Evstrop'ev [516]) by the formation of GeO_6 octahedra in the range of 0–15 mol% M_2O leading to densification, followed by the formation of GeO_4 tetrahedra and nonbridging oxygen above 15 mol% R_2O with the usual consequence of loosening the network. Scholze [70] explains the limitation of GeO_6 units by their mutual repulsion.

This phenomenon is sometimes referred to by the term "germanate glass anomaly" in analogy to the "boric acid anomaly." Caution should be exercised

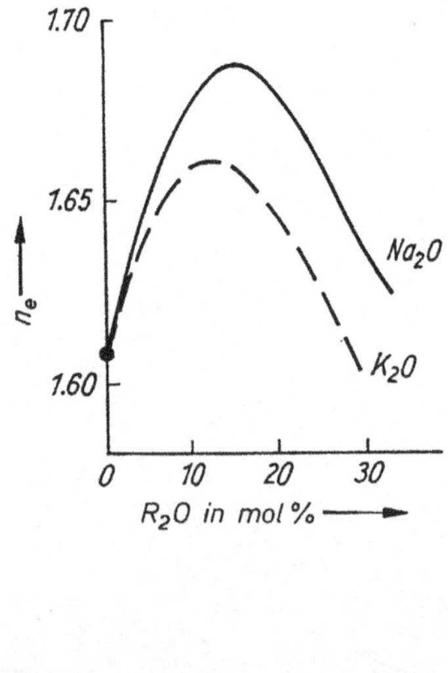

Fig. 7.56. Densities of binary alkali germanate glasses (Ivanov and Evstrop'ev [516]

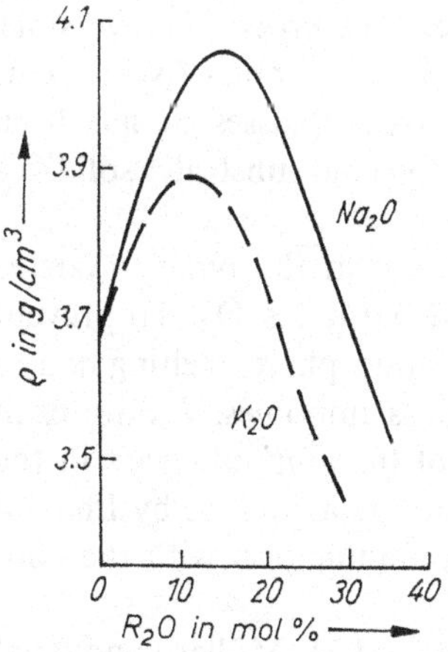

Fig. 7.57. Refractive indices of binary alkali germanate glasses (Ivanov and Evstrop'ev [516]

in accepting any interpretation at this time, considering that in the case of the "boric acid anomaly" any connection between property maxima or minima and coordination of B has been disproven. Much more experimental evidence will be required.

The exclusively scientific interest in germanate glasses prevailing in the past is beginning to be supplemented by some practical possibilities. High-lead germanate glasses transmit infrared to higher wavelengths than silicate glasses, while offering higher softening temperatures, hardness, and scratch resistance than chalcogenide glasses.

Germania is an important component in waveguide glasses, which can be obtained from $GeCl_4$ by, for example, plasma oxidation. With GeO_2 as a dopant in SiO_2, a higher index is obtained: conversely, it can be doped with lower-index components such as SiO_2 and B_2O_3 (Schultz [519]). Ferroelectric germanate glasses may serve as precursors for ferroelectric glass ceramics (Eysel et al. [520]).

7.11 Glasses Containing Arsenic Oxide

7.11.1 Glass Formation

For a long time, As_2O_3 has been added (in concentrations of 2 to 3 mass%) to commercial glass batches as a fining agent. Its effect is based on the release at about 1400 °C of oxygen from As_2O_5 which has been formed by the acceptance of oxygen at low temperatures:

$$As_2O_3 \xrightarrow[O_2]{1000\,°C} As_2O_5 \xrightarrow{1400\,°C} As_2O_3 + O_2$$

This oxygen is a major constituent of rising bubbles. The process is reversible: Residual small oxygen bubbles are resorbed by As_2O_3, reforming As_2O_5. This behavior plays an important role also in arsenic oxide glasses. It has been known for a long time that As_2O_3 vapor condensing on cold substrates solidifies as a glass.

Heaton and Moore [521] found glass formation in the binary systems K_2O–, PbO–, Sb_2O_3–, MoO_3–, SeO_2–, TeO_2– and V_2O_5–As_2O_3. To prevent the considerable evaporation of As_2O_3 in a normal atmosphere, melting had to be performed in closed fused silica or borosilicate glass ampoules. Variations in results were later traced to considerable solution of the ampoule glass in the melt. Areas of glass formation reported by these authors as well as by Imaoka [282] (Table 7.12), were doubtlessly affected by contamination with the container glass.

W. Vogel, H. Bürger, M. Weist, S. Seidemann, and H. Müller conducted extensive investigations in open corundum crucibles on glasses containing large amounts of arsenic oxide (Tables 7.12–7.14). It was found that Al_2O_3-free glasses were obtained where short melting times could be effected. Longer melting resulted in up to 20 mol% Al_2O_3 contents (Table 7.14).

All glasses contained arsenic in the form of As_2O_3–As_2O_5 equilibria (Tables 7.13–7.15). The older term "arsenite glasses" thus is misleading. Certainly "arsenate glasses" would have come closer to their composition.

Table 7.12. Limits of glass formation in binary As_2O_3 glasses (Imaoka [396])

System	Mol% oxide added
Li_2O–As_2O_3	0 . . . 48 Mol% Li_2O
Na_2O–As_2O_3	0 . . . 54 Mol% Na_2O
K_2O–As_2O_3	0 . . . 40 Mol% K_2O
CaO–As_2O_3	—
SrO–As_2O_3	0 . . . 48 Mol% SrO
BaO–As_2O_3	0 . . . 52 Mol% BaO

Table 7.13. Properties of As_2O_3 glasses: systems Na_2O–As_2O_5/As_2O_3 (A) and Na_2O–As_2O_5/As_2O_3–Al_2O_3 (B, C) (Vogel and Collaborators, 1973–1975))

Series	Impurity composition (in mol%)				n_d	ρ
	Na_2O	As_2O_3	As_2O_5	Al_2O_3		in $g \cdot cm^{-3}$
A	65.5	26.9	7.5		1.5820	3.154
	67.5	28.0	4.4		1.5733	3.148
	68.1	28.1	3.8		1.5705	3.139
B	57.2	33.7	5.8	3.3	1.6517	3.6620
	62.3	31.3	3.9	2.5	1.6520	3.5582
	68.0	28.0	1.7	2.3	1.6537	3.1005

Table 7.14. Properties of As_2O_5/As_2O_3 glasses: systems $BaO–As_2O_5/As_2O_3$ (A) and $BaO–As_2O_5/As_2O_3–Al_2O_3$ (B, C) (Vogel and collaborators, 1973–1975)

Series	Concentration (in mol%)				n_d	ρ in $g \cdot cm^{-3}$
	BaO	As_2O_3	As_2O_5	Al_2O_3		
A	42.0	48.4	8.6		1.7101	4.3745
	47.5	44.3	8.1		1.7197	4.4215
	50.3	41.7	7.8		1.7158	4.4524
B	41.4	50.8	2.5	5.3	1.7603	4.2253
	44.6	46.7	3.7	5.0	1.7644	4.2500
	48.1	42.4	3.7	5.6	1.7210	4.2990
	53.7	39.4	2.4	4.5	1.7240	4.4099
	56.2	35.3	2.3	6.1	1.7145	4.4305
C	41.0	32.2	3.5	23.4	1.6670	4.1262
	43.5	33.6	3.6	18.7	1.6700	4.1695
	44.8	31.9	2.7	20.7	1.6730	4.2283

Table 7.15. Properties of As_2O_5/As_2O_3 glasses (Vogel and collaborators, 1973–1975)

A. System $Na_2O–As_2O_5/As_2O_3–NaCl$

Concentration (in mol%)					ρ
Na_2O	As_2O_3	As_2O_5	NaCl	n_d	in $g \cdot cm^{-3}$
51.2	21.4	3.4	24.0	1.5683	2.949
56.6	22.8	3.9	16.7	1.5624	2.965
60.8	23.8	1.9	13.5	1.5012	2.952
64.9	25.8	1.4	7.9	1.5508	3.031
68.9	21.4	3.1	6.6	1.5634	3.050

B. System $BaO–As_2O_5/As_2O_3–BaCl_2$

Concentration (in mol%)					ρ
BaO	As_2O_3	As_2O_5	$BaCl_2$	n_d	in $g \cdot cm^{-3}$
42.2	29.7	2.3	20.9	1.7023	4.3475
49.2	32.1	3.9	14.8	1.7051	4.4391
51.5	36.6	4.6	7.5	1.7084	4.4714

C. System $BaSO_4–As_2O_5/As_2O_3–Al_2O_3$

Concentration (in mol%)					ρ
$BaSO_4$	As_2O_3	As_2O_5	Al_2O_3	n_d	in $g \cdot cm^{-3}$
35.7	53.9	5.4	5.0	1.6534	4.2957
31.7	55.7	7.2	5.4	1.6531	4.1850
26.3	59.4	9.8	4.4	1.6520	4.0300

As in tellurite glasses, considerable amounts of halides (fluorides, chlorides: not bromides or iodides) and sulfates can be incorporated (Table 7.15).

Glasses high in arsenic oxide are not very resistant to chemical attack. Like some phosphate glasses, some arsenic oxide glasses are hygroscopic but methods of stabilization are available.

Ternary glasses, including those with indium and gallium oxide, as well as transition elements were developed by Van Uitert et al. [522] and Grodkiewicz et al. [523].

7.11.2 Structure and Properties of Glasses of High Arsenic Oxide Content

Attempts to clarify the structure of arsenic oxide glasses must pay close attention to the As_2O_3/As_2O_5 equilibrium, which depends so much on melting conditions. Most structural considerations started from comparisons with the two crystalline modifications of As_2O_3 (Karutz and Stranski [524]). As_2O_3 occurs in a cubic modification, arsenolite and a monoclinic modification, claudetite (Fig. 7.58). The cubic arsenolite possesses a diamond-type lattice. It consists of intermolecularly saturated As_4O_6 building elements with the urotropine structure. Each As_4O_6 molecule is constituted from four each $AsO_{3/2}$ coordination polyhedra.

The monoclinic claudetite is also constituted from $AsO_{3/2}$ coordination polyhedra. Here, however, they are not assembled in closed molecule form, but from 2-fold infinite corrugated layers. The layers may be conceived as giant molecules and are held together by secondary valence forces. According to X-ray investigations by Böttcher et al. [525] and IR/Raman investigations by Heaton et al. [521] and by Adams [526] the As_2O_3-glass structure is claudetite-like.

The crystal structure of As_2O_5 has not been clarified sufficiently. Gmelin [527], Wells [528], and Pascal [529] assume a P_2O_5-like structure (Fig. 7.58), i.e., exclusive 4-coordination of As^V. Winkler, [530] however, discusses 4- and 6-coordinated As^V in crystalline As_2O_5. There are also contrasting interpretations of the structure of pure As_2O_5 glass (Krebs [534], Lin [532]). But the majority of authors interpret their experiments as confirming the AsO_4 tetrahedron as the structural unit, in analogy to those in phosphate glasses.

Considering the analytical results on glasses synthesized under ordinary conditions (see Tables 7.13–7.15), one is inclined to conclude that, besides a small portion of layer-like claudetite-type structural units with 3-coordinated As, a majority of arsenic ions in the form of AsO_4 units should be present.

It is not known whether, as in phosphate glasses, chain structures appear at increasing concentrations of network-modifiers. This would be a plausible assumption. At any rate, glasses with high arsenic oxide content will contain a variety of structural elements. Therefore, microphase separation phenomena are found only to a small degree or not at all.

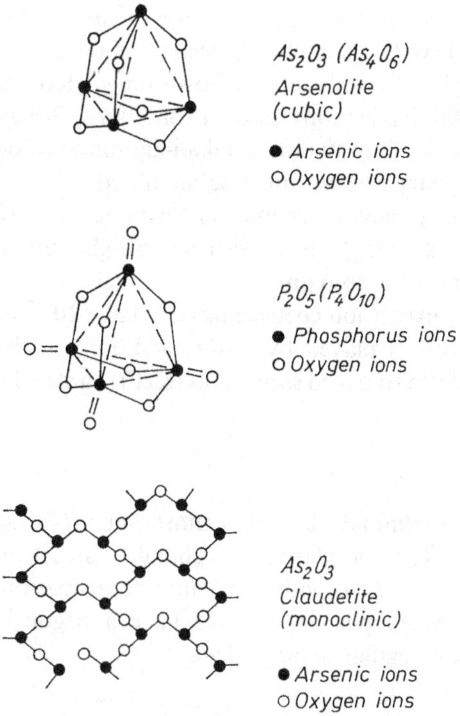

Fig. 7.58. Structural units in crystalline arsenic and phosphorus oxides

One of the most important properties of glasses high in arsenic oxide is that there is good light transparency between about 320 nm and 5.3 µm. Also, especially when halides or sulfates are incorporated, partial dispersions move toward unusual values. Considering the difficulties of preparation and reproducibility, it remains to be seen whether these valuable properties will lead to widespread applications.

7.12 Glasses Containing Antimony Oxide

7.12.1 Glass Formation and Some Important Properties

There are some analogies between As_2O_3 and Sb_2O_3. For instance, both show similarities in polymorphism. To obtain Sb_2O_3 in the vitreous state is difficult due its low field strength Z/a^2 (0.73), which must make it a poor glass-former. Even if, with Dietzel, one accounts for the high polarizability of Sb by allowing for a 20% increase, Sb_2O_5 would qualify much better since, in Sb_2O_5, the field strength is 1.76%. However, because of the high melting temperature of antimony oxide glasses (> 1000 °C), reduction of all Sb^{5+} to Sb^{3+} should be expected above 800 °C.

After much effort, Kordes [523] succeeded in the preparation of pure Sb_2O_3 glass by providing for fast melting and cooling conditions. Because of the strong polarizability of Sb^{3+}, the Sb_2O_3 glass is yellow to yellow-orange. Kordes determined an n_d value as high as 2.090. Under conventional conditions, Bishay and Askalanie [534] obtained clear binary (10–40 mol%) alkali-antimony oxide glasses. To obtain optical glasses, polynary systems must be mobilized.

Infrared windows can be based on systems 12–26 mass% Na_2O (K_2O) 7–22 Al_2O, 52–80 Sb_2O_3 (Hedden and King [535]), since, with just a light tint of lemon-yellow, they show good transparency to 6 μm.

Antimony oxide glasses have high expansion coefficients ($\alpha = 120 \times 10^{-7}$ to (with Pb, K) 212×10^{-7}). A special role is played by Sb_2O_3 additions in the control of partial dispersion, particularly in so-called short flints (KzF) (Table 1.1).

7.12.2 Structure

The structure of pure Sb_2O_3 glass resembles that of orthorhombic Sb_2O_3 (Bishay and Askalanie [534]). Most likely the glass has a chainlike structure corresponding to that of valentinite (Fig. 7.59) with Sb^{3+} in 3-coordination. Binary glasses between 0.05 to 2.0 Na_2O per Sb_2O_3 also have a trigonal-pyramidic structure in a somewhat less regular arrangement.

7.13 Glasses Containing Bismuth Oxide

Pure Bi_2O_3 glass cannot be obtained, in line with the low field strength of Bi^{3+}. However, very small additions (e.g., 1% SiO_2 or B_2O_3, but also of some CdO, BaO, ZnO, PbO) promote reasonably good glass formation (Rao [536], [537]). With larger additions, widespread areas of glass formation are found (Brekhovskich [538], Heynes and Rawson [539], Table 7.16).

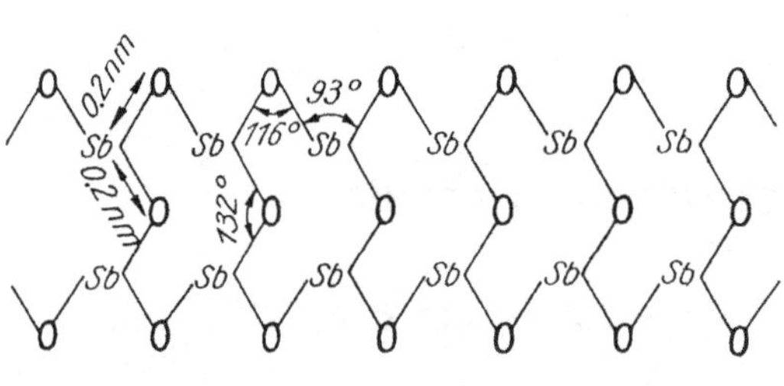

Fig. 7.59. Chain structure of crystalline Sb_2O_3 (valentinite) with bond distances and angles

Table 7.16. Limits of glass formations in binary Bi_2O_3 glasses

System	in Mol% Bi_2O_3
Bi_2O_3–SiO_2	9 . . . 57
Bi_2O_3–B_2O_3	9 . . . 75
Bi_2O_3–P_2O_5	12.5 . . . 29
Bi_2O_3–GeO_2	2

The existence and structure of high-bismuth silicate glasses may be compared to those of high-lead glasses because of the presence of electrons above a completed shell (3 in bismuth, 2 in lead) (Shabanova [540]). Among the very few applications are those to reflective highway paints and, using ease of chemical reduction, to channel multipliers (Gerhardt [541]).

7.14 Limited Glass Formation in Systems of Exclusively Scientific Interest

Limited areas of glass formation under unconventional conditions are discovered continuously. No immediate practical application is expected for such glasses nor are structures well established. Such systems are discussed in this section for the sake of completeness. Moreover, what is called an abnormal glass type today may some day attract specific applications.

7.14.1 Titanate Glasses

Pure TiO_2 will not solidify to a glass even at extreme cooling rates. Relatively small additions of SiO_2, B_2O_3, P_2O_5, even impurities (White [542]) suppress crystallization. Glass formation from 20 g melts of binary K_2O–, Rb_2O–, Cs_2O–TiO_2 systems in limited regions was found by Baynton et al. [425, 427]. Glass is obtained most easily in the system Cs_2O–TiO_2 (Marfels [543]). In 1–5 g melts, Rao [544, 545] found glass formations in the regions listed in Table 7.17.

During experiments to fabricate a barium titanate ceramic, it was found that a barium titanate glass could be obtained in minute quantities only, even at extreme undercooling (Smoke [546]) but 200 g samples were obtained when the area of glass formation was extended by adding > 20 mol% SiO_2. $> 15\%$ B_2O_3 or $> 25\%$ P_2O_5 (Herczog [547]). The system TiO_2–P_2O_5 (to 60% TiO_2) was explored extensively by Harrison and Hummel [548].

Glasses with a high TiO_2 content are useful when both high refractive index and light weight are desired, e.g., in eye glasses of high diopter (Faulstich [549]) or glass beads for road signs (Searight et al. [550]). In fluorophosphate- and fluorosilicate-based glasses, a high TiO_2 content brings about desirable extreme abnormal dispersion ("deep flints" TF, see Fig. 1.8). Morey [554] developed glasses of high refractive index and specific dispersion, starting from the system TiO_2–Ta_2O_5 (about 1:1) and adding La_2O_3 and ThO_2.

Table 7.17. Limits of glass formation in binary TiO_2 glasses

System	Glass-formation range	References
K_2O–TiO_2	29 ... 60 mol% K_2O	Rao [544, 545]
Cs_2O–TiO_2	25 ... 67 mol% Cs_2O	Rao [544, 545]
P_2O_5–TiO_2	24 ... 33 mol% P_2O_5	Harrison and Hummel [548]

The Ti^{4+} ion with a field strength of 1.04 is to be classified as an "intermediate" near the boundary of network-formers. There is no generally accepted model for its incorporation in glasses without the presence of classical network-formers. Usually TiO_2 complexes are assumed to be present, as in titanate crystals; however, TiO_4 groupings have also been considered, for example, by Marfels [543].

The equilibrium Ti^{3+}/Ti^{4+} must play an important role, especially for electrical properties in semiconductor glasses, as in the case of other variable valence ions, such as, for example, V, U, Mn.

7.14.2 Vanadate Glasses

V_2O_5 does not behave as much like P_2O_5 in regard to glass formation as one might expect. V_2O_5 melts are less inclined to evaporate, but are much more fluid, thus tending to crystallize on cooling: Pure V_2O_5 is not obtained as a glass. On rapid cooling, mg or g melts can form glasses with K_2O, MgO, CaO, SrO, ZnO, or UO_3 in limited ranges. Quite good glass formation in wide ranges for larger samples is achieved in the binary combinations of Table 7.18.

The greatest promise for utilization lies in the electrical properties of vanadate glasses. They have been recognized as electronic semiconductors (Denton and Rawson [552], Baynton et al. [426], Mackenzie [35], Adler [556]). Semiconduction is attributed to the facility of electron transitions between different states of valency, particularly V^{4+} and V^{5+}. These transitions (even V^{2+} has been confirmed) complicate structural analysis. For instance, both network-forming and -modifying positions must be considered.

In contrast to phosphate glasses, VO_5 groups – as in crystalline V_2O_5 – must be considered Fig. 7.60). But VO_4 groups, as well as chain and ring structures as in phosphate glasses, have been confirmed by paper chromatography (Ohashi and Matsuura [557]). Probably because of these similarities,

Table 7.18. Limits of glass formation in binary V_2O_5 glasses

System	Glass-formation range	References
$PbO-V_2O_5$	43.5 ... 49 mol% PbO	Denton and Rawson [552]
$BaO-V_2O_5$	30.5 ... 41.5 mol% BaO	Denton and Rawson [552]
$TeO_2-V_2O_5$	3 .. 91 mol% TeO_2	Denton and Rawson [552]
$As_2O_3-V_2O_5$	47.5 ... 100 mol% As_2O_3	Denton and Rawson [552]
$GeO_2-V_2O_5$	36.5 ... 94 mol% GeO_2	Denton and Rawson [552]
$P_2O_5-V_2O_5$	6 ... 100 mol% P_2O_5	Roscoe [553] Tammann and Jenkel [554] Rao [555]

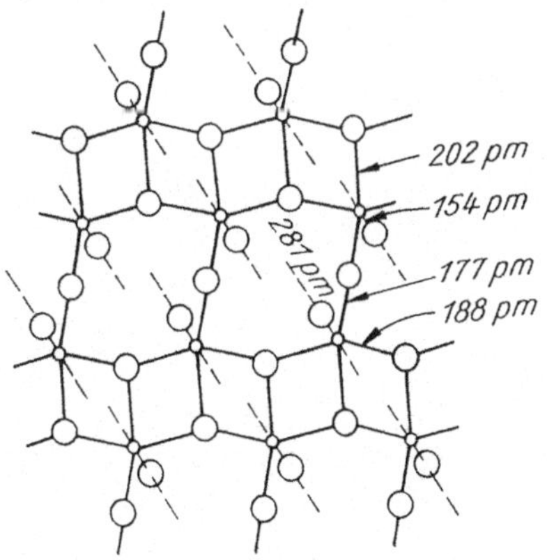

Fig. 7.60. Layer structure of crystalline V_2O_5

miscibility prevails in the system P_2O_5–V_2O_5 and mixed structural elements will form such as

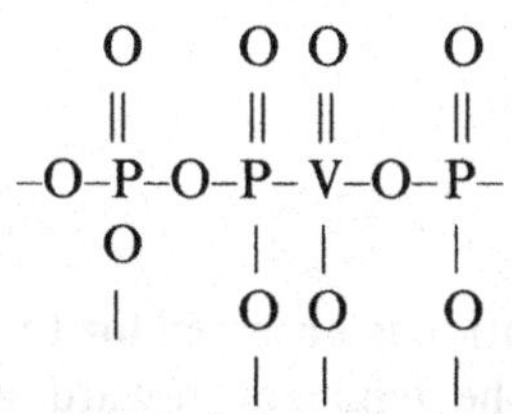

or

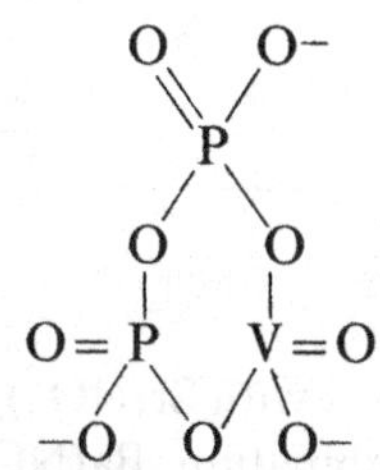

a polyphosphate vanadate chain a trimetaphosphate vanadate ring.

7.14.3 Nitrate Glasses

A series of nitrate glasses, discovered as early as 1930 by Rostkovsky [558] have purely scientific significance. If certain binary combinations of nitrates are cautiously melted and rapidly cooled in quantities of about 0.3 g, wide areas of glass formation are found in the systems listed in Table 7.19 (Thilo et al. [559]). Later investigators (Stevels [560], Dietzel and Poegel [561], Urnes [562], Van Uitert et al. [563]) agree more or less that a prerequisite of glass formation in these systems is the presence of alkaline earth nitrate complexes $(NO_3)^-$) which appear to be capable of connecting to a 3-dimensional network in analogy to $(BeF_4)^{2-}$ complexes. Such $[M^{2+}(NO_3)]^{2-}$ complexes can, however, originate in such systems only if the alkali ion introduced does not have too high a field strength, i.e., if it does not disturb too much, by counterpolarization, the coordination requirements of the alkaline earth ion. This is true if the difference in field strength between alkaline earth and alkali ion is $\Delta F < 0.14$. This is why glass formulation is observed in all $Mg(NO_3)_2$-alkali nitrate combinations. In combinations of $Ca(NO_3)_2$-alkali nitrate, no more glass formation is observed if the alkali is Li^+ (highest field strength).

Table 7.19. Limits of glass formation in nitrate glasses (Thilo et al. [559])

System	Mol%
$Mg(NO_3)_2$–$LiNO_3$	36 . . . 52
$Mg(NO_3)_2$–$NaNO_3$	21 . . . 47
$Mg(NO_3)_2$–KNO_3	24 . . . 61
$Mg(NO_3)_2$–$RbNO_3$	40 . . . 67
$Mg(NO_3)_2$–$CsNO_3$	47 . . . 75
$Ca(NO_3)_2$–$NaNO_3$	55 . . . 62
$Ca(NO_3)_2$–KNO_3	40 . . . 70
$Ca(NO_3)_2$–$RbNO_3$	32 . . . 75
$Ca(NO_3)_2$–$CsNO_3$	64 . . . 75
$Sr(NO_3)_2$–$RbNO_3$	69 . . . 83
$Sr(NO_3)_2$–$CsNO_3$	68 . . . 77
$Ba(NO_3)_2$–$CsNO_3$	78 . . . 79
$Cd(NO_3)_2$–$LiNO_3$	36 . . . 63
$Cd(NO_3)_2$–$NaNO_3$	28 . . . 64
$Cd(NO_3)_2$–KNO_3	21 . . . 67
$Cd(NO_3)_2$–$RbNO_3$	34 . . . 71
$Cd(NO_3)_2$–$CsNO_3$	27 . . . 75

In combinations with $Sr(NO_3)_2$, no glass formation is observed for Li, Na, or K. In the combination $Ba(NO_3)_2$–$Cs(NO_3)$, the tendency toward glass formation is minute.

More recently, Vinogradov et al. [564] found, under conditions of rapid cooling, glass formation in ternaries:

$$M^{I}NO_3 - M^{II}(NO_3)_2 - M^{III}(NO_3)_3$$

where $M^{I} = Na^+, K^+, Rb^+, Cs^+$; $M^{II} = Mg^{2+}, Ca^{2+}, Sr^{2+}, Zn^{2+}, Cd^{2+}, (UO_2)^{2+}$; and $M^{III} = Sc^{3+}, Eu^{3+}, Sm^{3+}, Tb^{3+}, Dy^{3+}$. The glasses containing rare earths are strongly luminescent.

Other systems include $Cd(NO_3)_2$–$TlNO_3$, $LiNO_3$–$AgNO_3$–NH_4NO_3 (Angell [565], Incher [566]). In nitrate glasses, glass formation reported by some authors (especially Van Uitert [563]) may have been aided by the presence of H_2O (Angell [565]).

No direct evidence for the structure of nitrate glasses by X-ray or neutron scattering is available to date. Indications are supplied by vibrational spectroscopy (Hester et al. [567]) and theoretical approaches (Furukawa et al. [568] and Van Weckem [569]).

For more detail on nitrate, nitrite, formate, acetate, thiocyanate, sulfate, chromate, iodate, and various hydrate glasses, reference is made to the extensive review by Angell [565], who with many coworkers, has significantly contributed to this field.

7.14.4 Carbonate Glasses and Glasses Based on $ZnCl_2$

Foundations similar to those for the formation of nitrate glasses appear to exist for K_2CO_3–$MgCO_3$. Eitel and Skaliks [149] first established that the compound $K_2Mg(CO_3)_2$ solidifies to a glass after melting under a pressure of about 100 MPa and cooling. Datta et al. [150] found a glass-forming region between 40 and 60 mol% $MgCO_3$, as well as compatibility with sulfate and fluoride systems, even in the absence of alkali. Glass formation in other pure carbonate systems is quite limited (one example: Li_2CO_3–$BaCO_3$). Not much is known about the structure, nor have carbonate glasses found any application. From the large series of the extraordinary ionic salt glasses, only one more type should be mentioned here, i.e., those based on $ZnCl_2$. Maier [570] found that a rapidly cooled $ZnCl_2$ melt solidifies to a glass. Schulz [571] found glass formation for $ZnCl_2$ melts containing up to 50 mol% KCl, KBr, or, especially, KI on rapid cooling. Although $(ZnCl_6)^{4-}$ as well as $(ZnCl_4)^{2-}$ groupings are present in the crystalline modification, it is believed that primarily a 3-dimensional network of $(ZnCl_4)^{2-}$, analogous to that of $(BeF_2)^{2-}$, is responsible for glass formation (Brehler [572], Imaoka et al. [573]). The T_g is, however, much lower (105 °C at 10 $deg \cdot min^{-1}$ by DTA) than that of BeF_2 (350 °C) (Angell et al. [565]).

Immiscibility was found in the large glass-forming region of $ZnCl_2$–KJ (Esteal et al. [547]). Very wide regions of glass formation exist with weak organic cations such as pyridinium (Esteal et al. [574]).

Iodine glasses, with a wide range of glass formation in the presence of 40–60 mol% CdI_2, have remarkably far IR transmission and moisture resistance (Cooper and Angell [575]).

7.14.5 Oxyhalide Glasses

Replacement of 0 by F in oxide (or F by 0 in fluoride) glasses is common, not surprising considering the almost equal sizes of O^{2-} and F^-. In commercial opal glasses partial replacement is used to obtain precipitation of fluoride crystals after heat treatment. In special optical fluoride glasses small amounts of oxides suppress undesirable crystallization. Stable and durable oxyfluoride glasses with extremely low T_g (down to slightly below 100 °C!) have been discovered (Tick [576]). A glass composed of about 60 Sn 6 Pb 35 P 60 F 120 O had a T_g of 95 yet exhibited a quite low solubility rate.

7.14.6 Oxynitride Glasses

Many binary and polynary oxynitride glasses containing typically 7–15 at% N have been developed and invariably found to have superior mechanical and chemical resistance. (Mackenzie and Zheng [577]). This feature is associated with N being bonded to three, not like oxygen to two network cations. A recent

MAS NMR study suggests that in Na–Si–O–N glasses (20, 30% Na_2O) half of N is bonded to 2, half to 3 Si (Unuma et al. [551]).

7.14.7 Oxycarbonate Glasses

Low C:O replacements increase mechanical properties more than N:O replacements presumably due to possible bonding to four network formers (disregarding possible effects of degree of covalency). (Mackenzie and Zheng [578]).

7.14.8 High-H_2O Glasses

Among those exotic glasses which may yet become technologically quite important are those containing a high percentage of H_2O ($> 10\%$). They are usually obtained by exposure to steam in an autoclave. The manner in which H_2O is accommodated differs from glasses of low "H_2O" content ($< 1\%$) where Scholze distinguishes various types of OH groups and hydrogen bonding (Scholze [87, 579–583]). The technology of high-H_2O glasses has been reviewed comprehensively by Bartholomew [584b].

7.15 Metal Glasses

In 1960 Klement and coworkers had for the first time reported that metals or alloys (Au_3Si for example) had solidified as amorphous solids, or glasses, after an extremely quick cooling of their melts. The development of a whole series of methods to obtain glasses from metallic melts or solids followed [586, 587]. In one case, a thin jet of metallic melt is driven under pressure onto a quickly rotating cooled disc and spun into a thin ribbon (Fig. 7.61a). In another case, a similar jet is rolled out into a thin ribbon between two quickly rotating cooled rollers (Fig. 7.61b). Metallic glasses are obtained only with extremely high

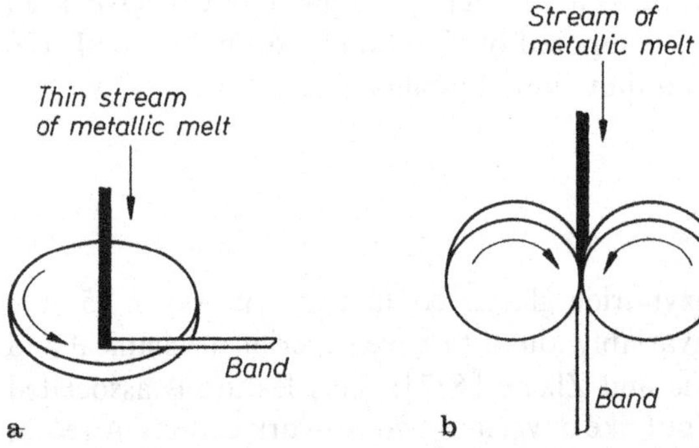

Fig. 7.61. Process for the production of metallic glasses. **a** Strolling away of the melt on quickly rotating discs; **b** Rolling out of the melt between quickly rotating rollers

cooling rates of the melts. The rates are approximately from 10^6 to 10^{10} $K \cdot s^{-1}$. This whole new field grew rapidly as is already proven by summarizing representations and monographs from Beck and Güntherodt [588], Cahn [589], Chaudari and Turnbull [590], Davies [591], Gaskell [592], Haasen [593], Hillenbrand and Hornbogen [594], Luborsky [595], Takayama [596], Van der Sande and Freed [597] or Wagner and Johnson [598]. Many questions remain nevertheless open in this young research area. Table 7.20 gives an abridged account of the enormous variety of element combinations which can solidify as glasses.

A complete series of theories exist for the cause of glass formation as well as the structure of metallic glasses. According to Egami and Waseda [600], the ratio of the radii of the atoms as well as the sterical hindrance resulting from the clusters formed while cooling, should play a big role in glass formation. One structural model presumes an irregular denser packing of spheres for the atoms of one element, while the atoms of the other metalloid element occupy the cavities (estimated to about 20%). The latter atoms give rise to the formation of a network due to their partly covalent character. A typical example would be: $T_{80}M_{20}$ where T = Fe, Co, Ni, Pd and M = B, C, Si, P. Phase separation has definitely been observed by Warlimont [601, 602] and Perlpezko, Galaup and Cooper [603]. The publications of Vind Nielsen [604, 605] and Donald and Davies [606] list the properties of about 400 different metallic glasses and show the extent and intensity of research in this field. Metal glasses possess a series of highly interesting properties which are partly derived from metals and partly

Table 7.20. Composition of metal glasses [599]

Binary glasses	Ternary glasses	Quarternary glasses
$Au_{81}Si_{19}$	$Au_{77}Ge_{14}Si_9$	
$Au_{73}Ge_{27}$	$Pd_{77}Au_5Si_{18}$	
$Pd_{82}Si_{18}$	$Pd_{77}Cu_6Si_{17}$	
	$Pd_{80}Si_{16}Sb_3$	
$Pd_{81}P_{19}$	$Pd_{60}Fe_{20}P_{20}$	
$Ni_{80}P_{20}$	$Ni_{80}Si_8B_{12}$	
	$Ni_{40}Pd_{40}P_{20}$	
	$Ni_{22}Pd_{62}Si_{16}$	
$Ni_{75}P_{26}$	$Ni_{75}Si_8B_{17}$	$Ni_{75}P_{16}B_6Al_3$
	$Ni_{75}P_{15}B_{10}$	$Ni_{68}Cr_{10}Si_{10}B_{12}$
$Co_{75}P_{25}$	$Co_{75}Si_{15}B_{10}$	$Ni_{68}Nb_{10}Si_{10}B_{12}$
$Fe_{83}P_{17}$	$Fe_{83}B_2P_{15}$	
$Fe_{83}B_{17}$	$Fe_{81}Cr_2B_{17}$	$Fe_{75}P_{15}C_6Al_4$
	$Fe_{80}P_{13}C_7$	$Fe_{75}P_{16}B_6Al_3$
	$Fe_{79}Si_{10}B_{11}$	$Fe_{40}Ni_{40}P_{14}B_6$
	$Fe_{75}P_{15}C_{10}$	
$Ti_{80}Si_{20}$	$Ti_{40}N_{40}Si_{20}$	
	$Ti_{62}Ni_{30}B_8$	
	$Ti_{72}Fe_{12}Si_6$	
$Ni_{60}Nb_{40}$		
$Cu_{57}Zr_{43}$		
$Cu_{50}Zr_{50}$		
$Gd_{67}Co_{33}$		

Table 7.21. Comparison of some properties of glasses and metals according to Güntherodt [586]

Property	Silicate glasses	Glassy metal	Crystalline metal
Formability	poor; brittle	good; ductile	good; ductile
Hardness	large	large	small
Break limit	low	high	high
Light transmission	good	non-transparent	non-transparent
Electrical Conductivity	poor	good	good
Magnetism	non-magnetic	various manifestations	various manifestations
Thermal conductivity	poor	good	good
Corrosion resistance	high	high	low

from classical glasses. A range of solid properties of classical silicate glasses are compared with properties of glassy and crystalline metals in Table 7.21 (Güntherod [586]).

Some metallic glasses are super conducting while others are suitable for the construction of magnet cores with a nearly loss-free hysteresis. Examples of such compositions are $Fe_{80}B_{20}$, $Fe_{29}Ni_{49}B_6Si_{12}$, $Fe_{82}B_{12}Si_6$, $Fe_{18}Mo_2B_{20}$ and $Fe_{40}P_{14}B_6$. Metallic glasses can only be produced as very thin ribbons or fibers. However thin ribbons have already been tentatively bound by laser or ultrasound welding methods into thicker samples suitable for use as magnet cores. Further potential applications in electrical engineering and other technical areas can not yet be assessed at this time.

7.16 Vitreous Carbon

The crystalline forms of carbon are diamond and graphite, the latter being the most common. Vitreous carbon is produced by solid pyrolisis of highly cross-linked aromatic polymers [607, 608].

Diamond is known to possess a three dimensional network whereas graphite has a layer lattice. The formation of the graphite's layer lattice is favored with solid pyrolysis. This graphitizing process is kinetically inhibited so that the irregular graphite regions remain small (300–400 nm). Closed pores which account for 25 to 40 vol% and have an average diameter of 1 to 5 nm occupy the space between the irregularly formed layers of the small graphite regions [609].

The diameter of the pores is, or course, dependent on the processing conditions. It is supposed to considerably vary it already at 700 °C. The densities of the three solid carbon forms also express the differences in structure: 3.516 $g \cdot cm^{-3}$ for diamond, 2.266 $g \cdot cm^{-3}$ for graphite and only 1.5 $g \cdot cm^{-3}$ for glassy carbon.

Vitreous carbon has a relatively low electrical conductivity of about $200\,\Omega\cdot cm^{-1}$. It shows conchoidal fracture like normal glass and is highly resistant to corrosion at high temperatures if no oxygen is present. One of its applications has thus been as crucibles for the production of highly corrosive glass melts under an inert atmosphere (e.g. pure fluoride or fluorophosphate glass melts). Other applications have already been found [610].

7.17 The Sol-Gel Method for Production of Glasses and Glass Ceramics

7.17.1 Introduction

The so-called "gel glasses" are produced according to a new unconventional process without going through a melt phase. The decisive breakthrough in this new method occurred in 1971 with the work of Dislich [512]. This new field grew considerably as proven by the relevant work, publications and conferences which followed Dislich's breakthrough [613–615]. There are currently three main methods:

- the alkoxide sol-gel method [617]
- the silica hydrosol method [611, 618, 619] and
- the Ormocer method [620]

A general overview of the methods follows. A more detailed study demands an extensive scientific literature survey, such as the (900 page) survey by Brinker and Scherer [621], (general); Brinker et al. [622] (films); Uhlmann et al. [623, 624] (films); Mackenzie and Ulrich [625], (topic: sol-gel optics, future); Hench [626, 627]. Schmidt [628] (topic: inorganic-organic, composites, optical), Reisfeld, Jöergensen [529], (topic: inorganic-organic, optical, large paper), Roy [630], (topic: history, future), Avnir et al. [631], (topic: inorganic-organic, optical; 85 references).

7.17.2 The Alkoxide Sol-Gel Method

The starting material for this method is an alkoxide as for example $Si(OC_2H_5)_4$ ("TEOS") for the production of silica glass. The alkoxide is at first dissolved in alcohol and hydrolyzed through successive water additions. Acids or bases act as catalysts for the reaction. The simultaneously occurring polycondensation processes lead to the formation of a sol [632–634] which solidifies into a wet gel (alcogel) in the course of maturation. The wet gel is at first transformed into a still porous xerogel through various drying steps. A further temperature treatment at 600–1200 °C leads under considerable shrinkage to a transparent glass [635, 636]. The gel-glass conversion occurs above the T_g of the resulting glass but clearly below its melt region. A transformation can also clearly occur by melting of the gels. The melt region of the gels definitely lies under one of the

Table 7.22. Schematic of the production of silica glass according to the alkoxide-gel method [613]

1 TEOS as starting material Addition of alcohol Addition of water and catalyst Heating of the mixture with mixing up to 60 °C HOMOGENEOUS CLEAR SOLUTION
2 Solution is left to stand at about 25 °C GEL
3 Drying of the gel at about 120 °C Temperature treatment GLASS

raw material oxide mix. A transformation can also take place with the simultaneous use of pressure and temperature. The chemical reactions are given here for the example of silica glass

$$\text{hydrolysis: } Si(OC_2H_5)_4 + 4H_2O \longrightarrow Si(OH)_4 + 4C_2H_5OH$$

$$\text{condensation: } Si(OH)_4 \longrightarrow SiO_2 + 2H_2O$$

Hydrolysis is a stepwise reaction: the (OC_2H_5) groups are replaced one at a time by (OH) groups. The condensation is analogous: water is continuously eliminated as three dimensional cross-linking occurs.

Table 7.22 gives a good schematic representation of the whole alkoxide sol-gel method (Beier, Meier and Frischat [617]). The sol-gel method is of great importance. Combinations such as tetraethoxysilane with all available metal alkoxides or with simpler compounds such as acetates and nitrates among others are possible. Nearly all glasses which were until now only produced from a melt, such as borosilicate glasses, can now be formed without going through the melt phase. The main problem of the gel glass production in "beaker glass" is that the sol-gel formation process is a time reaction. Catalysts are thus used to shorten the process [637]. Nevertheless the sol-gel process will naturally never be used for the bulk production of classical glasses but rather will find application in coatings and unusual shapes and will still expand (see Section 7.12.5).

7.17.3 The Silica Hydrosol Process

This process is very similar to the previous one. The difference between the two methods lies merely in the fact that the starting materials do not contain any organic groups but are silicate sol, colloidal silicates, amorphous silicates and metallic salts. A glass is produced by gel formation with addition of water and heating (see Table 7.23) [70]. A disadvantage in comparison with the alkoxide

Table 7.23. Schematic of the silica-hydrosol process

Starting material
- Silicate sol
- Colloidal silicate
- Amorphous silicate

↓ Addition of water

Homogeneous colloidal gel

↓ Destabilization

Homogeneous multi-component gel

↓ Thermal treatment and thickening begins

Reactive gel

↓ Heating at higher temperatures

Glass

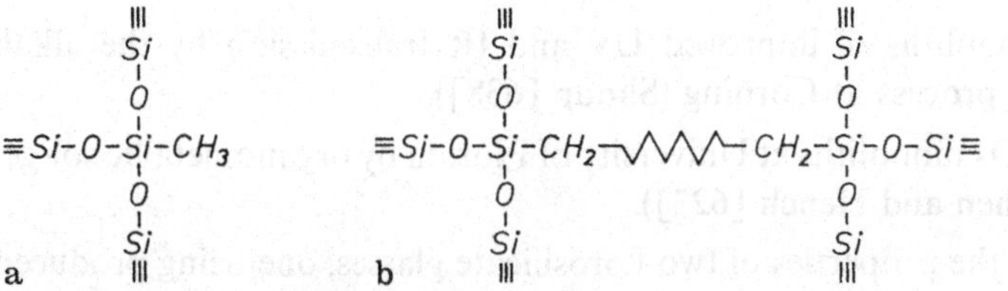

Fig. 7.62. Molecular building units of "ormocers"

sol-gel method lies in the fact that if water glass is used to supply SiO_2, the sodium ions must later be removed for many applications.

7.17.4 The Ormocer Method

Scholze has further modified the alkoxide sol-gel method and thus produced the new material group of the "ormocers" [620] (organically modified ceramics). The starting materials used in this method are not only such compounds as $Si(OR)_4$ which can be completely hydrolyzed but also compounds in which one or more (OR) groups are replaced by organic radicals R′. The starting products can hence be represented as $R'_nSi(OR)_{4-n}$. The organic radicals can be alkyl or aryl groups and still additionally contain functional groups such as –OH, –COOH, $-NH_2$ or $-CH{=}CH_2$. Scholze describes the materials obtained from the Ormocer method as inner composite materials which contain inorganic but also organic elements in their structure (see Fig. 7.62). Properties are thus obtained which can depend either on the inorganic or on the organic components. The conversion of the gel into a glass must obviously take place at lower temperatures than with pure inorganic gels due to the organic component of the gel.

7.17.5 The Importance and Application of Gel Glasses

Apart from the fact that gel glass production is possible at considerably lower temperatures than production from the melt phase, such microstructure formation processes as phase separation which occur in the melt are avoided. Thus completely clear glasses of the systems TiO_2–SiO_2 or ZrO_2–SiO_2 could in principle be produced. Only dull glasses are obtained from melts of these systems, due to the open mixture interstices in the melts. The production of bulk glasses by the sol-gel processes causes the biggest problems. Enormous stresses generally arise with drying of the gels thus causing the body to fracture. The problems here are obviously similar to those which occur in Vycor glass production. Considerable progress has been achieved by drying the gels under supercritical conditions [635] or by adding formaldehyde, diethyl formaldehyde, oxalic acid [465] or other compounds to the reaction products. Among recent achievements are the production of

- large SiO_2 monoliths of improved UV and IR transmission by the alkali silicate gelatin process at Corning (Shoup [638]).
- large SiO_2–TiO_2 monoliths at University of Florida by organometallic sol-gel technology (Chen and Hench [623]).

A comparison of the properties of two borosilicate glasses, one being produced from a melt and the other via the sol-gel method [612], is also scientifically interesting. Almost all properties of the analogous glass obtained via a melt process are achieved (see Table 7.24). Essentially simpler than bulk glasses, gel glasses can be produced and applied to substrates of all sorts. Among uses of

Table 7.24. Physical properties of borosilicate glasses [612]

Composition in mol% $SiO_2 = 86.85$; $B_2O_3 = 5.91$; $Al_2O_3 = 2.62$; $Na_2O = 3.92$; $K_2O = 0.66$	Gel-glass	Glass from the melt
H_2O content (% by weight)	0.039	0.023
C content (% by weight)	0.0002	
Density (g/cm^3)	2.28	2.27
Thermal expansion coefficient α (10^{-7}/°C)	32	32
Glass transition T_g (°C)	590	610
Refractive index n_e	1.477	1.474
Abbe's number ν_e	65	66

such films are

- planar optical waveguides and integrated components (Roncone et al. [640]; Zelinski et al. [641]; Chia et al. [642] including laser writing techniques.
- low temperature ion exchange
- porous membrane (Yazawa et al. [643]).

They are important as thermal insulation coatings due to their higher IR reflection [644], as corrosion protection, as anti-reflex coating [645] or also as ionic diffusion barrier strata [646]. Porous glasses [294], glass powder or glass ceramics of the most varied composition can be produced by the sol-gel process. While gel glasses are obtained by annealing at relatively low temperatures under pressure, glass ceramics can be produced from the same gel with treatment at a higher temperature [647, 648]. Glass ceramics compositions which were not possible via a melt process can be achieved via a gel process. Fibers can also be produced. A prerequisite is the spinning capacity of the solution as well as a fast consolidation after the drawing process [649]. Sols with one dimensional long molecules fulfil this requirement. They are above all obtained with a clean catalysis. The ormocers allow for a large range of properties. Philipp and Schmidt have described a development for production of ormocers in bulk for applications as contact lenses [650]. The development and application of new gel glass products will surely expand in the next years.

8 New Optical High-Performance Glasses

8.1 Fundamental Principles of the Dispersion Behavior of Glasses

As white light passes through a lens, a separate image is obtained for each wavelength or color of the spectrum due to the dispersion behavior of glasses. i.e. refraction of the glass is different for different wavelengths. These images overlay each other resulting in an erroneous and blurred image with colored edges. This effect can be corrected by placing a convergent lens and a divergent lens of different composition one after the other. The dispersion of the convergent lens must be exactly offset by the one of the divergent lens:

$$\Delta s_{\lambda F'} = ((P'_{\lambda F'} - P''_{\lambda F'})/(\nu_{e'} - \nu_{e''}))\cdot f$$

where $P'_{\lambda F'}$ is the relative partial dispersion of the first lens, $\nu_{e'}$ is Abbe's number and f is the focal length of the lens. The apochromatic condition (i.e. correction of the color distortion) is:

$$\Delta S_{\lambda F} = 0$$

If the relative partial dispersions (determined for all optical glasses at the same wavelengths), $P_{1,2} = (n_1 - n_2)/(n_3 - n_4)$, were to be plotted as a function of Abbe's number, all points would approximately lie on a straight line as observed in Fig. 8.1. This line is called "normal" line and the following relations apply for all glasses which lie on this line:

$$P_{1,2} = a_{1,2} + b_{1,2}\cdot\nu \quad \text{or} \quad \nu = -(a_{1,2}/b_{1,2}) + (1/b_{1,2})\cdot P_{1,2}$$

and for the other glasses:

$$P_{1,2} = a_{1,2} + b_{1,2}\cdot\nu + \Delta P_{1,2} \quad \text{or}$$

$$\nu = -(a_{1,2}/b_{1,2}) + (1/b_{1,2})\cdot P_{1,2} + \Delta\nu_{1,2}$$

The "normal" lines and thus the constants $a_{1,2}$ and $b_{1,2}$ are established by the coordinates $(P_{1,2}, \nu)$ of two glass types chosen such that the "normal" lines for all wavelength combinations lie in the center of the points in the respective diagrams. An example would be a crown glass with

$n_e = 1.521 \quad \nu_e = 60.12$
$n_d = 1.5192 \quad \nu_d = 60.36$

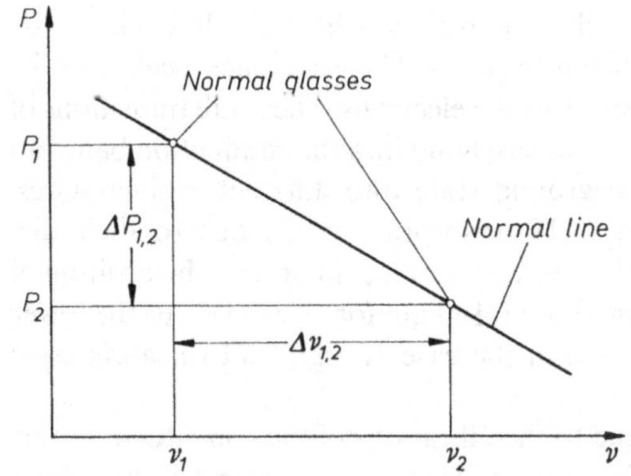

Fig. 8.1. Schematic of the "normal line" in the $P_{1,2}$-ν diagram

and a flint glass with

$n_e = 1.6241 \quad \nu_e = 36.09$
$n_d = 1.6200 \quad \nu_d = 36.35$

Brewster and coworkers have carried out detailed investigations on the partial dispersion relations in optical glasses. They compared the partial dispersion ratios: $P_{F,D}$ with $(n_F - n_D)/(n_F - n_C)$ and $P_{G',F}$ with $(n_{G'} - n_F)/(n_F - n_C)$ for silicate, borate and phosphate glasses. Silicate glasses, which are the usual optical glasses, were on the "normal" line while borate and phosphate glasses deviate. The secondary spectrum in silicate glasses can only be corrected by combinations with borate and phosphate glasses. The ratio of the partial dispersion to Abbe's number (that is the deviation from the normal line) for borate and phosphate glasses differ from the one for silicate glasses. Generally borate glasses possess lower partial dispersions at the same Abbe number than the silicate glasses and phosphate glasses have higher ones than the silicate glasses.

The dispersion of a glass is dependent on the position and intensity of the absorption peaks. Izumitani and Nakagawa [651] have summarized the different possibilities for the position of IR and UV absorption peaks and explained their effect on the partial dispersion. The empirical relation between partial dispersion and Abbe's number: $P_x = a_x + b_x \cdot \nu_e + c_x \cdot \nu_e^2$ is not fulfilled for unusual glasses. Izumitani and Nakagawa deduced that the unusual partial dispersion is linked with absorption of unusual glasses in the IR and UV region. They distinguish three cases:

a) The UV absorption peak occurs in the region of very short wavelengths as is the case for the unusual dispersion of fluor crown glasses for example.
b) The IR absorption band lies in the region of shorter wavelengths when compared with the normal glasses and therefore the ratio ν_e/P is smaller than for normal glasses. Short flint glasses and borate glasses which contain lanthanum have this type of unusual partial dispersion.
c) Glasses show an additional absorption in the UV region at longer wavelengths and the ν_e/P ratio is therefore larger than the one of normal glasses. Flint glasses which contain TiO_2 have this type of unusual partial dispersion.

The UV resonance peaks can generally be explained with the help of electronic energy levels. The absorption mechanism at the UV resonance peaks can be interpreted as a band to band transition of the electrons. These are transitions of optically activated electrons from the valence band into the conduction band (an "interband" transfer) and from the ground state into different excited states. Glasses possess overlain or blurred UV absorption peaks due to their disordered structure. As the content of heavy metal oxides increases, the position of the UV absorption edge, shifts from about 0.160 μm for silica glass to the lower limit of the visible spectral region e.g. in the case of high-lead silicate glasses (dense flint).

The IR resonance peaks are due to the vibrations of the constituent atoms which are related to a change in the electric dipole. The asymmetric stretching vibrations and the out-of-plane bending vibrations are important. For pure SiO_2 glass the asymmetric stretching vibrations lie at 9μm and the bending or deformation vibrations at 21.5 μm. These positions are assigned to vibrations of the SiO_4 tetrahedron. The introduction of network modifier ions changes the position of these bands. The band maximum of the asymmetric Si–O stretching vibration at 9 μm is displaced to longer wavelengths with the addition of alkali ions. An addition of PbO produces the same effect. The asymmetric B–O stretching vibration and the out-of-plane bending vibration peaks of borate glasses already occur at shorter wavelengths: at 7 to 8 μm and 13.5 μm respectively. Peaks which can be assigned to the BO_4 tetrahedrons appear with the addition of alkali ions or PbO. A displacement of the peaks to shorter

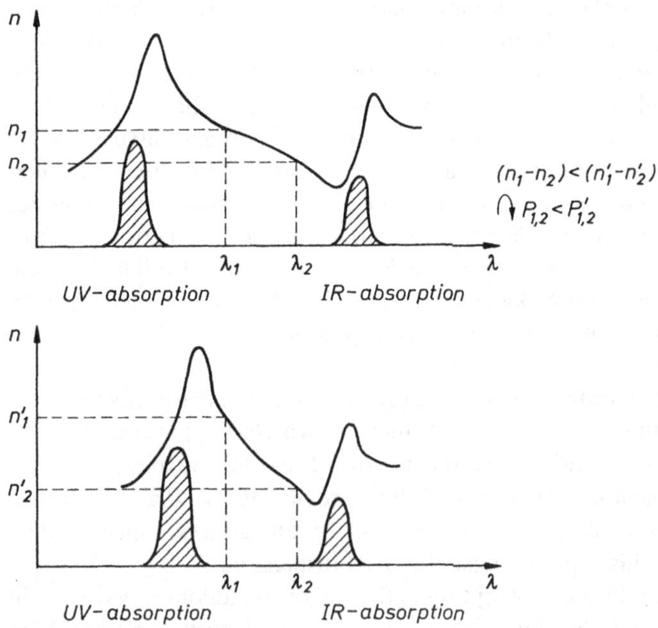

Fig. 8.2. Influence of the positions of the absorption peaks on the dispersion curve of a glass

wavelengths occurs. The P–O stretching vibration lies at about 10.5 μm for phosphate glasses.

A displacement of the peaks leads to an immediate change in the dispersion curve, in particular a change in the slope of the curve (see Fig. 8.2). A change of slope of the dispersion curve leads to a change in the refractive index difference which consequently raises the $P_{1,2}$ value (it tends to $(n_1 - n_2)/(n_3 - n_4)$ for an established $(n_3 - n_4)$ difference) or lowers it. The change in the $P_{1,2}$ value leads to a deviation of the position in the $P_{1,2} - \nu_e$ diagram since Abbe's number is also determined as $\nu_e = (n_e - 1)/(n_1 - n_2)$.

When the positions of the stretching vibrations of the individual base glasses are compared: Si–O 9μm, B–O, 7.5 μm, P–O 10.5 μm, the interpretations of Brewster and coworkers are confirmed: the substitution of SiO_2 by B_2O_3 leads to a negative deviation of the dispersion from the normal line and the substitution of SiO_2 by P_2O_5 to a positive one.

The position of the absorption bands can be influenced by the introduction of additional absorption centers. The production of glasses whose position in the $P_{1,2} - \nu_e$ diagram and to the normal line is completely controlled therefore becomes possible.

8.2 Change of the Dispersion with the Introduction of Additional Absorption Centers

Käs [652] investigated the modifications in the dispersion of glasses with an introduction of absorption centers in the IR and UV region. He assumed that normal glasses generally absorb in the UV region at wavelengths shorter than 0.3 μm. Additional absorption centers therefore had to be built with λ_0 0.35 μm in order to change the dispersion of these glasses: absorption bands with $\lambda_0 < 0.3$ μm would not change it and centers with $\lambda_0 > 0.4$ μm would color the glasses. The determination of the band position in the UV region with the introduction of ions in glasses is possible with the help of the ligand field theory.

The following ions possess absorption bands in the region $\lambda_0 = 0.3$ μm to $\lambda_0 = 0.4$ μm:

Fe^{3+} at 0.38 μm CN6 and 4
Ti^{4+} at 0.37 μm CN6
V^{5+} at 0.32 μm CN6
Cr^{6+} at 0.36 μm CN4
Ag^0 at 0.34 μm

These ions can be added as oxides. The Ag^0 center can be formed by irradiation. The use of the Fe^{3+} ion has disadvantages due to the Fe^{2+}/Fe^{3+} equilibrium. Fe^{2+} has an absorption band at 1.1 μm and consequently colors the glass. The Ti^{4+} ion is the most advantageous to use since no additional absorption bands in the IR or visible region are created. Cerium also presents this advantage (Kreidl and Davis oral comm).

If absorption centers are to be created for the IR region, molecule groups capable of vibrations must be introduced. The glass structure however will also change, therefore shifting the position of the absorption edge in the UV region as well as the position and dimension of the bands in the IR region. A determination of the influence of absorption centers in the IR region is therefore hardly possible.

8.3 Optical Glasses with Unusual Partial Dispersions

The use of crystals as an optical medium in the construction of high-performance optics have for a long time brought considerable advances in the field, due to their extreme partial dispersions. They were used for these advantages despite some unfavorable properties for the optics designer such as splitting of the crystals in preferential directions, low chemical resistance, double refraction in particular directions and low hardness.

A recent area of research has been the development of glasses with the same or still better partial dispersions than the aforementioned crystals. Abbe worked on a coarse characterisation of the capacity of an optical glass with its position in the n_e–ν_e diagram but he also developed a fine characterisation with the partial dispersion behavior. For the majority of optical glasses the linear Abbe relation applies

$$P_{1,2} = a_{1,2} + b_{1,2} \cdot \nu_e$$

where $a_{1,2}$ and $b_{1,2}$ are constants, ν_e is Abbe's number $\nu_e = (n_e - 1)/(n_{F'} - n_{C'})$ and $P_{1,2}$ is the relative partial dispersion $(n_G - n_{F'})/(n_{F'} - n_{C'})$. Most optical glasses lie on the so-called "normal" line:

$$P_{G,F'} = 0.57035 - 0.0014832 \cdot \nu_e$$

A prerequisite for an improvement in the correction of the color distortion in microscopes, telescopes or photo objectives is however a deviation from the "normal" line. The glasses have to possess as large a $\Delta P_{G,F'}$ as possible with a positive or negative sign: $\Delta P_{G,F'} = P_{G,F'} + 0.0014832 \cdot \nu_e - 0.57035$

New optical glasses have been designed with the same or better partial dispersions as crystalline Fluorspar and they can therefore replace these crystals in optics. This development was based on the discovery that very low quantities of metaphosphates influence the glassy solidification of a melt mixture of 70–98 mol% fluorides and monovalent, bivalent or trivalent metals. The principal building units of these glasses are chains of fluoride octahedra whose order or crystallisation is strongly suppressed by the introduction of mono or diphosphate building groups (see Fig. 8.3) [653].

Figure 8.4 gives an overview on the development state of the actual production of optical glasses with unusual partial dispersions [654]. Figure 8.5 describes in particular results of the development of a fluoroaluminate glass with low phosphate content. The optical properties of fluorspar were attained

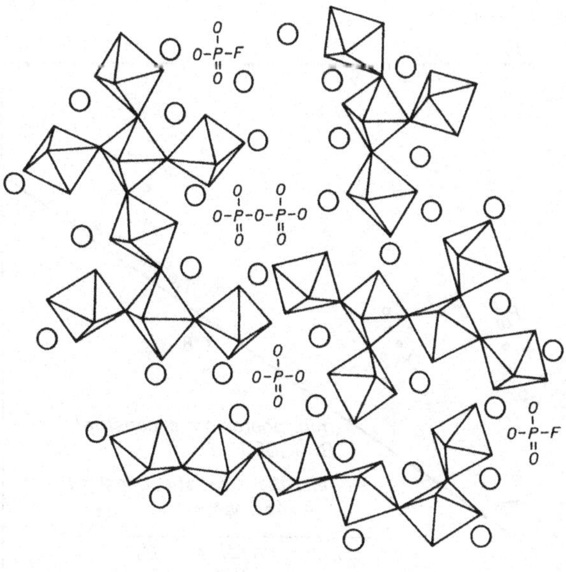

Fig. 8.3. The crystallization tendency of glasses is strongly reduced by the introduction of mono and diphosphate building units in the fluoride octahedra chains

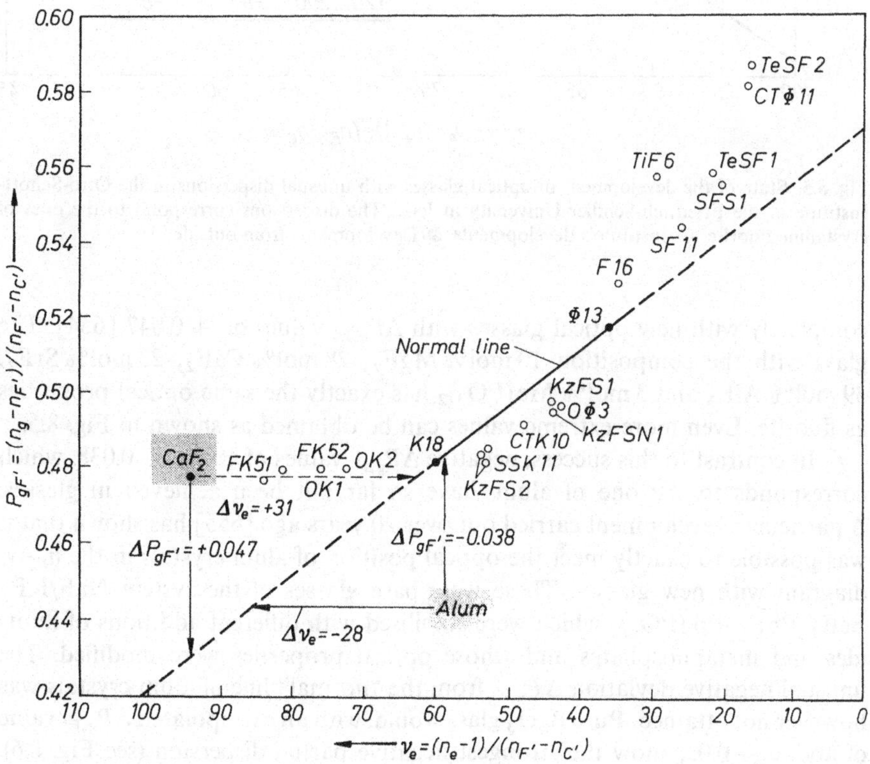

Fig. 8.4. Representation of the state of the development and production of optical glasses with unusual dispersions in the $P_{g,F'}$-ν_e diagram. CT, CTK, OK1, OK2: soviet glasses; FK51, FK52, KzFS1, KzFS2, TiF6: Schott Glass Works Mainz F16, SSK11, SF11: Jena Glass Works; TeSF1, TeSF2: developments from the Otto-Schott-Institute of the Friedrich-Schiller-University in Jena

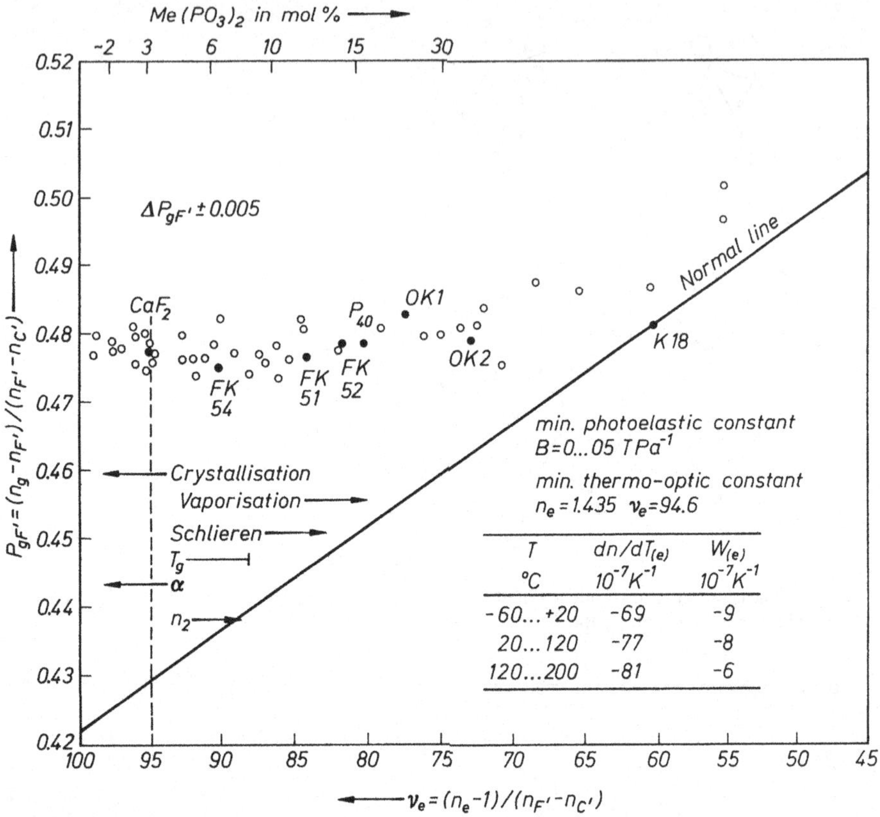

Fig. 8.5. State of the development of optical glasses with unusual dispersions in the Otto-Schott-Institute of the Friedrich-Schiller-University in Jena. The dispersions correspond to the ones of crystalline fluorite. ○ Institute's developments ● Developments from outside

completely with new optical glasses with $\Delta P_{g,F'}$ values of $+0.047$ [654]. The glass with the composition 10 mol% MgF_2, 28 mol% CaF_2, 23 mol% SrF_2, 39 mol% AlF_3 and 3 mol% $Me(PO_3)_2$ has exactly the same optical properties as fluorite. Even more extreme values can be obtained as shown in Fig. 8.5.

In contrast to this success, negative $\Delta P_{g,F'}$ values of about -0.038, which corresponds to the one of alum, have so far not been achieved in glasses. A particular development carried out over 20 years ago [655] has shown that it was possible to exactly meet the optical position of alum crystals in the $n_e - \nu_e$ diagram with new glasses. These were base glasses of the system NaF/KF–BeF_2–PbF_2–Pb $(PO_3)_2$ which were stabilised with different additions of fluorides and metaphosphates and whose optical properties were modified. The unusual negative deviation $\Delta P_{g,F'}$ from the "normal" line of alum crystals was however not attained. Pure B_2O_3 glass would, with an extrapolated $\Delta P_{g,F'}$ value of about -0.02, show the strongest negative partial dispersion (see Fig. 8.6). This glass however can only be produced with difficulty because of its strong hygroscopicity which is due to its structure. Nor it can be used as an

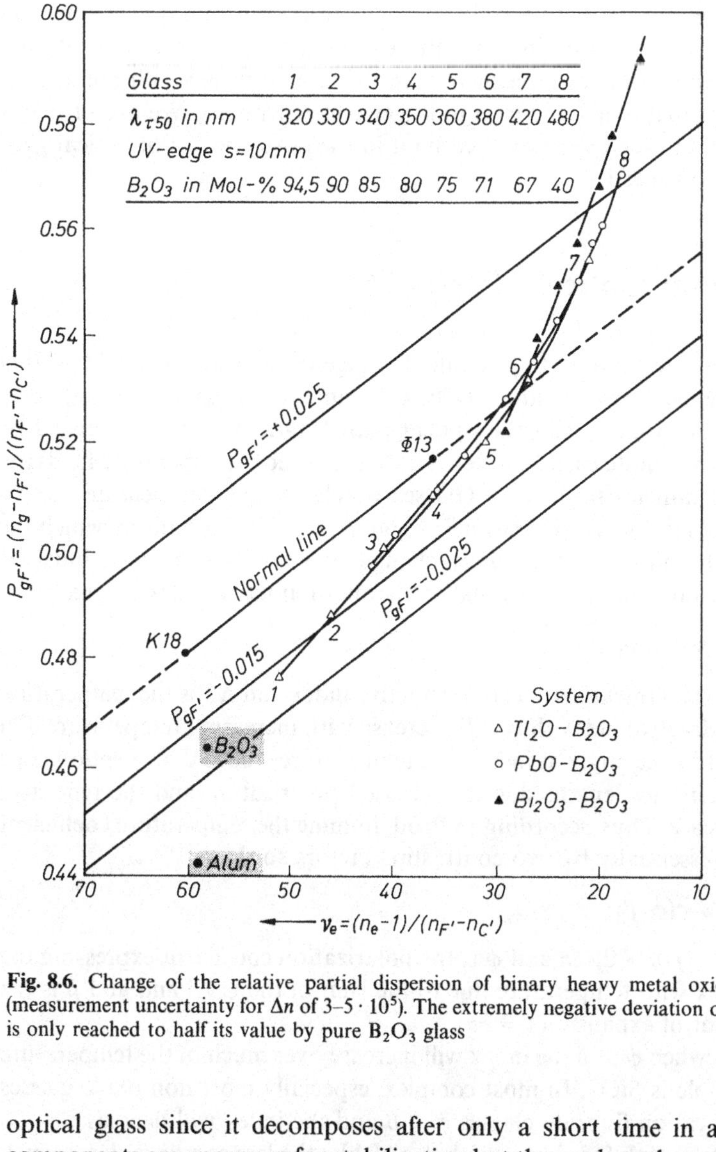

Fig. 8.6. Change of the relative partial dispersion of binary heavy metal oxide borate glasses (measurement uncertainty for Δn of $3\text{–}5 \cdot 10^5$). The extremely negative deviation of the alum crystal is only reached to half its value by pure B_2O_3 glass

optical glass since it decomposes after only a short time in air. Additional components are necessary for stabilization but they reduce the magnitude of the negative deviation. Alum crystals in comparison have a deviation of about -0.04.

Potassium ions as well as aluminum ions in alum ($KAL(SO_4)_2 \cdot 12H_2O$) are each coordinated with 6 H_2O molecules into an octahedron. Therefore 85 mol% of the structure of alum consists of H_2O molecules and these molecules also determine its unusual dispersion. A series of experiments has been carried out in the last years to intentionally introduce water into glasses in larger percentages. This is possible under pressure or also by hydrating them in water, acids, salt solutions in methanol/water mixtures among others.

Bartholomew's experiments deserve special mention: he could introduce up to a maximum of 20% H_2O (by weight) in alkali silicate glasses. Aside from the fact that such glasses would never be used as optical medium due to their low chemical resistance, the form in which the water is stored in the aforementioned silicate glass, more ionic or molecular as in the alum lattice, is still not known. The latter is probably not the case in larger concentrations. We are probably ready to cross a never before reached limit in the development of optical glass with alum-like properties.

8.4 Athermal Optical Glasses [656]

Optical systems are exposed to generally homogeneous temperature fields, that is temperature fields which are locally constant or posess a constant local gradient. A related change in the optical path length $\Delta l \cdot \Delta n$, experienced by a direct light ray is quite easily corrected. Inhomogeneous temperature fields are however more difficult to control. Glasses which actively compensate the so-called "thermal light wave aberration" are necessary. The amount by which the thickness of the glass approximately changes due to its expansion must be approximately inversely proportional to thermal refractive index change

$$\Delta d/\Delta T \approx -\Delta n/\Delta T$$

where d is the glass thickness, n is the refractive index and ΔT is the temperature change. The refractive index thus will decrease with increasing temperature. On the other hand increasingly at increasing temperatures the UV absorption edge moves to longer wavelengths due to increased polarization and the refractive index will increase. Thus according to Prod' homme the temperature coefficient is determined essentially by two contrasting factors so that:

$$\mathrm{d}n/\mathrm{d}T = c(\phi - \beta)$$

where $c = (n-1)(n+2)/6n$ and ϕ is the polarization coefficient expressing the change of index with temperature due to the shift of the UV band and β is the cubic coefficient of expansion ($= 3\alpha$).

In glasses when $\phi > \beta$ the index will increase over much of the temperature field. An example is SiO_2. In most complex, especially most non-oxide glasses such as heavy cation fluoride glasses $\phi < \beta$ and the index will decrease.

One has to search for glasses with $\phi \approx \beta$ like the less complex glasses such as fluorophosphate glasses. If the absolute amount of the refractive index is substituted, the following equation is obtained:

$$G = \alpha(n-1) + \mathrm{d}n/\mathrm{d}T$$

where α is the linear thermal expansion coefficient, G is the thermooptical constant and $\mathrm{d}n/\mathrm{d}T$ is the temperature coefficient of the refractive index. If the value of the term $\alpha(n-1)$ is the exact opposite of the one of the temperature coefficient of the refractive index $\mathrm{d}n/\mathrm{d}T$, the thermooptical constant G is equal to 0. Since the thermal expansion always experiences a positive change with increasing temperature, the refractive index must become smaller with increasing temperature. Normally glasses experience a positive change in their refractive index with an increase in temperature due to the strongly increasing polarisation capacity of the electron shells as they move further away from the nucleus. The terms $\alpha(n-1)$ and dn/dT consequently change in the same direction and G increases. In the case of inhomogeneous fields, a strong image distortion occurs due to locally different optical path lengths. The influence of the induced thermal stress which additionally increases the entire wave aberration of an optical system makes matters worse:

$$\Delta W_{\text{total}} = \Delta W_{\text{G}} + \Delta W_{\text{s}}, \quad \text{where}$$

$$\Delta W_{\text{total}} = d \cdot G \cdot \Delta T + d \cdot S \cdot \Delta T$$

d is the thickness of the optical medium and S is the photoelastic constant.

A small group of glasses show unusual refractive index-temperature behavior. These are mainly silicate, borate and phosphate glasses each containing fluoride as well as borosilicate, borosilicate aluminate, fluorophosphate glasses each containing fluoride, pure fluoride glasses as well as a few fluoride-free borate, phosphate and silicate glasses. If the temperature coefficient of the refractive index is sufficiently negative, these glasses can be thermostable. Therefore with a minimisation of ΔW_{G} (in the ideal case exactly opposite to the positive value of ΔW_{S}) the possibility exists to eliminate the total wave aberration caused by temperature changes.

According to Prod'homme the temperature coefficient of the refractive index is determined by essentially two factors: the polarisation coefficient phi and the cubic coefficient of expansion β ($\beta = 3\alpha$):

$$dn/dT = C \cdot (\phi - \beta)$$

$$C = (n-1)(n+2)/6n$$

(c is considered to be approximately constant)
If the polarisation properties outweigh the thermal expansion with a temperature increase (temperature dependent change of the polarisation capacity of the ions), the refractive index will increase. However if the polarisation coefficient phi is smaller and less temperature dependent than the cubic coefficient of expansion, the refractive index will decrease. Therefore in the development of athermal glasses. Figure 8.7 shows the empirical correlation between the thermo-optical constant G and the difference (ϕ–β) for different glass systems. The thermooptical constant G clearly first reaches values around 0 or smaller when the difference between the numerical data of the volume dilatation coefficient and the polarization coefficient ϕ amounts to approximately $100 \cdot 10^{-7}\text{K}^{-1}$ and the expansion effect is consequently greater.

If ϕ and β are approximately the same, the thermooptical constant G always has positive values ($\approx 50 \cdot 10^{-7}\text{K}^{-1}$). Since the numerical values of the polarisation coefficient of glasses of a very different composition are approximately of the same order of magnitude, the thermal expansion is clearly the decisive factor in the magnitude and the sign of the temperature coefficient of the

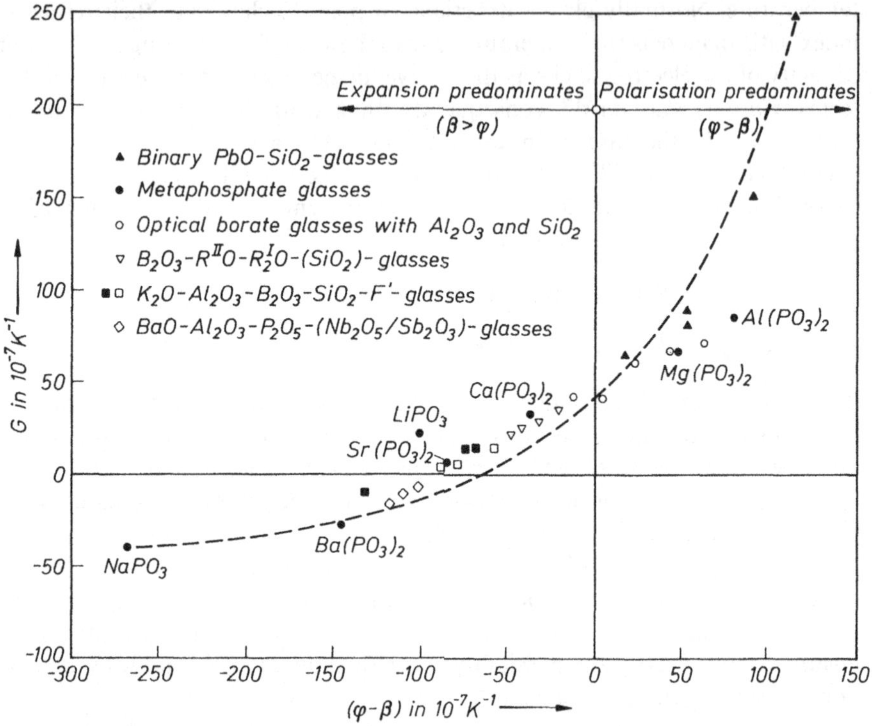

Fig. 8.7. Dependence of the thermal constant G on the polarisation coefficient ϕ and the volume expansion coefficient β for various glass systems

refractive index and therefore of the thermooptical constant. This accounts for the exceptional positive dn/dT of SiO_2 glass.

There is a qualitative correlation between the thermal expansion and the cation radius as well as between the polarisation coefficient of the refractive index and Dietzel's field strength. A series of network modifiers cations can be classified regarding their influence on the temperature coefficient of the refractive index in which the difference $(Z/a - r_c)$ (Dietzel's field strength Z/a and the radius of the cation r_c) is proportionally fixed to the difference $(\phi - \beta)$: $(\phi - \beta)$ $(Z/a^2 - r_c)$. (see Table 8.1 and Fig. 8.8).

The choice of a proper network modifier for a given glass is made according to the theory that high negative values for the difference $(Z/a^2 - r_c)$ are favorized at as negative a temperature coefficient of the refractive index as possible. The absolute quantities must also be considered. Table 8.2 shows a range of thermo optical data in relation to the aforementioned theory for some metaphosphate glasses. The following correlations are evident:

a) For the same base glass (P_2O_5 here) and the same quantity of network modifier oxide:

—the greater the ionic radius, and the smaller Dietzel's field strength, the more favorable is the influence of the ion on the thermooptical behavior of the glass

Table 8.1. Radii, field strengths and Z/a^2–r_K-values of cations in athermal glasses

Cation	r_K	Z/a^2	Z/a^2–r_K
Li^+	0,78	0,23	– 0,55
Na^+	0,98	0,19	– 0,79
K^+	1,33	0,14	– 1,19
Mg^{2+}	0,78	0,45	– 0,33
Ca^{2+}	1,06	0,35	– 0,71
Sr^{2+}	1,27	0,30	– 0,97
Ba^{2+}	1,43	0,26	– 1,17
Zn^{2+}	0,83	0,43	– 0,40
Pb^{2+}	1,32	0,29	– 1,03
Al^{3+}	0,57	0,84	+ 0,27
Y^{3+}	1,06	0,53	– 0,53
La^{3+}	1,22	0,46	– 0,76

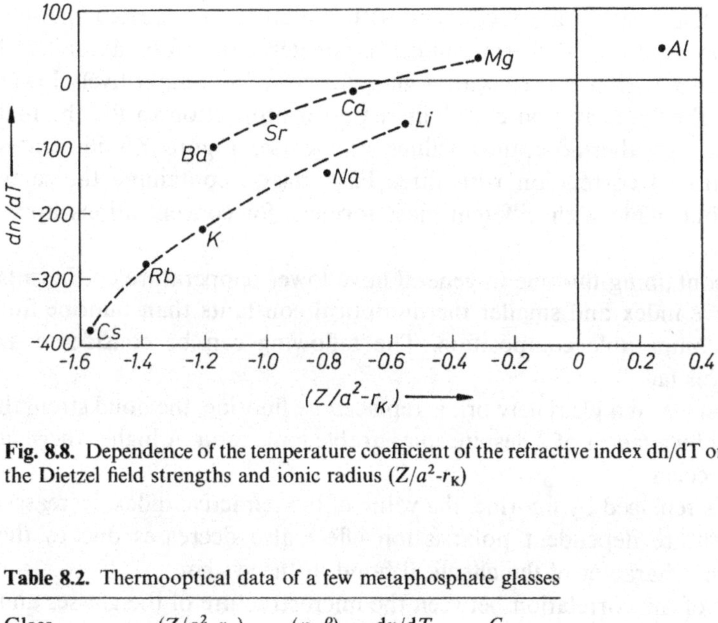

Fig. 8.8. Dependence of the temperature coefficient of the refractive index dn/dT on the difference of the Dietzel field strengths and ionic radius (Z/a^2-r_K)

Table 8.2. Thermooptical data of a few metaphosphate glasses

Glass	(Z/a^2–r_K)	(φ–β)	dn/dT	G
$LiPO_3$	– 0,55	– 101	– 61	+ 19
$NaPO_3$	– 0,79	– 266	– 150	– 44
KPO_3	– 1,19	– 429	– 233	– 107
$Mg(PO_3)_2$	– 0,33	+ 48	+ 29	+ 67
$Ca(PO_3)_2$	– 0,71	– 38	– 25	+ 31
$Sr(PO_3)_2$	– 0,97	– 86	– 59	+ 8
$Ba(PO_3)_2$	– 1,17	– 147	– 106	– 25
$Al(PO_3)_3$	+ 0,27	+ 120	+ 50	+ 83
$Y(PO_3)_3$	– 0,53	+ 114	+ 49	+ 75
$La(PO_3)_3$	– 0,76	+ 20	+ 9	+ 55
$Zn(PO_3)_2$	– 0,40	+ 124	+ 51	+ 83
$Pb(PO_3)_2$	– 1,03	– 183	– 104	– 9

—the empirical classification of the cations concerning their effectiveness is qualitatively correct.

b) Theoretical predictions of the degree of influence of a cation on the thermooptical parameter are in principle possible.

The choice of a proper base glass is of great importance for the achievement of the desired athermal properties. Table 8.3 lists the main refractive index n_e, the linear thermal expansion α, the temperature coefficient of the refractive index dn/dT, the thermooptical constant G as well as Dietzel's field strength Z/a^2 for the most important glass formers. If the basic glass former already possesses good thermooptical properties, thermal glasses can easily be synthesized by proper addition of network modifiers. Silicate and germanate base glasses are naturally not as promising as phosphate or borate glasses. The degree of influence of the network modifier cations is highly dependent on the field strength of the network formers. It can be qualitatively assessed that for a given cation and glass formers with high field strengths the polarization properties as a function of temperature are secondary with regard to the created thermal expansion. The effect on the thermooptical parameter is therefore favorable. If however the same cation is used with a glass former of a comparitively lower field strength, the "polarization effect" increases in proportion to the thermal expansion and the thermo-optical values are worse. Figure 8.9 illustrates the aforementioned correlation with three base glasses containing the same cation Pb^{2+} but made with different glass formers, for concentrations up to 15 mol% PbO.

Glasses containing fluorine in general have lower temperature coefficients of the refractive index and smaller thermooptical constants than fluorine-free glasses of a comparable composition. The following can be considered as a reason for this fact:

- If the oxygen ion in a glass network is replaced by fluorine, the bond strength is weakened by a factor of 2 despite comparable ionic radii. A higher thermal expansion occurs.
- As oxygen is replaced by fluorine, the value of the refractive index decreases. The temperature dependent polarisation effect also decreases due to the change in the character of the chemical bond in the system.

Investigations of the correlation between the microstructure of the glasses and their thermostable properties show that the thermo-optical constant is not

Table 8.3. Properties of the pure glass formers

Glass	n_e	α	dn/dT	G	Z/a	CN
SiO_2	1.4601	5.5	+ 102	+ 104	1.57	4
B_2O_3	1.4650	144	− 350	− 283	1.63	4
P_2O_5	1.4930	137	− 922	− 854	2.1	4
GeO_2	1.6056	71	+ 194	+ 237	1.29	4

(α, dn/dT and $G \times 10^{-7}\,K^{-1}$)
CN–coordination number

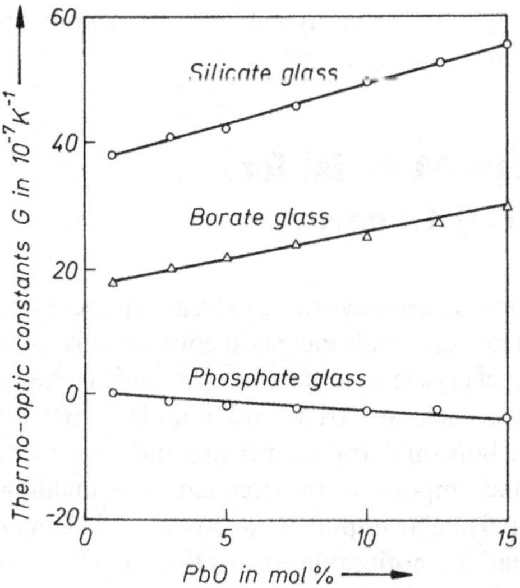

Fig. 8.9. Dependence of the thermooptical constant G on the PbO content in a SiO_2, B_2O_3 and P_2O_5 glass

sensitive to structure. The different stages of phase separation were also found to have no influence on the thermostable properties of the glass. Reitmayer and Schröder [657] postulated a correlation between the temperature coefficient of the refractive index and the change in the position of the UV absorption peak, λ_0, which still has not been explicitly examined. The first longer wavelength of UV absorption λ_0 is 8.5 eV (145 nm) for silicate glasses. All other measurements up to now of a temperature dependent change of the UV edge (200 . . . 400 nm) were simply the effect of impurities (e.g. Fe^{2+}/Fe^{3+}, Pt^{4+}/Pt^{2+} among others)

8.5 Non-Linear Refraction

The possibility of light-operated (photonic) switches in future information systems as well as the increasing development of light-operated sensors has caused intensive interest in non-linear optical glass materials. Although the non-linear refraction index defined by n_2 in the equation

$$n = n_0 + n_2 I$$

(where n is the resulting index, n_0 the linear index, and I the intensity of impacting light) is small (of the order of 10^{-13} esu) the effect above a threshold intensity is extremely fast (subpicoseconds). And, doping glasses with semiconductor crystals such as CdSe, or CdTe, has achieved non-linearities of 10^{-9} to 10^{-8}. (See e.g. Koyama et al. [658]. A typical scheme for using non-linearity in a sensor (Stegeman and Stolen [659] is represented by two fibers of exactly identical index touching, with the input in one fiber "cross-talking" to the output fiber. If a change in input intensity causes a slight change in index, the cross talk and thus the output is stopped. Precise control of size, shape and distance of particles as well as increased understanding of their function will have to precede expected applications. (Koizumi [660]). On the contrary, laser glasses have to have as small as possible n_2 to prevent focusing an resulting fracture.

8.6 Prerequisites on the Raw Material for the Production of Optical Glasses

As optical glasses found completely new and always more extreme applications, the requirements for the purity of the raw materials increased considerably. New process techniques for the production of classic as well as modern optical glasses have made earlier quality problems, such as striae and bubbles, virtually irrelevant. The use of platinum as a bulk material in melting and processing steps plays a decisive role. A second important requirement is a maximal transparency of the optical glass. In particular impurities in colored 3d element compounds in the raw materials lead to noticeable absorption or losses in transparency of the glasses produced.

Table 8.4 gives a general overview of the maximum heavy metal impurities still allowed in glass raw materials for the production of wave guide glass. If the new wave guide glass family of zirconium fluoride glasses is disregarded, the value obtained for the optical attenuation is only about 1 dB/km when iron compounds in a concentration of 10^{-8} to 10^{-11}% and such 3d elements as Ni, Cu, Cr, Co and V are completely absent. An optical attenuation of 1.1 dB/km at a wavelength of 1060 nm was obtained years ago by Bell Telephone. Traces of water in the glass structure also increase the optical attenuation losses. The microstructure of the glass plays a considerable role.

Turbidity losses due to phase separation reduce light transparency. As one goes to longer wavelength light, this phenomenon's effect on the attenuation losses decreases. In the classical melt processes for the production of optical glasses, the melting points of the different raw materials plays an enormous role on the melt drain, the volatility of certain components and therefore on the presence of striae and bubbles. A glass component can for example be introduced as an hydroxide, nitrate, carbonate, oxide or bound to other anions in the melt. The fewest technical problems can be expected when the melting points of all raw materials of a mixture do not have too large or too extreme a difference.

Table 8.4. Maximum amount of heavy metal impurities allowed in glass raw materials for the production of different optical glasses (in % by weight)

Impurities in the glass raw materials	Production of normal optical glasses (binoculars, photooptics ...)	High-performance optics	Fiber optic or wave conductor (modern telecommunications)
Fe_2O_3/FeO	max. $1\text{–}2 \cdot 10^{-3}$%	max. 2.10^{-6}%	$3 \cdot 10^{-8}$ bis $1 \cdot 10^{-11}$%
Ni, Cu, Cr, Co, V (as metal)	$\Sigma\, 1\text{–}2 \cdot 10^{-4}$%	max.each $1 \cdot 10^{-6}$%	—

9 Structure and Properties of Colored Glasses

9.1 General

The color of glass is simply the consequence of the absorption and emission of the electron clouds of its constituent ions. In other words: If a glass is irradiated by white light, mobile electrons of the outer electron shells of certain ions accept energy of a certain wavelength, while the remaining light, now deficient in certain components, no longer passes as white, but rather as colored residual light. A portion of the incident light is converted to thermal vibration; another part, however, contributes to the passing light by emission. This emission is caused by the return of excited electrons to their original level. This part does not conform exactly to the wavelength of the light absorbed.

Before discussing the chemistry of colored glasses, some basic parameters characterizing them will be briefly introduced. A simple and unequivocal characterization is by the transmittance τ_λ as a function of the wavelength λ of the irradiating light. The spectral transmittance τ_λ is the ratio of the light flux $(\phi_d)_\lambda$ leaving the glass to the light flux ϕ_λ of a monochromatic light radiation:

$$\tau = (\phi_d/\phi)_\lambda$$

Since, always, less light leaves the glass than enters it, this is always a true fraction. Analysis of the distribution of the incident light yields the following:

One portion of the incident light flux is weakened by absorption. This portion, termed ϕ_α is usually converted to heat. Another portion, ϕ_p, is lost by reflection at the surface. Thus, the incident light flux can be divided into three portions:

$$\phi = \phi_p + \phi_\alpha + \phi_d$$

where $\phi_d =$ the portion emerging, $\phi_\alpha =$ the portion absorbed, and $\phi_p =$ the portion reflected.

To calculate filter combinations and values for glasses of varying thickness, the only significant processes are those within the glass. It is therefore expedient to use parameters which disregard the reflection losses and refer only to the flux entering the glass. Thus, the "internal transmittance" θ_λ is defined as the ratio of the emerging light flux ϕ_d to that entering $\phi_i = \phi - \phi_p$:

$$\theta_\lambda = \frac{\phi_e}{\phi_i}$$

It is customary to measure τ_λ. By proper measuring techniques, e.g., immersion in a liquid of equal refraction, reflection can be eliminated and θ_λ can be determined directly.

Reflection losses depend on the quality of polishing, the angle of incidence, the index of refraction of the glass, and the surrounding medium as well as on wavelength. If the colored glass is in the shape of a parallel plane plate, bordering on air, with perfectly smooth surfaces without contamination (e.g., by dust or humidity), and the incidence is vertical (or not too far from vertical) the relation

$$\tau = P\Theta$$

The quantity P is defined as follows:

$$P = \frac{2n}{n^2 + 1}$$

where n = index of refraction and P is called the reflection (or Fresnel) factor, which changes only slightly with wavelength λ. Colored filter catalogs usually tabulate P values of various n values as a function of λ.

9.2 Absorption of Colorless Base Glasses

The transmission curve of simple colorless glasses in the UV is represented schematically in Fig. 9.1. At very short wavelengths (for SiO_2 glass at about 122 nm), a first very weak absorption (and reflection) maximum is observed, labeled λ_s. This maximum corresponds to the first eigenvibration of a colorless base glass.

More important is the UV absorption edge, conventionally designated λ_g, representing that wavelength at which the transmission has been reduced to 50%. Kordes [661] characterizes the absorption edge by λ_0 which is obtained by the linear extrapolation of the steep portion of the transmission curve. The

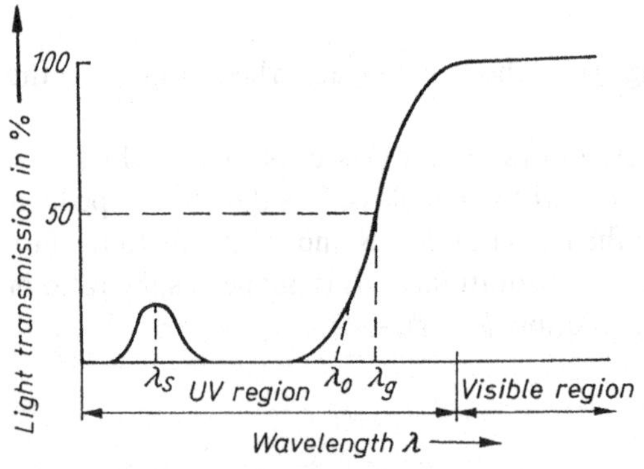

Fig. 9.1. Schematic of the light transmission of colorless glasses in the UV

advantage of this characterization is the relative independence on the quality of the sample (scattering by seeds or surface flaws).

Figure 9.2 and Table 9.1 show the shift to longer wavelength in the UV for base glasses in the order P_2O_5, SiO_2, B_2O_3, HPO_3, GeO_2. The P_2O_5 curve is hypothetical since it is hardly possible to obtain a P_2O_5 glass free of H_2O and since a very small H_2O content causes a shift of the edge toward longer wavelengths (compare the curves for P_2O_5 and HPO_3). Except for B_2O_3, the shift of the absorption edges corresponds with the decrease in field strength of the network-forming cations. Clearly the bond to oxygen plays a decisive role.

As to B_2O_3, here too the preparation of H_2O-free glass is difficult. Moreover, H_2O causes 4-coordination for some boron ions, resulting in a decrease of the field strength from 1.63 to 1.34, and a shift of the absorption edge to longer wavelengths.

A further shift of the curve for the pure network-forming glass results from the introduction of alkali and the concomitant production of non-bridging oxygen. As shown in Table 9.1, the UV absorption edge of SiO_2 glass may be shifted from 162 nm to 270 nm, that of B_2O_3 glass from 200 nm to about 360 nm. The shift increases in the order Li_2O, Na_2O, K_2O, Rb_2O, or Cs_2O. Basically, UV absorption is caused by the excitation of the electrons in oxygen

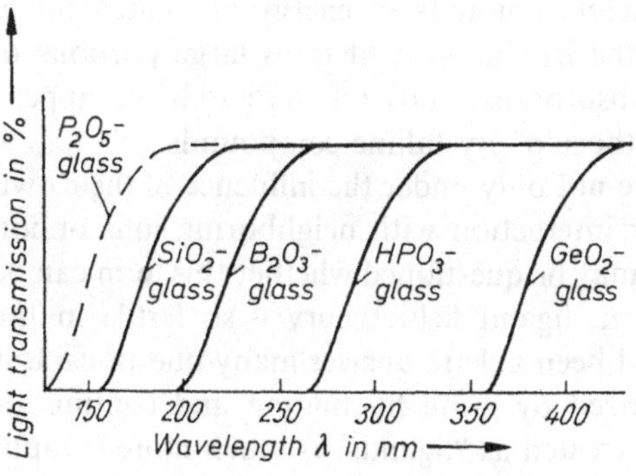

Fig. 9.2. Comparison of the UV absorption edges of various colorless glasses

Table 9.1. Limiting wavelength of absorption Λ_0 of various base glasses (Kordes [651])

Type of glass	Volume (in nm) (Limiting wavelength of absorption)	Field strength of the network-forming cation (= coordination)
P_2O_5	?	$P^{5+} = 2.1$
SiO_2	162	$Si^{4+} = 1.57$
B_2O_3	200	B^{3+} (CN = 3) = 1.63, (CN = 4) = 1.34)
HPO_3	273	$P^{5+} = 2.1$
GeO_2	363	$Ge^{4+} = 1.07$
P_2O_5–Me_xO_y	220	
SiO_2–Me_xO_y	270	
B_2O_3–Me_xO_y	360	

and will shift to longer wavelengths with a weakening of the oxygen bond. In detail, however, the absorption edge is strongly dependent on the concentration and valency state of small iron oxide impurities.

The glasses appear colorless since the shifts described do not reach the visible range of the spectrum. Only around 3000 to 5000 nm, i.e., in the infrared, the interaction glass radiation is resumed, as will be discussed later.

The interactions of glass with UV and IR are entirely different. The interaction with UV radiation involves the excitation of electrons, primarily of oxygen; that with IR radiation involves vibrations and rotations in entire groups of atoms, indeed of the entire network. Conventional colorless silicate glasses are usually not transparent above 5000 nm.

9.3 Glasses Colored by Ions

If one introduces 3d elements into the base glasses discussed in the previous section, e.g., Fe, Mn, Ni, Cu, Co, Cr, etc., one obtains the simplest type of colored glasses.

The electrons in the subgroups of the periodic system are particularly mobile, as expressed in their tendency to change valency. These elements, when introduced into glass, cause resonance absorption in their electron clouds under white-light irradiation. Usually discrete amounts of energy are taken up in certain wavelength ranges so that the irradiated light loses large portions of wavelengths in the form of defined absorption bands. Often these bands appear broadened and blunt compared to those of crystalline compounds.

But these excitable electrons are not only under the influence of their own nuclei, but also under that of their interaction with neighboring ions or ion groups, concisely termed ligands. It may be questioned whether this term can be applied properly to glasses; however, ligand field theory – so fertile in the chemistry of complexes – had indeed been able to answer many questions. But its application to glasses characterized by more fluctuating and continuous transitions of the glass constituents treated as "ligands" is much more complicated than in the chemistry of complexes. One is much more dependent on assumptions and hypotheses.

At any rate, absorption processes in glasses colored by ions are, if not identical with, clearly similar to those in aqueous solutions of the salts of 3d elements. This justifies the description of glasses colored by ions by the example of well-defined inorganic complex compounds. In the following paragraphs, some lucid basic principles for influencing the absorption properties of glasses colored by ions will be presented. Such principles may well be guides for the systematic development of colored glass types. But within the frame of a single chapter of a treatise on glass chemistry it will not be possible to treat all types and subtypes of colored glasses produced today nor their interpretation on the basis of ligand field theory. Reference should be made to specialized monographs from the classical text of Weyl [662] to the recent extensive treatment by Wong and Angell [18].

9.3.1 Dependence of Absorption on Network-Former

It is known from crystal chemistry that even a small change in the ligand field may cause enormous changes in absorption behavior, thus in coloration. Table 9.2 exemplifies the color changes in Co(III) complexes when merely one or two ligands in a cobalt hexamine complex are replaced by other ligands. Almost all hues of the spectrum are obtained as the luteocomplex is changed to the roseo-, purpureo-, violeo-, praseo-, flavo-, croceo-, or melanocomplexes. In glasses these changes are less pronounced, yet clearly apparent, when in a base glass at constant molar relations of its constituents a network-former is replaced, e.g., SiO_2 by P_2O_5, B_2O_3, BeF_2, etc.

Figure 9.3 gives an example from practice. Both color filters (BG1, BG24) (from the Schott catalog) are colored by Co^{2+} but BG1 is a silicate, BG24 a phosphate glass. The ligands are different. Thus, apart from the different UV and IR absorptions of the base glass, the absorption due to Co^{2+} is entirely different in the fields of SiO_4 and PO_4 ligands. Thus, since BG24 transmits more red light it appears reddish-violet compared to the pure blue of BG1.

Table 9.2. Colors of Co(III) complexes observed

Complex type	Color	Cation complex
Luteo-type	yellow	$[Co(NH_3)_6]^{3+}$
Roseo-type	pink	$[Co(NH_3)_5H_2O]^{3+}$
Purpureo-type	purple	$[Co(NH_3)_5Cl]^{2+}$
Violeo-type	red-violet	$[Co(NH_3)_4Cl_2]^{1+}$ *cis*-form
Praseo-type	green	$[Co(NH_3)_4Cl_2]^{1+}$ *trans*-form
Flavo-type	yellow-brown	$[Co(NH_3)_4(NO_2)_2]^{1+}$ *cis*-form
Croceo-type	yellow-red	$[Co(NH_3)_4(NO_2)_2]^{1+}$ *trans*-form
Melano-type	black	[Co-multiple core complex]

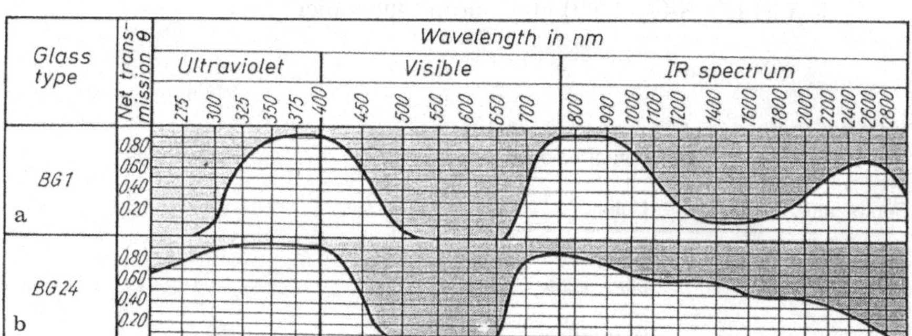

Fig. 9.3. Light transmission curves of some Schott colored glasses. **a** BG1: effect of CoO in a silicate base glass; **b** BG24: effect of CoO in a phosphate base glass

This effect may well be compared with two cases in classical complex chemistry where a changed ligand causes a significant change in absorption:

$[Cu^{2+}(H_2O)_4]^{2+}$	$[Cu^{2+}(NH_3)_4]^{2+}$
blue	violet

or

$[Co^{3+}(NH_3)_6]^{3+}$	$[Co^{3+}(C_2O_4)_3]^{3-}$
yellow	green

9.3.2 Dependence of Absorption on Modifiers

It is known from inorganic chemistry that mutual polarization effects between ionic partners have considerable influence upon the absorption behavior of a compound. Table 9.3 provides two examples. In one case the increasing polarization effect of the silver ion on halogen ions induces a color change from white (chloride) to yellow (iodide). In the second case the polarization by nickel induces a deepening of the color from a faint yellow (fluoride) to a deep black (iodide).

Now, if, for example KI confronts NiI_2, forming the complex compound $K_2(NiI_4)$, an enormous reduction in absorption and lightening of color results. Whereas NiI_2 is black, the $K_2(NiI_4)$ complex in solution is red. This is due to nothing else than the counterpolarization effect of K^+ on the first coordination sphere involving Ni^{2+} and I^+ ions.

To give another example, $K_4(Fe(CN)_6)$ (yellow) and K_2Zn_3 $(Fe(CN)_6)_2$ (white) are compared. Zn^{2+} ions have a stronger counterpolarizing effect vis-a-vis the central iron ion than K^+ ions. Analogous effects are found in glasses:

$Li_2O(PbO–SiO_2–NiO)$ glass: yellow

$Na_2O(PbO–SiO_2–NiO)$ glass: red-violet

$K_2O(PbO–SiO_2–NiO)$ glass: deep blue-violet

Table 9.3. Coloration of silver and nickel halides conditioned by the polarization of halogen ions

Compound	Color
AgCl	white
AgBr	faint yellow
AgI	yellow
NiF_2	faint yellow
$NiCl_2$	yellow-brown
$NiBr_2$	dark brown
NiI_2	black

The colored glass with the alkali ion of lowest field strength (K^+) exhibits the strongest absorption. This suggests that modification in the base glass without a change in colorant (chromophore) is an additional way to control the absorption behavior of colored glasses.

9.3.3 Dependence of Absorption on the Valency of the Chromophore

It is generally known that essentially all 3d elements tend to change valency easily. Changes in absorption behavior (color) of the crystalline compound or its aqueous solution result. Manganese, for example, occurs with a valency of 2, 3, 4, 6, or 7 in compounds, almost always characterized by significantly different coloration.

Under specific conditions, almost all these valencies of Mn are observed in glasses. It is, however, difficult to maintain just one of them: As a rule, equilibria are established. This is why the oxygen potential is of enormous importance when colored glasses are melted. It is out of the question to include in this text the absorption behavior of silicate-, borate-, borosilicate-, phosphate-, etc. base glasses containing all of the 3d elements. Reference is made for details to specialized texts such as those by Weyl [662] Wong and Angell [18] and Bamford [663].

A typical example of changes by both ligand and valency of a coloring cation is the behavior of iron in silicate and phosphate glasses.

Fe^{2+} in a silicate-based glass: deep blue-green

Fe^{3+} of equal concentration in the same base: deep yellowish-brown

Fe^{2+} in a phosphate-based glass: a slight greenish-blue

Fe^{3+} of equal concentration in the same base: slightly brownish.

In the two latter cases, P^{5+} has a significant counterpolarization effect on the Fe–O coordination. This fact has been used successfully in heat-absorbing protective glasses where the strong IR absorption of Fe^{2+} is combined with a high visible transmission. In addition, changes in the base glass composition affect the coordination of Fe^{2+} or Fe^{3+}. This is the cause of further color variations.

These are only examples for complications to be expected when colored glasses are melted. As a rule, base glass composition, chromophore concentration, furnace atmosphere, melting temperature and time, and cooling rate result in an equilibrium of Fe^{2+} and Fe^{3+}. Some control of this equilibrium is possible; however, Fe^{2+} or Fe^{3+} can hardly be excluded.

9.3.4 Dependence of Absorption on the Coordination Number of the Chromophore

Iron and cobalt are used as examples because of their dominant role in colored glasses.

9.3.4.1 Coordination Change Due to Change in Chromophore Concentration

The Fe^{3+} ion in acidic base glasses is usually present in the form of $(Fe^{III}O_4)$ complexes as long as its concentration is small. At much larger concentrations, however a change in absorption behavior (color) indicates a transition from $(Fe^{III}O_4)$ to $(Fe^{III}O_6)$ complexes. A similar effect is observed in glasses containing increasing amounts of Co^{II}. For silicate glasses the observed color changes can be summarized as follows:

$(Fe^{III}O_4)$ brown	$(Co^{II}O_4)$ blue
$(Fe^{III}O_6)$ yellowish-pink	$(Co^{II}O_6)$ pink

9.3.4.2 Coordination Change of Chromophore Due to Concentration Change of Network-Modifier

Adding modifier oxide to an acidic base glass containing constant small amounts of colorant tends to decrease the coordination of the chromophore. Thus, for instance, an Na_2O–B_2O_3–CoO (small amount) glass containing > 25 mol% Na_2O is colored a pure blue due to the $(Co^{II}O_4)$ complex. For lower Na_2O contents, however, the color shifts to pink due to the $(Co^{II}O_6)$ complex.

As is known from the study of the structure of borate glasses, non-bridging oxygen first reappears at a concentration of 25 to 30 mol% Na_2O, therefore making the higher coordination of the Co^{2+} ions possible.

9.3.4.3 Coordination Change of Chromophore Due to Changed Network-Former

Experiments show that equal concentrations of chromophore oxide, e.g., CoO, result in entirely different absorption properties (coloration) in different acidic base glasses:

P_2O_5	pink $(Co^{II}O_6)$
SiO_2	blue $(Co^{II}O_4)$
B_2O_3	pink $(Co^{II}O_6)$ low Na_2O, blue $(Co^{II}O_4)$ high Na_2O
B_2O_3–SiO_2	amethyst $(Co^{II}O_6)$ + $(Co^{II}O_4)$ low Na_2O, blue $(Co^{II}O_4)$ high Na_2O

In the case of P_2O_5, P^{5+} ions counterpolarize the Co–O coordination sphere to the extent of a loosening of the coordination from 4 to 6. A change to red-violet or blue requires large amounts of network-modifiers (alkali) which diminish the counterpolarization of P^{5+}.

Borosilicate glasses containing Co^{2+} are blue at high alkali contents and show an amethyst coloration at low alkali contents thereby indicating that a $[Co^{II}O_4]/[Co^{II}O_6]$ equilibrium exists in the glass.

9.3.5 Problems of Interpretation

The preceding section gave only simple examples for major influences on the transmission behavior of glasses colored by ions. The separation of these influences when interpreting an absorption curve is complicated, indeed problematic. It is not surprising that the results and interpretations of various authors are often contradictory when multicomponent glasses are investigated. For instance, an increasing alkali content may have the opposite effect when chromophore concentrations and network components are varied.

For this reason this book deliberately abstains from a detailed discussion of the numerous glasses colored by ions, including those of technological interest. In spite of the practical interest in such a survey, it would not easily contain useful generalizations. Instead, the following section will give examples of the technologically most important chromophores and selected typical absorption curves.

In fact, one potentially important influence on the transmission of a colored glass has, so far, not attracted sufficient attention. This is the influence of microheterogeneities, which may cause enrichment of the chromophore in a particular microphase. Basic experiments addressing these questions were treated more extensively earlier. But in this respect glass research now stands only at the beginning.

The practitioner will find extensive coverage of colored glasses in the text by Weyl [662]. As to more theoretical interpretations potentially valuable for design and control of colored glasses, the most pertinent work is not free of contradictions [664–670]. An excellent assessment including description as well as critique of the methodology, and extensive coverage of compositional fields is by Wong and Angell [18].

9.3.6 Technologically Important Chromophores and Selected Transmission Curves

In Table 9.4 the most important raw materials for the production of glasses colored by ions are summarized, including the effective chromophore and the usually resultant coloration. The coloration listed is typical, but may be altered considerably for the same chromophore due to the influences discussed in the previous sections. In practice, often more than one chromophore is used to obtain a specific color. In such cases the redox equilibria are not easy to resolve. For this reason, control often remains quite empirical. Concentrations generally vary from 0.5 to 5 mass%, in specific cases from 0.1 to 12 mass%.

Table 9.4. Chromophores for the production of ionically colored glasses

Raw material	Effective chromophore	Usual coloration
iron oxide Fe_2O_3 iron oxalate $Fe(C_2O_4)\cdot 2H_2O$	Fe_2O_3 FeO	yellow brown
cobalt (III) oxide Co_2O_3 cobalt carbonate $CoCO_3$	CoO	blue
copper (II) oxide CuO copper sulfate $CuSO_4\cdot 5H_2O$	CuO Cu_2O	blue red
chromium (III) oxide Cr_2O_3 potassium bichromate $K_2Cr_2O_7$	Cr_2O_3 CrO_3	green yellow
nickle (II) oxide NiO nickel (III) oxide Ni_2O_2 nickel carbonate $NiCO_3$	NiO	green
uranium oxide U_3O_8 sodium uranate $Na_2U_2O_7\cdot 3H_2O$	$U_2O_7^{2-}$ uranate ion UO_2^{2+} uranyl ion	yellow orange (no fluorescence) yellow green (strong fluorescence)
manganese dioxide MnO_2 potassium peramanganate $KMnO_4$	MnO Mn_2O_3 and higher oxides	faint yellow violet
rare earth oxides Ln_2O_3 mostly didymium $(Nd_2O_3 + Pr_2O_3)$	Nd_2O_3, Pr_2O_3	violet (with fluorescence)
cerium dioxide CeO_2	CeO_2 Ce_2O_3	colorless to yellow UV-absorbing

Figure 9.4 shows the pure transmission curves for 1–2 mm thickness (using θ_λ) for selected colored and filter glasses, containing the most used chromophores. The effective chromophores are listed for each glass, as are the base glass types.

The designations are those of Schott catalogs and characterize color families as follows:

BG	blue
GG	yellow
OG	orange
RG	red
VG	green
UG	violet (either IR-transmitting or not)
NG	neutral (gray)
SG	protective

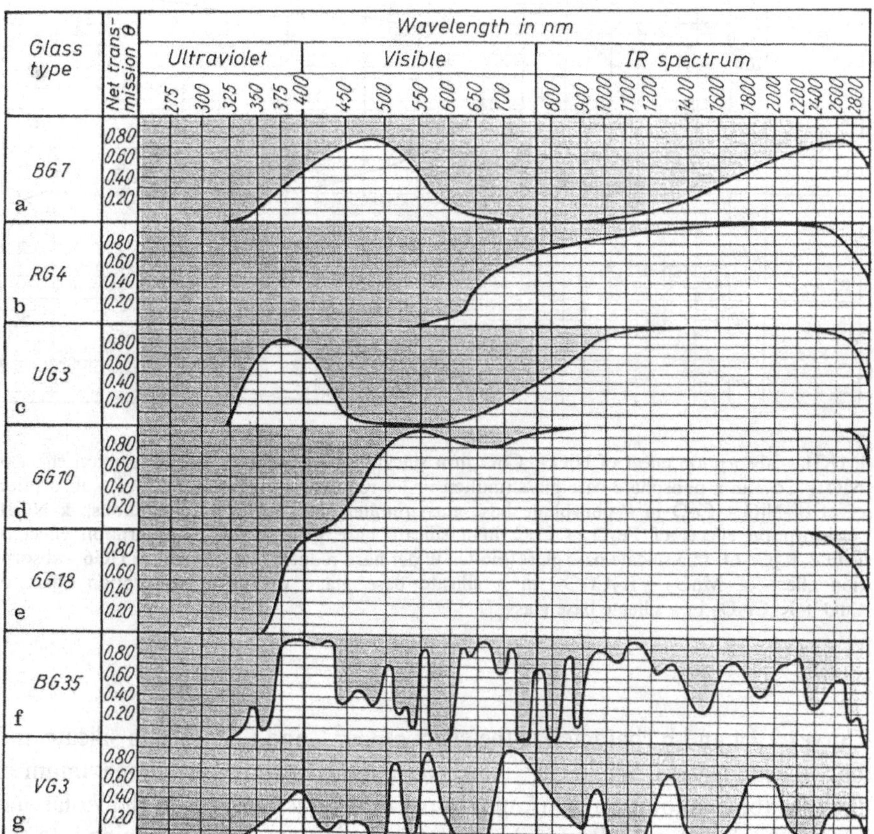

Fig. 9.4. Light transmission curves for some Schott colored glasses. **a** BG7 – absorption effect of CuO in a silicate base glass; **b** RG4 – absorption effect of Cu_2O in a borosilicate base glass; **c** UG3 – absorption effect of $KMnO_4$ in a silicate base glass (molten under oxidizing conditions); **d** GG10 – absorption effect of $K_2Cr_2O_7$ in a silicate base glass, **e** GG18 – absorption effect of CeO_2 in a silicate base glass; **f** BG35 – absorption effect of Didym (Nd_2O_3, Pr_2O_3) in a silicate base glass; **g** VG3 – absorption effect of $Na_2U_2O_7$ in a phosphate base glass;

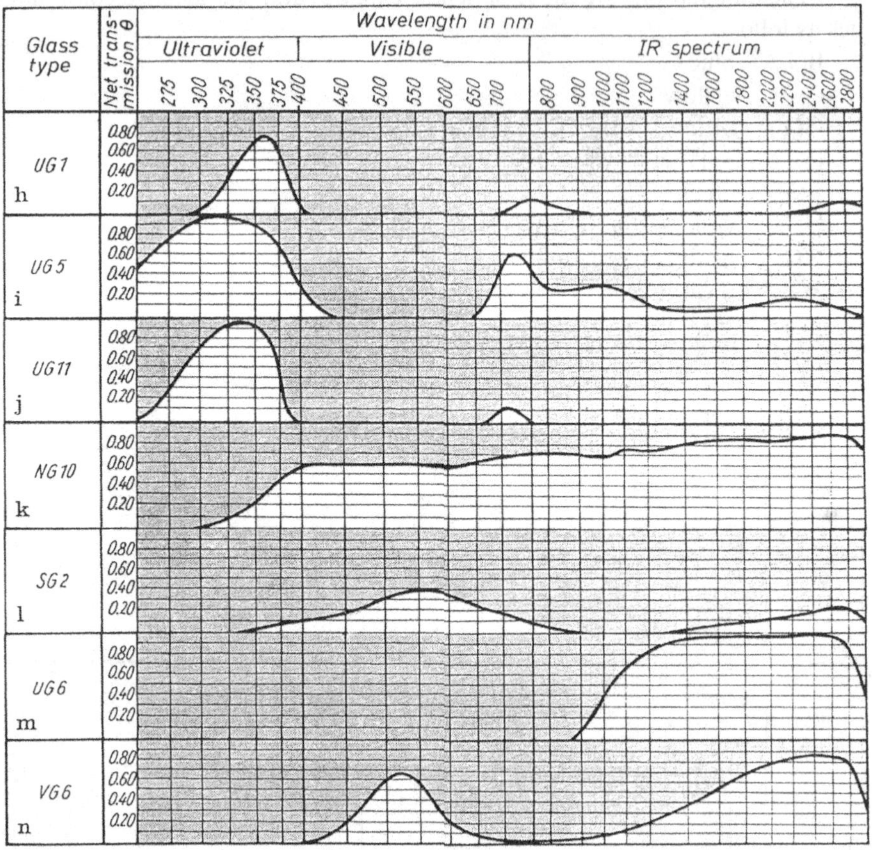

h UG1 – absorption effect of NiO + CoO in a silicate base glass; **i** UG5 – absorption effect of NiO + CoO in a phosphate base glass (molten under reducing conditions); **j** UG11 – absorption effect of NiO + CoO in a phosphate base glass (molten under reducing conditions); **k** NG10 – absorption effects of Fe_2O_3 + CoO in a silicate base glass; **l** SG2 – absorption effect of FeO + $K_2Cr_2O_7$ in a silicate base glass (molten under hard reducing conditions); **m** UG6 – absorption effect of MnO_2 + $K_2Cr_2O_7$ in a silicate base glass; **n** VG6 – absorption effect of CuO + $K_2Cr_2O_7$ in a silicate base glass

Curves 9.4 a and b characterize the effect of Cu^{2+} and Cu^{+} ions in silicate and borosilicate glasses. While Cu^{2+} ions cause a broad transmission maximum in the blue (blue coloration), Cu^{+} in a borosilicate glass absorbs in the violet and blue and exhibits a single, very broad transmission maximum reaching into the IR (to 3000 nm) (red).

Curves 9.4 c to g show absorption effects for Mn, Cr, Ce, Nd, Pr, U ions in silicate and phosphate glasses. The curves for glasses BG35 and VG3 are reminiscent of those for aqueous solutions of rare earth and uranium compounds. Combined with other filters, these two glasses are useful in narrow-band filter combinations.

Curves 9.4 h–j show the effects of Ni and Co ions in silicate and phosphate glasses. The difference is clearly apparent in the UV where it is determined by the base glass (compare Fig. 9.2). Since in a combination of Ni and Co nearly 100% of the IR and visible region is absorbed, this combination provides a filter passing only UV radiation.

Curve 9.4 k, typical for neutral (gray) glasses, is achieved by combining Fe and Co, so as to achieve a close to horizontal transmission curve.

The essential characteristic of a gray glass is that an absorption as even as possible occurs over the entire wavelength region in which the base glass is transparent transmission. There are gray glasses which reduce by 20, 40 or 60% over this entire wavelength region.

The curve 9.4 l results from the combination Fe–Cr in a silicate glass. The protective glass SG2 has to absorb the entire noxious UV and IR radiation while transmitting sufficiently (30–40%) in the visible range. Chromophore concentrations of up to about 6 mass% have been achieved.

The curve 9.4 m is achieved by combining Mn and Cr. All UV and visible radiation is absorbed. The glass UG6 transmits only in the IR (900–3000 nm).

The curve 9.4 n is based on the combination Cu–Cr in a silicate-based glass. Glass VG6 shows a single entirely symmetrical maximum in the visible and is used in monochromate filter combinations.

9.4 Striking Glasses

9.4.1 Composition, Preparation, and Absorption Behavior

So-called "striking" or cut-off colored glasses represent a second large family of colored glasses. They differ fundamentally both as to their way of production and their properties. All colors between yellow, orange, and red (GG, OG, RG) can be achieved.

Most "striking" colors are obtained in base glasses of systems K_2O (10–20 mass%)-ZnO (10–22 mass%)-SiO_2 (50–60 mass%). However, sometimes part of the K_2O is replaced by Na_2O, and ZnO by CaO and, in some types a few percent (1–4 mass%) B_2O_3 are added. The color is achieved by adding 1–3 mass% CdS, CdSe, or CdTe.

Melting must be under neutral or weakly reducing conditions, which is achieved by avoiding nitrates in the batch and maintaining proper gas:air mixtures in the fuel. Otherwise, S, Se or Te compounds would be oxidized to volatile SO_2, SeO_2, or TeO_2. Glassmakers call this undesirable occurrence "burning out." The remaining CdO would be a noncoloring component as is found in some novel optical glasses.

The chromophores are usually introduced as the cadmium chalcogenides (CdS, CdSe, CdTe) with an addition of excess elementary S or Se powder to counteract oxidation. Occasionally the reducing process is secured by additional Zn powder. Often Se is, in part, introduced as Na_2SeO_3.

After casting and fairly rapid cooling the striking glass melts solidify to colorless glasses. Only when reheated and held at 550 to 700 °C do they become colored, i.e., they "strike." During striking, the absorption edge steepens and moves toward longer wavelengths, depending on the additives, in the extreme case to the red, even IR region (Fig. 9.5).

This makes it possible to produce, from the same glass, several color filter types by controlled reheating. The "striking" glasses differ from those colored by ions mainly by the nearly quantitative absorption of white light from the UV to a defined wavelength, and the nearly quantitative transmission from that wavelength upward. The absorption and transmission regions are separated by a sharp edge, making it possible to obtain "pure" (saturated) colors. They make excellent cut-off filters for optical instruments, unlike filters based on ionic colors with their broad absorption bands. Figure 9.6 illustrates the absorption behavior of some typical "striking" or cut-off filter glasses of older Schott make. Chromophores and base glasses are identified. A comparison with the curves of Figs. 9.3 and 9.4 will demonstrate the difference in absorption behavior.

The clearest yellow glass GG5 still shows a distinct residual transparency of about 10% on the other side of the sharp absorption edge, in the near UV region from 425 to 360 nm (Fig. 9.6a). More recent developments [196, 671] have demonstrated that this disturbing residual transparency could be avoided with glasses of the system K_2O–ZnO–SiO_2 replacing a base glass of the system K_2O–CaO–SiO_2. Sharp absorption edges could be obtained even at 400 nm.

9.4.2 Base Glass Structure and Coloring Mechanism in "Striking" Glasses

The fact that, in contrast to glasses colored by ions, the "striking" glasses become colored only after a subsequent reheating process, suggests a completely different coloring mechanism. Indeed, it has been shown that the coloration of a "striking" glass (cut-off filter glass) is caused by the precipitation of an infinite number of submicroscopic cadmium chalcogenide crystals. This process will be

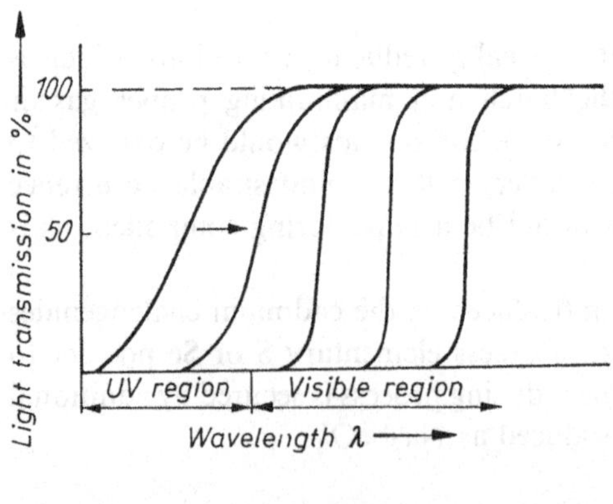

Fig. 9.5. Schematic of the shift of UV absorption edges into the long wavelength region of a "striking" glass during tempering

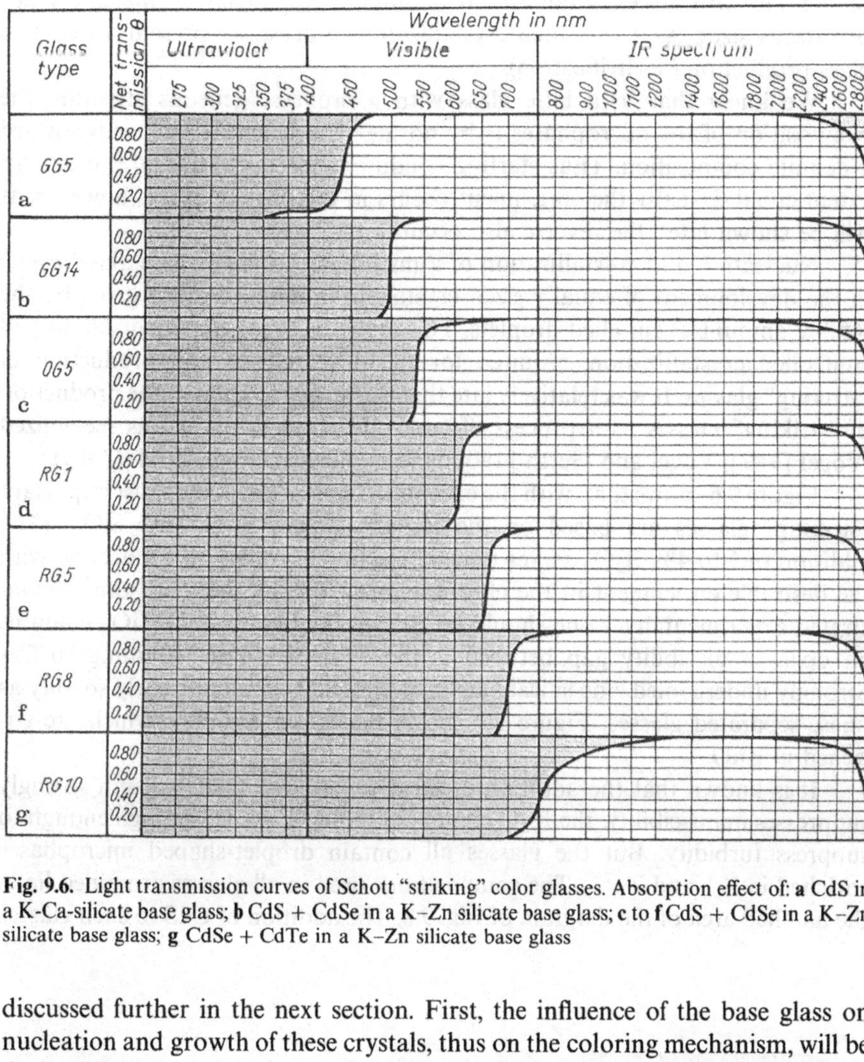

Fig. 9.6. Light transmission curves of Schott "striking" color glasses. Absorption effect of: **a** CdS in a K-Ca-silicate base glass; **b** CdS + CdSe in a K–Zn silicate base glass; **c** to **f** CdS + CdSe in a K–Zn silicate base glass; **g** CdSe + CdTe in a K–Zn silicate base glass

discussed further in the next section. First, the influence of the base glass on nucleation and growth of these crystals, thus on the coloring mechanism, will be dealt with.

The discussion of glass-glass phase separation included the observation that the small amount of 3d elements added to a base glass prone to unmixing will not be equally distributed in the microphases formed; rather they will be enriched in one phase offering optimal coordination conditions.

This enrichment in highly charged modifying cations must become very important for the nucleation and growth of crystals. Whereas in a homogeneous glass with a static distribution of cations nucleation is preceded by an ion diffusion process, this is not necessary in the separated phase in which they are enriched. In other words: Nucleation and growth of crystals is significantly facilitated in phase-separated base glasses. One more fact is decisive. In homogeneous base glasses, nucleation starts incidentally and sporadically in certain locations and at different times. For this reason, due to the different starting times for growth, large crystals form in addition to a spectrum of smaller crystals of various sizes. As a rule such crystallization processes represent a defect in glass production (devitrification).

We know that in a base glass with microheterogeneous structure the composition of the microphases is by no means accidental, but tends toward a definite composition. Thus starting conditions for nucleation are to a large extent equal. Usually the enrichment occurs in the droplet phase, however, as will be shown later, the reverse also occurs.

Nucleation and crystallization of a microheterogeneous base glass results in the development of equally sized crystals in infinite locations given by the infinite amount of enriched droplets. This is the basis for the understanding of controlled crystallization, required for the development and production of "striking" glasses. It was relatively late that this interpretation of the production of "striking" glasses as a process of controlled crytallization was recognized (Vogel [189], Vogel and Gerth [190]) and confirmed (Vogel et al. [196]).

Figure 9.6 shows that, with the exception of glass GG5, the most important "striking" glasses are based on the ternary system K_2O–ZnO–SiO_2. (The addition of 1 to 4% B_2O_3 serves the melting process only.) In connection with the desirable enrichment in the chromophore CdS – which has been demonstrated experimentally – one should recall that the binary ZnO–SiO_2 contains an open immiscibility gap between 2 and 35 mol% ZnO (Bunting [672]). Strongly undercooled zinc silicate melts of high SiO_2 content clearly solidify as phase-separated glasses. Figure 9.7 shows the 0.2 μm droplets which are enriched in ZnO.

It is known that the addition of a cation of weak field strength strongly suppresses immiscibility; the K_2O content of these glasses is just high enough to suppress turbidity. But the glasses all contain droplet-shaped microphases enriched in K^+ and Zn^{2+}. The compositions used in all glasses examined lie in the hatched area of the diagram of Fig. 9.8. The addition of CdS to such glasses

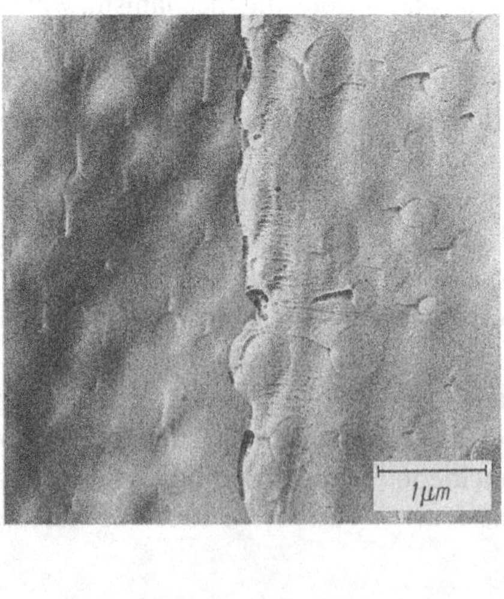

Fig. 9.7. Binary turbid zinc silicate glass of high SiO_2 content. Precipitation of a ZnO-rich droplet-shaped silicate glass phase (electron micrograph after replica preparation)

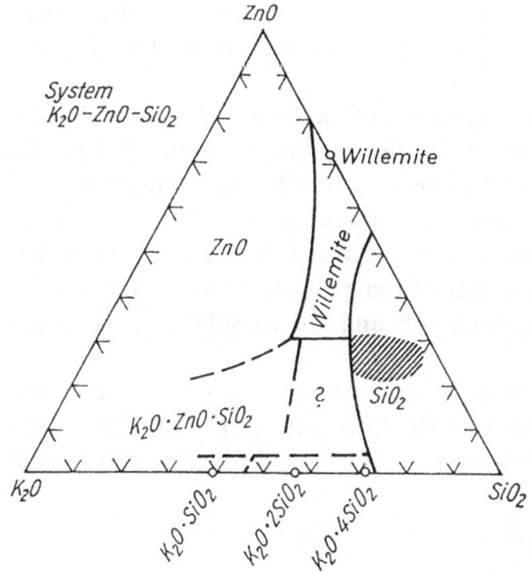

Fig. 9.8. Melt system K_2O–ZnO–SiO_2 [673]. The glasses investigated lie in the cross-hatched area

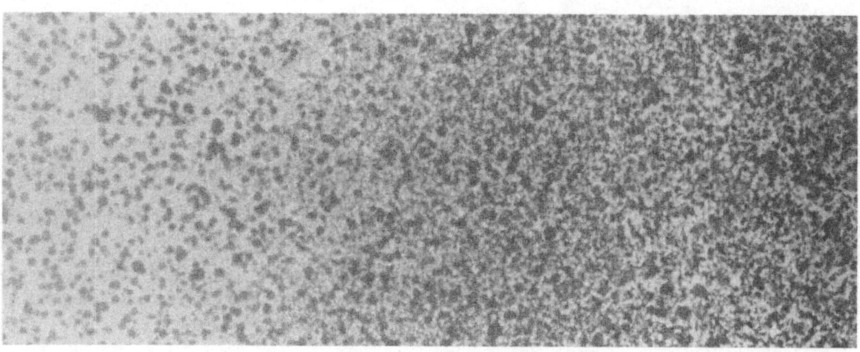

Fig. 9.9. Strip of a glass (system K_2O–ZnO–SiO_2–CdS) heat-treated for 45 h at 700–820 °C. Precipitation of yellow-brown droplet regions in a nearly colorless matrix phase (light micrograph (4:1)

represents the simplest case of a "striking" glass serving as a model for fundamental investigations.

It is, indeed, found that CdS is enriched in the Zn-rich phase, as expected from the relationship between Zn and Cd. Composition and cooling keep the droplets just small enough to suppress turbidity. But they grow rapidly on reheating, becoming large enough to be visible in the light microscope, thus revealing their color. After reheating for 45 h at 700 to 820 °C, a glass composed of 68 mol% SiO_2, 29 mol% ZnO, 3 mol% K_2O (1 mol% CdS added), changed in appearance, as shown in Fig. 9.9.

At $T = 720$ to 730 °C, the first yellowish-brown glassy droplet regions had been precipitated. At higher temperatures their color deepens and crystallization starts. The pictures prove clearly that CdS was enriched in the droplet microphases and by no means distributed evenly in the base glass. The matrix is indeed almost completely colorless.

Figure 9.10 again shows the glassy nature of the droplet. Figure 9.11 shows the start of crystallization within the droplet regions at about 740 °C. The crystallization droplets look almost like biological cells with their nuclei.

In the range of 780 to 800 °C a secondary phase separation occurred in the form of very small droplets which also are yellowish-brown (Fig. 9.12). This secondary separation is due to the increased ionic diffusion near 780 °C, leading to the precipitation of some remaining Zn and Cd as sulfides in a second enriched phase.

How much Zn and Cd remains in the matrix phase depends on the cooling rate. In these studies the composition of the CdS-containing K_2O–ZnO–SiO_2 glasses had been chosen deliberately so as to favor the separation of relatively

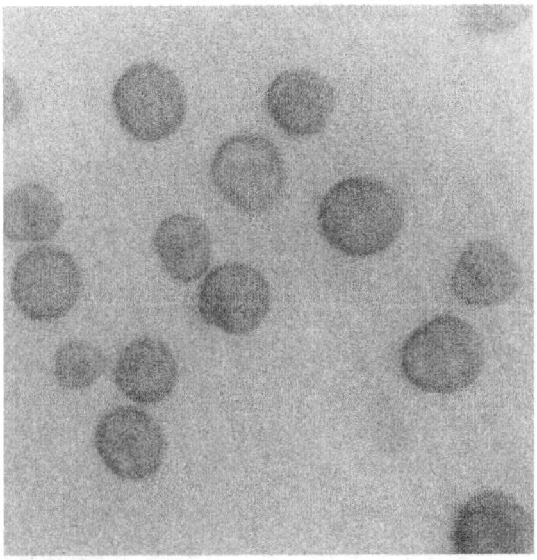

Fig. 9.10. Section from Fig. 9.9. In the temperature region to 730 °C the droplet regions remain clear (light micrograph ca. 45:1)

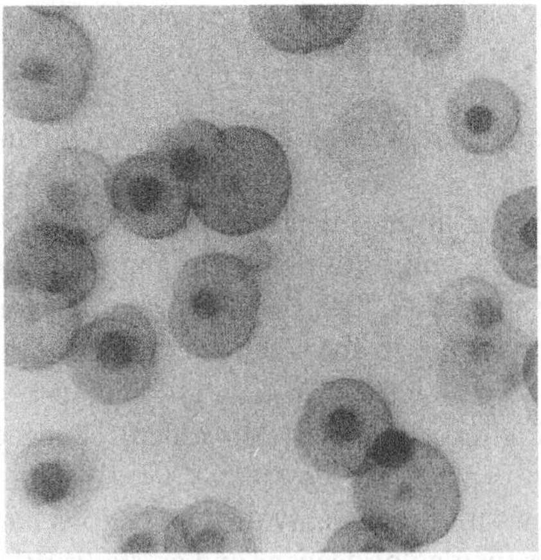

Fig. 9.11. Section from Fig. 9.9. Above 740 °C, crystallization starts within the droplets. Almost without exception, it starts in the center of the immiscibility regions (light micrograph ca. 50:1)

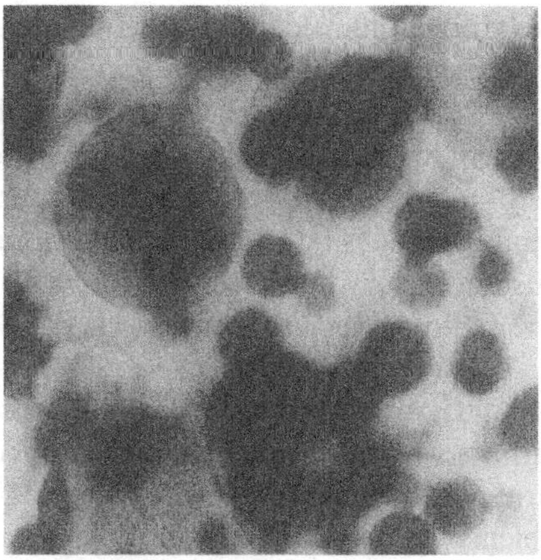

Fig. 9.12. Section from Fig. 9.9. Between 780 and 800 °C two droplet dimensions are observed. The droplets have been crystallized to a large extent (light micrograph ca. 80:1)

large primary droplets. This facilitated the investigation of composition, color, and crystallization processes.

This confirmation by the color differences observed in the light microscope of the suspected distribution of CdS chromophores was supplemented by electron microprobe evidence for the enrichment of Zn and K in the droplet phase. At the same time, the chief interferences of α-CdS and α-ZnS, much broadened due to the small dimensions of the crystal, could be observed. Apparently a reaction between CdS and ZnO had led to a concentration-dependent equilibrium and, because of the similar lattice dimension, the formation of mixed crystals. The mixed-crystal formation is confirmed by small shifts in the ZnS and CdS interferences which seem to depend on which sulfide forms the host lattice.

Even though these investigations must be considered as model investigations, they unequivocally prove the CdS enrichment in separated phases of the base glass. In production glasses (Fig. 9.6) the K_2O content must of course, be much larger (up to about 20 mass%). This considerably suppresses the tendency to phase separation so that the droplets become smaller in size and larger in number, avoiding opacification by excessive droplet growth during reheating ("striking").

The defect called "livery glass" in glassmaking practice (a brownish, more opaque appearance) is due to excessive concentration in chalcogenide or to impurities enhancing excessive phase separation. These disturbances may lead to crystallization and more or less 'ceramming" of the melt.

9.4.3 Coloring Mechanism in Striking Glasses

Striking glasses have been known for a very long time and their coloring mechanism has been the subject of numerous studies [196, 662, 671, 674–685]. A common factor of all these interpretations is the assumption that growth of the particles during reheating is responsible for the shift in the absorption edge. Some interpretations, however, sometimes still maintained, treat these colors as colloidal precipitates in analogy to the colloidal nature of gold ruby glasses. For this reason, one still finds the designation "Cd-S-Se ruby glass." Since chalcogenide and ruby glasses are dissimilar, except for their red color, this deceptive designation will be avoided. On the basis of many experimental results, it must be considered certain that colloidal precipitation does not cause the coloration of cadmium chalcogenide glasses. For instance, their fluorescence spectrum is more or less typical of that of the respective lattices. Metal colloids, however, are not fluorescent.

During the past 15 years, nucleation and growth processes have been the basis for the interpretations of several authors who have considerably enlarged our state of knowledge about the coloring mechanism in "striking glasses." Nevertheless there remain unsolved problems.

A change in color of "striking" glasses, involving a shift in absorption edge for polychalcogenides or larger crystals, cannot be explained simply by Mie scattering effects as can be done, e.g., for gold sols. Certainly the steep edge cannot be so interpreted. More recently, Rehfeld and Katzschmann [686] succeeded in attributing the absorption edge to electronic excitations using a band model for the crystalline cadmium chalcogenide precipitations.

The steepening of an absorption edge through reheating (Fig. 9.5) would correspond (Fig. 9.13) to an allowed electron transition from the valence to the conduction band according to

$$E_{\text{gap}} = h\upsilon_{\text{k}} = h\,\frac{c}{\lambda k}$$

where indices k refer to the steep absorption edge. The absorption edge thus is a measure for the width of the forbidden zone. Indeed limits for E_{gap} are:

$$E_{\text{gap}} = 1.61\ \text{eV}\ (= 13\,000\ \text{cm}^{-1}\ \text{or}\ 777\ \text{nm})$$

$$E_{\text{gap}} = 3.35\ \text{eV}\ (= 27\,000\ \text{cm}^{-1}\ \text{or}\ 370\ \text{nm})$$

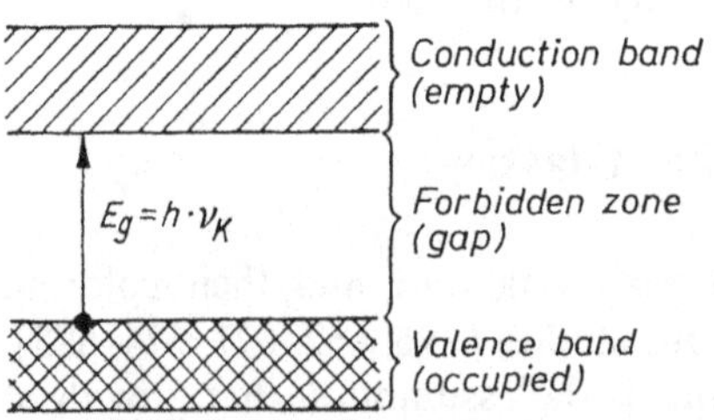

Fig. 9.13. Schematic of the electron transition from valence to conduction band

Values for typical chromophores in striking glasses are:

CdS 2.42 eV (= 19 500 cm^{-1} or 518 nm)

CdSe 1.73 eV (= 14 000 cm^{-1} or 717 nm)

CdTe 1.44 eV (= 11 600 cm^{-1} or 860 nm)

ZnS 3.53 eV (= 28 500 cm^{-1} or 340 nm)

Although the absorption edge for CdS and CdSe lies in the visible range, this is not the case for CdTe and ZnS, in agreement with practical experience.

Rehfeld and Katzschmann's [686] establishment of a relation between the location of the absorption edge and the size of the crystallites, exemplified for glasses containing CdSe, represents significant progress. Figure 9.14 illustrates the dependence of the UV absorption edge (thus the "striking" color) on the nuclear radius, as determined by small-angle X-ray diffraction and electron microscopy. The smallest radius corresponding to the start of striking ($E_{gap} = 2.75$ eV) was found to be 1.3 nm. This value at the intersection of the curve with the abscissa (the start of striking) was termed the critical radius of nucleation. Figure 9.14 also shows that the curve approaches a limiting value of 1.73 eV, which corresponds with the lattice absorption of crystalline CdSe. From this curve, after determining constants from a linear relation, one obtains between $r = \infty$ and $r = 1.3 \pm 0.2$ nm:

$$E_g = \left(\frac{1.70\,(\mathrm{nm})^2}{r^2} + 1.73\right)\mathrm{eV}$$

It should not be difficult to find an analogous relation to E_g to $1/r^2$ for glasses containing CdS and CdTe.

These results also dispose of the hypothesis that the shift in the absorption edge is related to the evolution of mixed crystals. No doubt, mixed crystals precipitate in glasses containing more than one chalcogenide. But clearly the shift occurs also in glasses containing only a single chalcogenide, where only particle size can account for the shift. On principle, the relation to particle size

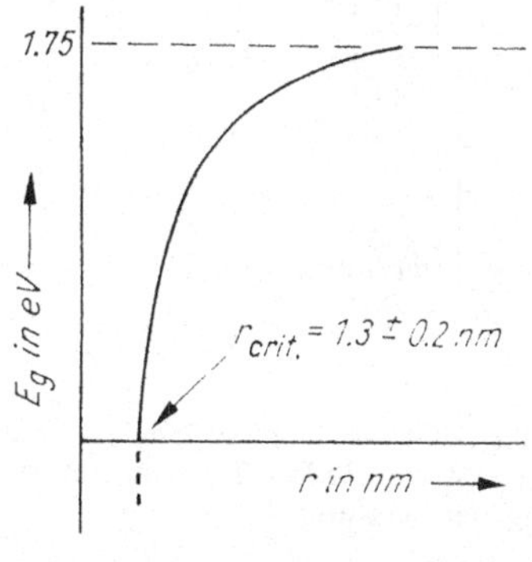

Fig. 9.14. Dependence of the location of the absorption edge in a "striking" colored glass of particle size [686]

will also account for the shift when mixed crystals are precipitated, provided that the lattice absorption of either component, e.g., CdS and CdSe, are on the same side of the absorption edge of the base glass. If this is not the case the shift may occur – against all previous experience – in the opposite direction, i.e., toward shorter wavelengths. This was observed in the case of glasses containing ZnS and CdS, which form a continuous series of mixed crystals. When CdS precipitates first, the normal shift toward the lattice absorption of 515 nm takes place (Fig. 9.15). At higher temperatures, however (620 to 680 °C), the steep edge travels in the opposite direction. The lattice absorption of pure ZnS lies at 340 nm, and the now-forming mixed crystal contains increasing amounts of ZnS, which is more soluble and precipitates later. On this basis, new shortwave cut-off glasses (Fig. 9.16) were developed which have no UV transmission beyond the absorption edge (compare curve a Fig. 9.6 with the curves in Fig. 9.16). The cut-offs are in regions where they could not have been attained in the older types.

Finally, it should be noted that, because of the strong affinity of Zn for S, a high ZnO content of the base glass suppresses losses of sulfur. Also, the solubility of CdS increases with K_2O content. Both of these factors are important in production. Equally important are phase-separation effects in the base glass since the cadmium chalcogenides are enriched in the K-rich, Zn-rich droplet phase. This influence remains, even when the phase-separation process has been suppressed to the point that droplet boundaries no longer can be ascertained and the glass has been frozen to a stage of swarmlike aggregations. It is also understandable that limited crystal growth during reheating is possible at

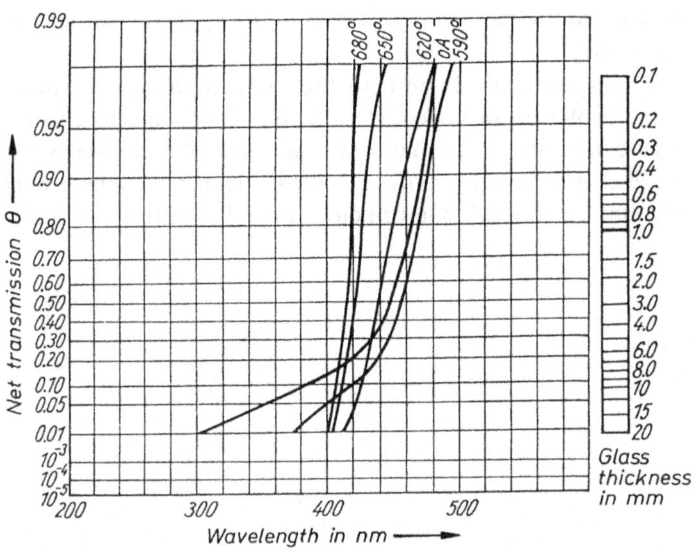

Fig. 9.15. Shift of the absorption edge with heat treatment of a CdS-containing "striking" colored glass (base glass: K_2O–ZnO–SiO_2) A: glass before treatment, no "striking" effect. The temperatures marked at all other curves correspond to the chosen "striking" temperatures

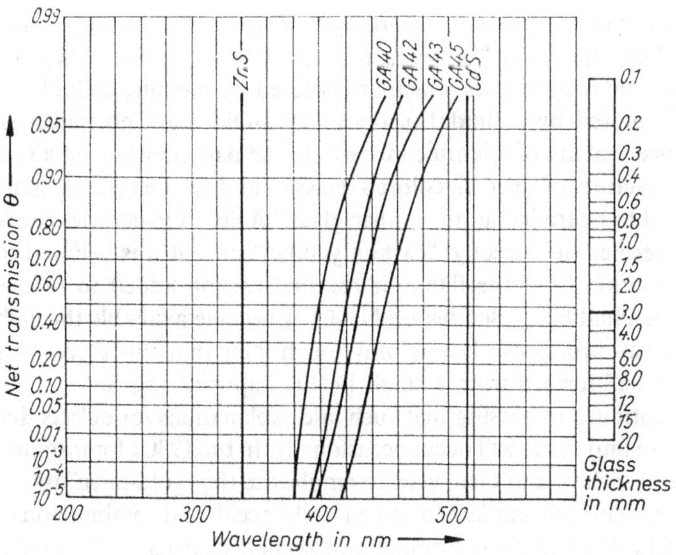

Fig. 9.16. Absorption edges of new "striking" colored glass types exhibit ideal steepness without any remaining transmission in the region of shorter wavelengths

constant particle number per unit volume since the primary crystal in the droplet (or aggregation) may still be increased by the residue of its kind in the matrix. In the model experiment of Fig. 9.12 this residue has been precipitated in a secondary phase-separation process in the form of a second, smaller droplet phase. When phase separation is strongly suppressed in the production glasses this residue may contribute to the limited crystallite growth through increasingly possible diffusion in the reheating process.

9.4.4 Related Glasses with Other Chromophores

Solubility conditions as described in the preceding sections are not limited to the chalcogenides of cadmium. Some colored glasses called "ruby" glasses are known, and have been investigated in regard to their chromophores (Zsigmondy [687]) which contain the following chalcogenides:

Sb_2S_3 ("antimony ruby")
FeS ("iron ruby," or amber glass)
PbS ("heavy metal ruby")
CuS ("heavy metal ruby")
NiS ("heavy metal ruby")
Mo_2S_3 ("heavy metal ruby")
W_2S_3 ("heavy metal ruby")
Ag_2S ("heavy metal ruby")

These glasses exhibit "striking" effects, i.e, light transmission changes on reheating. However, unlike the case of chalcogenide chromophores, the precipitated chromophores are not crystalline but, according to Zisgmondy, are truly colloidal dispersoids. Thus the term "ruby" is not entirely unjustified. They are a transition to the true ruby glasses which are colored by metallic colloids.

The "rubies" colored by colloidal heavy metal sulfides are no longer of practical importance, but are of scientific interest. In the past they played a role both in the manufacture of special colored glasses and as defects. A glass containing FeS used to be traded as an "amber-glass." A defect in colorless tank glasses used to be termed (incorrectly) "carbon yellow," since it was believed to be due to contamination from the flame or residual carbon added to reduce sodium sulfate (which had been used before Na_2CO_3 became available through the Leblanc or Solvay processes). It was only much later that this defect was found to be due to FeS formed from a small Fe_2O_3 impurity and S.

It might be mentioned in passing that such late explanations for defects due to the combination of impurities will occur occasionally. In the 1930s, for instance, the Jenaer Glasswork Schott and Gen. used to emphasize the "gold-olive" tint of their household ware. This tint was later ascribed to the accidental combination of trace impurities of Fe_2O_3 and TiO_2 forming a molecular dispersion of ilmenite ($FeTiO_3$). The attractive "honey-yellow" tint of Czechoslovakian CeO_2–TiO_2 in the 1930s was found to be due to Fe_2O_3 impurity in the purchased CeO_2.

For some time so-called "selenium-pink ruby glass" ("rosaline") played a certain role. It contained FeSe as the chromophore. Dietzel and Höfler [688], in particular, sought to determine the coloring processes involved in these glasses. However, conditions are far less clear than for the glass types described above. According to Dietzel and Hirsch [689], the partial oxygen pressure above the melt drastically controls equilibria between elemental selenium, selenides, selenites, and selenates, so that is often difficult to obtain reproducible production conditions.

It is very interesting that, when dissolved in glasses of high SiO_2 or B_2O_3 content, S imparts blue, oxidation-resistant colors.

The blue is attributed to S_3^-. The following states were identified from their absorption spectra: S_2 molecule for $Na_2O > 15$ mol%, S_2^- for > 20 mol% Na_2O, S_3^- 20–30 mol% polysulfides for 30–35 mol% (Ahmed et al. [690]).

Na_2S is much more soluble in some glasses. Polysulfides form, leading to colors changing via yellow to deep red with increasing chain length. In contrast to polyselenides and polytellurides, the polysulfides are relatively stable in glasses. None of these glasses has attained practical importance.

9.5 Glasses Colored by Metal Colloids (Ruby Glasses)

9.5.1 Composition, Fabrication, and Absorption Behavior

True ruby glasses originally were based predominantly on the systems K_2O–CaO–SiO_2 and K_2O–PbO–SiO_2, then the system K_2O–PbO–B_2O_3–SiO_2, finally replacing all PbO by Sb_2O_3. This brief survey demonstrates the relative

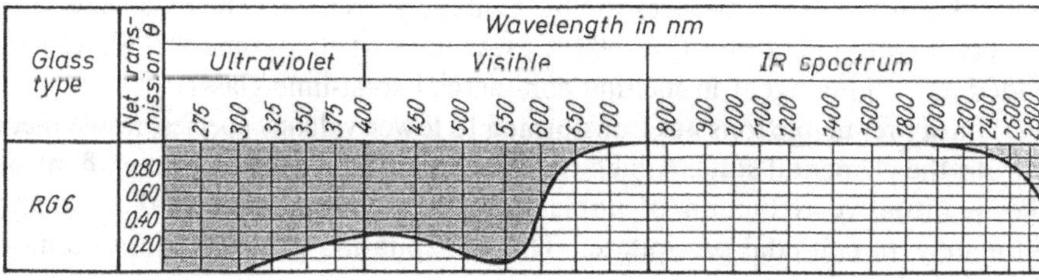

Fig. 9.17. Light-transmission curves of a Schott filter glass absorption effect of colloidal gold in a borosilicate base glass

independence of the coloring mechanism on the base glass, which thus clearly is chosen mostly to affect melting and working properties. All these base glasses, when doped with about 0.003 to 0.1 mass% of a copper, silver, or gold salt (preferably $CuCl_2$, $AgNO_3$, or $AuCl_3$) together with a reducing agent such as potassium hydrotartrate, $SnCl_4$, SnO_2, or Sb_2O_3, are usually still colorless or very weakly struck when conventionally cooled from the melt. Reheating results as in "striking" glasses, in "striking" to a more or less intense red for copper or gold doping, yellow to yellowish-brown for silver doping. Because red is obtained in many variations, true ruby glasses have been used almost exclusively in art glass and for casting (red layer over colorless glass).

Only exceptionally gold ruby glass has been used as a filter glass in optics. Figure 9.17 shows the transmission curve of a gold ruby glass. When compared with the transmission curves of true "striking" glasses on the basis of cadmium chalcogenides, which today can be fabricated with good reproducibility, the disadvantages of gold ruby glasses become clearly apparent (Fig. 9.17 vs Figs. 9.6, 9.16). The typical "striking" glasses (Fig. 9.6, 9.10) have steep absorption edges in the shortwave region corresponding to saturated colors: the gold ruby glasses have remaining transmission ranges up to 30%, reaching far into the UV, corresponding to a blue tint in the red color. Because of this disadvantage and the difficulties in reproducing a defined transmission property, gold ruby glasses have been practically eliminated from use as color and filter glasses in optics.

9.5.2 Structure and Coloring Mechanism in True Ruby Glasses

Whereas the coloring process on reheating of "striking" colored glasses containing cadmium chalcogenides is based on the precipitation of crystallites and the excitation of electrons in the respective crystal lattices, coloring of true ruby glasses is conditioned by the precipitation and growth of metal colloids. For this reason the reducing agents introduced into the batch are of considerable importance. In gold ruby glasses, mere heating may suffice to initiate coloration (thermal reduction), but this is not possible in glasses doped by silver or copper. One simply may use carbon compounds such as tartrates, oxalates, or sugar as reducing agents added to the batch. But Sb_2O_3 shows the same effect, since – as is well known – Sb_2O_3 changes to Sb_2O_5 in the glass melt. (The reverse reaction at very high temperatures (> 1300 °C) was used when 1–3 mass% Sb_2O_3 was added as a fining agent in melting commercial soda-lime glass).

Easily reducible ions are transformed to lower valency species, noble metal ions to the elemental state. A high Sb_2O_3 content in a ruby glass melt controls the quantitative reduction of noble metal ions to the elemental state and, ultimately, to colloidal precipitates. This precipitation process requires a more thorough description.

It is known in colloid chemistry that, in spite of the equal sign of their charges, particles will, under certain circumstances, continue to aggregate and will eventually coagulate, i.e., flocculate as macro particles; but there are certain protective additives which prevent complete flocculation. The cause of this protective effect is, for example, in the case of colloidal gold, combining of the gold particle and the protective colloid. This happens when there are more protective colloid particles, and they are smaller than the gold particles. One speaks of a protective envelope preventing approach and union of the gold colloid particles which would otherwise lead to coagulation. If the protective colloid particles are larger than the gold colloid particles, the gold particles will cling to the protective colloid particles ("protection by adhesion"). In either case the gold colloid particles – even if discharged by electrolytes – are kept at a distance to prevent flocculation.

In aqueous solutions, tin acid sols have been proved to be particularly advantageous protective colloids for metal sols such as those of Au, Pd, Pt, Cu, Ag, Hg, Bi, Se, or Te. The most striking example is the red-gold sol of the so-called "Cassius gold purple." Here envelope- as well as adhesion-type protective colloid was found to form.

It is true that entirely identical conditions could not be expected in the true ruby glasses based on the precipitation of copper, silver or gold colloids. But many modes of behavior are very similar. For instance, the addition of SnO_2 to a gold ruby glass batch may be considered to have an effect completely analogous to that in aqueous gold sols. In fact, the tin compounds may have a dual function in the glass – as a reducing agent for the noble metal ion, and as a protective colloid (in form of the colloidal SnO_2 formed in the reduction process) for the noble metal colloid.

At first, the colorless gold ruby glass formed from the melt contains an atomic solution of gold. Reheating causes the formation of colloidal gold under the stabilizing and protective colloid influence of the typical protective colloid additions which prevent coagulation (Heyman [693], Sigel [483]). The striking range can be extended by halides (e.g., 0.3–0.6% KCl) (Kaminskaya [694]).

The coloring mechanism parallels that of the structuring of the precipitation. It has been established that the color is chiefly due to specific absorption, and only to a lesser degree to scattering on very small particles. The absorption is due to the excitation of free electrons in collectively oscillating metal particles, depending on their size and shape (Doremus [285]). The absorption band peaks near 530 nm. Scattering and the effect of particles smaller than that of the wavelength of the irradiating light is covered by Mie's [683] theory.

If the degree of dispersion remains unaltered, Lambert's and Beer's laws remain valid for colloidal solutions. As described in the introduction to Chapter 9, irradiation by light causes, among other effects, partial absorption.

The intensity of light decreases along path x passed in the medium from the original value according to the exponential

$$J = J_0 e^{-ax} = J_0 \cdot 10^{-\alpha x}$$

or

$$\ln J_0/J = ax, \text{ or } \log J_0/J = \alpha x \text{ (Lambert's law)}$$

where a = absorption coefficient, α = extinction coefficient, and $\log J_0/J$ = extinction. In a solution, α is proportional to the concentration if the absorption centers do not change with a change in concentration:

$$\log J_0/J = \alpha x = \varepsilon c x$$

where ε is the percentual or molecular extinction coefficient, depending on whether concentration c is given in g/100 cm^3 or mol/L.

With increasing particle size, the extinction coefficient first rises and the absorption maximum is shifted toward longer wavelengths. The color changes from red via violet to blue. Scattering effects appearing with very strong particle growth can be neglected for the typical range of 50–60 nm.

For the sake of completeness, it should be noted that, on principle, colored glasses can be produced by the precipitation of other colloidal particles, e.g.,

"platinum ruby"	grayish-brown
"bismuth ruby"	grayish-brown
"antimony ruby"	grayish-brown
"lead ruby"	grayish-brown
"tin ruby"	brown
"cobalt ruby"	brown

For such glasses the reducing agents used in Cu, Ag, or Au ruby often will not suffice; it may even be necessary to resort to hydrogen. These glasses have not attained any practical importance.

9.5.3 Silver Stain

Often the content of glass apparatus or tubing must be protected against the effects of light, e.g., when photoemulsions are prepared for the film industry. On principle, this requirement may be satisfied by the use of the proper colored glass. In practice, a more rotational solution has been found. Conventional apparatus or tubing is used with the surface colored by producing within it a more or less brown silver ruby glass. The simplest method is to paint or spray the glassware with the so-called *silver stain.* Silver stain is a more or less consistent watery paste mixed from silver nitrate; kaoline, ocher, or clay; and a sulfur compound like a sulfide or sulfate (e.g., $BaSO_4$). After uniform coating, the glassware may simply be passed through a conventional annealing lehr a second time. Soft-coal briquets are added whose incomplete burning delivers CO gas.

After removal of the reheated and cooled ware, the residual paste is washed away. Depending on the chosen concentration of silver, annealing temperature, and time, the ware is colored from a bright yellow to a more or less deep brown. Similarly, gradations on measuring cylinders or beakers, other markings, and even lettering may be applied. Artists have used this process for centuries to apply decorations by masking or cutting. The underlying process may be understood as follows: Kaolin (ocher, clay) first functions as a carrier for the silver nitrate solution. During annealing glass components, mostly alkali ions, are exchanged for silver ions. At the same time, the CO gas reduces the silver ions, followed by the formation of coloidal silver.

It is assumed that the sulfur compounds lead to an intrusion of sulfur in the glass surface. Sulfur demanding more volume than oxygen widens the surface structure, facilitating silver diffusion. At any rate, silver staining in the absence of sulfur is found to result in irregular, spotty coloration.

Inhomogeneities become visible through the application of silver stain. The effect is analogous to that of enrichment of 3d elements in microphases. The silver ions are distributed differently in the matrix and the striae. In some cases this may constitute an interesting method of making striae more visible within the framework of quality control.

9.6 IR-Absorbing Glasses (Heat-Absorbing Glasses)

9.6.1 Application of Heat-Absorbing Glasses and Absorption Behavior of Glasses Containing Fe^{2+} and Fe^{3+} Ions

Glasses absorbing the IR wavelength range of the spectrum, but past visible light, are also called heat-absorbing glasses. Their application in modern science and technology is relatively manifold. A particularly important application is to cinema technology. Movie projection lamps not only emit considerably in the visible, but also emit a high portion of IR light (heat radiation). Since the latter must be kept away from the film, heat-absorbing glass filters are placed between the film and the light source. They absorb the entire impinging IR radiation while they, themselves, are heating. Therefore, one has to provide efficient removal of heat, either by metal mounting or cold air blowing. Otherwise the glasses themselves would become heat sources after functioning briefly.

The effect of the heat-absorbing glasses is based on a phenomenon discovered by Zsigmondy. He found that aqueous solutions of glasses, containing Fe^{2+} ions, pass visible light almost completely while almost completely absorbing the infrared region. Fe^{3+} ions, however, strongly absorb in the shortwave range of the spectrum, while noticeable transmission is maintained in the near IR. Figure 9.18 shows schematically the absorption behavior of silicate glasses containing Fe^{3+} and Fe^{2+} ions. Curves for Fe^{3+} ions glasses (Fig. 9.18(a)) cut

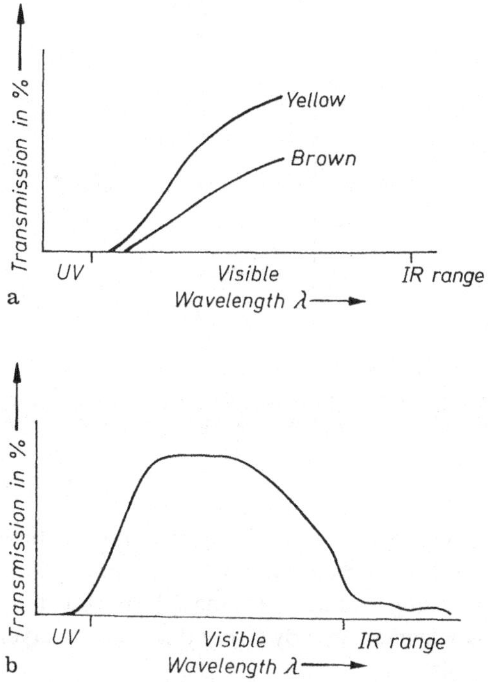

Fig. 9.18. Schematic of the absorption effect of Fe^{3+} and Fe^{2+} ions in a silicate glass. **a** Fe^{3+} in different concentrations in a silicate glass; **b** Fe^{2+} ions in a silicate glass

nearly diagonally almost one-half of the visible region. The resulting color is a deep brown due to $(Fe^{III}O_4)$ complexes, or a pinkish-yellow due to $(Fe^{III}O_6)$ complexes. Silicate glasses containing Fe^{2+}, however (Fig. 9.18(b)), show high transmission in the visible region, while transmission drops sharply toward the UV and IR regions. The resulting color is bluish-green. The absorption spectrum is due to $(Fe^{II}O_6)$ complexes (Fe^{2+} ions can only be 6-coordinate in glasses).

9.6.2 Development, Production, and Properties of Heat-Absorbing Glasses

The first heat-absorbing glasses according to Zsigmondy's discovery were based on the system Na_2O–CaO–SiO_2 doped with 1–3 mass% FeO. They still possessed noticeable absorption in the visible range (see Fig. 9.18(b)) and were colored a rather deep bluish-green.

Further development by Berger [695] represented a significant success since, for equal, even improved IR absorption, a considerable increase in the visible light transmission was achieved. Berger's heat-absorbing glass produced by Jenaer Glaswerk Schott u. Gen. was based on the system (Na_2O–CaO–SiO_2–P_2O_5–Al_2O_3)–FeO. The increased transmission in the visible was due to the partial replacement of SiO_2 by P_2O_5 (≈30 mass% SiO_2, 51 mass% P_2O_5). The interaction Fe–O in the $(Fe^{II}O_6)$ complex is weakened significantly by the stronger counterpolarizing effect of the P^{5+} ion (Fig. 9.19).

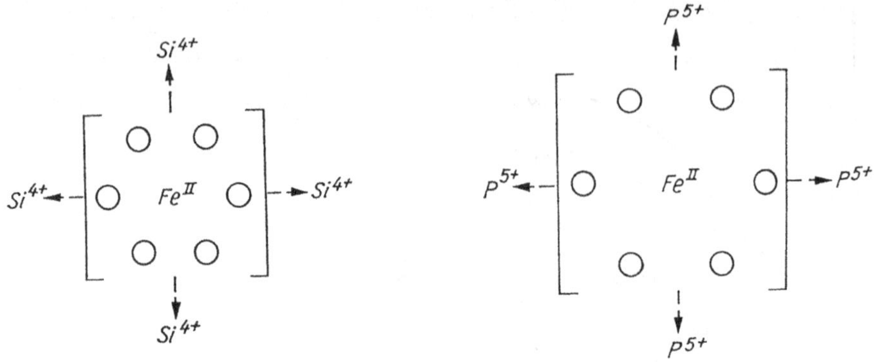

Fig. 9.19. The [$Fe^{II}O_6$] complex is much more loosened in a phosphate glass than in a silicate glass. The higher degree of transmission of Fe^{2+}-doped phosphate glasses is a result of this weakening of the Fe–O interaction

But the improvement in the optical properties of the new heat-absorbing glasses was accompanied by a deterioration in chemical resistance. Although the introduction of Al_2O_3 brought some improvement, the new heat-absorbing glasses could be used in practice only when covered by a very thin paraffin layer in an expensive process, or by SiO_2 glass.

In the mid-1940s the American Optical Company achieved further improvements in the invisible range transmission of heat-absorbing glasses of equal IR absorption by a further increase in P_2O_5 (5–10 mass% SiO_2, 70 mass% P_2O_5).

Such an increase in transmission had become necessary because colored film had conquered the world and glasses tampering with color rendering were not acceptable. Therefore, initially heat-absorbing glasses with improved visible range transmission were tolerated even if their chemical resistance was poor (German Standard: hydrolytical class 3).

An at least preliminary solution of the problem appears to have been offered by the development of new heat-absorbing glasses by Vogel and Heintz [696] based on the system (MgO–ZnO–Al_2O_3–P_2O_5)–FeO. The light transmission in the visible range is nearly optimal. They appear faintly blue-green. At the same time their IR absorption is excellent. A significant difference in relation to earlier developments is the complete absence of SiO_2, as well as all network-modifiers like alkali, calcium, or barium. Even though the network-former is entirely P_2O_5, the glass, otherwise composed of intermediate oxide like MgO, ZnO, and Al_2O_3, possesses excellent chemical resistance (German Standard: hydrolytic class I). No beneficiation by paraffin or SiO_2 layers is necessary.

The transmission curves of a new and an old heat-absorbing glass of equal thickness in the visible and the IR spectral regions are compared in Fig. 9.20(a) and (b). The almost perfect constancy of the high transmission in the visible range of the new glass is clearly seen in Fig. 9.20(a), the excellent absorption in the IR range in Fig. 9.20(b). In production everything must be done to prevent a transition from Fe^{2+} to Fe^{3+}, which would drastically impair the transmission

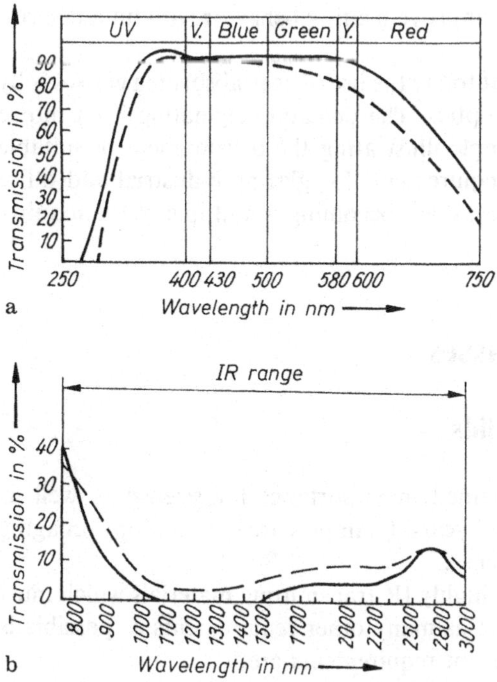

Fig. 9.20. Transmission curve of a silicophosphate heat protective glass as developed by Berger [696] (- - -); and of a pure phosphate heat protective glass developed by Vogel and Heintz [676] (—). **a** In the visible range (glass thickness 2 mm); **b** in the IR (glass thickness 1 mm)

characteristics according to Fig. 9.18. As a rule, an equilibrium Fe^{2+}/Fe^{3+} is established in a glass melt. In the past this was counteracted by using iron powder instead of iron oxide, with an addition of 0.5 to 1% sugar. Iron powder is used to produce the new heat-absorbing glasses and small portions of the MgO, ZnO, and Al_2O_3 content are introduced as metallic Mg, Zn, or Al powder.

From the viewpoint of glass chemistry the enormous improvement in chemical resistance achieved by the change from a silicophosphate to a phosphate glass is particularly interesting. The discussion of phase-separation phenomena (Chapter 6) has demonstrated and explained the strong tendency toward phase separation in SiO_2–P_2O_5 combinations which can be suppressed but not eliminated by certain additives, such as Al_2O_3. This is true even if the glasses appear perfectly clear. It has been proved that the poor resistance of these silicophosphate glasses is caused by phase-separation phenomena. Since, in the P_2O_5-rich phase of such silicophosphate glasses, alkali and other network-modifiers are also enriched, its solubility in water is quite high. The change to a pure phosphate base whose tendency to phase-separate is much lower (for reasons treated in Chapter 6) therefore brings about a strong increase in chemical resistance when adequate stabilization is provided, e.g., by Al_2O_3, MgO, or ZnO. Without sufficient insight into phase-separation phenomena, the improvement would be hard to understand since, in general, glasses become less resistant when one changes from SiO_2 or SiO_2–P_2O_5 to pure P_2O_5 systems. If phase separation was under proper control it was possible to add even small amounts of B_2O_3 (2–5 mass%) to improve melting behavior without a deterioration in optical properties.

Finally, it should be noted that today the use of heat-absorbing glasses is by no means limited to the field of optics. The demand originating from movie projectors serves only as an example illustrating the development of suitable filters. At present, however, the requirements for glazing industrial and office buildings, particularly in sunny areas, are becoming a multiple of the requirements for instruments in the past.

9.7 IR-Transmitting Glasses

9.7.1 IR Transmission of Solids

IR-transmitting materials have gained in importance for science as well as technology during the past 20 to 30 years. Examples are spectroscopy, imaging optics, and last, but not least, defense.

The systematic search for new, highly IR-transmitting materials which must simultaneously and optimally meet as many other requirements as possible is still active today. The most important requirements are:

(1) Highest transmission in the IR with an IR absorption edge as far as possible in the IR and a transmission free of absorption bands in other regions.
(2) Good resistance toward water and weathering
(3) Stability at high service temperatures (high melting point or softening point, in glasses)
(4) Not too low hardness.
(5) Small scattering losses throughout the transmission range.
(6) Radiation resistance.
(7) Minimal strain birefringence.

There are many materials which meet some of these requirements very well, but their application is precluded by their failure to meet some others, e.g., chemical resistance.

For imaging, dispersion behavior plays a critical role. While in classical light optics the dispersion behavior is known to be characterized by Abbe's number

$$\nu_d = \frac{n_{d(587)} - 1}{(n_{F(486)} - n_{c(656)})}$$

(note that more recently n_e and ν_e are used increasingly instead of n_d and ν_d), such a characteristic has not been standardized for the IR range. Schroder and Neuroth [697] propose

$$\nu_{1.529} = \frac{n_{1.529} - 1}{(n_{1.014} - n_{2.325})}$$

(λ in μm).

The absorption of a glass in the shortwave range (UV absorption edge) is generally conditioned on electron transitions in the range of the network-forming cation and its ligands. The absorption of a solid in the longwave range of the spectrum (IR absorption edge) derives from inner molecule or lattice vibrations. On the basis of a two-mass model one may estimate to a first approximation the eigenvibration of particles in dependence of their reduced masses and a force constant:

$$\bar{v} = (1/2\,\pi c)\sqrt{K/\bar{M}}$$

where $\bar{v}$ = eigenvibration of particles, c = speed of light, K = force constant, and $\bar{M}$ = reduced mass.

Progressing to heavier elements in the composition of an IR-transmitting material, the frequency of the eigenvibration of the particles decreases as their masses increase. This means that the IR absorption edge must move into longer wavelength regions.

Conductors are not transparent in the IR, but typical semiconductors are in the near-IR. In this case, the limiting wavelength corresponds with electron transitions between valence and conduction band. The energy levels are broadened. If the semiconductor is very pure the IR absorption edge is very steep. This is almost analogous to the short wavelength cut-off ("striking") glasses. Semiconductors with a steep longwave absorption edge are just as valuable cut-off filters. They have one disadvantage. With rising temperature the IR transmission decreases relatively as much as absorption due to lattice vibrations increases.

Table 9.5 represents a survey of various IR transmitting materials. It can be seen that a series of single crystals have excellent IR transmission (Table 9.5a). Their resistance to water and weathering, however, is quite low, so that they have to be protected by costly methods. Also, the art of growing large crystals free of defects is limited, and cost increases greatly with size. Often their low hardness is disadvantageous, as are their anisotropy and workability (cleaving).

A second class of IR transmitting materials which has gained practical importance and is available commercially are produced by "high-pressure hot-pressing" (Table 9.5b). Their fabrication avoids the problem of growing single crystals. These materials are transparent-to-opaque in the visible range, i.e., for the human eye, but quite transparent in the IR range. This is possible because scattering decreases with wavelength (Raleigh relation). For this reason, grain boundaries play a negligible role. High-pressure hot-pressed materials thus function well in the IR, but have disadvantages in the visible range.

Table 9.5c gives a survey of crystalline semiconductors. Table 9.5d contains IR-transmitting materials withstanding high temperatures.

The IR transmission of glasses is generally limited or entirely suppressed by the eigenvibration of the network. Borate and phosphate glasses of 5 mm thickness lose transmission at 3.5 μm. For silicate glasses the limit is at about

Table 9.5. Survey of IR-transmitting materials

a) Single crystals

Compound	IR limit	Refractive Index n_d	Water solubility $g/100\,g\,H_2O$
NaCl	15 μm	1.52	35.85
KCl	21 μm	1.47	34.35
KBr	27 μm	1.54	90.5
LiF	7.5 μm	1.36	0.27
MgF_2	8.0 μm	1.38	0.0076
CaF_2	10.5 μm	1.39	0.0017
CdF_2	11.5 μm	1.55	4.35
LaF_3	12.5 μm	1.41	
SrF_2	12.5 μm	1.43	0.011
BaF_2	14.0 μm	1.47	0.17
PbF_2	13.5 μm	1.76	0.064

b) Materials hot-pressed at high pressures

Material	IR limit	Refractive index at 3 μm	Trade name
Al_2O_3	6.5 μm	1.71	Lucalox
LiF	8.0 μm	1.38	IRG 10
MgF_2	9.2 μm	1.36	Irtran 1
MgO	9.4 μm	1.69	Irtran 5
CaF_2	11.5 μm	1.42	Irtran 2
BaF_2	14.2 μm	1.46	IRG 13
ZnS	1.7 μm	2.25	Irtran 2
ZnSe	2.18 μm	2.44	Irtran 4
CdTe	31.0 μm	2.69	Irtran 6

c) Crystalline semiconductors

Material	Transmission	Refractive Index at 3 μm
Si	1.2 ... 15 μm	3.43
Ge	1.8 ... 22 μm	4.04
CdS	0.8 ... 16 μm	2.27
GaAs	1.0 ... 16 μm	3.23
InAs	3.5 ... 9 μm	3.45
InP	1.0 ... 14 μm	2.53

d) High-tempterature resistant IR materials

Material	IR Limit	Refractive index at 3 μm	Melting temperature (° C)
Al_2O_3	5.8	1.7 (anisotropic)	2030
MgO/Al_2O_3 Spinel	5.5	1.72	2030
TiO_2	6.2	2.38 (anisotropic)	1825
$SrTiO_3$	6.8	2.22	2080
$BaTiO_3$	6.9	2.40	1600
MgO	8.5	1.69	2800
CaF_2	11	1.42	1360
BaF_2	14	1.46	1280

4.6 to 4.8 μm. Thus the classical glasses are not suitable for use as IR-transmitting materials. Also, even small contents of H_2O greatly influence IR-transmission. A fundamental clarification of the role of H_2O was provided particularly by Scholze [87, 579–583]. Generally, absorption bands due to H_2O in IR-transmitting glasses must be expected at the following wavelengths:

2.7–3 μm	O–H valence vibration
3.4–3.6 μm	H bridge bond to singly bonded O
4.5 μm	Strong H bridge
6.0 μm	O–H bending vibration

9.7.2 IR Transmission of Germanate, Tellurite, and Aluminate Glasses

The structure and properties of germanate and tellurite glasses have been discussed in preceding sections. The discussion has to be resumed in connection with their transmission in the IR. In particular, lead germanate glasses have good infrared transmission (as far as 5.8 to 6 μm), while being transparent (like flint glasses) in the visible range, exhibiting only a faint yellowish tint. They find practical application whenever 6 μm is sufficient, because of other positive properties such as high softening range, chemical resistance, hardness, and workability. Similarly, tellurite glasses are usable. The transmission may be extended to 7 μm by the insertion of halogens (Cl, Br, I) and sulfates. Tellurite glasses have excellent transmission between 0.4 and 7 μm, i.e., also in the visible range, just like normal optical glasses.

A very interesting group of glasses – also transmitting to about 6 μm – are the so-called aluminate glasses. As early as 1909, Sheperd et al. [698] found that micromelts in the system $CaO–Al_2O_3$ between 38 and 65 mass% Al_2O_3 freeze to a glass from melting temperatures between 1400 and 1600 °C. It wasn't until 1948/49 that Stanworth [382] and Sun [699] pursued this discovery and found that glass formation can be extended and larger pieces obtained if very small amounts (to about 5 mass%) of SiO_2 are introduced into $CaO–Al_2O_3$. An example is: 44.8 mass% CaO, 48.6 mass% Al_2O_3, 6.6 mass% SiO_2. Since such an addition of SiO_2 interferes with IR transmission, Hafner et al. [700] developed larger IR lenses for practical use to 6 μm by combining CaO and Al_2O_3, with other oxides such as BaO, La_2O_3, etc. An example is: 43.0 mass% CaO, 10.0 mass% BaO, 47.0 mass% Al_2O_3.

The pure aluminate glasses are also interesting from a structural view-point since they are oxide glasses containing no typical network-forming oxide. Perhaps the counterpolarizing effect of the Ca ion on the AlO_6 complex causes a transformation to an AlO_4 grouping (Fig. 9.21), which would signify one to a network-former. Apparently the effect of alkali is insufficient for such a transformation and alkali aluminate glasses are not obtained in conventional melting processes.

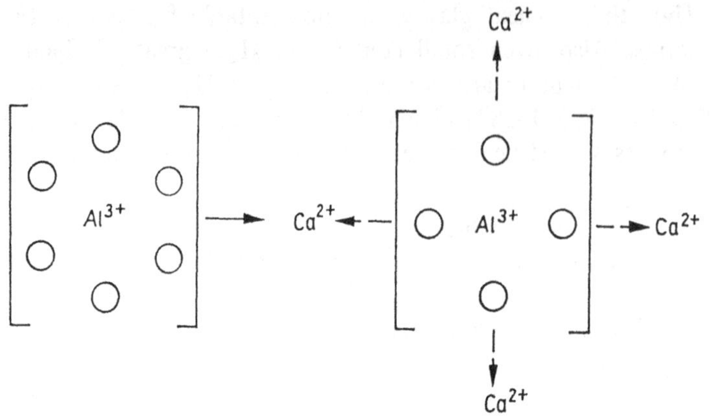

Fig. 9.21. Schematic of the coordination change of an Al^{3+} ion under the counterpolarizing effect of Ca^{2+} ions

9.7.3 IR-Transmitting Chalcogenide Glasses

Chalcogenide glasses are defined as non-crystalline solids solidified from melts of the chalcogens S, Se, and Te and their combinations with elements of groups IV, V, VII, including additions of many other elements (Table 9.6) [701–726]. In oxide or fluoride glasses, only stoichiometric ratios of cation to anion (1:3, 1:4) exist in their building elements SiO_4, BO_3, PO_4, BeF_4, etc., especially since excess oxygen would form volatile O_2. The cation:anion ratios, however, in chalcogenide glasses can be varied within very large limits, since excess S, Se, or Te remain liquid in their melting ranges.

To avoid easily occurring oxidation of chalcogens, chalcogenide glasses must be melted under an inert gas atmosphere, usually in closed containers, e.g., silica or Vycor® ampoules. This necessity obviously limits the volume of glass attainable in production.

Under the impetus of their potential uses as IR transmitting and semiconducting materials, research and development have expanded to an extent that no adequate coverage is possible within the scope of this book. Pioneering early work by the school around Kolomiets [727–729] and early development for potential use in switching devices leading to much insight into structure and properties by Ovshinsky and Bienenstock [730–739] deserve special mention. Structural studies by Krebs [740–742], Feltz et al. [243, 704, 718–720, 726, 743–746] and Linke et al. [681, 682, 684, 692, 715] are other examples of earlier studies. The interested reader is also referred to the following specialist papers: [244–247, 701–703, 705, 706, 708, 712, 713, 715, 716, 721, 723, 724, 740].

A recent survey from the viewpoint of glass chemistry rather than electronics and physics of the immense body of work on contemporary studies since the appearance of the first original edition of this book is by Kreidl [285].

The exploration and interpretation of the electronic states in these glasses has provided impetus to consider models extending to oxide and other glasses.

Table 9.6. Overview of glass formers in binary and ternary chalcogenide systems

	Glass system	Glass former	References
1.	As–S	35 ... 90 Mass-%S	Frerichs [701]
2.	As–Se	40 ... 100 Mass-%Se	Fraser [702]
3.	As–Te	40 ... 50 Mass-%Te	Tsugane, Haradome, Hioki [703]
4.	Ge–S	10 ... 30 Atom-%Ge. 39 ... 43 Atom-%Ge	Kawamoto, Tsuchihashi [246]
5.	Ge–Se	0 ... 30 Atom-%Ge. 38 ... 42 Atom-%Ge	Feltz, Lippmann [704]
6.	P–S	5 ... 25 Atom-%P	Heyder, Linke [705]
7.	P–Se	0 ... 52 Atom-%P	Heyder, Linke [705]
8.	As–S–Se		Flaschen, Pearson, Northover [706]
9.	As–Se–Te		Borisova, Panus, Obrazcov [707]
10.	As–S–Tl		Flaschen, Pearson, Northover [706]
11.	As–Se–Tl		Flaschen, Pearson, Northover [706]
12.	As–S–Br		Pearson, Northover [708]
13.	As–S–J		Flaschen, Pearson, Northover [709]
14.	As–Te–J		Pearson, Northover [708]
15.	As–S–Sb		Kawamoto, Tschuchihashi [710]
16.	As–Se–Cu		Asahara, Izumitani [711]
17.	Ge–As–Se		Webber, Savage [711]
18.	Ge–Sb–S		Linke, Böckel [713]
19.	Ge–Sb–Se		Linke, Heyder [714]
20.	Ge–P–S		Hilton, Jones, Bran [715]
21.	Ge–P–Se		Linke, Heyder [714]
22.	Ge–S–Tl		Linke, Eberhardt: Dietz [716]
23.	Ge–Se–Tl		Linke, Eberhardt: Dietz [716]
24.	Ge–As–S		Borisova [717]
25.	Ge–S–Pb		Feltz, Voigt [243]
26.	Ge–Se–Pb		Feltz, Senf [718]
27.	Ge–S–Sn		Feltz, Schlenzig [719]
28.	Ge–S–Ag		Feltz, Thieme [720]
29.	Ge–Se–Hg		Burkhardt [721]
30.	Ge–Se–Te		Pazin, Obrazcov, Borisova [722]
31.	Ge–As–Te		Krebs [531]
32.	Si–As–Te		Hilton, Brau [723]
33.	Si–P–Te		Hilton, Jones, Brau [715]
34.	P–Sb–Se		Linke, Hey [724]
35.	P–As–Se		Egorova, Kokorina [725]
36.	Ge–Si–As–Te		Feltz, Maul, Schönfeld [726]

Mott [773] was among the first to modify the classical band model for crystalline solids to suit the non-repeating arrangements in glasses. Mott is the only "non-crystalline person" to have obtained the Nobel prize for his work. An excellent general review of non-crystalline semiconductors is by Adler [556].

The most recent critique of the many models rather closely following Mott's is by Emin [774]. It is not possible to discuss the situation within the framework of this book. Suffice it to say that the most significant concept of semiconduction in non-crystalline solids is the replacement of a gap in the existence of electronic states in crystals by one merely in the mobility of states ("mobility gap, mobility edges"). The development of both theory and practical applications remains extremely active and is documented in the proceedings of large annual conferences on "amorphous semiconductors" which must be consulted for more up-to-date information.

Sulfur, Se, and Te form glasses themselves. One typical example for the structure of binary chalcogenide glasses is given in the following.

9.7.3.1 Arsenic Sulfide Glasses

As_2S_3 glass may be considered a "classical" chalcogenide glass. As early as 1870, Schultz-Sellack [775] described the glass-forming tendency of As_2S_3 melts. It may have been after further fundamental studies by Frerichs [701] that the first chalcogenide glass was produced commercially and, by about 1960, applied as an optical material.

Pure As_2S_3 glass is prepared quite easily, since it can be distilled at 703 °C to eliminate impurities. Even minute amounts of H_2O must be excluded to eliminate OH bands interfering with IR transmission (Fig. 9.22). Figure 9.22 also demonstrates its good transmission up to 12 μm. The low total transmittance (70%) caused by the high refractive index ($n = 2.47$ to 2.38 for $\lambda = 1$ to 10 μm) can be improved (to 90% and more) by special antireflection coatings.

The structure of As_2S_3 glasses is closely related to that of crystalline As_2S_3 (orpiment) (see, e.g., Frerichs [701], Petz et al. [776], Leadbetter [777]), which in turn resembles that of As_2O_3 (Fig. 9.23) with O replaced by S. This structure consists of ($AsS_{3/2}$) groupings connected in zig-zag layers.

Glasses form both with excess S (to pure S) and some excess As consisting of various groups (Myers and Felty [778]). At very low As content, depolymerized S chains lead to very low T_g, and S_8 rings may be present up to an As:S ratio of about 1:10. The appearance of $AsS_{3/2}$ groupings crosslinking S chains leads to increasing T_g with a maximum at the composition As_2S_3. At higher As content,

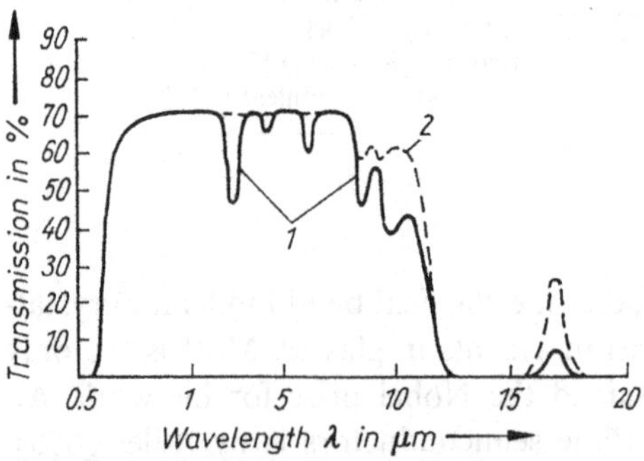

Fig. 9.22. Transmission curves of two As_2S_3-glasses. (*1*) containing some H_2O; (*2*) containing no H_2O

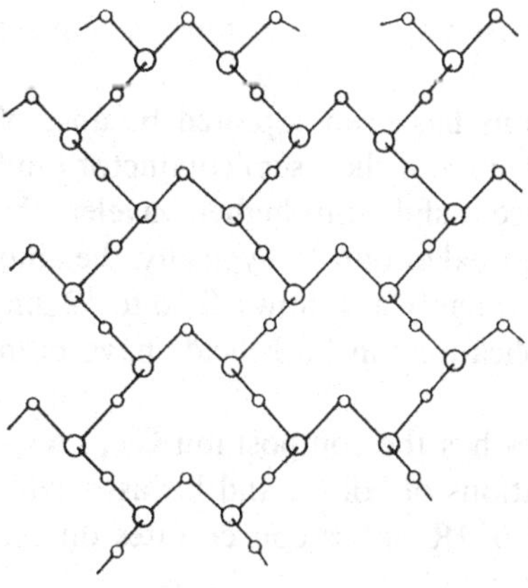

Fig. 9.23. Schematic of the structure of As_2S_3. It is analogous to that of As_2O_3, with just the O ions replaced by S ions

the structure is broken by As_4S_4 groups, similar to those in crystalline As_2S_4 (realgar) and after some decrease in T_g glass formation ceases.

The presence of these quite different groupings, which between S and As_2S_3 constitute a co-polymer, is expected to cause phase separation which has, indeed been verified [244–247, 762–764, 779]. Maruno and Noda, for example, find a T_g plateau (176 °C) between $As_2S_{2.6}$ and $As_2S_{2.5}$, characteristic of immiscibility. IR light guides have been based on the variation of refraction with addition of S to As_2S_3.

9.7.3.2 Other Chalcogenide Systems

Chalcogens (S, Se, Te) form a vast class of non-oxide glasses. Unlike oxygen, which is a gas at the temperatures of oxide glass synthesis, chalcogens are solid or liquid in the range if which they therefore can form glasses way beyond the stoichiometry of chalcogenides. And unlike oxygen they form elemental glasses. Structures in these wide ranges of chalcogen and chalcogenide glasses are based on rings, chains or three-dimensional networks as well as their combinations. While obviously the network former and modifier concepts cannot be applied to chalcogen glasses, similar depolymerization functions can be assigned to certain low valency substitutes like Tl, Ag, Cl, I. These ions play a role in photosensitive (index-change) integrated optics. A first use was in IR optics, indeed continued and expanded into the field of sensors. The constitution of chalcogenide glasses has been successfully related to electronic structure and electronic properties such as photoluminescence, photochromium, photostructural changes, photoconductivity, switching, etc. The basic concept is based on the classical modified valence alternate-pair model (Kastner [780]) which postulates defects (in which anions, e.g. Se, may occur with 3 or 1 instead of 2 coordination; cations, e.g. As, with 2 or 4 instead of 3 coordination) and their proximity. In photostructural changes irradiation changes the arrangement in the valence alternate pair and may be reversible.

This extensive field of glass forming systems and their structures and properties cannot be covered adequately in this text. Reference is made to reviews such as: Elliott [459] (general), Fritzsche [782] (photostructural change), and Kreidl [783] (general). The most important cations in applied systems are As, Sb, Ge.

A vast area of chalcogenide compositions has been explored because of their transmission in the infrared beyond 10 μm and their semiconductor (and related electronic) properties. IR transmittance is shifted to higher wavelengths as vibrations occur between heavier ions with weaker bonds. Typically, the kind of chalcogenide used affects IR transmission roughly as follows: S 10 to 12 μm, Se 14 to 16 μm, Te 18 to 20 μm. Early applications in both fields have been reviewed by Cornet [772].

A commercial U.S. IR-transmitting glass has the composition Ge_{33} As_{12} Se_{55} and is based on the extensive investigations of Hilton and his associates [968]. At present, most activity in the field of IR optics concentrates on the fluoride glasses described in an earlier chapter.

An extensive survey of other binary and polynary glass-forming regions end their structural relations has been made by Kreidl [286]. Some references to earlier work follow: As–S, Frerichs [701], As–Se, Fraser [702], As–Te, Taugane et al. [703], Ge–S, Kawamoto and Tsuchihashi [246], Ge–Se, Feltz and Lippmann [704], P–S, P–Se, Heyder and Linke [705], As–S–Tl, Flaschen et al. [706, 709], As–Se–Tl, Borisova et al. [707], As–S–Se, Flaschen et al. [706, 709], As–Se–Te, Borisova et al. [707], As–S–Br, Pearson and Northover [708], As–S–Sb, Kawamoto and Tsuchihashi [710], As–Se–Cu Asahara and Izumitani [711] Ge–As–Se, Webber and Savage [712], Ge–Sb–S, Linke and Boeckel [713], Ge–P–Se, Ge–Sb–Se, Linke and Heyder [714], Ge–P–S, Hilton et al. [715], Ge–S–Tl, Ge–Se–Tl, Linke et al. [716], Ge–As–S, Borisova [717], Ge–S–Pb, Feltz and Voigt [243], Fe–Se–Pb, Feltz and Senf [718], Ge–S–Sn, Feltz and Schlenzig [719], Ge–S–Ag, Feltz and Thieme [720], Ge–Se–Hg, Burkhardt [721], Ge–Se–Te, Pazin et al. [722], Ge–As–Te, Krebs [531], Si–As–Te, Hilton and Brau [723], Si–P–Te, Hilton, Jones, and Brau [715], P–Sb–Se, Linke and Hey [724], P–As–Se, Egorova and Kokorina [725], Se–Si–As–Te, Feltz et al. [726].

Since the preparation of Kreidl's review [286], some newer important findings on chalcogenide glasses include:

(1) A thorough detailed analysis of the structure of pure Se glass based on computer simulation (Wei et al. [784] and interpretations by Corb et al. [785]) of many experimental and theoretical approaches has become possible. Essentially, the structure is characterized by disordered chains coiled around each other, with *cis* symmetry resembling rings to the point that previous disagreement on rings can be appreciated.
(2) The structures of As–Se binary glasses were reinterpreted by Feltz et al. [786]. The main difference from the case of As–S glasses is the disproportioning of realgar-like $AsSe_{4/2}$ groups, leading to immiscibility, yet the possibility of obtaining high-As quenched glasses.
(3) Very high conductivity analogous to that in fast ion-conducting oxide glasses exists also in some Li-containing chalcogenide systems [787–789].
(4) Tough IR-transmitting glass-ceramics were based on $GeSe_2$ nucleated by $ZrSe_2$ (0.25 mol%) (Cheng et al. [790]).

9.8 Opacified Glasses

9.8.1 Mechanisms of Opacification

Opacification, in all its variations, is always caused by small particles having a refraction different from that of the matrix in which they are dispersed. On principle, it does not matter whether these particles are crystalline or not. According to Ostwald [791], four main kinds of opacification – two of which usually overlap – may be distinguished on the basis of the relation of particle size (d) to wavelength (λ).

(1) Opacification by refraction ($d > \lambda$).
(2) Opacification by reflection ($d > \lambda$).
(3) Opacification by diffraction ($d \approx \lambda$).
(4) Opalescence ($d < \lambda$).

Opacification by refraction is brought about by the difference in refractivity between matrix and dispersed phase: The passing light ray is refracted as it enters and exits the dispersed particle and thus becomes deflected (Fig. 9.24(a)). Similarly, the ray is deflected when it is reflected upon entering and exiting a dispersed phase of different refractivity. (Fig. 9.24(b)). Both phenomena overlap, appearing whenever the dispersed particles possess dimensions larger than the wavelength of light.

When the particle size diminishes and becomes of the order of magnitude of visible light, opacification by diffraction becomes dominant over that by refraction and reflection.

The intensity of scattering caused by such particles is described by Rayleigh's relation (Fig. 9.25):

$$I_{\alpha,\lambda} = \frac{\pi^2 V^2}{\lambda^4 r^2}\, \varepsilon^2(\Delta\varepsilon - 1) \sin^2 \alpha$$

where I = intensity of scattering, ε = relative dielectric constant of particles, V = volume of particles, r = distance of particles from point of measurement, and α = scattering angle.

This relation shows that, in the first approximation, for equal particle size the scattering intensity

$$I_{\alpha,\lambda} \approx V^2/\lambda^4$$

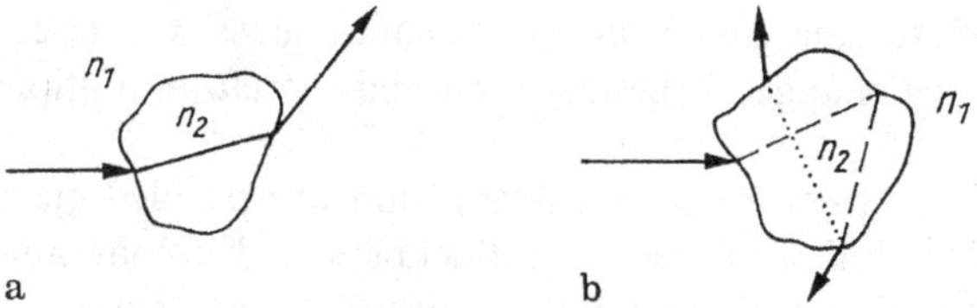

Fig. 9.24. Schematic of opacification by refraction (**a**) and reflection (**b**)

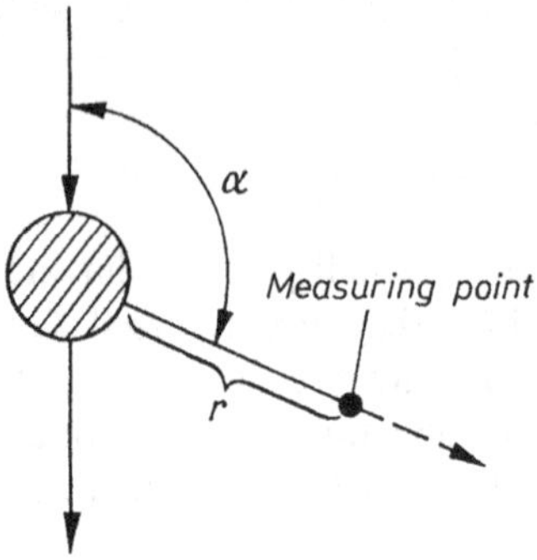

Fig. 9.25. Schematic of a scattering particle illustrating the Rayleigh relation

is proportional to the square of the volume of the particle, and inversely proportional to the 4th power of the wavelength of light. This means that the shortwave rays of the incident white light are scattered more strongly than the longwave rays; e.g., for blue versus red light about 10 times more strongly. This is why an observer on earth, where sunlight is scattered by the gas particles of the atmosphere, sees the "sky" as blue. The "sky" appears black to an observer in a space ship in the absence of scattering. In the morning and the evening the sun appears red since the blue part of the light has been lost through scattering on the long path (Fig. 9.26).

The maximum for opacification by diffraction occurs for particles of the dimension of the wavelength of the incident light. If the opacification is very strong, one speaks of "diffusion". As the dimensions of the dispersed particle fall significantly below the wavelength of incident light, "diffusion" changes to "opalescence".

Opalescence is the name for the type of coloration (dichroism) of a highly dispersed system with no or faint opacification. The material usually appears yellowish-red to brown when one is looking through it, and purplish-blue when one is looking upon it. The dichroism is caused by the prevalence in scattering of the blue part of the spectrum.

It is the contamination of the environment and water droplets that make the sky look opacified. In a clean atmosphere it should be opalescent.

In addition to refraction, reflection, and diffraction, the function of an opacified glass is also affected by absorption. This is of particular importance if it is used in illumination. Other uses of opacified glasses are for example, thermometer scales, sanitary ware, and decoration.

The higher light absorption of an opacified glass in comparison with that of a clear glass used to be assumed to be just due to the longer path length which a light ray has to cover before emerging from an opacified glass. Countless particles produce a continuous change in the direction and therefore multiple scattering.

Gehlhoff and Thomas [792] compared the absorption of opacified glass bulbs with that of clear glass bulbs of the same thickness and colour and calculated the average free path length of light in opacified glass from the absorption values. The path length thus determined was 3.94 times longer than

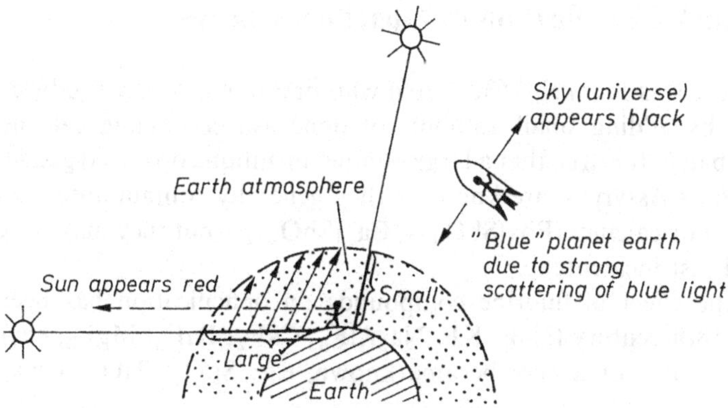

Fig. 9.26. Color phenomena in nature due to light scattering

in clear glass. They also compared the absorption of a clear glass bulb with a thickness 3.94 times greater with that of the opacified glass. If only the longer path length was responsible for the higher absorption of opacified glass, then both values should agree. However this was not the case. The absorption of the clear glass was 3.8%, that of the opacified glass is 14%. The opacifying particles had to be assumed to absorb the light more intensely. This is much better understood today due to a better knowledge of phase separation in glasses.

In the model of Gehlhoff and Thomas no light was lost due to scattering. This is however not the case when an opacified glass is illuminated by one light ray or a pencil of parallel rays and the light emerging in the direction of the incident ray is measured. Therefore the question of light transparency for the described case has considerable interest. Rötger [231] has clarified the situation. The absorption due to light scattering in accordance with the Rayleigh effect must be separated in the calculation of the absorptivity of opacified glasses. It is not possible to calculate the absorptivity of an opacified glass by the simple determination of an extinction coefficient. Rötger [232] determined the scattering factor q_{RO} which exactly gives the scattering proportion from a particle (see Fig. 9.25):

$$q_{RO} = \frac{\lambda^4}{S} \log \frac{100}{\theta_R\,\%} = \frac{2\pi^5}{27\,M} (\varepsilon - 1)^2 D^6 c$$

where S is the thickness, θ_R is the retransmission for pure Rayleigh scattering, M is 2.303 (conversion factor), ε is the dielectric constant, D is the diameter of the particles, c is the concentration of particles and λ is the wavelength of light. The scattering factor q_R is a measure of scattering after the elimination of the multiple scattering of a light ray on countless particles. This factor also takes into account the retransmission and therefore the absorption ratio.

9.8.2 History and Classification of Opacified Glasses

It is believed that it was Kunkel (1630–1705) who first intentionally produced opacified glasses by adding small amounts of bone ash containing calcium phosphate to the batch. It is true that a large number of antique opacified glasses (such as Egyptian or Assyrian) are known – they generally contain antimony compounds of lead or calcium ($Pb_3(SbO_4)_2$, $Ca_3(SbO_4)_2$ – but they may have become opacified just incidentally.

In Europe the effect of fluorine compounds on opacification has been known since the 18th century (NaF, KF, Na_3AlF_6, CaF_2, BaF_2, MgF_2, etc.). Later, so-called "white" opacifiers became known, e.g., SnO_2, TiO_2, ZrO_2, CeO_2, ZnO, etc.

Also, arsenates, particularly lead arsenate ($Pb_3(AsO_4)_2$), and other phosphates (e.g., Na_2HPO_4) were used to produce "opaque" and "milk" glasses. Usually about 1–5 mass% was sufficient, but occasionally much more was added.

Considering this variety in opacifiers, it will not surprise anyone that not just one mechanism is operative. A large body of literature (see for example, Jochmann [793], Springer [794], Knizek [795], Schönborn [796]), addresses this question.

Originally, opacification was attributed almost without exception to the precipitation of numerous microcrystals. In antique glasses, undissolved batch components might have been responsible. More recently, however, phase separation – precipitation of droplets of non-crystalline microphases – was beginning to be considered coresponsible. What has been learned about phase separation since about 1960 shows that a well-opacified glass may be produced in the absence of any of the above-mentioned opacifiers. Many opacified glasses thus just represent an advanced stage in the formation of the microstructure in many glass systems.

It has become necessary, in spite of some overlap, to delineate two mechanisms: opacification by minute crystalline particles, and opacification after phase separation based on droplet-shaped regions of immiscibility. Sometimes older concepts must be revised.

Today one should classify all glasses containing phosphates, arsenates, and fluorine compounds as opacified by droplet-shaped immiscibility regions. While the observed precipitation of crystalline phases may not be of secondary importance, it is certainly of secondary origin.

It is true that glasses opacified by the rest of the above-mentioned opacifiers, such as SnO_2, TiO_2, ZrO_2, CeO_2, ZnO, could be classified as prevalently opacified by microcrystalline phases. But even here crystallization is usually preceded by unmixing. In a few cases crystallization may not even occur. In the following sections some examples of opacified glasses are discussed.

9.8.3 Phosphate Opal Glasses

After early studies of glasses opacified by phosphates by Inwald [797] and Zschimmer et al. [798], it was particularly Schönborn [799] who thoroughly investigated the dependence of opacification on heat treatment and additives. Even he did not come to any final conclusion about all details of the mechanism. Andrews et al. [800] assumed the formation of $Ca_3(PO_4)_2$ on the basis of X-ray work, but Weyl [801] considers that F or OH must be present to form fluorine-, or hydroxyapatite ($3Ca_3(PO_4)_2\ CaF_2$, $3Ca_3(PO_4)_2\ Ca(OH)_2$). Schönborn had proposed $Ca(PO_3)_2$ as the opacifying phase. A common factor is the need for the simultaneous presence of Ca and P.

Vogel [151, 208] carried out investigations on a perfectly transparent glass of the composition: $SiO_2 = 70.7\%$, $B_2O_3 = 17.0\%$, $Al_2O_3 = 2.5\%$, $Na_2O = 6.5\%$, $As_2O_3 = 0.3\%$, $Sb_2O_3 = 0.1\%$ (by weight), by adding in a series of six experimental melts (a to f) the following components:

	a	b	c	d	e	f
mass % P_2O_5	2.5	2.5	2.5	2.5	2.5	2.5
mass % CaO	—	0.5	1.0	1.5	3.0	4.5

The glasses were observed in the electron microscope. Glass a was opacified very weak glass due to the very small droplet size (see Fig. 9.27). The droplet size increased about 20 times with an increase in the content of CaO at constant content (2.5%) of P_2O_5 in glasses b to f. Secondary immiscibility and precipitation of small droplets inside the larger droplets occurs with a content of CaO

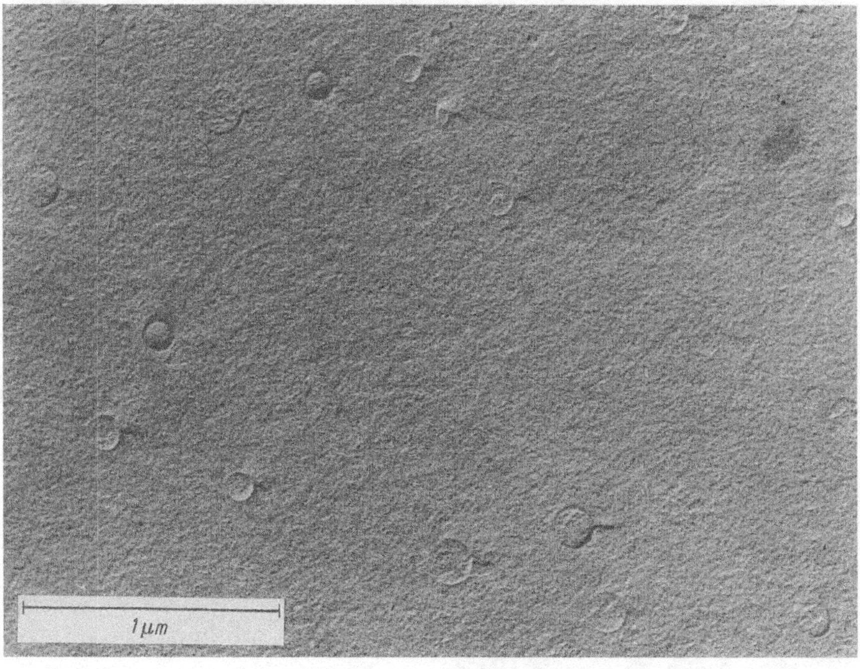

Fig. 9.27. CaO-free borosilicate glass opacified by phosphate. Low opacity due to a few very small P_2O_5-rich droplet regions (electron micrograph after replica preparation)

greater than 3.0% (see Fig. 9.28). At higher contents, crystallization of apatite in the primary droplets occurs while the secondary droplets remain glassy (Fig. 9.29). This result shows that the calcium ions accumulate in the glassy phosphate droplet phase and provide for an increase of the droplet sizes.

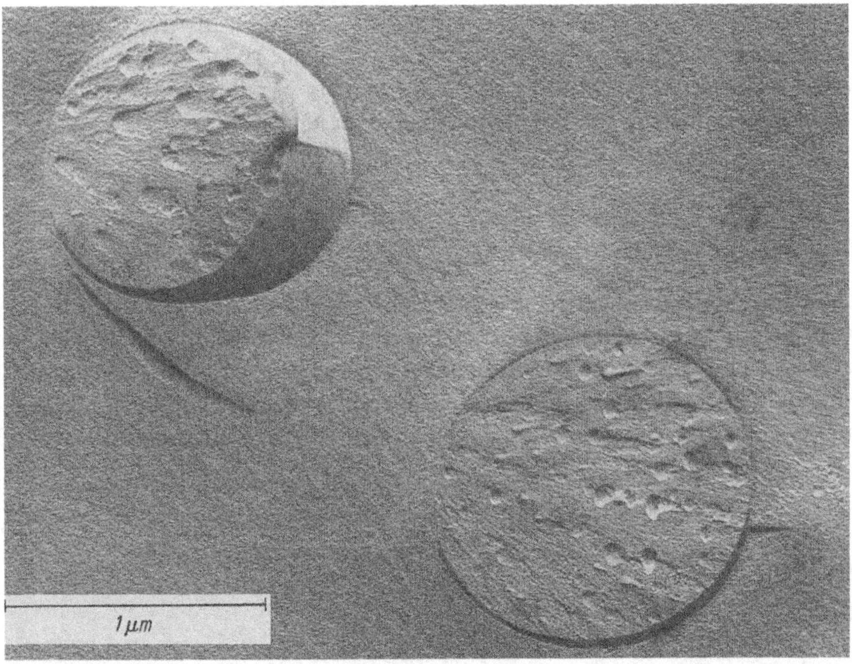

Fig. 9.28. Borosilicate glass containing 3 mass% CaO opacified by phosphate. High opacity due to large droplet regions (droplet phase rich on CaO and P_2O_5). Secondary precipitation of SiO_2-rich droplets in the macrodroplet phase (electron micrograph after replica preparation)

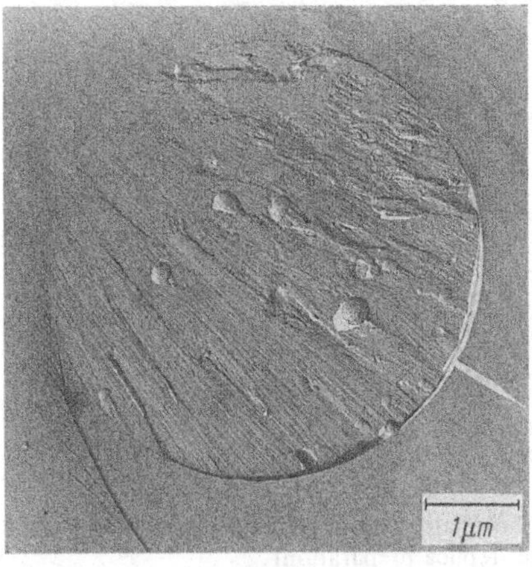

Fig. 9.29. Borosilicate glass containing 3 mass% CaO opacified by phosphate. Now the macrodroplet phase is partially crystallized

The known immiscibility processes in borosilicate glasses help to understand the observations. Alkali borosilicate glasses with low alkali contents tend to a definite phase separation. One even reaps the benefits from it with Vycor glasses. An alkali rich borate phase precipitates in the form of droplet shaped immiscibility regions in an SiO_2 matrix glass phase for glasses of the system Na_2O–B_2O_3–SiO_2 with a high SiO_2 content. If P_2O_5 is added to such a base glass, the glass presents a combination of all three main glass formers B_2O_3, SiO_2, P_2O_5. The experimentally observed phase separation can be theoretically predicted. Due to the field strength difference between the main glass former cations $F_{P^{5+}}$ (= 2.1), $F_{Si^{4+}}$ (= 1.57), $F_{B^{3+}}$, CN3 (= 1.63), $F_{B^{3+}}$ CN4(= 1.34), all partial systems SiO_2–B_2O_3, SiO_2–P_2O_5 and B_2O_3–P_2O_5 should and do indeed phase separate. The cation with the highest field strength dominates the unmixing process in a combination of all binary systems. In the present case, the cation P^{5+} gathers alkali from the already unmixed borosilicate phase into an additional droplet phase rich in P_2O_5. An alkali-rich phosphate glass droplet phase is therefore clearly observed in a B_2O_3–SiO_2-matrix glass phase (see Fig. 9.27). The B_2O_3–SiO_2 glass phase is also unmixed and contains extraordinarily small droplets (as a rule a B_2O_3–SiO_2 glass is therefore wrongly viewed as an homogeneous glass). A small addition of Al_2O_3 will dampen the unmixing process. If CaO is added, it too enters the phosphate phase droplets: the oxygen ion contributes to the shielding of the network former ion with the highest field strength (P^{5+}). An accumulation of CaO in the phosphate phase also favors the crystallization of the calcium phosphate compound: apatite which has a relatively low solubility. The addition of CaO also increased the solubility of SiO_2 at high temperatures. With cooling, an SiO_2-rich phase precipitates with the P_2O_5-rich droplets. It is interesting that the equilibrium diagram of the system $3CaO \cdot P_2O_5$–$2CaO \cdot SiO_2$ also allows mixed crystals only at high temperatures (Trömel, Harkort and Hotop [802]). At lower temperatures all the mixed crystals decompose into two crystalline phases, a P_2O_5-rich and an SiO_2-rich crystal phase. The precipitation of apatite (Fig. 9.29) does not decrease the opacity of the glass. A particle is known to scatter more light as the refractive index difference with the matrix glass phase is larger. The increased refractive index of the crystalline apatite phase consequently increases the scattering effect and further contributes to the opacity.

An unlimited increase of the content of CaO does not increase opacity any further. At times, excessive corrosion of pots or tanks cause an uncontrolled increase in CaO in a phosphate opal melt. The consequences are shown in Figs. 9.30 to 9.36.

Figure 9.30 shows the normal case in the production of phosphate opal glasses: countless P_2O_5-rich microphases in a B_2O_3–SiO_2 matrix glass phase. In Figs. 9.31 and 9.32, the droplet structure starts degenerating due to too high a content of CaO. The tendency for crystallization is too high in the droplet phase. The original spherical particles are converted in the crystallization process into small barrel-shaped entities. "Branches" will grow out on these until they finally become ideally shaped crystallites (Figs. 9.33 and 9.34). In Fig. 9.35 the process goes further and seriously hinders production. The droplet

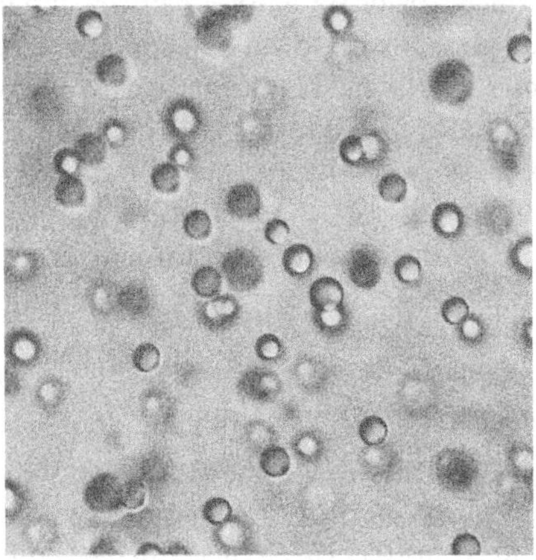

Fig. 9.30. Borosilicate glass opacified by phosphate. Uniform droplet dimension conditions uniform opacity (light micrograph ca. 500:1)

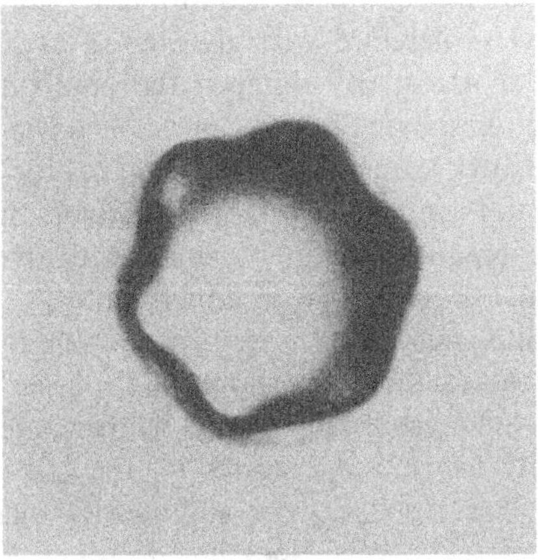

Fig. 9.31. Borosilicate glass opacified by phosphate. Because of excessive CaO content of the glass, the droplets degenerate (starting crystallization). As crystallization progresses, the droplets change shape, showing first six symmetrical protrussions (light micrograph ca. 416:1)

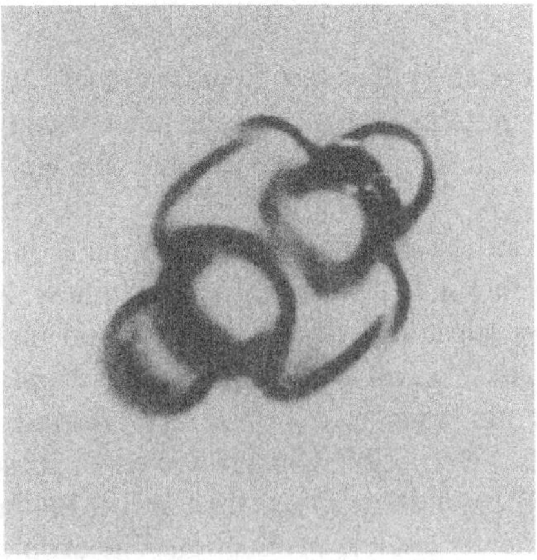

Fig. 9.32. Borosilicate glass opacified by phosphate. The change in shape has progressed

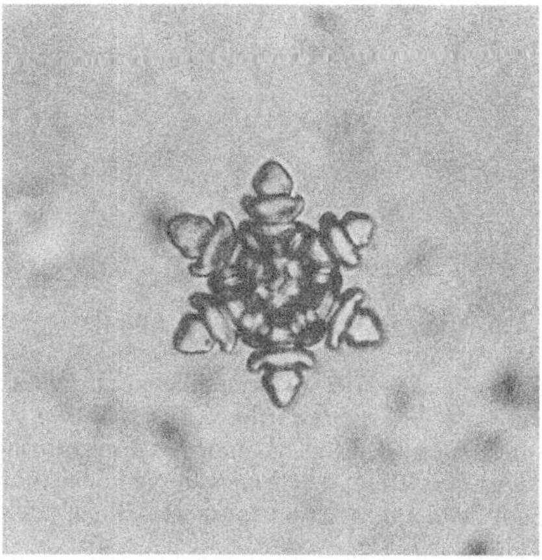

Fig. 9.33. Borosilicate glass opacified by phosphate. Crystallite of $Ca_3(PO_4)_2$ originating in a droplet-shaped immiscibility region (light micrograph ca. 250:1)

Fig. 9.34. Borosilicate glass opacified by phosphate. Progressing growth of the crystallite (light micrograph ca. 350:1)

Fig. 9.35. Borosilicate glass opacified by phosphate. Production problems due to entirely nonuniform opacification. Partial excessive CaO content causes spontaneous growth at $Ca_3(PO_4)_2$ crystals. Thus droplet-free halos form and the glass is transparent around those crystals (light micrograph ca. 200:1)

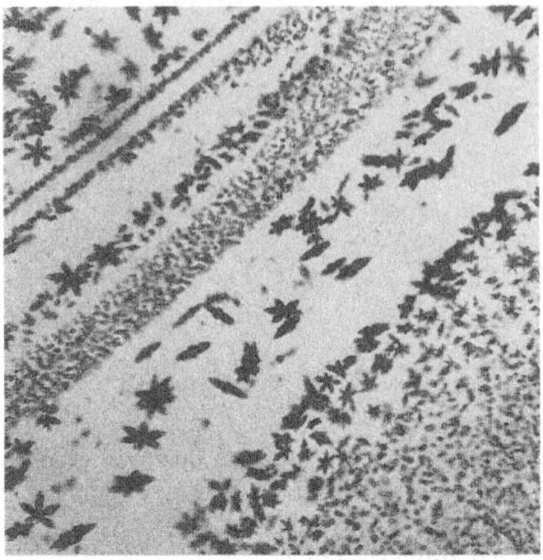

Fig. 9.36. Borosilicate glass opacified with phosphate showing CaO-rich striae. Here the growth of $Ca_3(PO_4)_2$ crystallites is particularly advanced. Because of the withdrawal of P_2O_5 from the microdroplets originally providing uniform scattering, they have vanished once more. All striae appear clear (transparent) (light micrograph ca. 85:1)

shaped glass microphases clearly disappear around the large crystallites and clear glass halos have developed. P_2O_5 and CaO have been extracted from the droplet regions by the large crystallites for their own growth and the light scattering effect decreases enormously due to the lower number of particles. Fig. 9.36 shows a particularly pronounced case with CaO-rich striae in an opacified glass. Such a glass appears to the human eye as an opacified glass with countless completely transparent threads running through it and the glass is therefore useless.

The critical CaO content must be established for each base glass and be exactly monitored during production. The proportions for arsenate opal glasses are analogous to those for phosphate opal glasses.

9.8.4 Fluorine Opal Glasses

The primary role of phase separation in opacification is also recognized for systems opacified by fluorine compounds. It is known that, because of the comparable ionic radius, some O^{2-} can be replaced by F^- in nearly all oxide glasses, with a drastic effect on properties. Since F^- is monovalent, it cannot act as a bridge between two Si^{4+} like O^{2-}; it may be said to break the network, to depolymerize the silicate system. This must lead to a significant decrease in viscosity, which in turn encourages processes whose kinetics are impeded at high viscosity, such as phase separation, nucleation, and crystal growth. Besides, surface tension is significantly decreased. Conditions have been modeled by experiments conducted in the system Na_2O–SiO_2 (Vogel et al. [151, 196]).

Although, in contrast to the system Li_2O–SiO_2, no visible immiscibility had been reported in the system, Na_2O–SiO_2 observations soon led to the postulate (Vogel [205]) and, finally to clear verification (Zarzycki and Mezard [803], Porai-Koshits et al. [804], Hammel [257]) of phase separation in this system, too.

The smallness of the droplets, which lets the glass appear clear, made the study of the influence of fluorine convincing. When 3.4 to 6.6 mol% F replaced O the dimension of droplets increased from about 80 nm (800 Å) to 600 nm (6000 Å) (Table 9.7, Fig. 9.37). Above 6.6 mol% F there is no significant further increase, but spherulites of crystalline NaF precipitate.

That is also the reason for the abrupt deflection in the curve of Fig. 9.37.

The model may be applied to all fluorine opal glasses. It does not matter through which compound F is introduced, but polyvalent cations (Ba, Ca, Mg, Al, Ti, Zr) do influence the beginning of crystallization, hence opacity.

It should be noted that phase separation can be controlled not only by fluorine additions. Fig. 9.38 represents a case of multiple phase separation (six phases or more, five spherical), which should reinforce scattering effects. A very high degree of whiteness (higher than 96%) can be obtained with new opacified glasses developed by Gerth, Rehfeld and Vogel [805]. This type of glass is very important and is used as a standard for optical whiteness measurements or as equipment in that type of optical instrumentation. These glasses are formed by

Table 9.7. Influence of the replacement of O^{2-} by F^- ions in a base glass 18 mol% $Na_2O \cdot 82$ mol% SiO_2 on the droplet dimension

Replacement of O by F (mol%)	Droplet diameter (Å)
3.4	800
4.3	1460
4.8	1800
5.4	2670
6.6	6000
10.4	6250

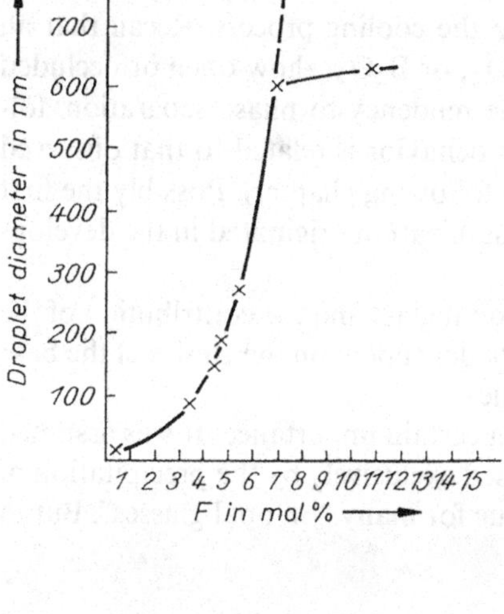

Fig. 9.37. Influence of F content on droplet dimension in a binary sodium silicate glass (18 mol% Na_2O, 82 mol% SiO_2) in which O ions have been replaced by F ions

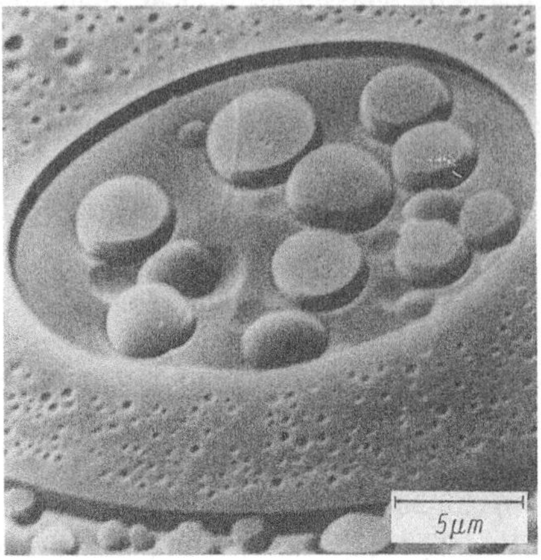

Fig. 9.38. Multiple phase separation in a barium borosilicate glass (at least 6 microphases are present) (electron backscattering picture)

combining opacifiers and the main components: (SiO_2–B_2O_3–Al_2O_3–KPO_3–KF–NaF–AlF_3–PbO–$CdWO_4$). The addition of heavy metals ensures an optimal refractive index difference with the particles which separate out (microphases and crystals).

9.8.5 Opal Glasses Based on SnO_2, TiO_2, ZrO_2, CeO_2, ZnO, and Other Compounds

Opal glasses containing the so-called opacifiers SnO_2, TiO_2, ZrO_2, CeO_2, ZnO, or other compounds today have relatively little importance compared to the two main types (fluorine and phosphate) described above. Their properties in a sense represent a transition to those of enamels, porcelains, and glass-ceramics.

Added in small concentrations, they promote the tendency toward phase separation of the base glass and are usually enriched with the network-modifiers in the droplet phase. Because of their low solubility they usually are precipitated in the form of nuclei or crystallites during the cooling process. Recall that all these compounds, in combination with SiO_2, or B_2O_3, show open or occluded immiscibility gaps. There is a considerable tendency to phase separation, followed immediately by crystallization. This behavior is related to that observed in controlled crystallization (treated in the following chapter). Possibly the first developments in the field of controlled crystallization originated in the development of opal glasses.

Obviously the scattering power of an opal glass and the contribution of the crystalline phases precipitated are also quite dependent on the choice of the base glass. There is a great variety of possibilities.

For some time "tin opal glasses" had a certain importance. It was assumed that their strong scattering power was caused exclusively by the precipitation of crystallites of tin oxide. This is certainly true for many "tin opal glasses". But in simple alkali glasses up to 10 mol% of SnO_2 are easily soluble and, moreover, suppress phase separation. This is just to exemplify the fact that, depending strongly on the base glass, these compounds need not promote opacity.

9.8.6 Light Scattering and Color of Microdisperse Two-Phase Glasses

The phenomenon of light scattering accompanied by coloration in microdispersed two-phase glasses will be introduced using a "macromodel."

Around 1936 the dispersion light filters produced by the Jenaer Glasswork Schott und Gen [806] were quite important. Their function was based on an effect discovered, and described by Christiansen [807]. They consisted of a cuvette filled with a mixture of a glass powder of defined grain size and a liquid. The glass and the liquid (our "macromodel" phases) possess different dispersion but the same refractive index for a certain wavelength λ_0 of white light. If white light passes the cuvette, only the fraction of wavelength λ_0 will pass both phases without being scattered, while light of any other wavelength is scattered sidewise, more so the larger the difference in refractive index (Fig. 9.39). In this way a monochromatic light filter ("Christiansen filter") is formed.

Since the temperature dependence of dispersion is quite different for glass and the organic liquids used (e.g., methyl benzoate, glycerol, sugar solution, etc.), it is possible to change the wavelength λ_0 of the light transmitted without scattering over the entire visible region (see Fig. 9.40). While for the liquids a family of dispersion curves exists for temperatures between 0 and 70 °C, the dispersion of the glass is constant in this interval. Thus, the wavelength of the light passed without scattering (λ_0) at a particular temperature can be read as the intersection of the glass curve and the curve of the liquid. It can be seen that the intersection (and thus the wavelength λ_0) migrates toward shorter wavelengths with increasing temperature (Fig. 9.41). At the same time, Fig. 9.41 shows that such filters may be designated as practically monochromatic. In a further development of these "dispersion filters" (in contrast to absorption filters), it was particularly Knudsen [808, 809] who tried to minimize excessive and disturbing temperature sensitivity by embedding the glass grain in lacquers, celluloid, rubber, etc. Finally he proposed the preparation of two-phase glasses, where color changes would require temperature variations of several hundred degrees. Such filters would be quite temperature-independent.

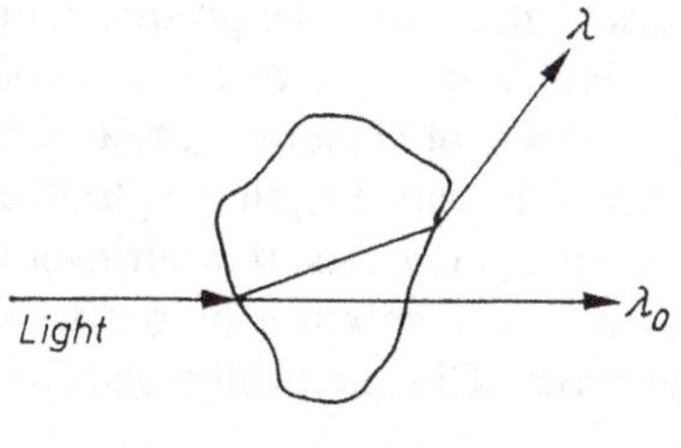

Fig. 9.39. Schematic of the function of a classical Jena dispersion filter. Embedding medium and glass grains possess the same refractive index for wavelength λ, so that light of this wavelength will pass the two-phase system without deviation

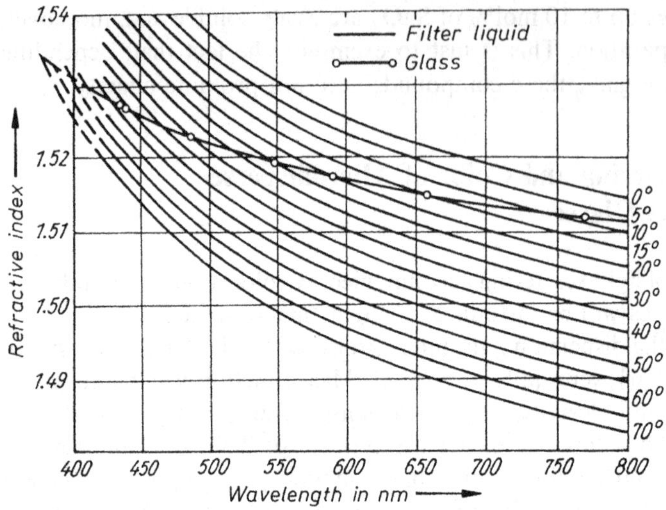

Fig. 9.40. Dispersion of a glass and a dispersion liquid at various temperatures

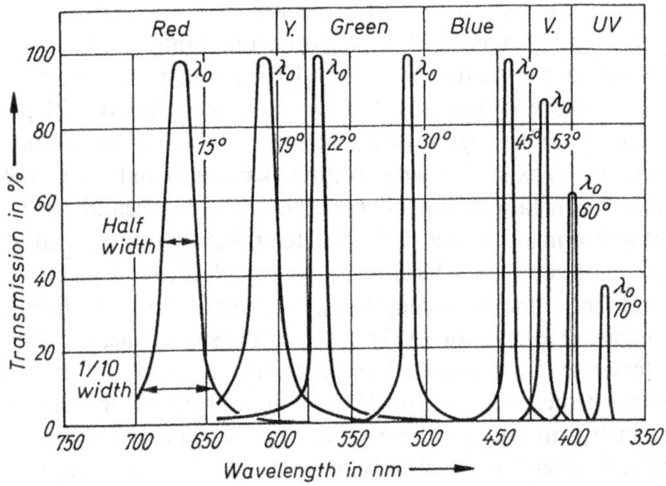

Fig. 9.41. Transmission curves of a dispersion filter for various filter temperatures

This proposal by Knudsen brings us back to the dimensions which are so important for recent concepts about glass structure. True, on a laboratory scale such "Knudsen glasses" exhibiting the Christiansen effect could be produced in Jena, but an industrial production proved impossible because of the lack of uniformity in droplet size and the sporadic appearance of crystals. For, as the droplet size decreases approaching the dimensions of the wavelength of light, the filter effect decreases and the Rayleigh diffraction effect increases. If birefringent crystals appear, this is equivalent to the presence of a mixture of particles of different refraction, and the filter effect decreases. Chromatically disperse

Knudsen two-phase glasses must be considered coarsely disperse, and it is only under such conditions that their filtering behavior could be optimized.

In the section introducing opal glasses, mention was made of the appearance of color based on the diffraction and opalescence behavior of very small particles. It was shown that the color of the blue sky and the red morning and evening sun could be understood through the Rayleigh scattering relations.

An example will now be given showing that the same cause may lead to serious trouble in the glass-production process. Only complete understanding of the mechanisms described here will ensure the prevention or repair of such disturbances.

During the production of some extreme optical glasses, sometimes a considerable reddish-brown discoloration is observed, which depends strongly on temperature. When lenses are pressed from such glasses, there is an upper size limit because the slower cooling of the larger sizes causes a yellowish-brown discoloration. This defect could be traced unequivocally to the diffraction at finely dispersed phases (opalescence). The color was due to the very small dimensions of a dispersed K_2TiF_6-rich fluoride phase which moves the wavelengths of the reflected light toward the UV. Besides, no Tyndall effect can be observed because of the smallness of these droplets.

Figure 9.42 shows the light-transmission curves of such glasses which are free of any Tyndall effect, and are colored more or less yellowish-brown as a consequence of variations in thermal history. These varying colors originate in the precipitation of droplet-shaped glass phases of different size, as can be demonstrated by electron microscopy.

Figure 9.43 shows a colorless pressed glass. As shown by the comparison with the molybdenum oxide crystal plane, there are no droplet-shaped immiscibility regions present (or their size is below the resolution of the replica process). In Fig. 9.44, phase separation and the formation of droplet-shaped regions is just beginning. This glass shows just a faint yellowish-brown color. In Fig. 9.45, innumerable finely dispersed droplets can be distinguished: This glass is brown, but shows no Tyndall effect. By adequate heat treatment, the dimensions of the droplets can be enlarged to the point that, besides the brown color in transmission, the blue-violet color in reflection becomes apparent.

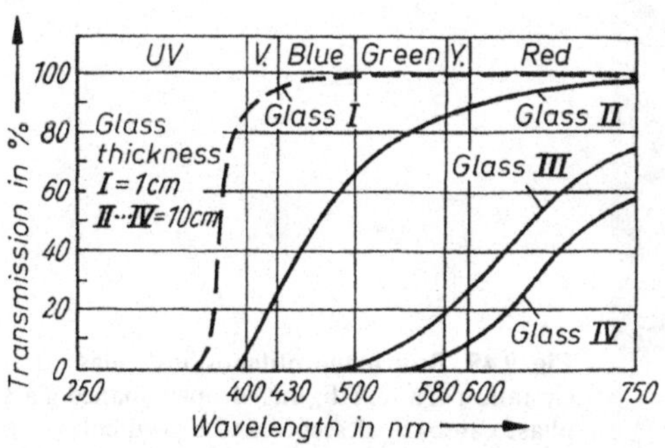

Fig. 9.42. Light-transmission curves of 4 fluorotitanosilicate glasses of equal composition but different cooling rates. Difference in resulting colors caused by opalescence. *I* strongly quenched (lens blank) *II* less quenched (colorless) *III* normal annealing (faint yellow-brown) *IV* slow annealing (dark brown)

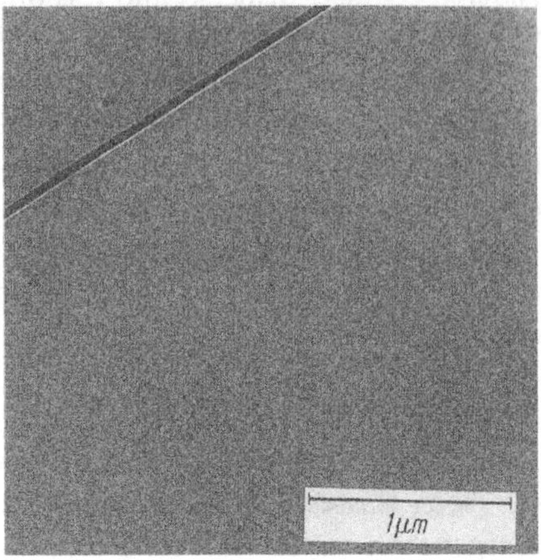

Fig. 9.43. Electron optical picture of a fluorotitanosilicate glass which had been quenched rapidly, thus becoming colorless. Comparison with the molybdenum oxide crystal test plane shows that no droplet-shaped immiscibility regions can be demonstrated since – if they are present at all – their size would lie below the limit of evidence by the method of preparation used

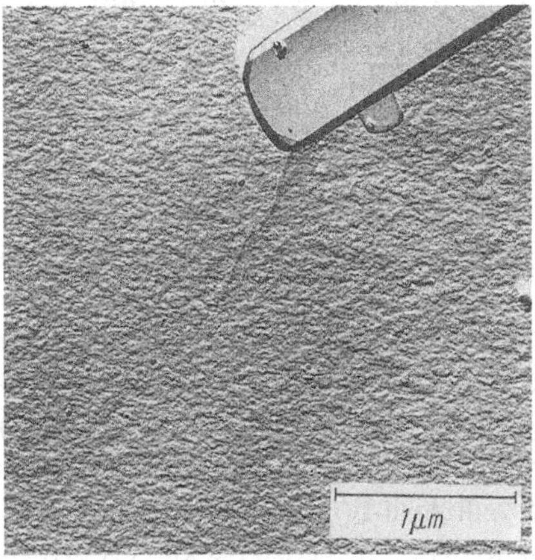

Fig. 9.44. Tyndall-free fluorotitanosilicate glass, faintly yellow-brown. Phase separation just starting (top right: molybdenum oxide crystal test plane)

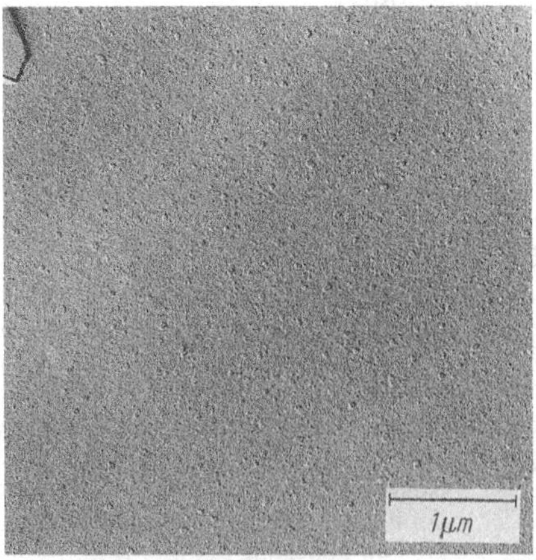

Fig. 9.45. Brown fluorotitanosilicate glass. Precipitation of a K_2TiF_6-rich droplet-shaped glass phase causes opalescence and brown color

Rötger and Besen [232] have investigated quantitatively the dependence of droplet dimension on cooling rate for these optical glass types. They found that the number of the droplet-shaped immiscibility regions in 65 times larger if, instead of holding at a certain temperature during cooling, the same holding temperature is reached after passing the immiscibility region from below. Growth as a function of holding time is proportional to the volume, not to the diameter of the droplet.

The disturbance in production was completely removed by the proper chemical interference with the phase-separation process (Vogel et al., [6] Vogel [151]). In practice, such effects have not only been observed in optical glass production, but also in the production of technical glasses.

In summary, it should be stressed that some color effects treated in this section are not based on the absorption of 3d element complexes (the usual case), but on the precipitation and the scattering behavior of small particles in the glass.

10 Crystallization of Glasses

10.1 General

One of the most important causes of defective glass is crystallization, which is, after all, the equilibrium stage of the solidifying melt. From this viewpoint, the manufacture of glass is based on tricking the melt so that it will not attain the low-energy state it strives to assume. Whether one wishes to avoid crystallization or to control it, it is necessary to master theoretically the phenomena of nucleation and crystal growth.

As indicated in Chapter 2, the crystallization of glasses is dominated by two factors: nucleation frequency and crystalline growth (Tammann [11]). Crystallization is predicted on the formation of nuclei, which occurs in the important state of transition between the liquid melt and the crystalline solid state.

Kleber [810] and Roy [811] have given a lucid description of the conditions for this transitional stage. In organic polymers, as well as in silicates (i.e., typical glass melts), nucleation and crystallization will not occur spontaneously, as in salt melts, even when the saturation or freezing temperature is passed slowly. The processes are slowed for reasons discussed earlier. As shown in Fig. 10.1 it is necessary to pass a second curve ("curve of supersaturation") at temperatures below those of the saturation or freezing curve before nuclei and crystals will form. This second limit is blurred; the area between the two limits, which cannot be determined exactly but is critical for the existence of glasses, is called the Ostwald-Miers region. A crystal introduced into this region could grow, but nucleation is excluded by its limit. The nucleation frequency (see in Fig. 2.2) is essentially determined by two factors (Kleber [810]): (1) the work of nucleation proper, which is defined as the energy required for nucleation if its building blocks are already present, and (2) the activation energy required to bring the building blocks of the nucleus to the location of its origin.

The work of nucleation proper consists of a difference of two energy contributions. One is the work required to form a new surface, overcoming the surface tension of the environment. The second contribution is energy gained – as is well known – on ordering an amorphous region to a crystalline lattice. This portion is volume-dependent. Neglecting factor (2), i.e., assuming that the building blocks are already in location, the work of nucleation will be small if the stoichiometric composition of the amorphous region is close to that of the ordered crystal to be formed. In that case the difference in surface tension is conditioned only by the difference in ordering, and nucleation will be much easier than in the case of a compositional difference between environment and nucleus.

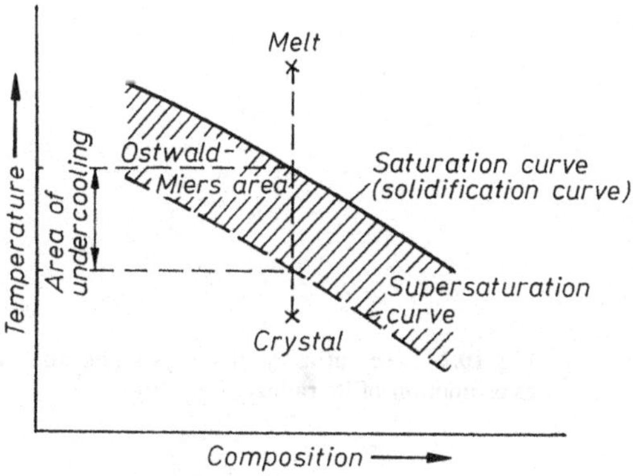

Fig. 10.1. Schematic of the transition of a highly viscous melt to the crystalline state

10.2 Theoretical Considerations

10.2.1 Homogeneous Nucleation

The thermodynamic treatment of nucleation starts from the requirement that a process of phase change must be associated with a decrease in the free enthalpy ΔG. The total free enthalpy consists of a volume contribution and a surface contribution.

$$\Delta G = -\Delta G_v + \Delta G_0$$

The volume contribution ΔG_v corresponds to that portion of the energy which is gained in the nucleation process; it thus becomes negative. Assuming a spherical shape.

$$-\Delta G_v = -\frac{4}{3}\pi r^3 \Delta g_v$$

where r is the radius of the nucleus and Δg_v the change in free volume enthalpy during phase transition.

The surface contribution ΔG_0 corresponds with the energy to be provided when a nucleus is formed or enlarged against the opposing surface tension of the environment. It is thus positive:

$$\Delta G_0 = 4\pi r^2 \sigma(\Delta g_0)$$

where σ = surface tension. The free enthalpy of the interface Δg_0 can be neglected.

Thus the free total enthalpy becomes

$$\Delta G = -\frac{4}{3}\pi r^3 \Delta g_v + 4\pi r^2 \sigma$$

Figure 10.2 shows the free enthalpy of nucleation as a function of r. As the resultant of ΔG_v and ΔG_0 it first rises, but decreases from a maximum ΔG^* corresponding to a critical radius r^*.

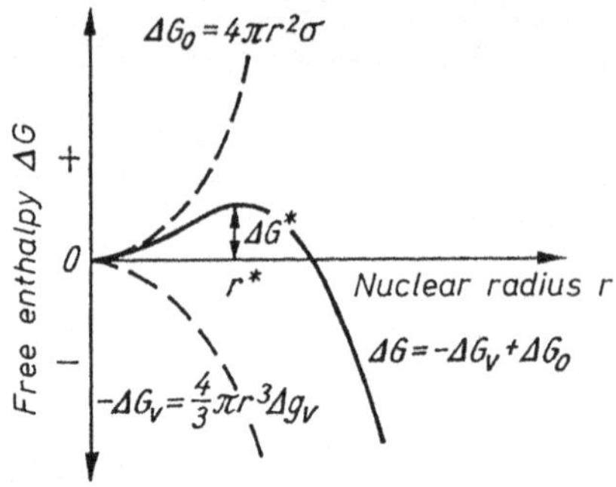

Fig. 10.2. Free enthalpy of a crystalline nucleus as a function of its radius ($T < T^m$)

This maximum is found to be

$$\Delta G^* = \frac{16\pi\sigma^3}{3(\Delta g_v)^2}$$

$$\left(\text{from } \sigma\Delta G/\mathrm{d}r = 0 = -4\pi r^{*2}\Delta g_v + 8\pi r^*\sigma\right.$$

$$r^* = 2\sigma/\Delta g_v$$

$$\left.\Delta G^* = -\frac{4}{3}\pi\frac{8\sigma^3}{g_v^2}\Delta g_v + 4\pi\frac{4\sigma^2}{g^2}\sigma\right)$$

It is important that the critical radius r^* also decreases with undercooling ΔT (Fig. 10.3).

The change of the free enthalpy of nucleation with temperature is described by the following general equation:

$$\Delta G^* \approx \frac{1}{(T_m - T)^2}$$

i.e., as the temperature T approaches the melting temperature $T_m(\Delta T \to 0)$ ΔG^* becomes infinitely large and no nucleation occurs.

The nucleation frequency (nuclei $s^{-1} \cdot cm^{-3}$) is according to thermodynamic considerations (Volmer and Weber [812]).

$$I = A\exp(-\Delta G^*/KT)$$

where I corresponds to Tammann's "I" (Fig. 2.2), A is a constant, ΔG^* = the free enthalpy of nucleation, K = Boltzmann's constant, and T = absolute temperature.

Taking into account the enthalpy of activation for the diffusion of ions from a homogeneous solution to the location of nucleation ΔG_D, Volmer's relation becomes:

$$I = A\exp[(-\Delta G^* + \Delta G_D)/KT]$$

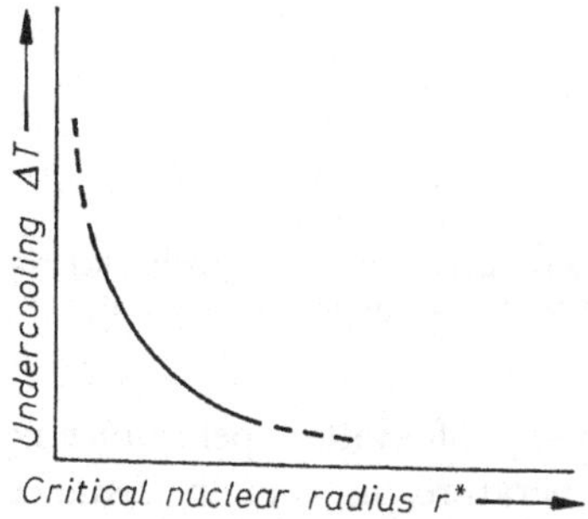

Fig. 10.3. Change of critical nuclear radius with the undercooling of a melt

10.2.2 Heterogeneous Nucleation

The term heterogeneous nucleation refers to the facilitation of the nucleation of the main phase by the presence of some alien crystalline nucleus. Frequently, epitaxial interaction between the primary alien phase and the main phase cannot be excluded in this process. Because of the direct influence on the atoms of the main phase deposited on the primary alien nucleus, one may talk about the growth of the primary alien nucleus by means of a different species. In such an extreme case there would be no need to overcome any nucleation barrier. Normally the critical free enthalpy of nucleation is just decreased by a considerable amount. Accordingly, the Volmer equation must be modified.

The strength of interatomic interaction between the alien nucleus and the liquid main phase is best expressed by the measurable contact angle (Fig. 10.4).

$$f(\theta) = \frac{(2 + \cos\theta)(1 - \cos\theta)^2}{4}$$

Thus, heterogeneous nucleation occurs if the interatomic forces between liquid phase and substrate (alien nucleus phase) are larger than those within the liquid phase ($\theta < 180\,°$). There will be no heterogeneous nucleation if they are smaller ($\theta = 180\,°, f(\theta) = 1$).

In this way heterogeneous nucleation occurs if

$$\Delta G^* \text{ (heterogeneous)} < \Delta G^* \text{ (homogeneous)}$$

According to Turnbull, heterogeneous nucleation can thus be described by

$$I = A'' \exp\{[-\Delta G^* f(\theta) + \Delta G_D]/KT\}$$

10.2.3 Crystal Growth

For the mathematical description of crystal growth u (Fig. 2.2) various relations have been proposed which match one or the other experimental curves.

According to Turnbull,

$$u = \frac{fD_u}{a_0}(1 - e^{\Delta G/KT})$$

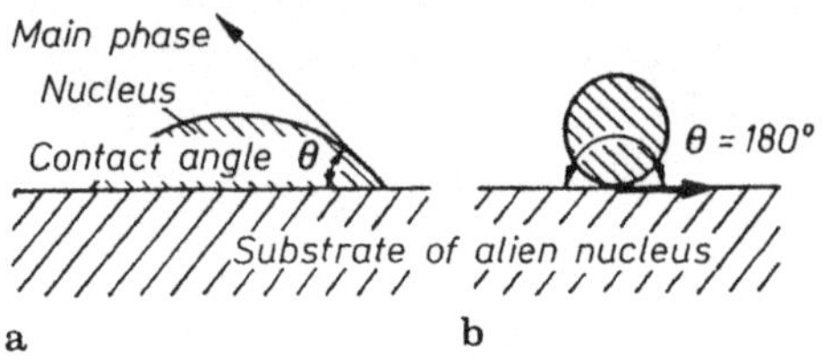

Fig. 10.4. Schematic of a beginning heterogeneous nucleation on alien substrate

where u = crystal growth rate, G = free enthalpy of motion per molecule, K = Boltzmann's constant, T = absolute temperature, a_0 = jump distance, f = fraction of interface available for the deposition of molecules, and D_u = kinetic constant for crystal growth.

Taking into account the viscosity η of the melt, this relation transforms to

$$u = \frac{fKT}{3\pi a_0^2 \eta}(1 - e^{\Delta G/KT})$$

A different relation, proposed by Frenkel, [77] has been found fitting for crystallization data, particularly for alkali silicate glasses:

$$u = Ae^{-1/KT(\Delta U + CT_0/\Lambda \Delta T)}$$

where u = crystal growth in $\mu m \cdot min^{-1}$, A, C = constants, ΔU = activation energy of self-diffusion in the melt, T_0 = equilibrium parameter in K, ΔT = undercooling, T = temperature, and Λ = heat of condensation for two-dimensional condensation.

In this manner both Turnbull and Frenkel interpret and describe crystal growth by thermodynamic as well as by kinetic parameters.

For a better understanding of many crystallization phenomena in glasses, the steps involved in the deposition of ions on a growing crystalline lattice are discussed in the following.

Modern molecular theories (Kossel [814–816] and Stranski [817–820]) have contributed decisively to the understanding of crystal growth (also see [821, 822]).

Recall that in heteropolar crystals the energetic interactions between ions of opposite charge can be described, in the first approximation, by Coulomb's law:

$$K = Z_1 e Z_2 e / a^2$$

where K = force of attraction, e = elementary charge, and Z = valence, or for equal valence:

$$K = (Ze)^2 / a^2$$

Imagine the formation and growth of an NaCl crystal. If 1 g·mol Na^+ and Cl^- ions approach each other from an infinite distance (i.e., as an ideal gas), a crystal forms through the steps ion → molecule → Chain → lattice plane → lattice with a gain in the lattice energy U. The energies of formation for each of these steps will represent fractions of this energy.

The energy of formation (or work of dissociation) for a molecule is, in the first approximation:

$$\phi_M = -(e^2/a)$$

(a) Consider a simple ionic chain (Fig. 10.5). Ions 1 and 2 will attract each other, ion 3 is repelled by 1 at a distance $2a$, and ion 4 is again attracted. The total energy of formation then will be:

$$\begin{aligned}\phi_1 &= -\frac{e^2}{a}+\frac{e^2}{2a}-\frac{e^2}{3a}+\frac{e^2}{4a}\dots\\ &= -\frac{e^2}{a}\left(1-\frac{1}{2}+\frac{1}{3}-\frac{1}{4}\dots\right)\\ &\approx -0.6932\dots e^2/a\end{aligned}$$

$$\Phi_1 = -\frac{e^2}{a}+\frac{e^2}{2a}-\frac{e^2}{3a}+\frac{e^2}{4a}\dots = -\frac{e^2}{a}\left(1-\frac{1}{2}+\frac{1}{3}-\frac{1}{4}\dots\right).$$

$$\Phi_1 = -0.6932\frac{e^2}{a}.$$

(b) Consider the formation of a 2-dimensional lattice plane, i.e., the start of a second chain next to one or several completed chains (see Fig. 10.6). Two ions at distance a attract each other, ions at distances $a\sqrt{2}$ and $2a$ are repelled because of their equal charge. The resulting energy of formation is:

$$\phi_2 = \underbrace{-e^2+\frac{2e^2}{a\sqrt{2}}-\frac{2e^2}{a\sqrt{5}}+\frac{2e^2}{a\sqrt{10}}\dots}_{\text{1st chain}}+\underbrace{\frac{e^2}{2a}-\frac{2e^2}{a\sqrt{5}}+\frac{2e^2}{a\sqrt{8}}\dots}_{\text{2nd chain}}$$

$$= -0.1144\frac{e^2}{a}$$

(c) For the formation of a new chain on a completed lattice plane a similar summation yields:

$$\phi_3 = -0.0662\frac{e^2}{a}$$

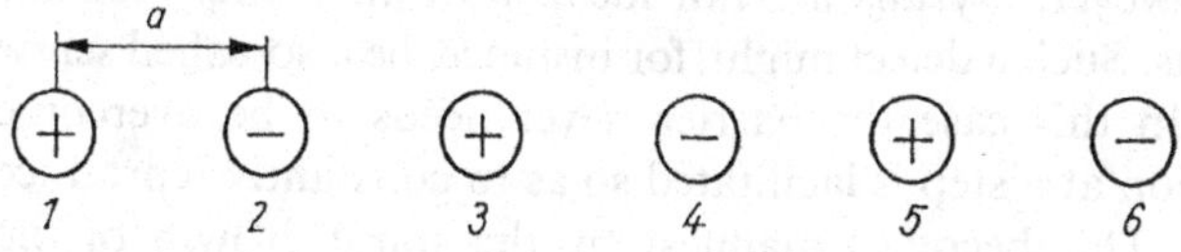

Fig. 10.5. Approach of an ion to an ion chain

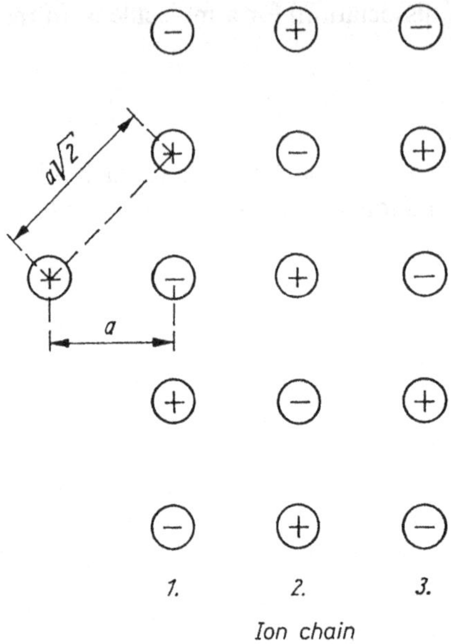

Fig. 10.6. Approach of an ion to several completed ion chains

(d) For the formation of a chain on a finished lattice plane,

$$\phi_4 = \phi_3 + \phi_1 = (-0.0662 - 0.6931)\frac{e^2}{a} = -0.7593\frac{e^2}{a}$$

(e) For the formation of a second chain on a finished lattice plane,

$$\phi_5 = \phi_3 + \phi_2(-0.0662 - 0.114)\frac{e^2}{a} = -0.1806\frac{e^2}{a}$$

(f) For the continuation of a second chain on a complete lattice plane,

$$\phi_6 = \phi_3 + \phi_2 + \phi_1 = -0.8738\frac{e^2}{a} \approx -\frac{U}{2}$$

(i.e., with the deposition of an ion at this stage, one-half of the lattice energy has been gained). All these steps are presented schematically in Fig. 10.7.

It can be seen that, to build a new chain on a complete lattice plane, a certain barrier must be overcome. Here the gain by depositing an ion is the smallest. In practice, however, crystals are not ideal, with an entirely regular lattice, but possess defects. Such a defect might, for instance, be a so-called screw dislocation (Fig. 10.8). In this case the barrier never needs to be overcome entirely because deposition at a step is facilitated so as to constantly reproduce the step (Fig. 10.8 a–c). This becomes manifest by the spiral growth of the crystal.

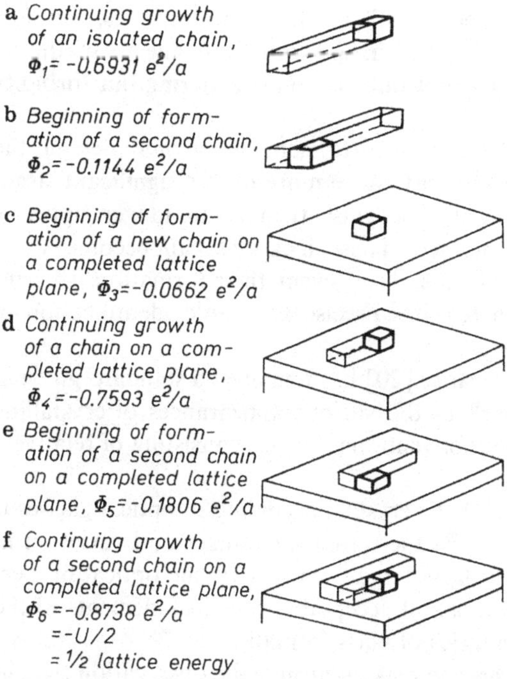

Fig. 10.7. Growth stages of a crystal lattice (a)–(f) and resulting energies of formation ϕ (Kossel and Stranski [814–820])

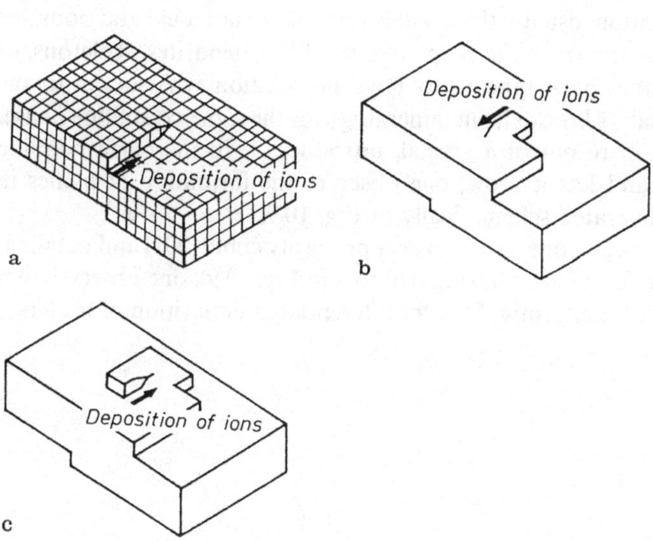

Fig. 10.8. Schematic of spiral growth on crystals starting at a screw dislocation (Kleber [810]). **a** Screw dislocation; **b** and **c** stages of ion deposition on constantly recurring screw dislocations

10.3 Crystallization as a Defect in Glass

The precipitation of crystals impairs glass quality, mainly through two phenomena. Very small crystals ("crystallites" – but not those postulated by the crystallite theory), sometimes called "dust" in the optical glasses, interfere considerably with light transmission and image formation due to scattering effects. If they are somewhat larger, not only Tyndall scattering but turbidity may result.

If the crystals grow to still larger sizes (dimensions in millimeters), the difference in coefficient of expansion may cause more or less significant strain around the crystalline precipitation. The glass then is no longer optically isotropic and birefringence is observed. These locations can become stress centers at which fracture will originate. To prevent their formation, efficient management of glass production requires precise and rapid identification of crystalline precipitations in glasses.

Jebsen-Marwedel's classical treatise [20] has remained a standard guide in this domain. Many defects, as well as the various appearances of crystalline phases are treated, particularly for conventional glasses consisting of relatively few components.

Identification and cataloging of the crystals occurring in optical glasses is still in a beginning stage. Yet it is just there that the danger and variation in crystalline precipitation are particularly large because extreme optical properties tend to demand the use of almost all acceptable chemical elements of the periodic system to the very boundaries of glass formation.

An attempt should be made here to make plausible the observation that, in the case of devitrification, usually the crystals formed are not ideal and bounded by planes, but rather appear in the shape of crystallites, dendrites, skeletons, or spherulites. What one finds appears to have no relation whatsoever to the crystal shapes and habits known from mineralogy for the same crystalline phase.

If one cuts a sphere out of a crystal, introduces it in the corresponding saturated solution, and lets it grow, one observes the formation of planes in certain places, as illustrated schematically in Fig. 10.9.

In this particular case, one would expect primarily cubic (001) and octahedral (111) planes. After some growth stages (0 to 3 in Fig. 10.9), one observes that the planes do not shift uniformly. Due to differentiated deposition of material,

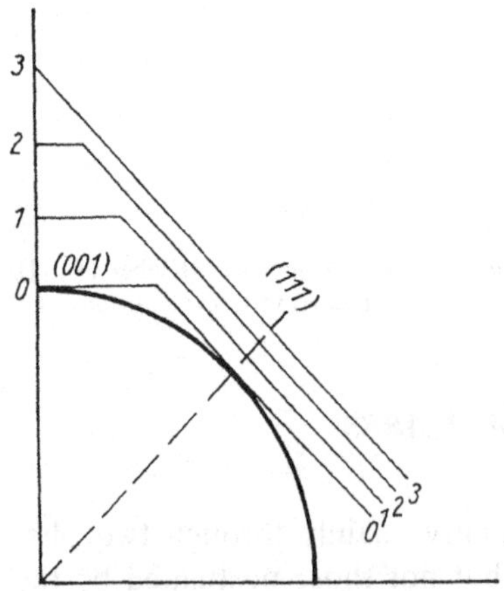

Fig. 10.9. Schematic of the growth of a sphere, cut out of a cubic crystal, in the corresponding saturated solution (0–3: growth stages) 001 cube plane, 111 octahedron plane

one plane moves faster than another. In Fig. 10.9 the cubic plane (001) grows more rapidly. That means, however, that it becomes smaller and smaller until it vanishes, so that just the octahedral plane (111) remains, at which less material is deposited and which therefore shifts at the slowest rate. Corners and edges of crystals thus may be defined as planes of maximum rate of moving.

In this connection, the well-known phenomenon should be mentioned that, in undercooled glass melts exclusively, crystallites and dendrites are precipitated. This phenomenon is based on the fact that, at the beginning of spontaneous crystallization of an undercooled melt, growth clearly occurs in the preferred direction. Since conventional glass melts are always very viscous, the transport of new growth material by diffusion is too slow. Thus planes with the smallest rate of shifting at first cannot form, and skeletons, as pictured in Figs. 10.10 and 10.11, result.

Now one might ask for the basic reasons for the difference in growth between different planes in the same crystal, the difference resulting in the formation of different habits.

The main causes are external circumstances, particularly the influence of impurities. Apparently impurities are absorbed preferentially at certain locations in a growing crystal without being incorporated into the lattice, but blocking the deposition of the pure material.

In this manner the rate of shifting can vary so much that the habit changes. In the process of crystallizing NaCl, the influence of urea causes the change in habit from cubic to octahedral (scheme of Fig. 10.9). But other external

Fig. 10.10. Ideally uniform crystallite of cristobalite in a high-SiO_2 silicate glass (300:1)

Fig. 10.11. Dendrite precipitation of crystal skeletons resembling water plants in a BeF_2 glass (250:1)

circumstances (e.g., pressure, temperature, rate of cooling) also have an influence on the formation of a crystalline habit, just as the atmospheric climate influences the growth of a plant.

Although the phenomenon of changing habit is well understood in the fields of mineralogy and petrography, its understanding is still in a beginning stage in the field of glass technology. The establishment of four or five different habits and appearances of $Al(PO_3)_3$ crystals in phosphate glasses as a function of glass composition and structure (Vogel [823]) has shown that in glasses, too, changes in crystal habit are probably not an exceptional phenomenon.

10.4 Controlled Crystallization

10.4.1 Principles of Controlled Crystallization

In both science and technology there are many cases where defects, undesirable reactions, or frightening natural phenomena become important once their essence has been recognized and mastered. Crystallization used to be one of the most prevalent disturbances in glass production, but it was later controlled and utilized. The novel material "glass ceramic" ("Vitroceram") originated. This nomenclature indicates that such ceramics are not processed in the usual way, i.e., by firing with only partial melting of certain components, but via the primary step of a glass melt. It is only in a secondary step that the glass becomes the "glass ceramic" ("Vitro ceram").

What are the significant differences between controlled and uncontrolled crystallization?

Unintentional crystallization phenomena during the production of technical or optical glasses (i.e., crystallization as a defect) are usually characterized by the presence of an entire spectrum of crystals of various sizes. In a relatively homogeneous base glass, often sporadically and accidentally temporal and local conditions occur, at which the critical nuclear dimension is surpassed and a nucleus becomes capable of growth. If these conditions might occur later at another location, the nucleus that grew first might already have grown to a considerable size. The characteristics of controlled crystallization differ fundamentally from such nucleation and growth phenomena occurring sporadically at different times and locations.

Criteria of controlled crystallization are: (1) high nucleation frequency, uniform throughout the entire glass volume, (2) very uniform crystal size, and (3) very small crystallite dimensions (usually only a few micrometers).

The primary process of controlled crystallization in glass is controlled microphase separation. This is achieved by the specific selection of the base glass, specific additives, and a suitable temperature schedule. As is known from the exploration of microphase separation phenomena in glasses, the composition of the microphases in the glass tends toward that of a stable chemical compound. The degree to which this is achieved depends on the degree of undercooling.

In this manner a microphase enriched in network-modifiers approaches a "precrystalline condition." In practice, phase separation is controlled in such a way that this phase is the droplet phase. The presence of such well-defined, though still disordered, structural elements in a droplet region means that the work needed for nucleation is greatly reduced since it is no longer necessary to bring the constituent ions to the location of nucleation (the diffusion enthalpy ΔG_D). In other words, nucleation proceeds more easily since, in Turnbull's equation, $(\Delta G^* + \Delta G_D)$ is reduced to ΔG^* only.

Nucleation is also facilitated by the formation of a heterogeneous nucleus which further encourages crystallization of this microphase by epitaxial interaction. In this case Turnbull's term further reduces to $\Delta G^* f(\theta)$ which is smaller than $\Delta G_*(f(\theta) + \Delta G_{D\Delta})$, as well as just ΔG^*, since $f(\theta) < 1$. This epitaxial interaction plays an essential role in controlled crystallization. Epitaxy is defined in crystal chemistry and crystallography as the phenomenon of directional deposition of a substance on a crystalline substrate, or, in our special case, the growth of a nucleus with foreign matter. Such phenomena of oriented deposition of matter play an important role within the framework of a large area of crystallography in which much progress has been attained, particularly by Neuhaus [822] and Kleber [810].

A crude rule is: Epitaxial interaction between two crystalline phases is always likely if at least two lattice parameters do not differ by more than $\pm 15\%$.

It is characteristic for a glass-ceramic batch to contain small concentrations of nucleating additives (catalysts, mineralizers) such as fluorides, phosphates, TiO_2, ZrO_2, compounds of W, V, Ni, Cr, Mn, Fe, Cu, Ag, Au, the platinum metals, etc. According to their function they can be placed into two categories:

(1) First are those additives which primarily enhance phase separation, i.e., affect nucleation indirectly. Fluorides and phosphates should be included in this group.
(2) The second group contains almost exclusively additives which are soluble in the glass at high temperature but precipitate at reduced temperatures, because of strong oversaturation, i.e., via secondary phase separation in a pure form, e.g., as TiO_2. Typical nucleation catalysts such as TiO_2 form melt systems with the main glass-former (e.g., SiO_2) characterized by even, open immiscibility gaps (see Fig. 10.12). According to the characterization of various nucleation phenomena given before, the primary precipitation of, for example, TiO_2 nuclei, must be considered a homogeneous nucleation process. In most cases these primary nuclei act via epitaxial interaction as what might be called initial fuses for the main crystallization of the entire droplet phase. This effect then is to be considered under the term of heterogeneous nucleation. The entire nucleation mechanism in glass ceramics thus represents combined homogeneous-heterogeneous nucleation.

This appears the main reason for the difficulty, even impossibility, of determining the crystalline phase of the primary nucleus by X-ray diffraction. For if the primary nucleus attains its critical size it continues to grow with foreign matter (epitaxy), escaping X-ray determination by remaining "X-ray-amorphous."

When controlled microphase separation (influencing droplet number and dimension) and primary precipitation of heterogeneous nuclei have established the prerequisites for ceramizing the glass, the existing conditions and further reactions may be characterized as follows:

(1) Equal starting conditions for nucleation prevail in all droplet microphases (this is never possible in wild crystallization).
(2) Crystal growth at first proceeds to the droplet boundary which presents a barrier against further growth. This ensures size equalization of the crystal,

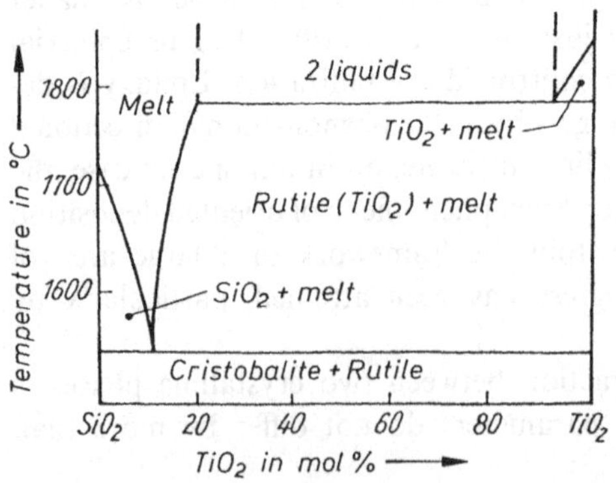

Fig. 10.12. Phase diagram of the system TiO_2–SiO_2 (DeVries et al. [824])

since droplet regions of phase separation in glasses are known to possess relative uniformity of dimension (Fig. 10.13).

(3) At higher temperatures, reappearing diffusion processes lead to a simultaneous crossing of the former spherical phase boundaries by the growing crystals, until the spherulitic growth fronts collide (Fig. 10.14, 10.15).

Because of the very large number of nuclei this condition is reached very soon, resulting in the very small crystallites of the final product. As a rule this causes the mechanical strength of the product to be much increased rather than much decreased, as in the case of macroscopic crystalline inclusions.

The first patents representing a realistic basis for the production of glass-ceramics prescribe a heat-treating schedule as exemplified in Fig. 10.16. Forming and annealing of the molten clear glass occur during step I. Subsequent reheating to temperature T_1 (nucleation temperature) leads to nucleation. During a further temperature increase to T_2 (crystallization temperature), complete

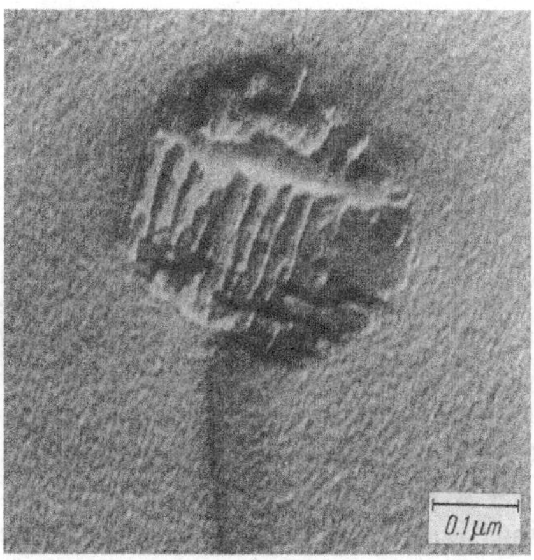

Fig. 10.13. Droplet-shaped immiscibility region with fracture "flag." The droplet-shaped phase is just starting to crystallize. Precipitation of a crystallite whose growth is stopped at the phase boundary because of the vast difference in composition between droplet and surrounding glass

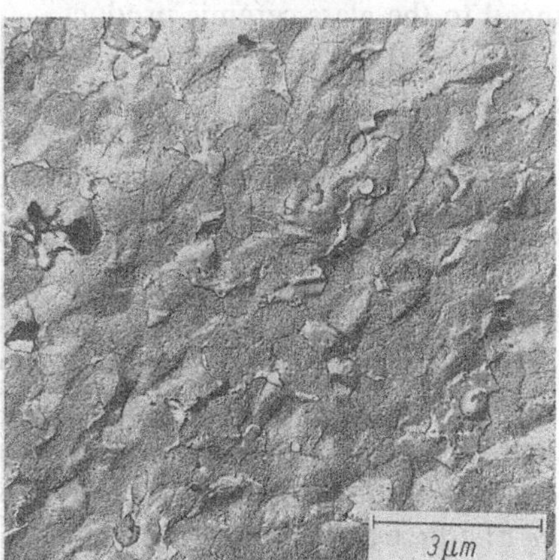

Fig. 10.14. Progress of phase separation and crystallization in the Pyroceram® glass with heat treatment: Many immiscibility regions touch each other. This also stops the growth of crystals. The preparation was slightly etched with HF (electron micrograph after replica preparation)

Fig. 10.15. Final stage of the crystallization of a Pyroceram® glass. The single-crystal regions remain recognizable. Within the regions the primary heterogeneous TiO_2 nuclei formed via secondary phase separation (very small droplets) can be identified

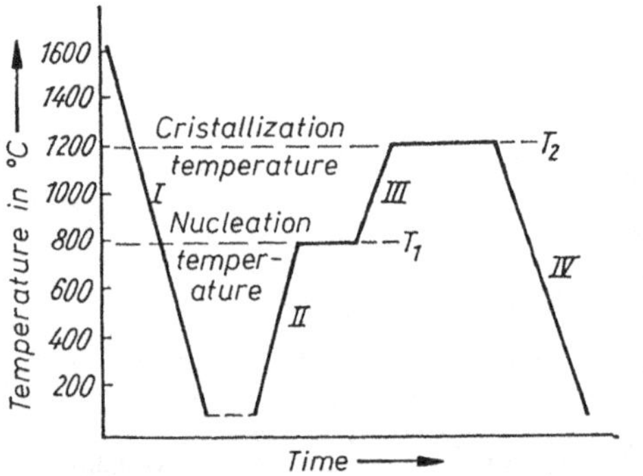

Fig. 10.16. Schematic of the formation of a glass-ceramic as a function of time and temperature in stages I to IV (Stookey [829]). *I* Glass processing *II* nucleation *III* crystallization *IV* annealing of final product

crystallization ("ceramizing") occurs. Step IV represents the annealing of the final product.

Today it is known that the first holding (step II) represents merely the further development of phase separation, where nucleation of crystals may, but need not, occur. But all the prerequisites for nucleation are obtained in this step. During the second holding (step III) the glass is changed to the glass-ceramic product.

The development of glass ceramics doubtless starts with the first patents of Becker [825] and is based on later investigations by Albrecht et al. [826, 827].

But the earliest attempts to make porcelain from glass (Reaumur [828]) had failed because the heat treatment of ordinary soda-lime glass started crystallization on the surface. In France, between about 1895 and 1920, street pavements had been made from heat-treated scrap glass (Doremus [285, p. 49]) or basalts (the latter containing TiO_2 which inadvertedly helped nucleation and growth), controlled on the basis of Tammann's papers. This idea was not pursued, for economical reasons.

Certainly, the first development toward mature production was made possible by the directive work of Stookey [829, 830] (for patents see [831]) at the Corning Glass Works, who marketed the first commercial glass ceramics products since 1957.

Today more than 1200 patents cover the field of glass ceramics, as do many scientific and technological publications. For this reason, it is clearly impossible to treat, or even cite, all pertinent results. What follows comprises selected lines of development which seem essential at the present time and pertinent examples. A much more detailed treatment is by McMillan [832] and by Beall [833] and Duke [834].

10.4.2 Pioneering Developments at Corning Glass Works

10.4.2.1 Glass Ceramics with Minimal Coefficients of Thermal Expansion

10.4.2.1.1 Composition, Production, and Application

Corning Glass Works effected a worldwide breakthrough in the development and production of glass ceramics based on Stookey's fundamental studies. The first products reaching the international market carried the trademark "Pyroceram." Later this designation was often used as a general term for the entire new group of materials, the glass ceramics, in spite of their highly differentiated variations.

Since it is impossible to present the development and modification of glass ceramics in its entirety, the characteristics of glass ceramics with a minimal coefficient of thermal expansion are explained using products "Pyroceram 9606 and 9608" as examples.

Stookey's basic patents [831] disclose that the standard types Pyroceram 9606 and 9608 are built on base glasses in the system Li_2O–Al_2O_3–SiO_2–TiO_2, usually containing some MgO and ZnO. A whole series of other components, such as BeO, CaO, SrO, CdO, BaO, PbO, MnO, FeO, CoO or NiO, may also be present, just to modify the primary properties. The standard Pyroceram types consist of about 90–95 mass% SiO_2, Al_2O_3, MgO, Li_2O, and TiO_2, with 2–20 mass% TiO_2 being designated as nucleation catalysts.

Melts of these compositions, as well as their manifold modifications, solidify when cooled to normal glasses. The sequential heat treatment, involving two holding steps, transforms the glass to a glass ceramic. The holding steps are characterized in the patent literature as follows:

(a) Holding at the so-called nucleation temperature. Since this temperature (800 °C) lies below the crystallization temperature, the glass remains practically transparent.
(b) Subsequent holding at the crystallization temperature (1100 to 1200 °C). The glass is transformed to the microcrystalline product. (For (a) and (b) see Fig. 10.16.

In comparison to conventional ceramics, the Pyrocerams are distinguished by a series of properties unattainable before in normal ways. The chief advantage appears to be the possibility of endowing the ceramic end product with any desired shape obtained by glass processing (tubing, drawing, pressing, blowing, automatic machine forming etc.). Volume changes during holding and crystallizing treatments are less than 1%. Although most Pyrocerams soften at much higher temperatures than ordinary glasses and ceramics (1200–1350 °C), they possess extraordinarily low coefficients of expansion.

Depending on the chosen relation of $Li_2O/Al_2O_3/SiO_2$, more β-spodumene or more β-eucryptite is formed as the main crystalline phase. Since β-eucryptite has a negative coefficient of expansion, and β-spodumene one of about zero, one succeeds in producing ceramics of weakly positive, weakly negative, or zero expansion. Certain glass ceramics can be produced in opaque or transparent modifications, the latter being the consequence of a minimal difference in refraction between the precipitated crystals and the remaining glass phase. The surface hardness of some glass ceramics equals that of hardened tool steels. Other extraordinary applications are based on high abrasion resistance, mechanical strength, chemical durability, and low porosity. A special role is played by the electrical properties of new glass-ceramics. The best conventional porcelains are surpassed in insulating power. A brief survey of the most important properties is given in Table 10.1.

The new ceramic materials made from glass were first applied in radomes. Other remarkable and extraordinary applications include materials for radar antenna housings, airplane parts, exhaust valves, and turbine wings for compressors and motor pistons. An extremely important application of glass-ceramics of minimal expansion was the manufacture of large astromirrors for mirror telescopes. Because of the low coefficient of expansion and high mechanical strength, mirrors could be made much thinner. Large astromirrors (2–5 m in diameter) remain unchanged under the load of their own weight or as a consequence of temperature variations, thus avoiding disturbances limiting the efficiency of telescopes using glass mirrors. The minimal coefficient of expansion also provides excellent temperature shock resistance. This property introduced the wide field of application to stove tops, replacing steel tops.

10.4.2.1.2 Structure and Properties

Analyzing the base glass underlying Pyroceram, Li_2O–MgO–ZnO–Al_2O_3–SiO_2–TiO_2, in regard to phase separation, one finds that its constituent silicatic binaries such as Li_2O–SiO_2, ZnO–SiO_2 (Fig. 10.17), Al_2O_3–SiO_2, and TiO_2–SiO_2 exhibit a pronounced tendency toward immiscibility.

The immiscibility behavior of binary lithium silicate glasses was studied extensively and proved by Vogel and Byhan [275] (Fig. 6.1, 7.6). The binary MgO–SiO_2 even shows an open immiscibility gap between 0 and 40 mol% (Kracek, [414] Fig. 7.9).

The often-doubted tendency toward immiscibility of glasses in the system Al_2O_3–SiO_2 was proved unequivocally by McDowell and Beal [237]. Figure 10.18, for example, shows an electron micrograph of a binary Al_2O_3–SiO_2 glass (15 mol% Al_2O_3, 85 mol% SiO_2) demonstrating this tendency toward immiscibility. The system TiO_2–SiO_2 exhibits an open immiscibility gap over the entire range of composition (De Vries et al. [824] Fig. 10.12). As we know, in combinations of several systems immiscibility decreases, but does not entirely suppress the tendency toward immiscibility. This provides the prerequisites for

Table 10.1. Some important properties of classical "Pyroceram" types

Properties		Pyroceram 9606 V	Pyroceram 9608 V	Starting glass for Pyroceram 9608
Density g/cm (25 °C)		2.60	2.50	2.4035
Softening temperature °C (10^6 Pa·s)		1350	1250	1050
(Glass-transition transition temperature °C (10^{12} Pa·s)				682
Coefficient of expansion ($\alpha \cdot 10^7 \cdot K^{-1}$) (0 to 300 °C)		56	4 . . . 11	33.6
Specific heat $J \cdot Kg^{-1} \cdot K^{-1}$ (25–400°C)		0.963	0.984	
Modulus of elasticity $Pa \cdot 10^{-12}$		11.9	8.6	
Thermal conductivity $W \cdot m^{-1} \cdot K^{-1} \cdot 10^2$		0.0087	0.0047	
Bending strength $Pa \cdot 10^6$		203.07	34.335	41.202
Poisson's number		0.0245	0.25	
Breaking strength (ground surface) $Pa \cdot 10^{-9}$		–13	11 . . . 16	
Knoop hardness (100 g) $Pa \cdot 10^6$		5.58	6.78	
Dielectric constant ε′				
At 1 MHz	25 °C	5.58	6.78	
	300 °C	5.60		
	500 °C	8.80		
At 10,000 MHz	25°C	5.45		
	300 °C	5.51		
	500 °C	5.53		
loss angle tan δ				
At 1 MHz	25°C	0.0015	0.0030	
	300 °C	0.0154		
	500 °C			
At 10,000 MHz	25°C	0.00083		
	300 °C	0.00075		
dielectric loss	500 °C	0.00152		
At 1 MHz	25°C	0.009	0.020	
	300 °C	0.086		
	500 °C			
At 10,000 MHz	25°C	0.002		
	300 °C	0.004		
	500 °C	0.008		
Specific resistance	Ω·cm			
	250 °C	10	8.1	
	350 °C	8.6	6.8	

controlled crystallization. In the case of Pyroceram glasses, Vogel and Gerth [189, 190] demonstrated by electron micrography that the process of controlled crystallization corresponds to the preceding theoretical description, using a glass of the composition (in mass%):

69 SiO_2	3.0 MgO
18.5 Al_2O_3	2.5 Li_2O
5.0 TiO_2	1.0 ZnO

remainder: Na_2O, K_2O, CaO, Fe_2O_3, As_2O_3.

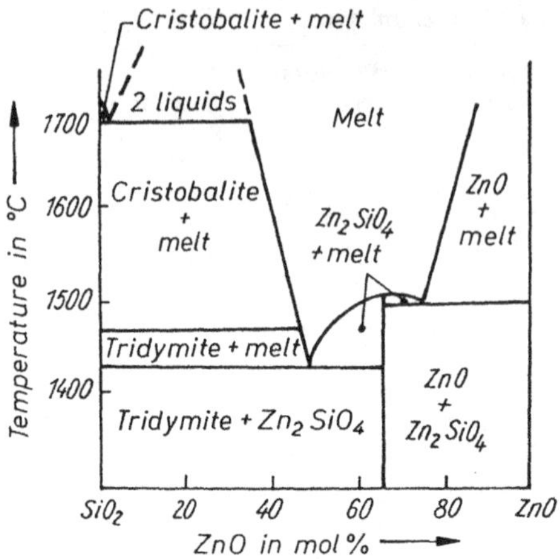

Fig. 10.17. Phase diagram of the system ZnO–SiO_2 (Bunting [672])

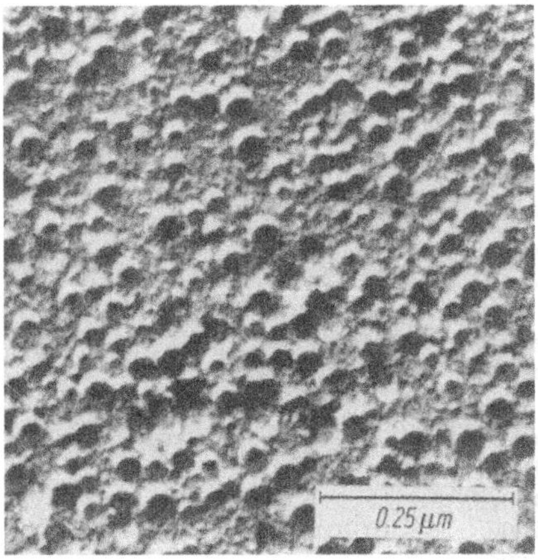

Fig. 10.18. Electron micrograph at a quenched Al_2O_3–SiO_2 glass (15 mol% Al_2O_3–85 mol% SiO_2) (McDowell and Beall [237])

During the first hold; phase separation initially progresses (Fig. 10.19). At the end of the first hold or during the second hold, nucleation and crystallization will occur in the droplet silicate phase enriched in Li, Mg, Zn, Al, and Ti ions.

In Fig. 10.14 crystallization is already crossing the droplet boundaries, in Fig. 10.15 ceramizing has been terminated: The growth fronts of the crystallites have collided and growth has stopped.

In Fig. 10.15 one may just recognize the secondary phase separation in the form of very small TiO_2-rich droplets wthin the former large-droplet region.

Fundamental studies of controlled crystallization of glasses or systems suited for the development of glass ceramics with very low coefficients of expansion were carried out by many authors [151, 233, 835–883]. For more detail the reader is referred to McMillan [832].

The most important crystalline phases, essentially determining the properties of a glass ceramic, have been identified as β-eucryptite ($LiAl(SiO_4)$), β-spodumene ($LiAl(Si_2O_6)$), cordierite ($Mg_2Al_3(AlSi_3O_{18})$), and rutile (TiO_2).

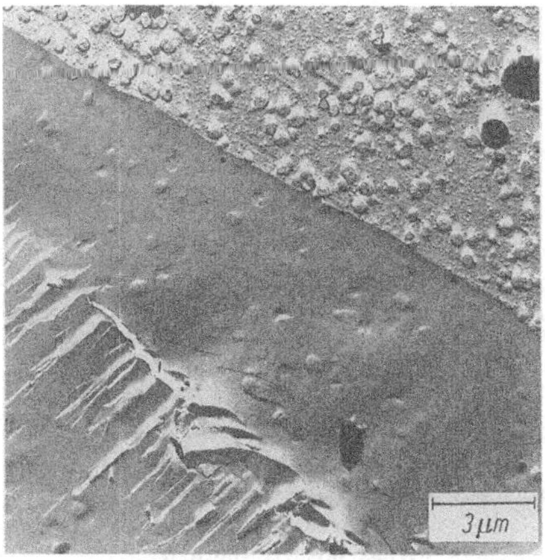

Fig. 10.19. Heat-treated Pyroceram glass; droplet-shaped immiscibility regions. Glass region top right etched by hydrofluoric acid (electron micrograph after replica preparation)

Table 10.2. Lattice constants of the main phases occurring in "Pyroceram" and their deviations from the constants of the heterogeneous nucleus (TiO_2)

Crystalline phase	Lattice constant in Å *a*	*b*	*c*	Deviations of lattice constants in relation to nucleation agent (TiO_2)
Rutile TiO_2 Nucleation agent	4.58		2.95	
β-spodumene $LiAl[Si_2O_6]$	9.5	8.3	5.24	$a_{Spod.} = 2 \cdot a_{Rutile} \approx +3.7\%$ $b_{Spod.} = 2 \cdot a_{Rutile} \approx -9.4\%$ $c_{Spod.} = 2 \cdot c_{Rutile} \approx -11.2\%$
β-eucryptite $LiAl[SiO_4]$	5.2		11.17	$a_{Eucr.} = 2 \cdot a_{Rutile} \approx +12.6\%$ $c_{Eucr.} = 4 \cdot c_{Rutile} \approx -5.3\%$
Cordierite $Mg_2Al_3[AlSi_5O_{18}]$	17.0	9.78	9.33	$a_{Cord.} = 4 \cdot a_{Rutile} \approx -6.7\%$ $b_{Cord.} = 2 \cdot a_{Rutile} \approx +6.8\%$ $c_{Cord.} = 2 \cdot a_{Rutile} \approx +1.9\%$

Additionally, in modified compositions the following crystalline phases show up: cristobalite (SiO_2), quartz (SiO_2), magnesium titanate ($MgTiO_3$), willemite (Zn_2SiO_4), aluminium titanate (Al_2TiO_5), sphene ($CaTi(OSiO_4)$), wollastonite ($Ca_3(Si_3O_9)$). Stookey [829] has already pointed out that the catalytic effect of TiO_2 on nucleation could be understood on the basis of epitaxial interaction. Table 10.2 confirms this suggestion. As shown in the table, the lattice constants of the nucleation catalyst, TiO_2, and the main crystalline phases (β-spodumene, β-eucryptite, and cordierite) agree within ± 15%, thus fulfilling the condition for directional deposition.

Possible modifications in expansion in some glass-ceramics based on the system Li_2O–Al_2O_3-SiO_2 have been surveyed by Mehmel [866] (Table 10.3). The coefficient of thermal expansion α may increase to $11 \cdot 10^{-7}\ K^{-1}$.

As a rule, crystallite size in glass-ceramics is about 1 μm or somewhat less. But it is not only the size of the crystallites and their mutual penetration, but

Table 10.3. Crystalline phases in glass ceramics from the system Li_2O–Al_2O_3–SiO_2 determining thermal expansion behavior

Molar ratio in base glass Li_2O	Al_2O_3	SiO_2	Main crystalline phase	Thermal expansion behavior
1	1	2	β-eucryptite $LiAl[SiO_4]$	Strong contraction (negative coefficient of expansion)
1	1	3	Solid solution of SiO_2 in β-Eucryptite	Small contraction (negative coefficient of expansion)
1	1	3.5	Like for molar ratio 1:1:3	Zero thermal expansion
1	1	4	β-spodumene $LiAl[Si_2O_6]$	Weak positive thermal expansion ($\alpha \approx 9 \cdot 10^{-7} K^{-1}$)
1	1	6	Lithium orthoclase $LiAl[Si_3O_8]$	Weak positive thermal expansion ($\alpha \approx 5 \cdot 10^{-7} K^{-1}$)
1	1	8	Petalite $LiAl[Si_4O_{10}]$	Weak positive thermal expansion ($\alpha \approx 3 \cdot 10^{-7} K^{-1}$)

also the amount of the remaining glass phase which determine the properties of Vitrocerams. This amount can be varied within wide limits.

Glass ceramics containing eucryptite and spodumene were further developed in a very interesting way by Beall [884] and by Sack and Scheidler [876, 885]. Their low-expansion glass ceramics are perfectly transparent. According to Sack [886], transparency is achieved by the simultaneous use of two catalysts (TiO_2, ZrO_2) in the system Li_2O–MgO–Al_2O_3–SiO_2. Apparently the double doping has resulted in a considerable diminution of the crystallite size below 1 μm, analogous to the suppression of phase-separation phenomena by the combination of two or more binaries with a strong tendency toward immiscibility (competition principle). Of course, the assimilation of the refractive index of the crystalline phase to that of the remaining glass also enhances transparency.

10.4.2.2 Machinable Glass Ceramics

10.4.2.2.1 General, Composition, Production

Machinable glass ceramics are a young, fast-spreading field of investigation. Since, on principle, any glass or glass ceramic can be worked by machines, e.g., ground, polished, cut, diamond-sawed, etc., machinable glass as a new material should be characterized more specifically.

A machinable glass ceramic is defined as a glass ceramic which can, like metals, be turned, drilled, milled, or threaded. These glass ceramics thus are characterized by permitting a more or less far-reaching application of the same machines which are generally used to work metals. The application of such machines to the working of normal glasses or glass ceramics would cause the fracture of the workpiece.

The development of such glasses is based on the controlled precipitation of mica phases from certain base glasses. Such systems were studied by Daniels [887] at the University of Missouri-Rolla. Machinable glass ceramics were first developed at Corning (patents by Stookey [888], Beall [889, 890] and Grossmann [891]) and described in publications by Beall et al. [833] and Grossmann [892, 893]. The first product offered on the international market under the name "Macor" may be characterized by the following base compositions (in mass%):

44 SiO_2, 16 MgO
16 Al_2O_3, 10 K_2O
8 B_2O_3, 6 F.

Glassmaking and ceramming processes correspond to those typical for all glass ceramics. The precipitated crystalline phase is potassium phlogopite ($KMg_3(AlSi_3O_{10}F_2)$), i.e., a mica phase which must constitute more than two-thirds of the total volume to effectuate machinability.

This material is developing rapidly, leading to many applications, particularly the replacement of metals. While, of course, these glass ceramics can never compete with metals for ductility, they offer entirely new combinations of properties. Table 10.4 lists the properties of some original Corning machinable glass ceramics.

Table 10.4. Properties of machinable glass ceramics of Corning Glass Works

Properties of glass ceramics	Type of glass ceramic			
	9650	9652	9654	9656
$\alpha \cdot 10^7\ K^{-1}$ (25 ... 400 °C)	97	74	64	63
Maximum thermal shock resistance in °C	800	800	800	800
Tensile strength in $Pa \cdot 10^6$ at 25 °C	86.33	92.21	49.05	59.84
Tensile strength in $Pa \cdot 10^6$ at 400 °C	59.84	77.50	46.11	47.09
Tensile strength in $Pa \cdot 10^6$ at 600 °C		70.63	45.13	45.13
Bending strength in $Pa \cdot 10^6$	225.06	343.35		196.2
Modulus of elasticity in $Pa \cdot 10^6$	60822	63165	56898	58860
Volume resistance in Ω cm at 500 °C	10^7	10^{11}	10^{12}	10^{12}
Dielectric constant (10 kHz, 25 °C)	6.1	5.7	5.6	5.6
Dielectric loss $\tan\delta \cdot 10^{-4}$ (10 kHz, 25 °C)	30	60	60	20
Excellent chemical resistance against water, acids, bases.				

The structure-determining and crystallization processes involved in the fabrication of machinable glass ceramics are by no means fully understood. Beall considered a certain amount of B_2O_3 necessary; other authors believe that one may omit B_2O_3.

10.4.2.3 New Mica-Containing Glass Ceramics

The development of machinable glass ceramics introduced by Beall [890b] and described so far was based on the properties of the phases K-phlogopite ($KMg_3(AlSi_3O_{10}F_2)$) and borofluorophlogopite ($KMg_3(BSi_3O_{10}F_2)$). Grossmann [891b] developed glass ceramics based on tetrasilicofluoromicas ($K_3Mg_3(Si_4O_{10}F_2)$) which, though less chemically resistant, are said to be suited for the production of "artificial marble" after introduction of coloring cations. Beall [890a] describes machinable alkaline earth fluoromicas which exhibit swelling properties in contact with water.

10.4.2.4 Chain Silicate Glass Ceramics

Beall and coworkers who first developed silicate glass ceramics containing silicate with chain structure. The products obtained had very different properties [900, 901]

- A glass ceramic was developed from the glass systems ZrO_2–MgO–SiO_2 and ZrO_2–Li_2O–Al_2O_3–MgO–SiO_2. It contained enstatite ($MgSiO_3$), a mineral from the pyroxene group with single chain units.
- A glass ceramic was produced from the melt system K_2O–Na_2O–CaO–MgO–SiO_2–F to which Al_2O_3, P_2O_5, LiO_2 and BaO were added. The main crystal phase was potassium fluorine richterite ($KNaCaMg_5Si_8O_{22}F_2$), a mineral from the amphibole group with double chain units.
- A glass ceramic was also obtained from the melt system Na_2O–K_2O–CaO–SiO_2–F. Its main crystal phase was canasite ($Ca_5Na_4K_2Si_{12}O_{30}F_4$), a mineral which is formed from multiple chain units.

Canasite glass ceramics have a bending strength of about 200 MPa and a breaking toughness $> 3\ MPa^{1/2}$. The very unusual properties open possibilities for use of the chain silicate glass ceramics as sustrate material in electronics, as electroporcelain, as porcelain for household use, as very thin household dishes, as house paneling and even as roofing tiles.

10.4.2.5 Strengthening of a Special Glass by the Chemcor Process

The first strengthening processes for glass objects were based on the prevention of the influence of mostly macroscopic surface lesions. The process of tempering, known for a long time and described in detail in a recent review by Gardon [902] can be understood in this sense. A fairly thick-walled glass object is heated above the glass-transition temperature, and quenched by cold air or cold metal plates. The surface glass solidifies rapidly while the inner layers continue to contract. In this way the surface ends up in compression, while the interior is in tension, compensating for the surface compression. Fissures in the surface are compressed and, under load, resist their opening and the introduction of the fracture process. Fracture occurs only after the load overcomes surface compression. A minimum wall thickness is required. Otherwise the differential in cooling is insufficient and the compression layer is too thin, so that fissures may reach into the tension zone and carry out their weakening effects.

A novel process for strengthening a special glass based on the same principle was described by Stookey and Olcott [903] and became known under the name of "Chemcor Process".

Objects fabricated from a Na_2O/K_2O–Al_2O_3–SiO_2–(TiO_2) base glass–differing from a typical Pyroceram base glass only by the substitution of Na_2O and K_2O for Li_2O–are immersed in a lithium salt (preferably Li_2SO_4) melt. In this manner the K^+ and Na^+ ions of the surface are replaced by lithium ions, converting a thin layer to a Pyroceram glass with all its typical nucleation and crystallization characteristics. A subsequent heat treatment (as in Fig. 10.16) transforms this thin Pyroceram glass layer to Pyroceram, i.e., in this very thin layer β-spodumene and β-eucryptite crystals grow, just as in the entire bulk of Pyroceram: The extremely low, or even negative, expansivity of this surface layer gives rise to enormously high compression in this surface zone. Since the surface layer is extremely thin, the strengthened glassware remains entirely clear and transparent.

The strength achieved in this manner by a controlled surface crystallization following ion exchange is about 590–690 MPa. The strength of thin specimens might be considerably higher. The strain distribution and the behavior under bending load of a conventionally strengthened glass plate and of one strengthened by the Chemcor process are illustrated in Fig. 10.20.

A slowly cooled glass plate shows no strain along the entire cross section, corresponding to the zero line of Fig. 10.20 a. If a strain-free plate is subjected to a bending load, as indicated by the arrows in Fig. 10.20 a, tension is induced in the surface (to the amount a), and on the bottom equal compression (to the amount a′), while in the center a neutral zone remains unaffected. The strain distribution in the entire plate is reproduced by the straight line C.

A conventionally tempered glass plate shows compression on the surface as well as, obviously, on the opposite face (about 200 MPa in Fig. 10.20 a, represented by curve A). Toward the center, compression decreases and attains maximum tension in the center after crossing the zero line. Starting from the opposite face, conditions are mirrored in the same way. Now the strain distribution is no longer represented by a straight line, but by a parabola. If now a bending load, as indicated by the arrows, is applied, the strain distribution is shifted. Compression at the front surface is decreased by the amount of the bending load (a), while compression at the opposite face is increased by the same amount (a′). As shown by Fig. 10.20, a so-tempered glass plate may be subjected to a considerable bending load before the surface compression is compensated and finally replaced by tension, inducing the danger of failure.

Figure 10.20 (b) illustrates analogous conditions for a glass plate strengthened by controlled crystallization according to the Chemcor process. As

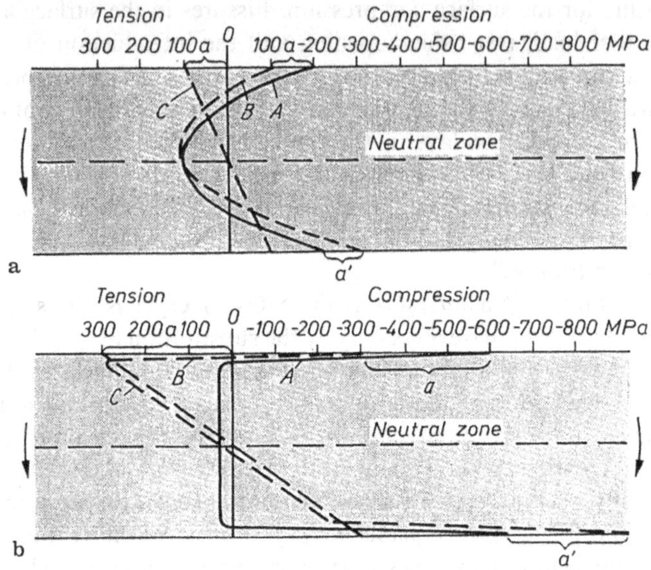

Fig. 10.20. Schematic of strain in strengthened glass plates (Stookey et al. [903]) **a** Cross section of a glass plate strengthened by tempering; **b** Cross section of a glass plate strengthened by the Chemcor process. In bot cases the surface layers have been put under strong compression while the interior is in tension. Under bending stress compression decreases on one, increases by the same amount on the other side. The load resistance of a glass plate depends mostly on the compression of the surface

can be seen in Fig. 10.20 b, surface compression is much higher than in the conventionally tempered glass. Thus, a much greater bending load is possible before tension occurs in the surface, inducing the danger of failure. In the United States the Chemcor process was used primarily for the strengthening of flat glass, particularly for motor vehicles and aircraft.

10.4.3 Fundamental Investigations in the Development of Glass-Ceramics at the Otto Schott Institute of the Friedrich Schiller University in Jena [904]

The first developments of glass ceramics by Stookey [831] have given a decisive impulse to this entire new field. Today there are well over 1,000 patents on glass ceramics and even more publications. Glass ceramics represent a new material class with some property combinations unknown until now and they consequently find applications in the most different areas of science and technology. This new field continues to broaden today. A fundamental reason for this is the extensive knowledge of the microstructure of glasses and of the microstructure formation processes which occur in all glass melts, including nucleation and crystallization.

The particular recognition that practically each controlled crystallization must be preceded by a controlled microphase separation has brought considerable advances in this area [151, 186–190, 196, 321]. A great variation of properties can therefore be observed in glass ceramics even when the base glass used is the same. Additives can decisively influence the microprocesses occurring in the base glass and therefore confer new properties on resulting glass ceramics.

10.4.3.1 Nucleation and Crystallization Kinetics of a Base Glass from the $MgO–Al_2O_3–SiO_2$ System [322, 879, 880]

Melts from the SiO_2-rich portion of the ternary system $MgO–Al_2O_3–SiO_2$ solidify easily as a glass. The glasses obtained demonstrate a distinct crystallization tendency with thermal treatments. The base glasses chosen are from a relatively narrow area of the ternary $MgO–Al_2O_3–SiO_2$ melt diagram (see Fig. 10.21.)

The base glass with the following composition: 15.5 mol% MgO, 12.9 mol% Al_2O_3 and 71.6 mol% SiO_2 is just beside the eutectic gap between the tridymite and cordierite fields but still in the tridymite field in the ternary diagram. This indicates that the chosen base glass is still easily melted and at the same time shows a sufficient tendency for phase separation. If the glass was directly in the eutectic gap, the latter would no longer be the case.

Figure 10.22 shows phase separation in the base glass after the normal cooling process (the cooling rate in the transformation region was 3–4 K/min^{-1}). This degree of undercooling was maintained for all the following experimental glass melts. A precipitation of droplet shaped microphases which

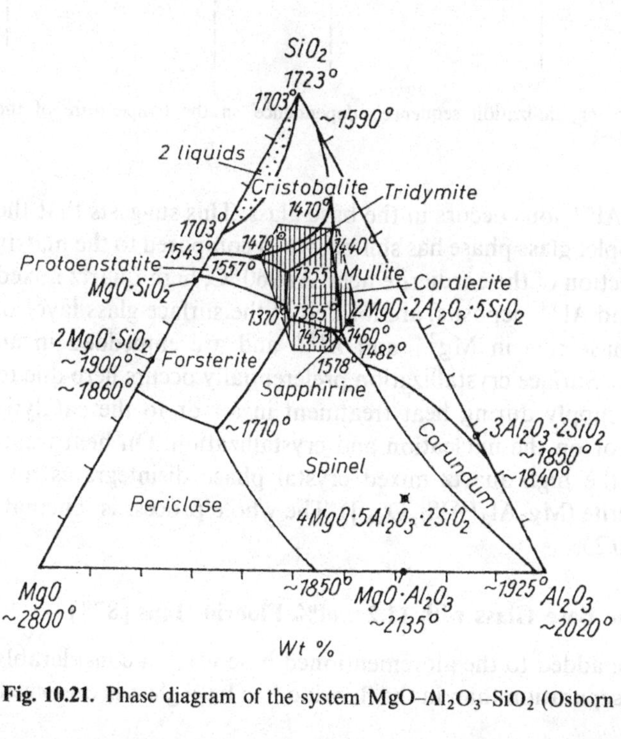

Fig. 10.21. Phase diagram of the system $MgO–Al_2O_3–SiO_2$ (Osborn and Muan, in Levin [905])

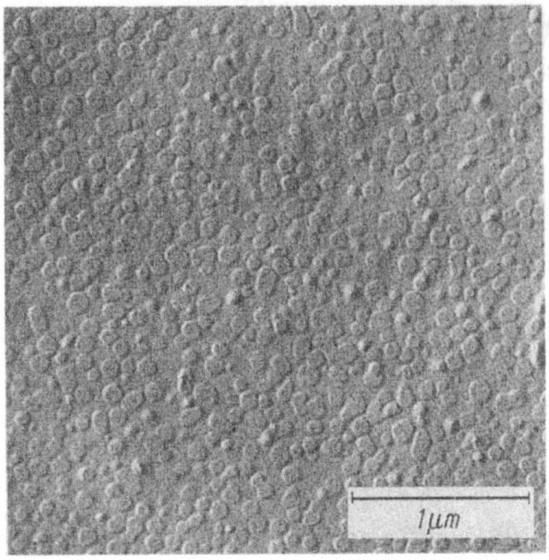

Fig. 10.22. Magnesium aluminosilicate glass ($MgO = 15.5$ mol%, $Al_2O_3 = 12.6$ mol%, $SiO_2 = 71.9$ mol%). Precipitation of droplets rich in MgO and Al_2O_3 in a matrix glass rich in SiO_2 (replica method)

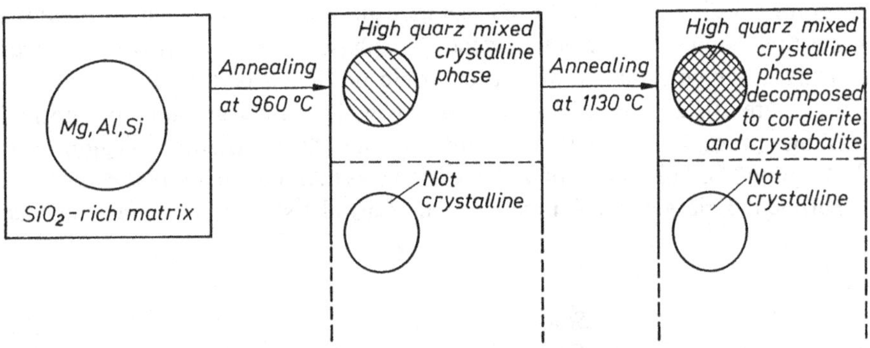

Fig. 10.23. Schematic of the crystallization sequence's dependence on the temperature of the MgO–Al_2O_3–SiO_2 initial glass

are rich in Mg^{2+} and Al^{3+} ions occurs in the base glass. This suggests that the composition of the droplet glass phase has shifted, when compared to the matrix glass phase, in the direction of the cordierite field. At 960 °C, high-quartz mixed crystals (with Mg^{2+} and Al^{3+} deposits) are formed at the surface glass layer of the silicate droplet phase rich in Mg^{2+} and Al^{3+} and are embedded in an SiO_2-rich matrix phase. Surface crystallization preferentially occurs here due to the increased oxygen supply during heat treatment in air or to the catalytic influence of water vapor on the nucleation and crystallization. On heat treatment up to 1130 °C, the high-quartz mixed crystal phase disintegrates into cristobalite and cordierite ($Mg_2Al_3[AlSi_5O_{18}]$). The whole process is schematically shown in Fig. 10.23.

10.4.3.2 Doping of the Base Glass with 11.2 mol% Fluorine Ions [878]

When fluorine ions are added to the aforementioned base glass, a considerably more advanced phase separation than in the fluorine-free base glass is observed.

This occurence is a well-known fact. If the dimensions of the droplets in the two glasses are compared (Figs. 10.22 and 10.24) the glass containing fluorine ions has droplets about three times larger than the ones of the fluorine-free glass. When the fluorine enriched glass is annealed at 960 °C, cordierite and cristobalite precipitate immediately inside the droplets at the glass surface. The high-quartz mixed crystal precipitation shown in Fig. 10.23 was avoided. The process which is first achieved at a temperature of 1130 °C with a fluorine-free base glass was possible here at a lower temperature (960 °C) due to the addition of fluorine ions. This addition caused the viscosity to decrease and therefore increased ion mobility.

10.4.3.3 High-Strength Glass Ceramics Containing Spinel [882, 906, 907]

The strengthening procedure described in Section 10.4.2.5 is effected through a thin compression layer whose violation would cancel the achievement. Therefore, no surface finishing can be applied.

It would seem more attractive to obtain strengthening throughout the glass volume. To start with, the base glass should be free of ions with relatively low bond strength, such as alkali ions. A further increase in so-called volume strength should be possible, if the macroscopic compression concept could be transferred to the microscopic range. If primary precipitation of crystals with a higher thermal expansion than the surrounding glass is achieved, the latter will be placed under compression. The effect will be the larger, the larger the difference in the coefficients of expansion between glass and crystalline phase.

McMillan [832] stated that, on principle, the strength of glass ceramics decreases with decreasing coefficient of expansion. On the basis of this concept, Vogel et al. [906] were able to achieve further progress.

If about 10 mol% NiO is added to a ternary base glass, containing 68.9 SiO_2, 21.1 Al_2O_3, and 10 mass % MgO, an interesting change in the crystallization kinetics is observed (see Fig. 10.25) with an additional heat treatment (in comparison to the base glass in Fig. 10.23). Phase separation is

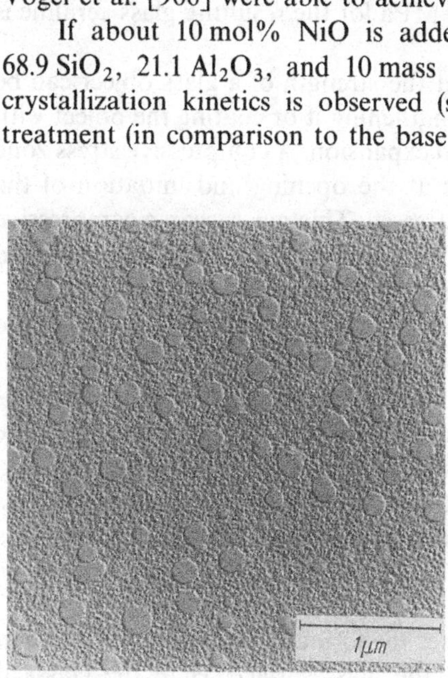

Fig. 10.24. Magnesium aluminosilicate base glass (like in Fig. 10.22) in which 11.2 mol% oxygen ions are replaced by fluorine ions. Clear increase of the immiscibility tendency recognizable from the increase in the droplet sizes Fig. 10.22

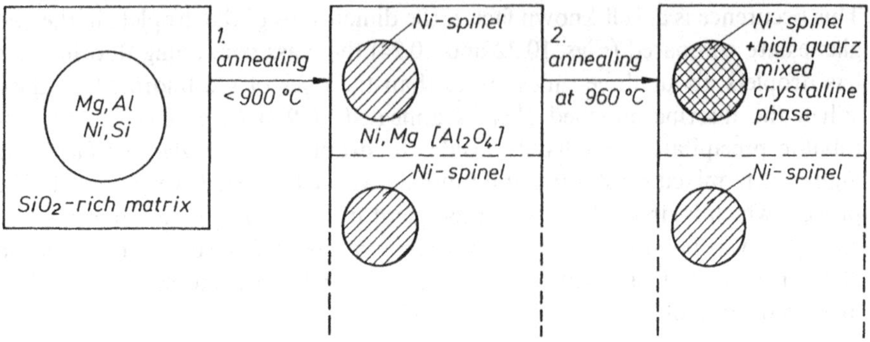

Fig. 10.25. Schematic of the crystallization sequence in an NiO-doped base glass in function of the temperature. The base glass (Fig. 10.22) has been doped with 10 mol% NiO

favoured here like (with the introduction of fluorine ions). The droplets are rich in Mg^{2+}, Al^{3+} and Ni^{2+} ions. Whereas the crystallization kinetics of a ternary base glass after an additional heat treatment will not change with an addition of less than 8 mol% NiO, this is no longer the case if 10 mol% NiO is added (Fig. 10.25). Already with an additional heat treatment at less than 900 °C, a nickel-spinel precipitation (Ni, Mg, Al_2O_4) occurs in all droplet phases. A second heat treatment at 960 °C leads to the formation of high-quartz mixed crystals from the droplets in the glass surface. The earlier spinel precipitation caused the droplets to be depleted in Mg^{2+} and Al^{3+} ions. The residual concentrations are however sufficient for the formation of high-quartz mixed crystals.

The crystallization order described here is of significant importance for the applications of this type of glass ceramic. An increase of the bending strength from 90 MPa (initial glass) to about 590 MPa for the resulting glass ceramic is observed.

It is known in glass technology that the strength of a glass object can be considerably increased by thermally strengthening it or coating the object with a thin foreign glass layer of lower thermal expansion. A compressive stress zone in the glass surface hinders notch tears at the opening and initiation of the fracture when the glass is mechanically stressed. This macroscopic compressive stress concept for the increase of the strength of a glass was carried over to microdimensions and used in the described glass ceramic containing NiO.

Figure 10.26 is a schematic representation of the expansion ratios in the glass ceramic containing NiO and Table 10.5 lists the linear expansion coefficients of particularly interesting crystal phases. If Ni-spinel crystals are precipitated in the entire glass volume, the nickel-spinel crystals will contract more with a decrease of temperature than the surrounding glass matrix since $\alpha_{\text{Ni spinel}}$ $80 \cdot 10^{-7}/K^{-1}$ is greater than α_{glass} $33 \cdot 10^{-7}$ K^{-1}. A strong compressive stress zone therefore exists around all Ni-spinel crystals and the entire ceramic is strongly stressed in microdimensions. The strength of the glass ceramic is therefore increased.

The second annealing at 960 °C causes the formation of high-quartz mixed crystals with an α value of about 0 at the glass surface. Here the classical

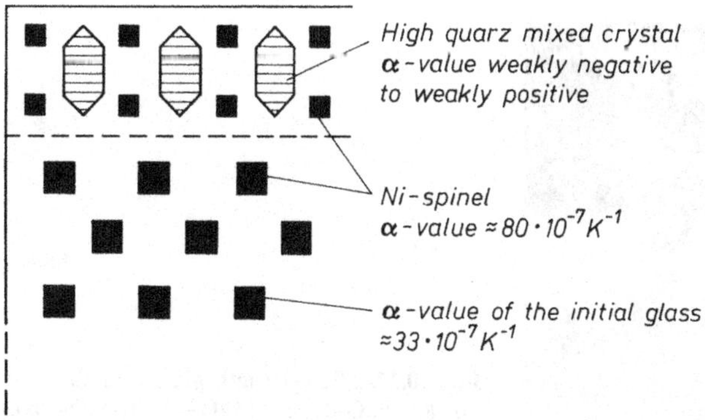

Fig. 10.26. Schematic representation of the expansion ratios after the second tempering step in an NiO-doped base glass and after precipitation of the crystal phases in the area near the surface and in the centre of the glass sample

Table 10.5. Coefficient of thermal expansion of crystal phases and glasses

Crystal phase	Coefficient of thermal expansion $\alpha \cdot 10^7\ K^{-1}$
Cordierite $Mg_2Al_3[AlSi_5O_{18}]$	14 (25 ... 400 °C)
Cristobalite SiO_2	500 (20 ... 300 °C)
t-Quartz SiO_2	132 (20 ... 300 °C)
Nickel spinel $(Ni, Mg)Al_2O_4$	80 (20 ... 300 °C)
End glass $[MgO–Al_2O_3–SiO_2]$	33
Ni-spinel-glass-cermic (core) $[MgO–Al_2O_3–SiO_2–NiO]$	44
High-quartz mixed crystals (with deposits of Mg and Al)	from slightly negative values to about 35
Spinel $MgO \cdot Al_2O_3$	80 (20 ... 800 °C)
Mullite $3Al_2O_3 \cdot 2SiO_2$	40 (20 ... 500 °C)
Mg-titanate $MgO \cdot TiO_2$	79 (25 ... 1000 °C)

macroscopic compressive stress concept can be applied. Whereas the bending strength increases from 90 MPa (starting glass) to 290 MPa after the first heat treatment, it reaches values of about 590 MPa after the second heat treatment. Fig. 10.27 shows the crystal precipitation after two tempering treatments and Fig. 10.28 shows the resulting stress ratios on the original product.

It must be mentioned that such a highly solid glass ceramic should not be heated to temperatures over 1100 °C. A decomposition of the high-quartz mixed crystals into cordierite and cristobalite would ensue as shown in Fig. 10.23. Since the α value for cordierite and cristobalite are higher than that of the high quartz, the glass ceramic's strength would decrease.

Note that strength data on glasses and glass ceramics must be taken as relative. Measuring technique, sample size, shape, and surface condition must be given. Unfortunately, this condition is disregarded in much of the literature.

Fig. 10.27. High-strength glass ceramic based on an NiO-doped $MgO–Al_2O_3–SiO_2$ base glass. The glass ceramic exhibits in its interior a precipitation of nickel spinel ($(Ni, Mg)Al_2O_4$) in the shape of small cubic crystallites, and on its surface formation of high-quartz mixed crystals

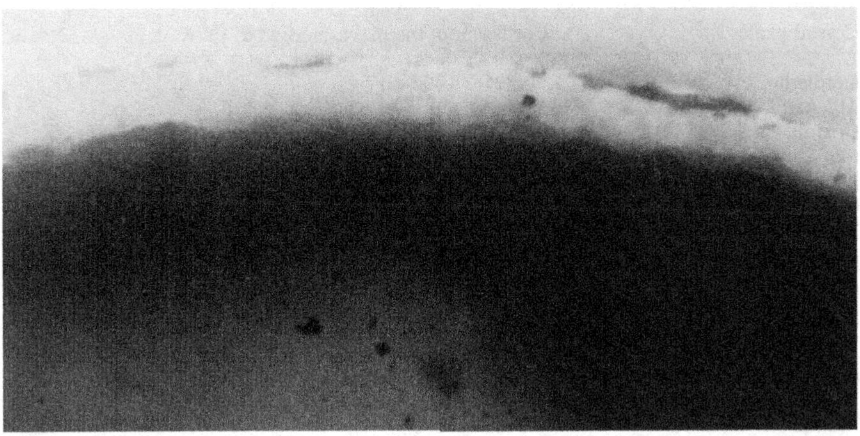

Fig. 10.28. Strain-optical investigation of the glass of Fig. 10.27. Light micrograph of a foil 0.4-mm thick in polarized light. Since the surface layer containing high-quartz mixed crystals possesses a lower thermal expansion than the core containing Ni spinel, a surface compression layer forms and strength increases

In the present case, bending strength was determined on 3-mm-thick rods after systematic abrasion with silicon carbide of defined grain size in a rotating drum. This procedure ensures that the data are true mean values, not top values (which are much higher for unabraded samples.)

Figure 10.27 shows the precipitation of spinel throughout the bulk as well as the β-quartz solid-solution skin on the surface. Figure 10.28 shows the strong compression effect of the surface layer conductive to this further increase in strength.

The structure-property relations exemplified here for the case of a glass ceramic containing nickel spinel confirm the first hints of McDonnell and Beall [237] and of Müller and Scheidler [908] as to possible precipitation of spinel in

$MgO–Al_2O_3–SiO_2$ glasses with alkaline earth, zinc, or transition element ion [237] additions. They also document the possibility brought up by Tucker and Stuart [909] of increasing the strength of glass ceramics by the specific development of a surface crystallization skin, promoting compression.

The method described here has the advantage over the Chemcor process that it is possible to further shape the glass by grinding, polishing, sawing, etc. after the first heat treatment, since the strengthening skin (30–60 μm thick), which cannot be violated, originates only during the second heat treatment, in which the shape is fully preserved.

10.4.3.4 Single Doping of the Base Glass with 2–10 mol% TiO_2 (Ti_2O_3) also Leads to High Strength Glass Ceramics [904, 910, 911]

The crystallization kinetics of an $MgO–Al_2O_3–SiO_2$ base glass to which TiO_2 was added, are very versatile and complicated. Not only the influence of the more or less high portions of Ti^{3+} ions must be considered here. Ti^{3+} ions are known to considerably favor more nucleation and crystallization than Ti^{4+} ions. Apart from an alteration of the nucleation at lower temperatures, a fine crystalline product is also maintained. A practical consequence is the fact that the effective addition of TiO_2 can be considerably reduced with a reduction of the Ti^{4+} ions to Ti^{3+} ions in the melt.

Two different base glasses were chosen for the experimental investigations of the complicated crystallization process of a $MgO–Al_2O_3–SiO_2$ glass containing TiO_2/Ti_2O_3.

Base Glass of a Composition from the Cordierite Field

($MgO = 25.4$ mol%, $Al_2O_3 = 12.6$ mol%, $SiO_2 = 57.0$ mol% enriched with 5.0 mol% TiO_2)

In such a glass, strong phase separation is observed: a silicate droplet phase rich in MgO, Al_2O_3 and titanium dioxide is embedded in an SiO_2 rich matrix glass phase (see Fig. 10.29). This is the case with normal as well as redox melting of the glass.

A thermal treatment of the normal molten glass at 810 °C leads to the precipitation of magnesium aluminotitanate $[x(MgO \cdot 2TiO_2) \cdot y(Al_2O_3 \cdot TiO_2)]$ in the droplets. High-quartz mixed crystals appear in the matrix glass phase after treatment at ca. 900 °C. At ca. 980 °C, cordierite ($2MgO \cdot 2Al_2O_3 \cdot 5SiO_2$) is formed from the reorganisation of the high-quartz mixed crystal. Finally at ca. 1010 °C only cordierite is present.

If the same glass was reduced while melted and subjected to thermal treatments, the crystallization behavior would be different. At about 810 °C, Mg–Al titanate is formed just as for the normal glass. However some of the Ti^{4+} ions are replaced by Ti^{3+} ions. An SiO_2-rich petalite-like phase is also observed at this temperature. The high-quartz mixed crystals appear at 900 °C. Cordierite however is not formed at 980 °C but rather as high quartz disappears, low-temperature quartz mixed crystals and spinel ($MgO \cdot Al_2O_3$) are formed. A complete conversion to cordierite occurs at about 1100 °C as with normal molten glass.

All the crystal phases mentioned here represent main crystal phases. This does not exclude the presence of very few portions of secondary crystal phases such as sapphirine and spinel among others.

Electron spin resonance analyses (ESR) have demonstrated that 25% of the Ti^{4+} ions are reduced to Ti^{3+} ions in the process of melting the glass. When

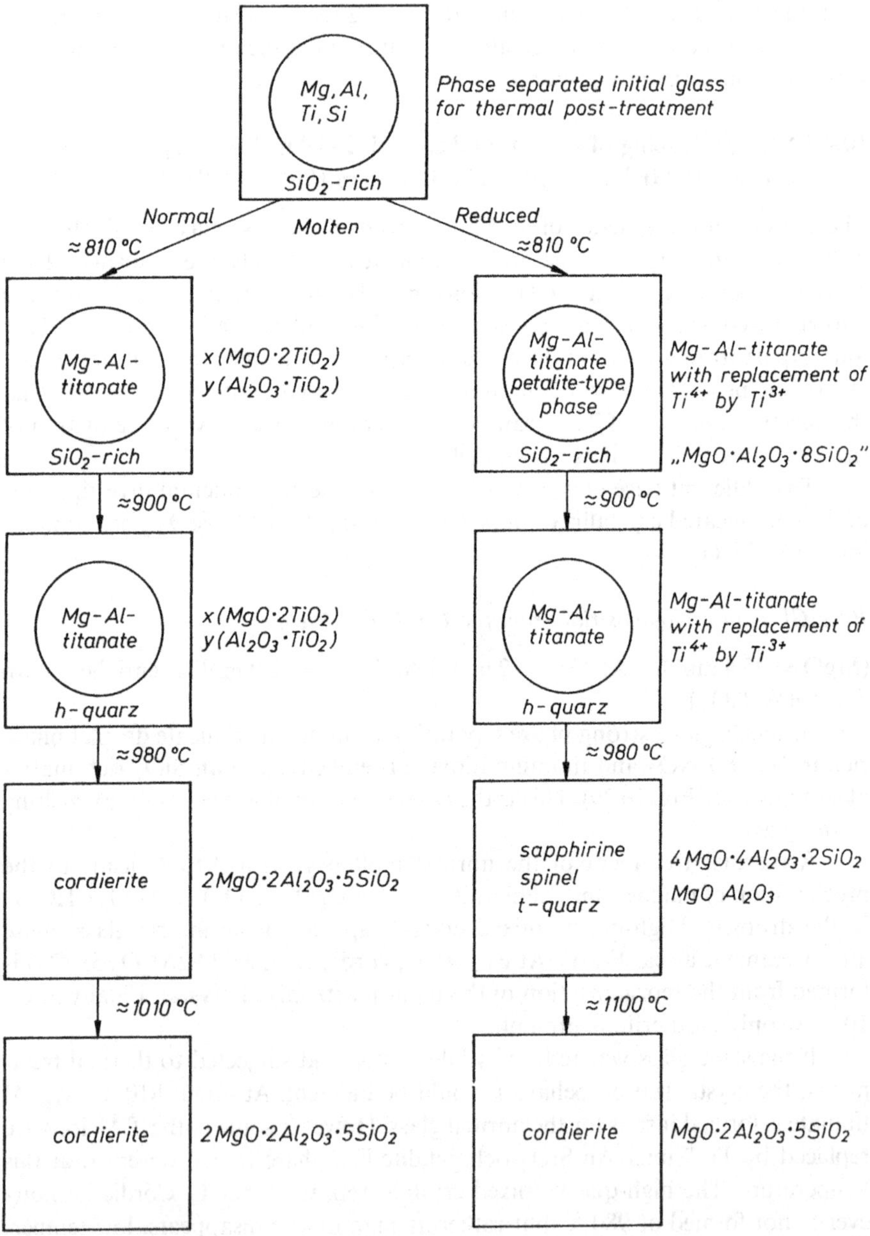

Fig. 10.29. Schematic of the crystallization sequence of a TiO_2-doped $MgO–Al_2O_3–SiO_2$ base glass ($MgO = 25.4$ mol%, $Al_2O_3 = 12.6$ mol%, $SiO_2 = 57.0$ mol%, $TiO_2 = 5.0$ mol%) from the cordierite precipitation field

compared to the initial glass, the glass ceramics thus obtained are remarkable for their increased mechanical strength: from 90 MPa in the initial glass to a maximal value of about 400 MPa. The latter value was obtained after treatment at 950–1000 °C and was obviously due to the low-quartz precipitation. The cordierite conversion initiated at higher temperatures causes a drop in the mechanical strength of the glass ceramic.

Base Glass of a Composition From the Mullite Field

(Mgo = 14.8 mol%, Al_2O_3 = 19.2 mol%, SiO_2 = 57.6 mol% enriched with 8.3 mol% TiO_2)

Phase separation is also observed here: a silicate droplet phase rich in MgO, Al_2O_3 and titanium dioxide is deposited in a SiO_2 rich matrix glass phase. The crystal formation order for a glass melted under normal and under reducing conditions is shown in Fig. 10.30. The following differences are striking for glasses from the cordierite and mullite fields (compare Fig. 10.29 and 10.30):

- Spinels appear in the base glass from the cordierite field as the main crystal phase, only in the presence of Ti^{3+} ions.
- Spinels are also present in the base glass from the mullite field, free of or poor in Ti^{3+} ions.

The presence of Ti^{3+} ions in the base glass (from the mullite field will stabilize and increase the number of low-quartz crystals.

The reduction of 25% of the introduced TiO_2 to Ti_2O_3 is of decisive importance for the resulting mechanical strength of the glass ceramics. The favored spinel and low-quartz precipitation causes the mechaical strength of the initial glass (90 MPa) to increase by a factor of 4–8. If the annealing temperature is chosen such that cordierite is already starting to be formed, the mechanical strength of the sample will increase by a factor of 9–11. Cordierite crystals develop first at the surface. They cause a low macroscopic effective compressive stress zone at the surface. They later grow into the entire glass volume with the decomposition of the primary crystal phases. The mechanical strength at this point is considerably reduced. The strength behavior of the investigated glass ceramics is always understandable in view of the thermal expansion coefficients listed in Table 10.5 for all the crystal phases present.

10.4.3.5 Double Doping of the Base Glass with 11.2 Mol% F and 5.2 Mol% Na_2O Yields Machinable Glass Ceramics [846, 847, 894, 911, 912]

In this special case a base glass from the ternary system $MgO–Al_2O_3–SiO_2$ (see Fig. 10.21) was chosen: 10 wt% MgO, 21.1 wt% Al_2O_3, 68.9 wt% SiO_2. As shown in Fig. 10.21, this composition is still located in the tridymite field, but close to that of cordierite. This location was considered optimal because the high SiO_2 content promised high chemical resistance for the glass-ceramic. It is true that a location on the eutectic line between the tridymite and cordierite fields would have brought technological advantages because of a lower melting temperature. On the other hand, a decreased tendency toward phase separation might have been disadvantageous for the structure-forming processes primarily responsible for the properties of the end-product.

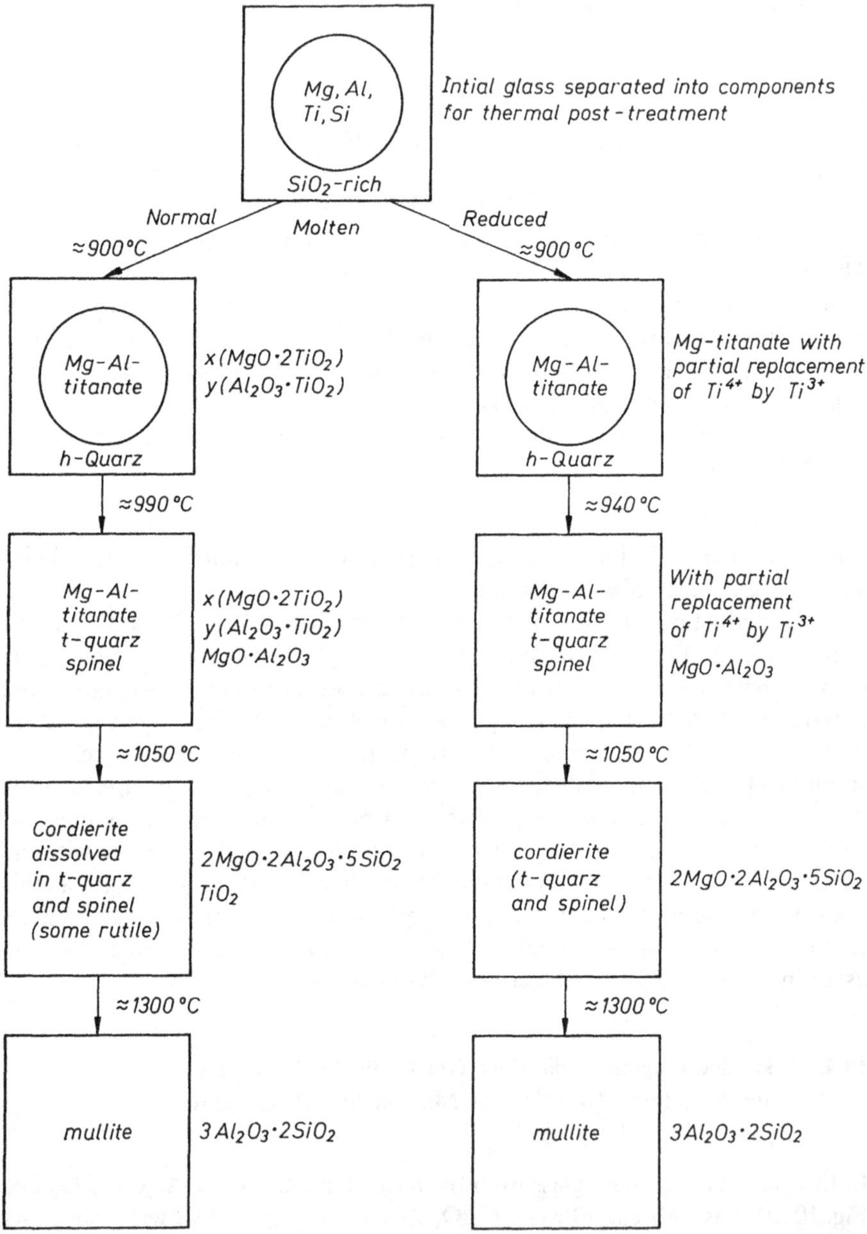

Fig. 10.30. Schematic of the crystallization sequence of a TiO_2-doped MgO–Al_2O_3–SiO_2 base glass (MgO = 14.8 mol%, Al_2O_3 = 19.2 mol%, SiO_2 = 57.6 mol%, TiO_2 = 8.3 mol%) from the mullite precipitation field

According to Fig. 10.3, this base glass still possesses a clearly demonstrable tendency toward phase separation, a prerequisite for controlled crystallization. The recognizable droplet regions constitute a silicate phase enriched in MgO and Al_2O_3.

It is well known that the replacement of oxygen by fluorine ions enormously enhances the tendency toward phase separation (Fig. 9.37). Indeed, 3.7 mass% F (= 11.2 mol% F) in the base glass (achieved, for example, by replacing some MgO by MgF_2) causes an enormous increase in the droplet regions of Fig. 10.31 (see Fig. 10.32).

Crystallization

If the F-free base glass is heat-treated, first an unstable high-quartz solid solution phase containing Mg and Al precipitates up to about 960 °C. This solid solution phase is stable at room temperature only if more than 8 mol%

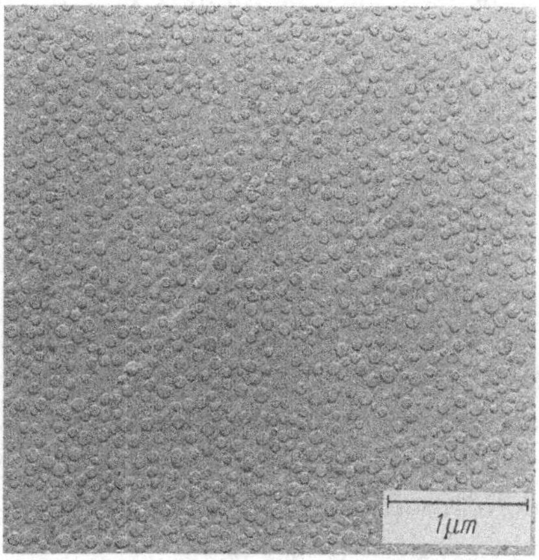

Fig. 10.31. Magnesium aluminosilicate glass precipitation of a silicate droplet phase enriched in MgO and Al_2O_3 (electron micrograph after replica preparation)

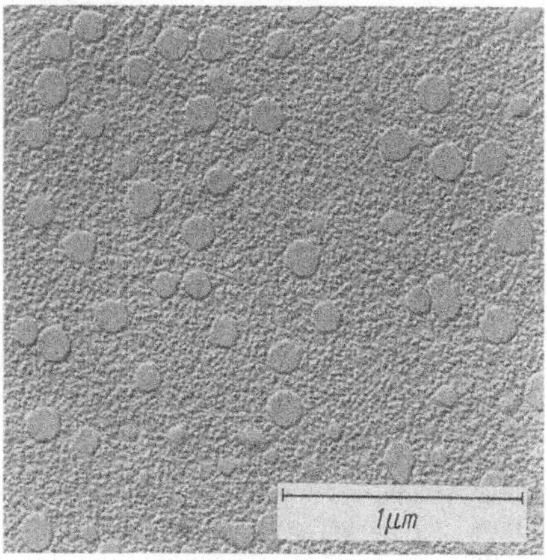

Fig. 10.32. Magnesium aluminosilicate glass as in Fig. 10.31 but with 11.2 mol% oxygen ions replaced by fluorine ions. Noticeable increase in tendency toward unmixing, documented by the increase in droplet size (electron micrograph after replication)

$(MgO + Al_2O_3)$ are embedded in the high-quartz lattice. On further increase in temperature, at about 1130 °C (holdng for two hours) the stable cordierite phase $(Mg_2Al_3\ (AlSi_5O_{18}))$ appears. This effect is clearly conditioned by progressing phase separation. This leads to a further enrichment in Mg and Al ions in the droplet phase, thus moving its composition into the cordierite field.

If F-containing base glass is subjected to the same heat treatment, the cordierite phase is precipitated at 960 °C, the temperature at which, in the F-free glass, the high-quartz solid solution phase had appeared. What in the F-free glass had been achieved by increasing the temperature, namely an enhancement of immiscibility, now has been achieved by the substitution of fluorine for oxygen ions, which influenced the nature of the precipitated crystalline phase.

One circumstance, however, is remarkable. The crystallization process, both in the F-free and the F-containing glass, occurred only in the surface. There was no crystallization in the bulk. Apparently, the greater availability of oxygen on the surface (i.e., the usual water skin) played an essential role for the coordination of the higher valency ions, thus for crystallization.

If 5.2 mol% Na_2O is introduced into the base glass containing 11.2 mol% F (system $MgO–Al_2O_3–SiO_2F$)) to increase oxygen availability, spontaneous bulk crystallization occurs. But now the precipitated phase is a mica phase: sodium phlogopite $(Na_{0.5-1}\ Mg_3\ (AlSi_3O_{10}F_2))$ (Fig. 10.33).

Nucleation and Crystallization Kinetics

As indicated in the preceding sections, nucleation and crystallization phenomena leading to phlogopite precipitation in $MgO–Al_2O_3–SiO_2$ glasses containing various additives as a function of heat treatment are by no means simple. Investigations of the kinetics of these processes are therefore quite important.

If the base glass from the system $Na_2O–MgO–Al_2O_3–SiO_2–F$ is rapidly undercooled from its melt, it exhibits a microstructure characterized by numerous, very small, droplet-shaped immiscibility regions (Fig. 10.34).

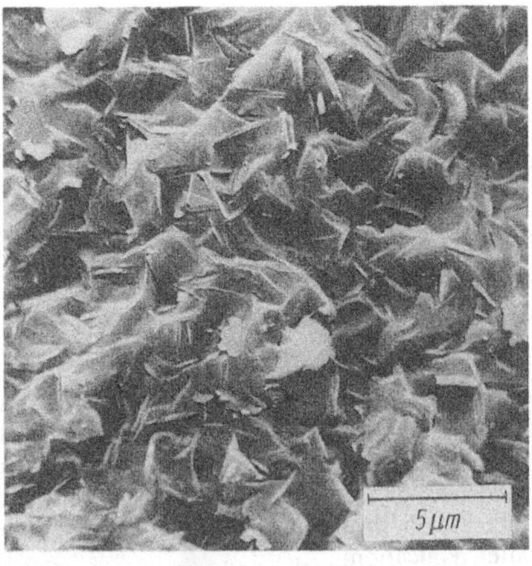

Fig. 10.33. Magnesium aluminosilicate glass containing fluorine as in Fig. 10.32, but with 5.2 mol% Na_2O added. Reheating leads to volume crystallization with precipitation of sodium phlogopite (scanning electron micrograph)

Fig. 10.34. If the glass of Fig. 10.33 is quenched very rapidly from the melt it exhibits numerous very small droplets (in the center of the picture is the MoO_3 crystal test plane)

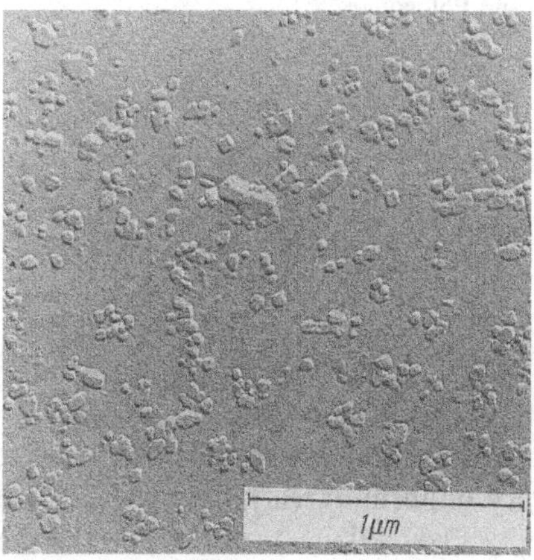

Fig. 10.35. Careful reheating of the glass of Fig. 10.34 at 780 °C promotes crystallization of sellaite (MgF_2) and norbergite ($2MgO \cdot SiO_2 \cdot MgF_2$) (both identified by X-ray diffraction) within the small droplets

After reheating from ambient temperature to 780 °C and holding for 48 h, the droplet regions are found to be entirely (or to a large extent – this is hard to determine) crystallized, as shown in Fig. 10.35 and confirmed by X-ray investigation. Sellaite (MgF_2) and norbergite ($2MgO \cdot SiO_2 \cdot MgF_2$) have formed simultaneously. (See also Dalal and Davis [913] and Takusawaga and Saito [914]).

An increase in temperature from 780 to 980 °C and a 15-min hold causes the conversion of sellaite and norbergite to Na-phlogopite (Fig. 10.36). Certain components are withdrawn from the very small sellaite and norbergite crystals to form the main phlogopite phase (Fig. 10.36) while they are redissolved. The process is quite analogous to crystallization in glasses opacified by phosphates (Figs. 9.35 and 9.36). An epitaxial interaction sellaite-phlogopite cannot be excluded (i.e., Table 10.6).

It is important to note that, for slow reheating through the temperature region favoring the formation of sellaite and norbergite, some uniform, very

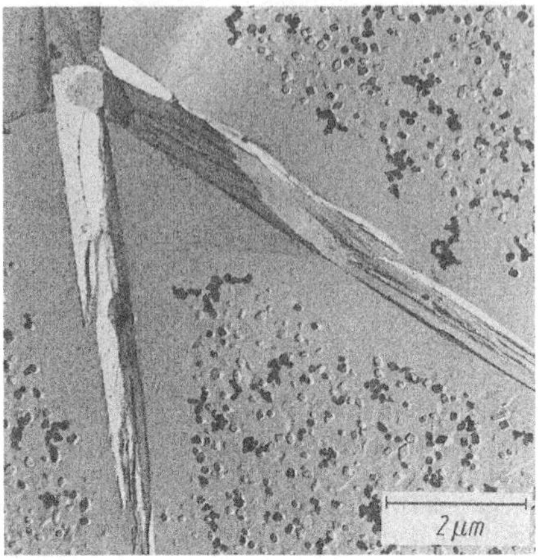

Fig. 10.36. Further reheating of the glass of Fig. 10.35 up to 980 °C causes the transformation of the sellaite and norbergite crystals to large sodium phlogopite mica crystals (see. particularly around the large mica crystals, halos free of the small sellaite and norbergite crystals)

Table 10.6. Comparison of Parameters for Sellaite and Phlogopite

Parameter	Sellaite (ASTM) (Å)	Phlogopite [915]	Parameter deviation (in %)
a_0	4.66	5.299 Å	$a_{0.S} : a_{0.P}$ $\Delta a_0 = 11.8\%$
b_0	3.08	10.135 Å	$3b_{0.S} : b_{0.P}$ $\Delta b_0 = 9\%$

small phlogopite crystals will appear (Fig. 10.33). If, however, the sellaite-norbergite region is passed rapidly, or if one cools from the melting temperature to 980 °C and holds at this temperature, very uniform, large phlogopite crystals appear (Fig. 10.37). In these latter cases only a few sellaite or norbergite crystals seem to have formed, presenting heterogeneous nuclei for the formation of considerably larger phlogopite crystals.

Machinability

Machinability, as defined previously, is without any doubt primarily based on the controlled precipitation of mica crystal phases in the glass. Analysis in more depth reveals a series of important controlling parameters which must be considered in production.

It is well known that the micas are layer silicates with (usually) foliform habit and excellent cleavage in the 001 plane. Their structure is characterized by the loose connection via (usually) K^+ or Na^+ ions of layer packets (Fig. 10.38). These layer packets in turn consist of two tightly connected $(Si_2O_5)^{2-}$ layers. Characteristically, each fourth tetrahedron (in Fig. 10.38 a cross section vertical to the 001 plane) in such a layer of six rings of tetrahedra is an AlO_4 tetrahedron. The firm connection of a layer packet is achieved by Mg^{3+} and F^+ ions in the form of a brucite layer.

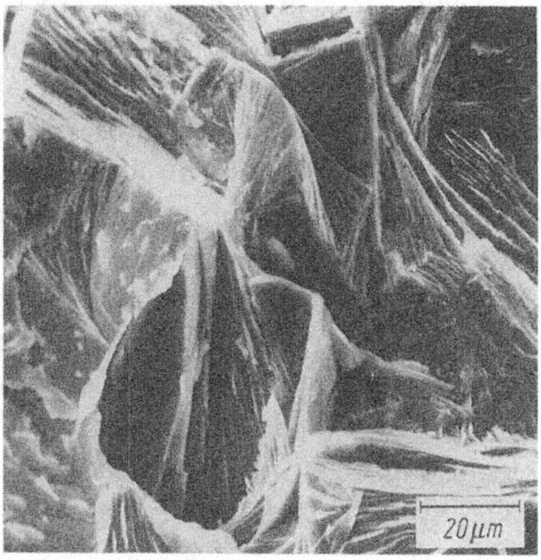

Fig. 10.37. Large sodium phlogopite crystals ($Na_{0.5-1}Mg(AlSi_3O_{10}F_2)$) in a base glass of the system Na_2O–MgO–Al_2O_3–SiO_2–F (scanning electron micrograph)

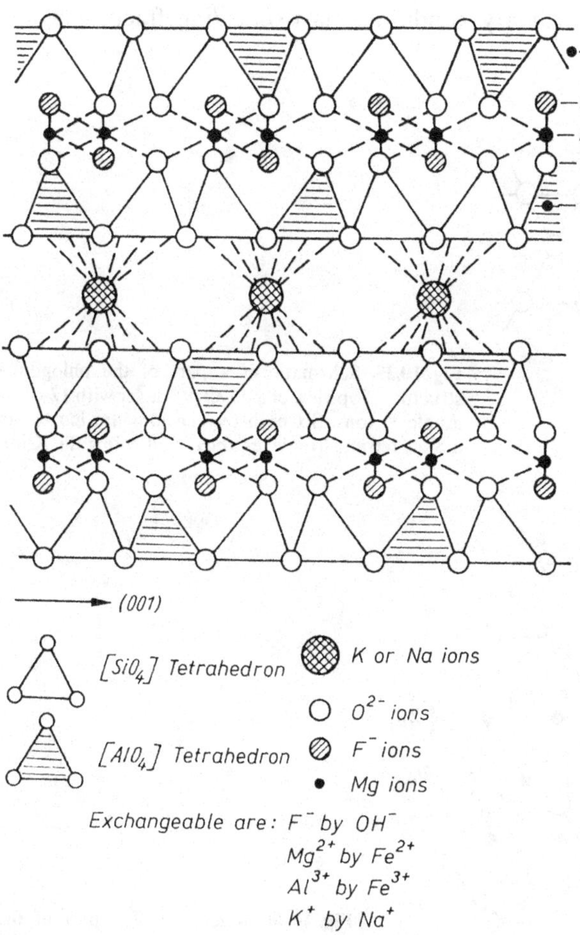

Fig. 10.38. Schematic of phlogopite structure

As shown schematically in Fig. 10.38, the Mg^{2+} ions are bonded directly to O^{2-} ions, which represent corners of the tetrahedra. The second valence of the Mg^{2+} ion is satisfied by the F^- ion, which is always located in the center of six-ring tetrahedra. In this way the two tetrahedron layers appear interlocked. The Mg^{2+} ion may be replaced by an Fe^{2+} ion, the F^- ion by an OH^- ion.

As also shown in the scheme of Fig. 10.38 all Mg^{2+} ions connecting the two tetrahedron layers are in 6-coordination. Such three-layer packets are loosely connected by K^+ or Na^- ions. These alkali ions function as a charge balance for the unsatisfied oxygen valences of protruding SiO_4 or AlO_4 tetrahedra (see Fig. 10.38). According to Fig. 10.39, all alkali ions are in 12-coordination with six O^{2-} ions, each belonging to one layer packet. While the schematic of Fig. 10.38 serves to clarify these arrangements. Fig. 10.40 represents the exact projection of the (001) plane of the phlogopite layer: the two tetrahedron layers of packets are somewhat staggered.

The significance of the mica structure for machinability rests primarily with the extraordinarily fast spread of fracture within the alkali-ion layer (001) plane (good cleavage!). Under ordinary conditions, fast-spreading fracture would

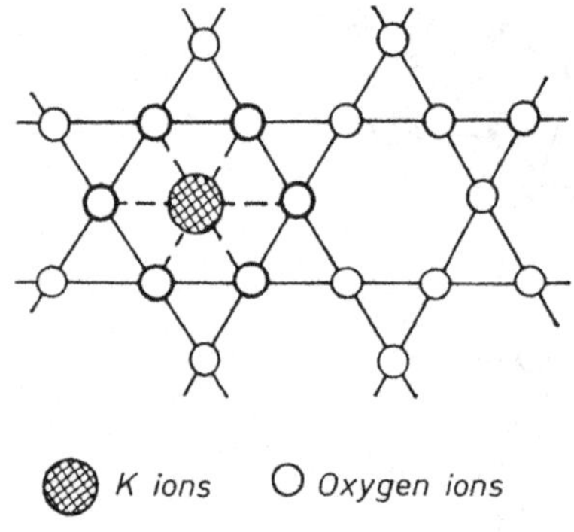

Fig. 10.39. Schematic of a part of the phlogopite structure. Top view of a $(Si_2O_5)^{2-}$ layer with 12-coordinate K ions. Six each oxygen ions are located in a SiO_4 tetrahedron layer above and below the K ion

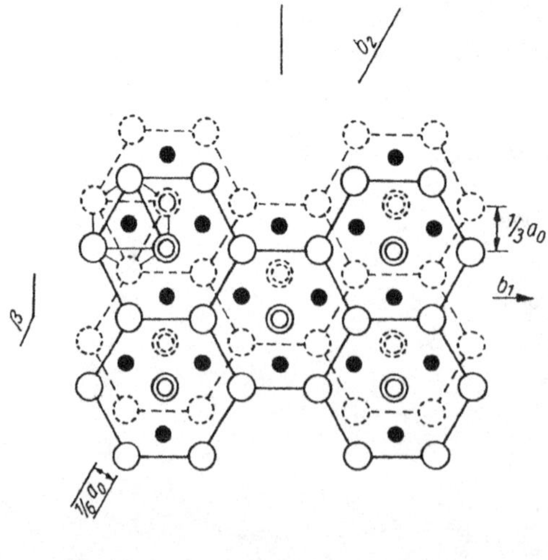

Fig. 10.40. Schematic of a part of the phlogopite structure. Projection of the (001) plane in the phlogopite layer

cause failure. In the special case of a phlogopite ceramic this is prevented by a "card house" structure (Fig. 10.33). A fracture spreading always in the (001) direction, jumping from one to another neighboring phlogopite crystal, has to change direction constantly and will fail quickly. Thus, the glass-ceramic will not fail.

Another very important influence on machinability is crystal size. It has been found that a glass ceramic with very small phlogopite crystals can be machined, but not as well as one with large crystals. This might be understood in terms of repeated energy requirements for new cleaving processes if the fracture spread fails too quickly. It has been established that there is a certain optimal crystal size.

The lack of optimal control of reheating schedules may also cause impaired machinability. As shown in Fig. 10.41, long holds often cause the transformation of fine, interlocked phlogopite foils into bulkier, rectangularly bounded rough disks, usually leading to a loss in interlocking. This prevents easy fracture propagation of neighbor crystallites, which in turn affects machinability. Figure 10.42 illustrates machining of a phlogopite glass-ceramic on a lathe.

Doping of a Different Base Glass with 11.5 mol% F^- and 6.4 mol% Na_2O/K_2O [904, 916, 917]

(MgO = 21.2 mol%, Al_2O_3 = 19.5 mol%, SiO_2 = 59.3 mol%)
Optimization of the machinability of a glass ceramic containing mica is generally dependent on the base glass composition, the thermal treatment of the glass, the crystal portion of the total volume and the dimension and configuration of the crystals [322]. The external appearance and crystal aggregation were also found to exert a very decisive influence on the machinability of a glass-ceramic [904]. In a base glass only slightly different from the one described in the previous section, the phlogopite mica crystal phase can be precipitated in a form not observed until now with the proper temperature treatment. At a temperature of 750–1000 °C, the phlogopite crystals will appear in a spherical lamellae

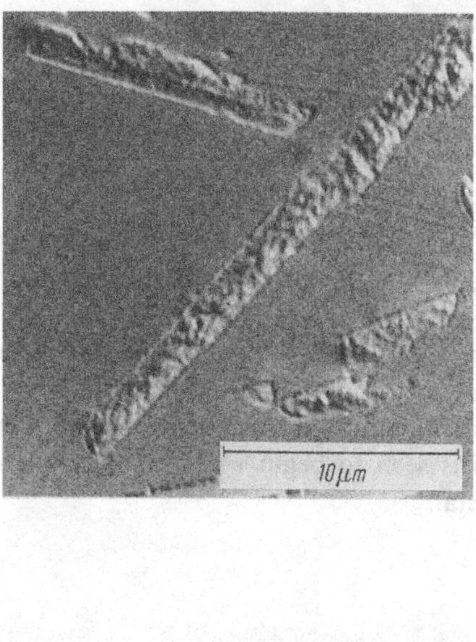

Fig. 10.41. Sodium phlogopite crystals in a glass, as in Figs. 10.33; or 10.37, but after long reheating times. The crystals have become bulkier, connectivity with other crystallites has become lost

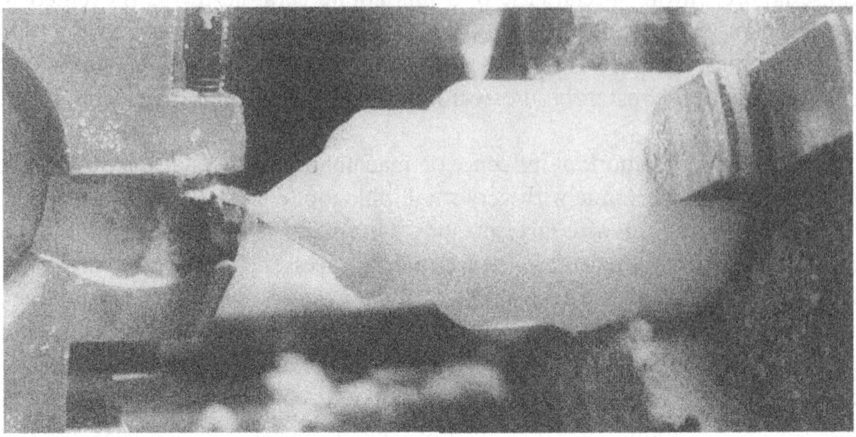

Fig. 10.42. A piece of a machinable sodium phlogopite glass ceramic being worked on a lathe

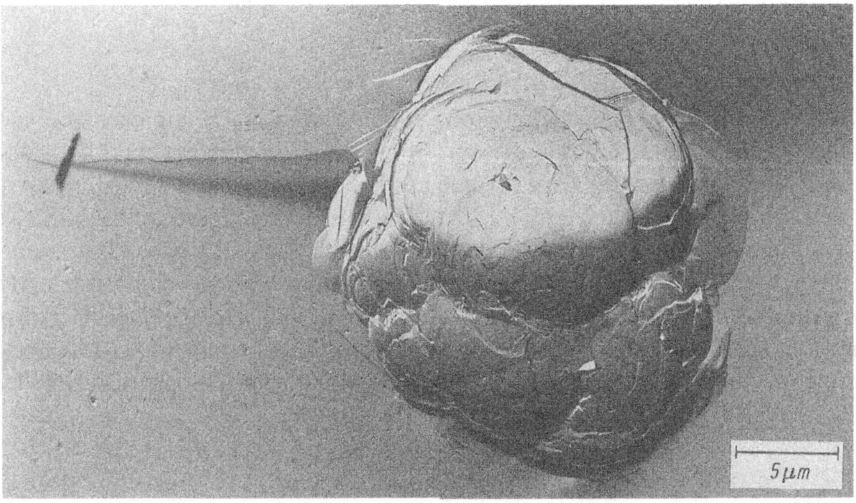

Fig. 10.43. Phlogopite crystals in a glass of modified base composition (MgO = 21.2 mol%, Al_2O_3 = 19.5 mol%, SiO_2 = 59.3 mol% and F^- and Na_2O/K_2O doping); spherical lamella arrangement of the phlogopite crystals (replica method)

configuration (similar to a cabbage see Figs. 10.43–10.47) rather than in the form of flat sheets (house of cards formation, see Fig. 10.33). The machinability can therefore be improved by a factor of 4–5.

Figure 10.43 is a replica electron micrograph of phlogopite crystals in the typical spherical lamellae configuration.

Figure 10.44 is a cross-section cut through the spherical crystal aggregates. The spherical formation of the crystals is seen to begin with the start of the precipitation of the first phlogopite crystal.

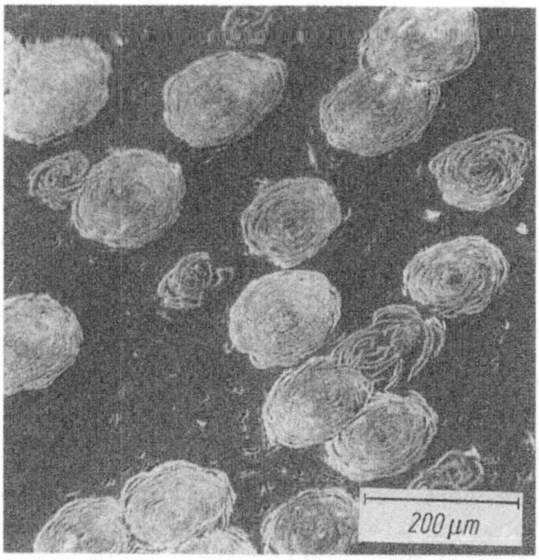

Fig. 10.44. Spherical lamella arrangement of phlogopite crystals; cross section through the spherical aggregate. At the beginning of the phlogopite crystallization, the aggregates are still separated from one another. (scanning electron micrograph)

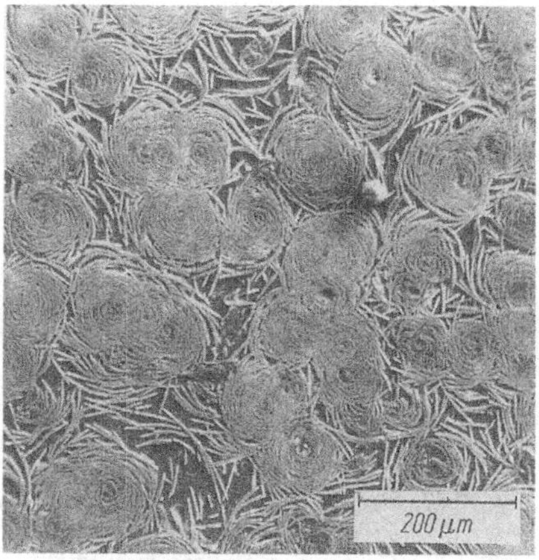

Fig. 10.45. Advanced stage of the phlogopite crystallization. Many spherical aggregates touch each other (scanning electron micrograph)

Figure 10.45 shows an advanced ceramic stage. The "cabbage"-like crystal aggregates and single crystals come into contact.

Figure 10.46 is a cross-section cut through a single aggregate. A certain bridge formation or growth between the single lamellae can be recognized.

As the glass is further tempered in the temperature range of 1000 to 1200 °C, the phlogopite crystals are drawn again. Figure 10.47 shows an intermediate stage in such a process. The stretched state is maintained with a fast undercooling of the sample. Preliminary structure analyses indicate that the reason for the new appearance of the phlogopite crystals lies in modifications of the concentrations of Na^+ and K^+ as well as in modifications of the dioctahedron-trioctahedron character of the crystals (see Fig. 10.38). This process is very complex in nature. It can be controlled by modifying the concentrations of each single component of the base glass.

A systematic increase of the total (Al_2O_3 + MgO) content in comparison to SiO_2 induces a transfer from the phlogopite crystals (spherical lamellae

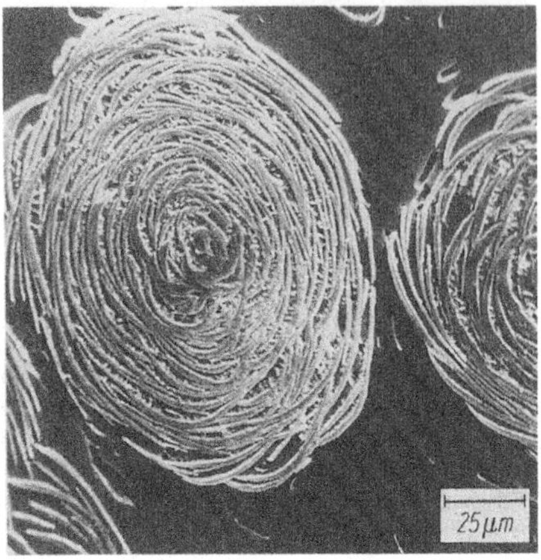

Fig. 10.46. Single spherical aggregate of phlogopite crystals. Definite growths can still be observed between the single lamellae (scanning electron micrograph)

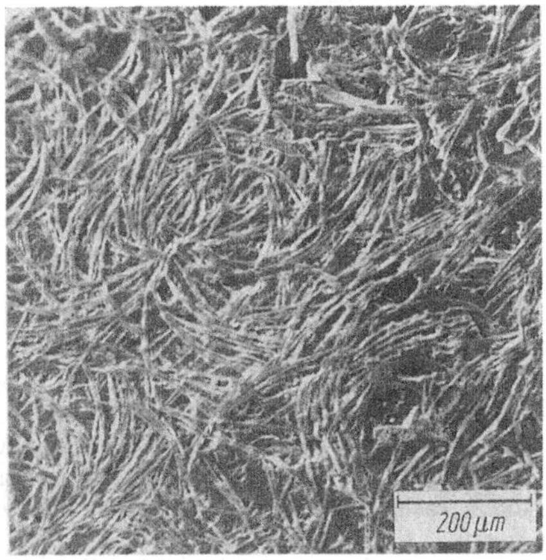

Fig. 10.47. Stretching of the original spherically ordered phlogopite sheets with heating to 1000–1200 °C (scanning electron micrograph)

configurations) to the cordierite crystals. If sodium and potassium ions are no longer available in sufficient quantity for the stabilization of the tetrahedron layers, cordierite is formed in tetrahedron rings. The $[AlO_{4/2}]^-$ groups in cordierite are stabilized by the Mg^{2+} ions.

All crystals seen in Figs. 10.43 to 10.47 were identified by X-ray to be phlogopite crystals and are under no circumstance identical to chrysotile crystals which are known to be found in small roll shapes. Many questions regarding the spherical arrangement of phlogopite crystals have not yet been answered in a satisfying way and remain open.

10.4.3.6 Ferrimagnetic Glass Ceramics

Replacement of MgO and Al_2O_3 in a base glass doubly doped leads to ferrimagnetic glass ceramics [904, 918–927] (MgO = 20.7 mol%, Al_2O_3 = 19.6 mol% and SiO_2 = 59.8 mol% with 11.0–12.6 mol% F^- and 5.6–7.0 mol% K_2O).

The phase separation tendency will increase in a base glass from the mullite field doped with 11–12 mol% F^- and 50–70 mol% K_2O if MgO and Al_2O_3 are partly or completely replaced by FeO or Fe_2O_3 (see Fig. 10.38).

Three glasses with different iron oxide contents were chosen for thermal treatment and investigation of the crystallization kinetics. The Fe^{2+}/Fe^{3+} ratio varied as follows:

Glass 1	Glass 2	Glass 3
FeO = 1.6 mol%	FeO = 4.6 mol%	FeO = 10.5 mol%
Fe_2O_3 = 0.8 mol%	Fe_2O_3 = 2.2 mol%	Fe_2O_3 = 3.1 mol%

These Fe^{2+}/Fe^{3+} ratios were obtained from the arithmetical analysis of the Mößbauer spectra shown in Fig. 10.48.

Nucleation and crystallization processes have been demonstrated to start inside the droplet regions, already upon cooling of the liquid glass (the

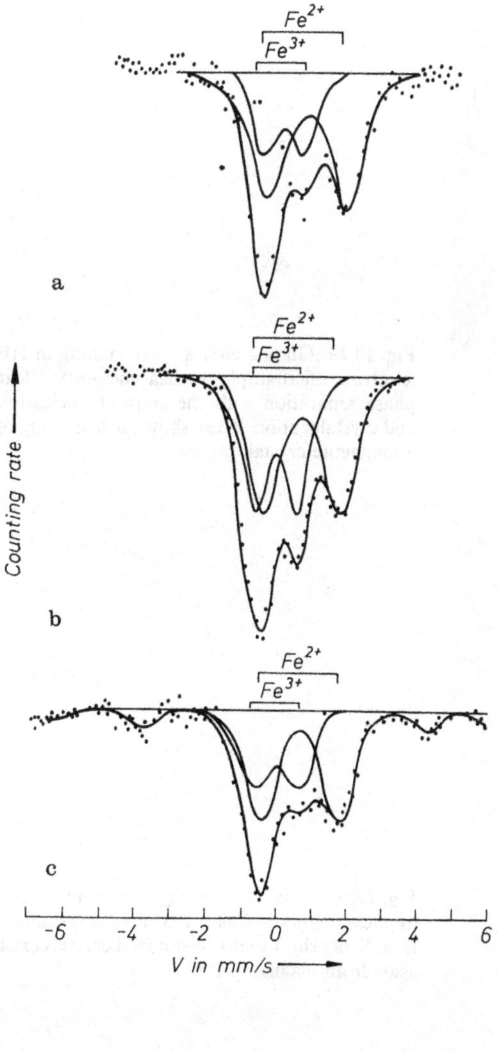

Fig. 10.48. Mößbauer spectra of glass 1 (**a**), glass 2 (**b**), and glass 3 (**c**). Calibration doublets for the quantitative determination of the Fe^{2+}/Fe^{3+} ratios in the glass are simultaneously included for certain Fe^{2+} and Fe^{3+} concentrations

undercooling rate in the transition region 3–4 $K \cdot min^{-1}$). While in glass 1 only phase separation occurred, nucleation has clearly begun in glass 2 (Fig. 10.49) and crystallization can be observed in glass 3 (Fig. 10.50).

Magnetite (Fe_3O_4) is exclusively precipitated in glass 2. Magnetite and spinel mixed crystals of the Mg, $Fe^{II}(Fe_x^{III}Al_{1-x})_2O_4$ type are formed in glass 3. A thermal treatment of the three initial glasses leads to the crystallization sequence schematically representated in Fig. 10.51. One must realize that the three initial glasses are subject to different initial conditions. Glass 1 corresponds to the state in Fig. 10.51a, glass 2 to 10.51b and glass 3 to 10.51c.

Heat treatment of glass 1 starting at 650 °C will first lead to the precipitation of magnetite in the droplets (state in Fig. 10.51b). Next formation of two spinel mixed crystal phases I and II of the type Mg, $Fe^{II}(Fe_x^{III}Al_{1-x})_2O_4$ will occur. The lattice constant in type I is $a = 0.827$ nm and in type II $a = 0.813$–0.823 nm. Finally as the temperature is further increased, magnetite

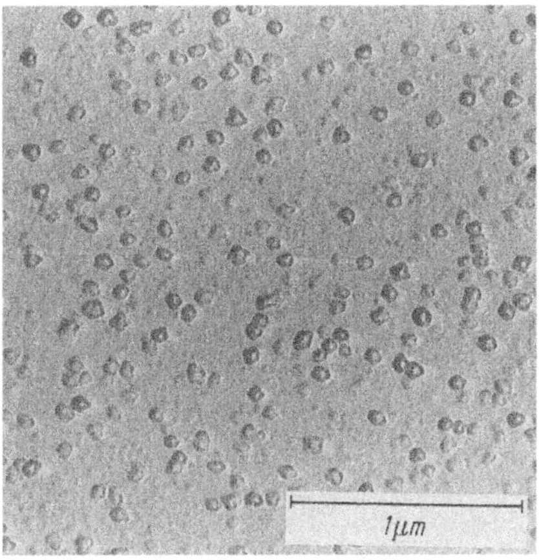

Fig. 10.49. Glass 2 after a weak etching in HF (electron micrograph, replica method). Clear phase separation with the start of nucleation and crystallization. X-ray show the beginning of a magnetite crystallization

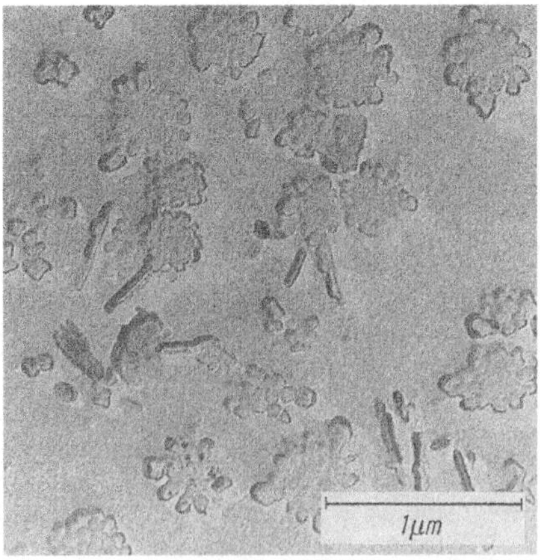

Fig. 10.50. Glass 3 after slight etching in HF (replica method). Clear advanced crystallization. X-ray show that spinel mixed crystals exist aside from magnetite

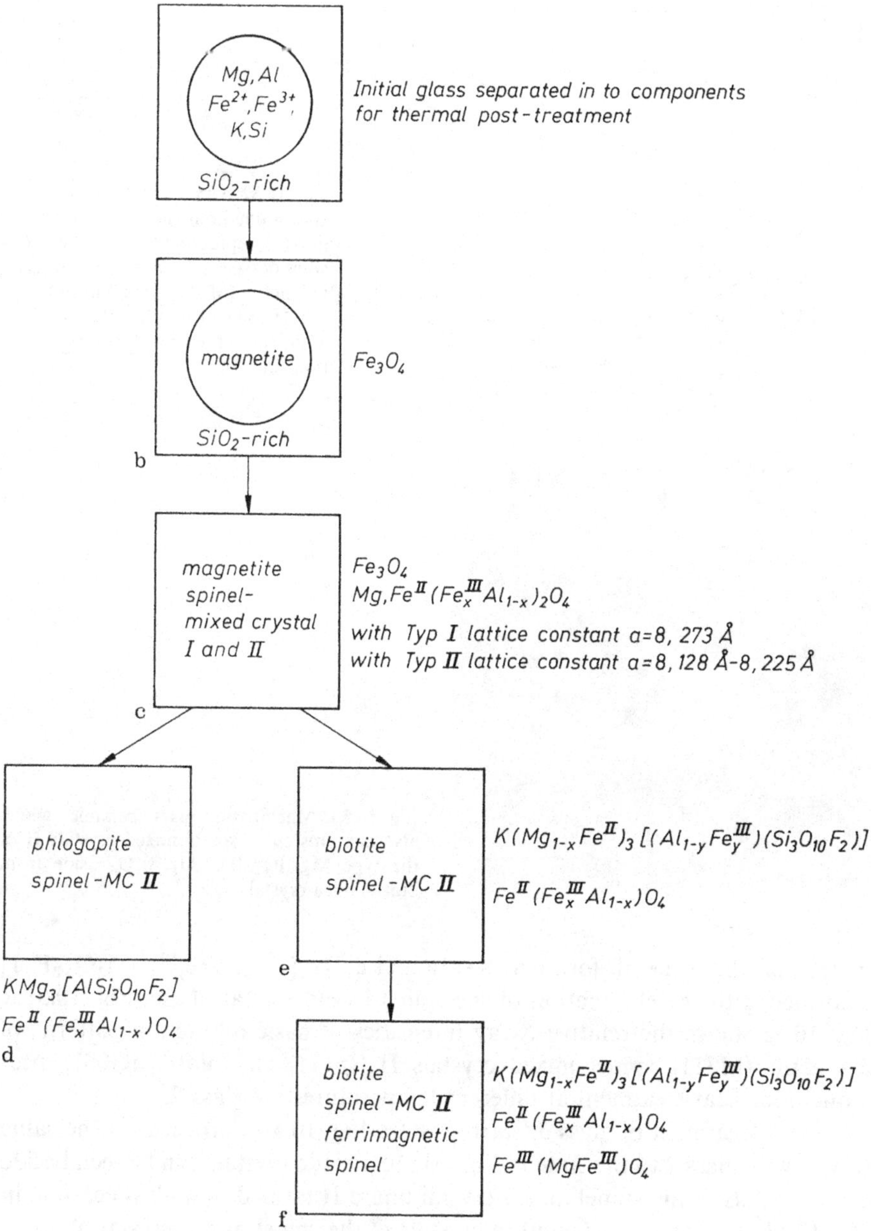

Fig. 10.51. Schematic of the crystallization sequence of a [MgO–Al_2O_3–SiO_2–K_2O–F^-] base glass, in which MgO and Al_2O_3 are replaced by FeO or Fe_2O_3

and the spinel mixed crystal phase I disappear with the new formation of phlogopite crystals.

Heat treatment of glass 2 (starting state of Fig. 10.51b) will lead, at even lower temperature than glass 1, to the additional precipitation of two spinel mixed crystal phases I and II (Fig. 10.51c). As the temperature is further increased,

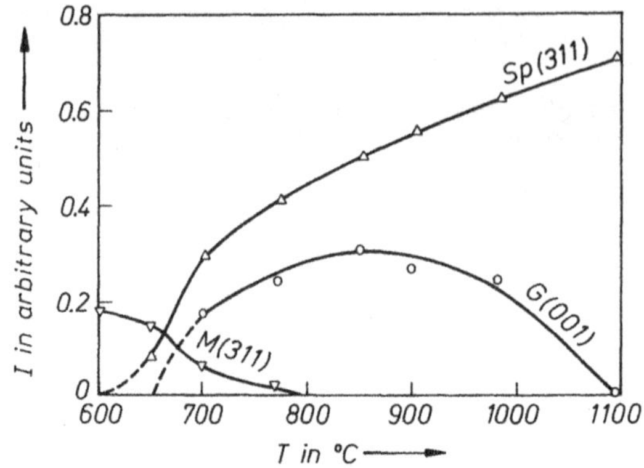

Fig. 10.52. Crystallization sequence with a one-hour thermal treatment of glass 2 represented by X-ray intensities of base diffraction reflections for magnetite (M311), spinel mixed crystals II (Sp311) and biotite (G001) as a function of temperature (crystallization time 1h)

Fig. 10.53. Machinable glass ceramic which also contains cubic spinel mixed crystals II of the type Mg, Fe^{II} $(Fe^{III}Al_{1-x})_2O_4$ aside from biotite mica crystals

biotite of the general formula $K(Mg_{1-x}Fe_x^{II})_3$ $[(Al_{1-y}Fe_y^{III})$ $(Si_3O_{12}F_2)]$ is formed with the elimination of the spinel mixed crystal phase (Fig. 10.51c). Fig. 10.52 shows the relative X-ray intensities of basic diffraction patterns of magnetite (M311), spinel mixed crystals II (sp311) and biotite (G001) after a one-hour heat treatment at different temperatures for glass 2.

Heat treatment of glass 3 (starting state Fig. 10.51c) proceeds in the same way as with glass 2 to the state in Fig. 10.51c. Biotite crystals can be seen beside cubic crystals of the spinel mixed crystal phase II formed in a glass ceramic in Fig. 10.53. The iron enrichment in crystals of the spinel mixed crystal phase II and in biotite crystals respectively can be observed in Figs. 10.54 and 10.55. If relatively high quantities of burn iron and magnesium oxide are available in the base glass (e.g. $FeO/Fe_2O_3 > 6$ mol% and $MgO > 15$ mol%), ferrimagnetic spinels of the type $Fe^{III}(MgFe^{III})O_4$ are formed additionally with a thermal treatment above 1000 °C (Fig. 10.54). This type of glass ceramic is ferrimagnetic. All glass ceramics described which contain magnetite or ferrimagnetic spinel (Fig. 10.51b, c and f) are ferrimagnetic as can be seen in the hysteresis curve in Fig. 10.56.

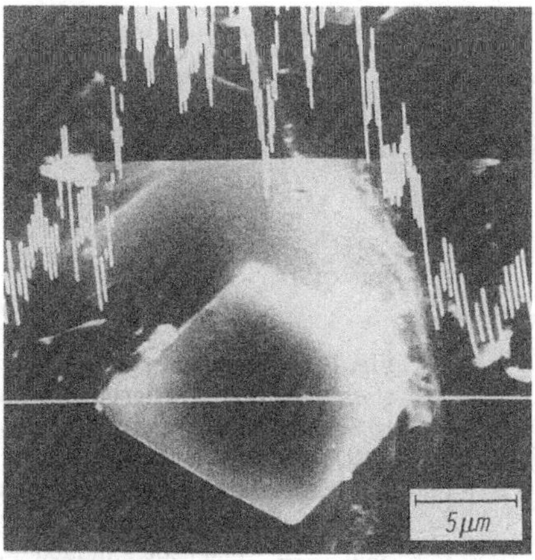

Fig. 10.54. Scanning electron micrograph of the base glass with 1.0 mol% FeO and 0.5 mol% Fe_2O_3 after a three-hour thermal treatment at 1200 °C with the X-ray intensity curve for the $Fe_{K\alpha}$ radiation. The electron beam was run along the white horizontal line on the sample. The intensity curve shows that the iron content in the cubic spinel crystals sharply increases in comparison to the matrix

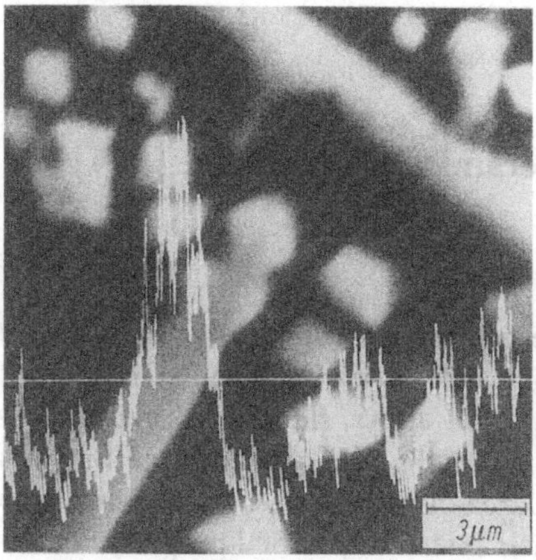

Fig. 10.55. Scanning electron micrograph of base glass 3 (similar conditions as for Fig. 10.54). As shown by the X-ray intensity curve for the $Fe_{K\alpha}$ radiation, the sheet-like mica crystals represent biotite containing iron. Cubic spinel crystals are also observed

These results prove that it is possible to produce glass ceramics with ferrimagnetic properties on the basis of magnetite precipitation (Fig. 10.51b and c). This is achieved by the simultaneous presence of biotites, which provide machinability and of ferrimagnetic spinel of type $Fe^{III}(MgFe^{III})O_4$, which provide ferrimagnetic properties.

The most varied properties or property combinations can be obtained for glass ceramics by using almost the same base glass changing addition. Such properties as very high mechanical bending strength, very good machinability, ferrimagnetic properties, assets along with ferrimagnetism can be achieved. Other favorable secondary properties such as high chemical resistance or high surface quality are also characterisitic of these new materials [880, 928–933].

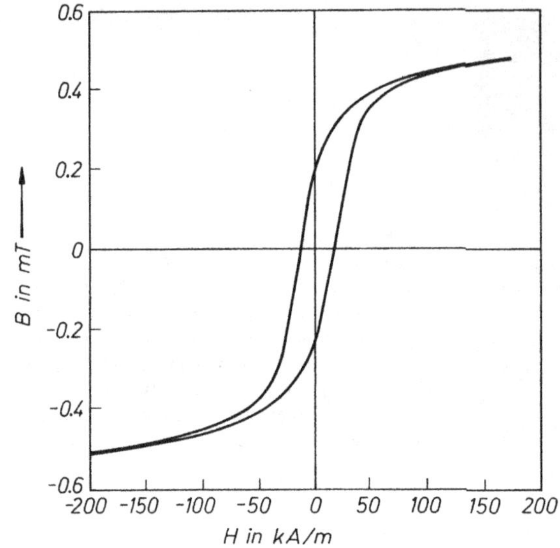

Fig. 10.56. Hysteresis curve of the glass containing magnetite with 10.5 mol% FeO and 3.1 mol% Fe_2O_3 with no thermal treatment. The material is characterized by the shape of the curve as magnetically "soft" (B magnetic induction, H magnetic field (strength)). The same curve would be obtained with a precipitation of ferrimagnetic spinel of the type Fe^{III} $(MgFe^{III})O_4$

10.4.4 Development of Bioglass Ceramics for Medicine [934]

10.4.4.1 Introduction

The enormous advances achieved in the field of medicine in the last decades were not only due to new instruments but also to new materials. One cannot imagine modern medicine without metals and alloys, sintered corundum, organic high polymers (also as composite material), glassy carbon etc.. Bioglass ceramics open new chances in a research area of natural sciences and medicine which is only at the beginning. Due to their extensively varied property combinations, bioglass ceramics can adapt better to the advances of medicine than the conventional implants. The two main properties of bioglass ceramics are: biocompatibility, the introduction of the material in the tissues of the human body without rejection or a toxic effect, and bioactivity, or the property of properly growing into the tissues of the human body. None of the classical biomaterials show this property. The range of application of the bioglass ceramics, which are still partly at the animal or clinic testing stages, extends from the replacement of vertebrae, replacements in the middle ear, throat, nose, eyes, the whole head area, as well as the shoulder and leg up to the replacement of the roots of teeth and the use of tooth superstructure (hard tissue substitute in the broadest sense). The question of the long term behavior of a bone-bioglass-ceramics contact on hardening under varying loads has not yet been answered satisfactorily. The whole interdisciplanary research area is still too young for that. However negative effects due to the material has not been observed so far.

10.4.4.2 Development of Bioglass Ceramics, Present State, Requirements and Target

The developments achieved until now have already brought considerable advancement in the field of medicine. Silicophosphate glasses are usually applied as coating and thin films on metals, sintered corundum and other implant materials [935–937]. These glasses are very similar to the bone substance which has a high apatite content. Also thin glass films which were partly converted to crystalline apatite by heat treatment, have been used. These bone-related glasses or glass ceramic films already exhibit a fair bioactivity and intergrowth with the bone tissue. Thin films as a rule will however completely dissolve and consequently be ineffective. Since this property is advantageous for other applications fully absorbable glasses and glass ceramics have been developed [938–941]. Sintered ceramics were also developed as a solid material. They are as a rule sintered products from crystalline apatite [941]. Phosphate glasses in powder form in which crystalline apatite is formed by heat treatment have also been introduced as sintered products [942, 943]. Phosphate glass powder and a silicate glass powder, in which crystalline silicate such as wollastonite and devitrite, are precipitated were also sintered together in order to further modify the properties of the product [944, 945]. The products obtained are more or less bioactive and porous due to the presence of apatite. They can be worked on with hard metal tools. Porosity and relatively low mechanical strength often prove to be a large disadvantage.

An important advancement was made by producing glass ceramic products from solid glasses by heat treatment. Hench et al. the first decisive step in the way to a solid intergrowth between inorganic materials and bones, by using glass film or partly crystalline glasses [946–948]. The composition of his base glass was (% by weight): SiO_2 45%, CaO 24.5%, Na_2O 24.5%, P_2O_5 6.0%.

In the 70s, a significantly higher chemical stability had been obtained with the material "Ceravital". Ceravital is made from the base glass system SiO_2–Na_2O–CaO–$Ca_3(PO_3)_2$–MgO–K_2O–CaF_2 [949, 950]. A series of consecutive developments [951–953] brought further advances. Very extensive investigations, especially regarding end uses, showed that a good bioactivity was obtained. However the intergrowth area between the bones (or other tissues) and the glass ceramic widens due to the relatively high solubility of the latter. This process obviously does not reach an equilibrium state after a long term contact of the material with the bones. The danger of the implant coming loose under mechanical strain at the boundary zone therefore exists. For other applications the boundary zone will however be sufficiently stable. The main crystal phases of the aforementioned new material are apatite $Ca_5[F/(PO_4)_3]$, wollastonite $Ca_3[Si_3O_9]$ and/or devitrite $Na_2O \cdot 3CaO \cdot 6SiO_2$.

It has always been our goal to develop improved bioglass ceramics with new properties. The knowledge of the microprocesses involved in the solidification of a glass melt or of the transformation of a glass into a glass ceramic by a controlled crystallization is fundamental for this development. The demands on the cooperators in the medical field are very diverse. The current work in the field of bioglass ceramics can already be compared to the development of the range of high-grade steel for various applications. The required primary properties are:

- complete biocompatibility or
- high bioactivity with

  as high a mechanical strength as possible
  higher chemical resistance in particular
  against body fluids and
  good machinability of the material as much
  as possibly by the physician himself.

When good biocompatibility is achieved, the results of cell compatibility tests both in vitro and in vivo must show no rejection or toxicity and additionally a direct boundary with the bone free of connective tissue must take place. A good bioactivity implies that the implant activates human tissues to participate in a real biochemical intergrowth with the implant according to a partial solution and ion exchange process.

In 1982, Vogel et al. [954] and Höland et al. [899] succeeded for the first time in developing a machinable bioactive glass ceramic, containing both fluorophlogopite and apatite, suitable for bone implantation in humans.

10.4.4.3 Development of Biocompatible and Machinable Glass Ceramics

Machinable Glass Ceramics from the System Na_2O–MgO–Al_2O_3–SiO_2–F with Planar, Sheet-Shaped Mica Crystals

A glass ceramic is regarded as machinable when it can be spun, moulded, drilled or threads can be put on it with metal processing tools or with hard metal tools without the tool breaking as is the case with a normal ceramic. A test for comparing machinability consists of measuring the time required for a hard metal drill to reach a known drill depth with a constant load.

Beall, Grossmann et al. [846, 847, 955] in 1970 found that a controlled precipitation of mica crystals, which have an optimal size for the achievement of good machinability, will occur in a base glass. The crystals will touch each other and finally occupy 2/3 of the total volume of the ceramic.

A model example is shown in Fig. 10.33. An alkali metal oxide (5.2 mol%) was added to a standard base glass (15.5 mol% MgO; 12.6 mol% Al_2O_3 and 71.9 mol% SiO_2) in which 11.2 mol% oxygen ions were replaced by fluorine ions. A controlled phase separation of the base glass leads to a specific separation of sodium phlogopite crystals ($Na_{0.5-1}Mg_3[AlSi_3O_{10}F_2]$) and to the required mica [956, 957] from sodium phlogopite (see Section 10.4.3.4). This ceramic which was developed for special purposes in the manufacturing of scientific instruments and for mechanical engineering purposes is also biocompatible, and can be used in several ways in medicine. Certain physical properties are listed in Table 10.7.

Table 10.7. Properties of the machinable glass ceramics with planar fluorophlogopite mica crystals composition (mol%): Na_2O 3–15, MgO 7–23, Al_2O_3 10–30, SiO_2 45–70, F 3–6 [934]

Density	2.5 g/cm^3
Linear thermal expansion coefficient (20–450 °C)	75.10^{-7} K^{-1}
Bending strength	90 MPa
E-modulus	50 GPa
Compressive strength	450 MPa
Impact toughness	2.0 kN/m
Hydrolytic grade	1–2
Acidic grade	3
Basic grade	1–3
Surface roughness	0.15 m
Machinability	very good

Glass Ceramic of Improved Machinability from the System Na_2O/K_2O–MgO–Al_2O_3–SiO_2–F with Curved Mica Crystals [958, 959]

A new mica phase can be made to precipitate in a modified standard base glass (21.2 mol% MgO; 19.5 mol% Al_2O_3; 59.3 mol% SiO_2) enriched with 11.2 mol% F^- and 6.4 mol% Na_2O/K_2O. The mica crystals are not flat sheets but rather curved and grouped in cabbage-like aggregates (see Fig. 10.46). The X-ray spectra from energy dispersion X-ray emission microanalyses (see Fig. 10.57) demonstrate with a comparison of the Mg and Al peaks that the Al concentration is higher in the curved mica sheets than in the flat ones [960]. With an increase of the total (Al_2O_3 + MgO) content in comparison to the SiO_2 content in the glass, Mg^{2+} ions are partly replaced by Al^{3+} ions in the new type mica crystal (1.5 Mg^{2+} corresponds to 1 Al^{3+}). Stress is therefore experienced in the octahedral layer of the mica structure and this in turn leads to a bending of the mica crystals. This is connected with modification of the dioctahedral-trioctahedral character of the crystals evidenced by X-ray diffraction. If the concentration of (MgO + Al_2O_3) is further increased, cordierite crystals ($Mg_2Al_3[AlSi_5O_{18}]$) precipitate; the $[AlO_{4/2}]$ groups in cordierite are stabilized by the Mg^{2+} ions. The glass ceramic with the curved mica crystals can be 4 to 5 times better in terms of machinability [958, 959] than the one with flat mica sheets mentioned in Section 10.4.3.4. This machinable glass ceramic is biocompatible and can be used for many purposes in medicine like the one described above. Its main properties are listed in Table 10.8.

Machinable Glass Ceramic from the System Na_2O/K_2O–MgO–Al_2O_3–SiO_2–F with Mica and Cordierite Crystals

Machinable glass ceramics which only contain mica crystals were already described as mere glass ceramics which only contain cordierite crystals ($Mg_2Al_3[AlSi_5O_{18}]$) [956] The latter particularly demonstrate high mechanical strength and toughness. Mica and cordierite crystals can be simultaneously

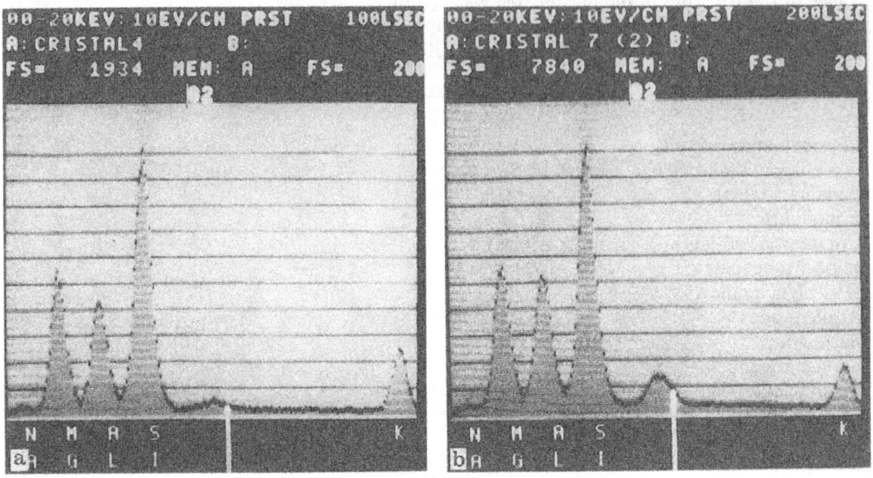

Fig. 10.57. **a** EDAX of flet fluorophlogopite mica crystals. Mass ratios Al/Si = 0.43, Mg/Si = 0.71. **b** EDAX of bent fluorophlogopite mica crystals. Mass ratios Al/Si = 0.51, Mg/Si = 0.63

Table 10.8. Properties of the best machinable glass ceramics with bent fluorophlogopite mica crystals composition (mol%): M_2O 5–12, (Na_2O 0–8, K_2O 0–6) MgO 8–17, Al_2O_3 21–36, SiO_2 34–60, F 1–7 [934]

Linear thermal expansion coefficient (20–400 °C)	$50–75 \cdot 10^{-7}\,K^{-1}$
Bending strength	up to 110 MPa
Hydrolytic grade	1–2
Basic grade	1–3
Machinability	excellent

precipitated in controlled phase separation (in the way described in paragraph 10.4.4.3) from a specifically modified base glass of the following composition (% by weight): SiO_2 43–50%, Al_2O_3 26–30%, MgO 11–15%, Na_2O/K_2O 7–10.5%, F^- 3.3–4.8%, Cl^- 0.01–0.6%, CaO 0.1–3%, P_2O_5 0.1–5%. The two advantageous properties, good machinability and high toughness, are combined in the resulting glass ceramics. These properties and also the full biocompatibility, good polishing capacity, very low surface roughness and dyability enable the material to be used for dental protheses. Fig. 10.58 a shows a typical cordierite crystallite in a technical glass as a glass defect. For comparison, the cordierite mica crystallization in the described glass ceramic is shown in Fig. 10.58 b. The close interrelations also expressed in the illustration (transformation from curved phlogopite mica crystal to cordierite crystal). Prominent properties of the biocompatible, machinable mica cordierite glass ceramic are given in Table 10.9.

All three types of glass ceramics described here (see Tables 10.7 to 10.9) and also their sub-types (which fulfill special demands) can be used in science and technology as well as in medicine as biocompatible material.

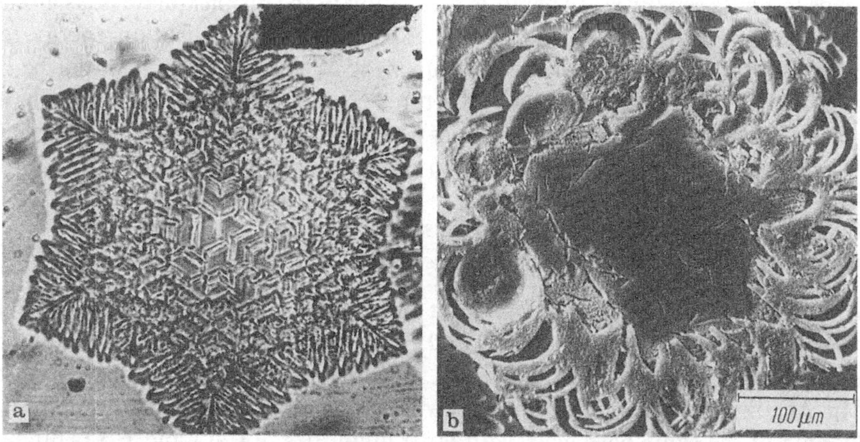

Fig. 10.58. a Cordierite crystal as a defect in a technical glass (light micrograph); **b** cordierite crystal in the new machinable glass-ceramic with bent phlogopite mica crystals (scanning electron micrograph)

Table 10.9. Properties of the machinable mica (phlogopite) cordierite glass ceramics [934]

Density	2.5 g/cm^{-3}
Linear thermal expansion coefficient (20–400 °C)	75–125 · 10^{-7} K^{-1}
Bending strength	90–140 MPa
E-modulus	70 GPa
Compressive strength	450 MPa
Breaking toughness K_{IC}	up to 1.9 Pa · $m^{1/2}$
Vickers hardness	up to 8000 MPa
Hydrolytic grade	1–2
Acidic grade	3
Basic grade	1
Surface roughness	0.1 m
Machinability	good to very good

10.4.4.4 Development of Bioactive Glass Ceramics

Machinable, Bioactive Glass Ceramics as well as the Kinetic Processes Involved in Production

The machinable glass ceramics described in Section 10.4.4.3 have proved to be biocompatible which means they cause no disruption in living cells. An implant material becomes bioactive when it contains apatite crystals, the natural building substance of the bone [849, 950, 961]. Therefore the problem was to simultaneously precipitate from a glass melt, mica (phlogopite) and apatite crystals in variable proportions. The foundation and prerequisite for a controlled crystallization in a glass is a controlled phase separation. The formation of stable molecular building units and their concentration in particular regions leads to phase separation. Therefore nucleation in the droplet phase is, as a rule, considerably facilitated: since the entire amount of the activation energy required to form a nucleus of the critical size is not necessary as it would be in a base glass with a homogeneous structure. The immiscibility structure of base glasses in the machinable glass ceramics described so far has been demonstrated. Which further steps are needed now in order to achieve bioactivity?

CaO and P_2O_5 Doping of the Base Glass from which Fluorophlogopite Mica Crystals are Precipitated

If the base glass from which flat sheet phlogopite crystals are precipitated is doped with small quantities of CaO and P_2O_5, the immiscibility structure of the initial glass does not change. Figure 10.59 shows a silicate droplet phase rich in Mg, Al, alkali metal and fluorine ions whereas the matrix phase is rich in SiO_2. Doping of the same base glass with large quantities of CaO and P_2O_5 fundamentally changes the immiscibility behavior: the original silicate droplet phase disappears and relatively large droplet regions rich in P_2O_5 are formed in the silicate matrix phase (see Fig. 10.60). If the two base glasses, to which CaO and P_2O_5 were added, are subjected to a special temperature treatment, only the well known flat phlogopite crystallites are precipitated from the glass, with small additions of CaO/P_2O_5. Only apatite crystals are precipitated in the glass with larger quantities of CaO and P_2O_5, phlogopite crystals are not simultaneously formed. This first negative result can be explained by the different kinetics of the immiscibility processes in both glasses (see Fig.10.61).

In case A, the small CaO/P_2O_5 quantities are mostly homogeneously absorbed by the droplet phase without any considerable change in the known crystallization behavior. In case B with the large CaO/P_2O_5 addition, a pure phosphate droplet phase rich in Ca^{2+}, Mg^{2+}, Al^{3+}, K^+/Na^+ and F^- is formed in the solid base glass. Since the field strength, Z/a (Z is the valence and a the distance between anion and cation), of P^{5+} (2.1) is higher than that of Si^{4+} (1.57), a higher shielding will result from the presence of P^{5+}. A greater portion

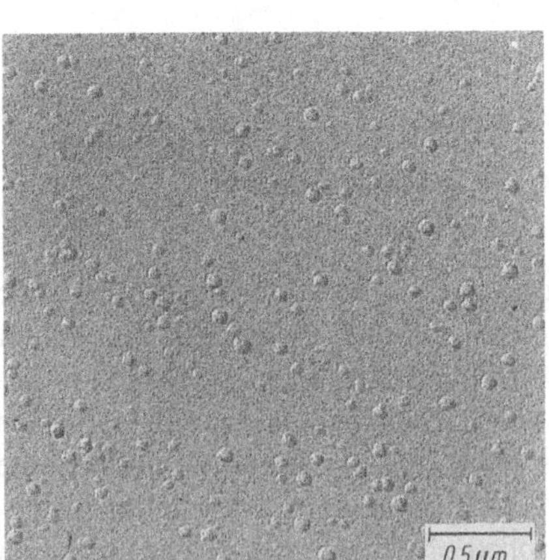

Fig. 10.59. Immiscibility structure of the base glass doped with a small quantity of CaO/P_2O_5 after rapid quenching. Small silicate droplet regions

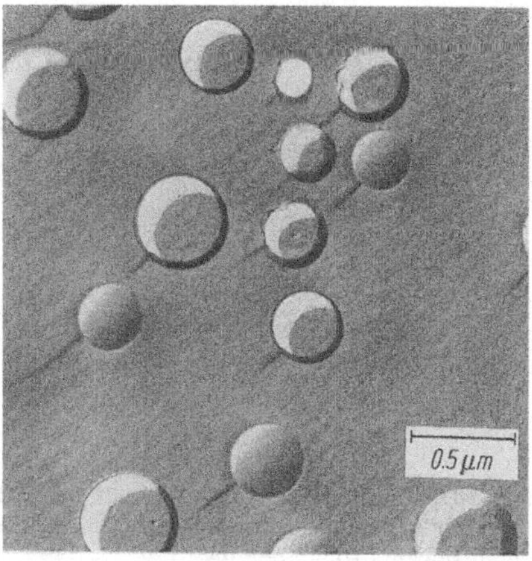

Fig. 10.60. Immiscibility structure of a base glass doped with a large quantity of CaO/P_2O_5. Large phosphate droplet regions; the small silicate droplet regions have disappeared (replica method)

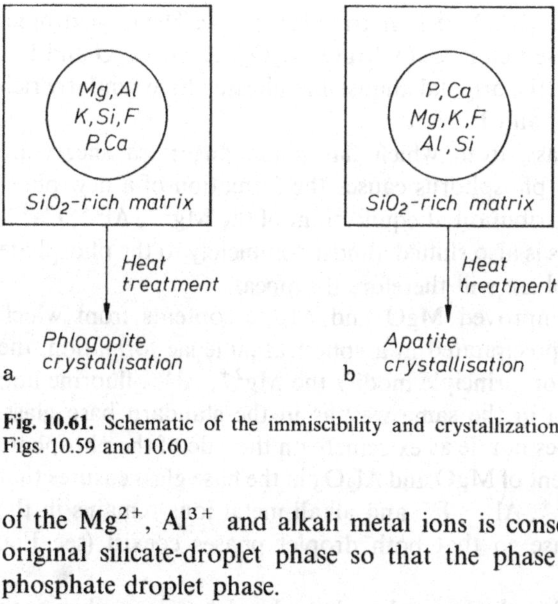

Fig. 10.61. Schematic of the immiscibility and crystallization behavior of glasses according to Figs. 10.59 and 10.60

of the Mg^{2+}, Al^{3+} and alkali metal ions is consequently extracted from the original silicate-droplet phase so that the phase disappears in favor of the phosphate droplet phase.

An increased fluorine content in the base glass favors this process due to a strong viscosity decrease. When this base glass is heat treated, only apatite will crystallize. The prerequisites for a simultaneous precipitation of phlogopite are lost.

CaO (10–19 mol%) and P_2O_5 (2–9 mol%) Doping of the Base Glass from which Fluorophlogopite Mica Crystals Are Precipitated in Doping Within Spherical Lamellae Formations

The addition doping within CaO and P_2O_5 of the base glass with the higher Mg^{2+} and Al^{3+} ion content (see Section 10.4.4.3) leads primarily to a three phase glass (see Fig. 10.62). Two droplet phases different in dimension and

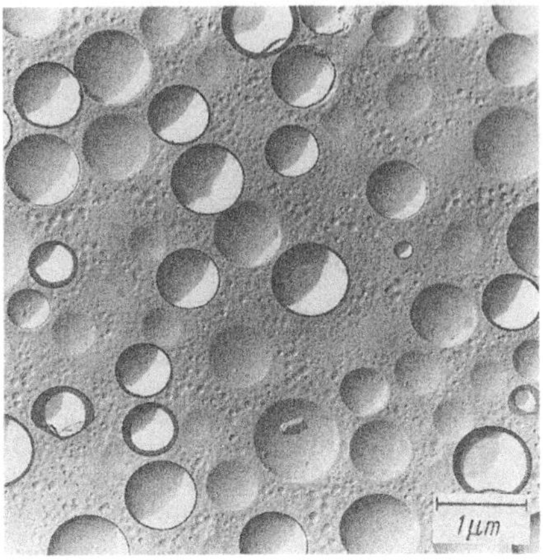

Fig. 10.62. Base glass with three phases. The large phosphate glass droplets coexist with the small silicate droplet regions

composition are embedded in a SiO_2 rich matrix glass phase. The large droplet regions are a phosphate phase rich in CaO, MgO, Al_2O_3, K_2O/Na_2O and F^-. The small droplet regions are the original immiscible silicate phase similarly rich in MgO, Al_2O_3, Na_2O/K_2O and F^-.

In a standard base glass, from which flat phlogopite mica sheets are precipitated, the presence of phosphorus causes the formation of a new phosphate droplet phase. The distribution at equilibrium of the Mg^{2+}, Al^{3+}, Ca^{2+}, fluoride and alkali metal ions is also shifted almost completely to the phosphate phase. The silicate droplet phase will therefore disappear.

In a base glass with improved MgO and Al_2O_3 contents from which phlogopite mica sheets are precipitated in a spherical lamellae formation, the addition of CaO/P_2O_5 will on principle modify the Mg^{2+}, Al^{3+}, fluorine and alkali metal ion distribution in the same way as in the standard base glass. However the equilibrium does not lie as extremely on the side of the phosphate droplets. The increased content of MgO and Al_2O_3 in the base glass ensures that a sufficient quantity of Mg^{2+}, Al^{3+}, F^- and alkali metal ions remains in the original silicate-droplet phase so that both droplet phases coexist (see Fig. 10.62).

Fluoroapatite is formed in the large phosphate droplet regions when such a glass is heat treated. Since the small silicate droplet regions still contain Mg^{2+}, Al^{3+}, alkali metal and fluorine ions, phlogopite mica crystals are also precipitated. The mica crystals are however formed as flat sheets again because some Mg^{2+}, Al^{3+}, alkali metal and fluorine ions are withdrawn from the silicate droplet phase by the higher field strength of P^{5+}. Figure 10.63 shows how the large P_2O_5-droplet regions start transforming into apatite crystals. Surfaces of apatite crystals can already be recognized in the droplets. Figure 10.64 shows apatite crystals beside fluorophlogopite ones. The ratio of both crystal phases as well as the ratio of the residual glass phase to the crystal phases can be adjusted. The formation of apatite crystals reduces the machinability of the product but it remains sufficient. Consequently a fundamental partial goal has been reached

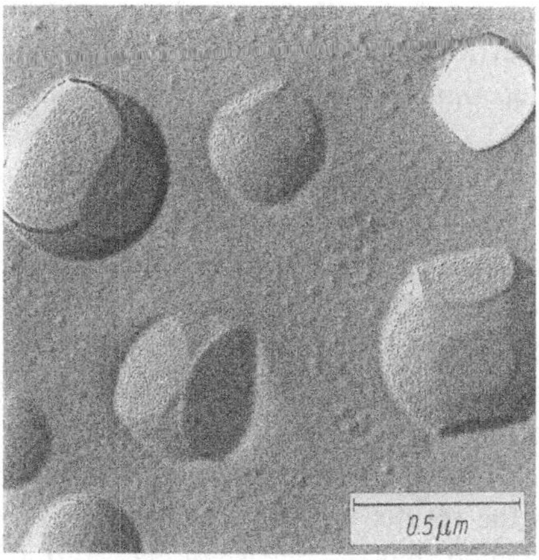

Fig. 10.63. The large phosphate glass droplets start transforming into crystalline apatite with annealing (scanning electron micrograph)

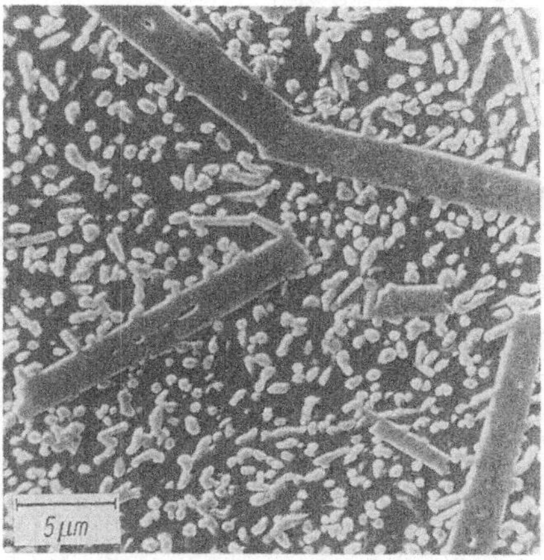

Fig. 10.64. Scanning electron micrograph of a machinable bioactive glass ceramic. After thermal treatment of the glass of Fig. 10.63 at 1050 °C, phlogopite and apatite crystals form side by side

with a base glass of the following composition (mol%):

SiO_2 19–54%	F^- 3–23%
Al_2O_3 8–5%	CaO 10–34%
MgO 2–21%	P_2O_5 2–10%
Na_2O/K_2O 3–8%	

Chemical and Mechanical Properties

The behavior of the surface of bioglass ceramics towards chemical agents is of particular interest concerning their applications. The water resistance of the bioglass ceramic which was optimized for medical applications was measured according to the standard procedure and a hydrolytic grade of 1 to 2 was obtained. Glass ceramics which are machinable and have a high apatite content have the hydrolytic grade 2 whereas machinable glass ceramics with a low apatite content but a high phlogopite content have the grade 3. Two internationally known bioglass ceramics which are however not machinable have the hydrolytic grade 5. The high p_H grade obtained according to standard procedures amounts to 1, the acidic grade to 3. These two properties are however of little importance for medical applications since the test procedures cannot be compared to body conditions.

The Ringer solution [950] is a model liquid which resembles the human body fluids (aqueous solution which contains Na^+, Ca^{2+}, Cl^- and CO_3^{2-} ions). After cooking the developed machinable bioglass ceramic for 40 hours in the Ringer solution, the surface roughness amounts to only 0.5 μm. A test for an increase of the Na^+ and K^+ ions concentrations in the Ringer solution was negative. The treatment of bioglass ceramic powder with tris-buffer solution (2-amino-2 hydroxymethyl-1,3 propanediol-HCl–H_2O mixture) is another very meaningful method for the determination of the ion release of an implant under body-like conditions. Table 10.10 shows the results.

Further investigations have demonstrated that the release of Na^+ and K^+ has almost come to a standstill after 14 days and definitely after 4 weeks. The properties of the machinable glass ceramics containing apatite can be varied to a wide extent by changing the proportion of apatite to mica crystals. The values obtained with a drill test range from machinable to excellently machinable. The mechanical bending strength is measured on a glass-ceramic sample with a three-point support. It ranges from 140 MPa with glass-ceramic samples having a low apatite content to 220 MPa with samples having a high apatite content. In order to keep the differences in the measured values low, the surface of the test body was systematically attacked before the measurement by rotating it in a drum in the presence of SiC grains (see Table 10.11 for mechanical properties).

Biocompatibility and Bioactivity

The following well known laboratory tests: the INT reduction test, the LDH release test and in particular culture cell tests have given very positive results. In the latter one, the materials were added as a dust to the cell cultures and in concentrations from 0.4 to 3.2 mg/ml. Neither the glass-ceramic tested nor the

Table 10.10. Ion Emission of bioactive machinable glass ceramics after treatment with tris-buffer solutions. Experimental conditions: pH of the Buffer solution 7.4; experimental temperature 37 °C; 3 g of glass ceramic powder of grain size 0.16–0.315 mm; 100 ml solution; continuously stirred sample. t = reaction time. The ion concentrations obtained are given [934]

Bioglass ceramic	t(h)	ion concentration (mg/l) Na^+	K^+	Al^{3+}
High apatite content	168	1.0	0.7	—
(apatite 40 vol%	336	1.1	0.7	—
phlogopite 20 vol%)	672	1.8	0.7	—
High phlogopite content	168	1.8	2.1	—
(apatite 20 vol%	336	1.9	2.1	—
phlogopite 70 vol%)	672	2.1	2.1	0.08

Table 10.11. Mechanical properties of the machinable bioactive glass ceramics [934]

Density	2.8 g/cm^3
Thermal expansion coefficient (20–300 °C)	$80–120 \cdot 10^{-7}$ K^{-1}
Bending strength	140–220 MPa
E modulus	77–88 GPa
Compressive strength	500 MPa
Breaking toughness	0.5–1.0 MPa·m$^{1/2}$
Vickers hardness	up to 5000 MPa (500 H V 10)

reference substance (sintered corundum) influences cell reproduction. The glass ceramic can therefore be considered as biocompatible.

Gummel and Schulze [962, 963] carried out some experiments on animals. Double implants of glass ceramic and sintered corundum were systematically carried out. Sintered corundum is known to be absolutely biocompatible.

Bioactivity was evaluated with a series of implants in the head of the lower leg (head of tibia) of guinea pigs. The heads of the tibias were investigated after leaving the implants for 8, 12, 16 weeks and up to 2 years. Mechanical ejection tests proved that the shear strength of the implant bone intergrowth was 8 times larger for glass-ceramic implants than for sintered corundum ones. It reached maximum values of 5 N/mm. Bone remains could still be observed by electron microscopy on the glass-ceramic implants after their removal.

Light and electron microscopy give further insight into the situation. Figure 10.65 is the micrograph of a direct bone-glass-ceramic intergrowth. Already after 8 weeks, the bone substance (dyed with alizarin complexon) grows directly into the glass-ceramic implant. Bone cells and a blood vessel are even observed in direct contact with the implant. Optimal bioactivity has therefore been achieved. For a comparison, Fig. 10.66 is the micrograph of a bone-sintered corundum boundary zone also dyed with alizarin complexon. Even after 16 weeks no virtual intergrowth with the bone is observed. The gap between the bone and the implant was filled out with a binder tissue layer. The implant can consequently therefore be pulled out like a cork out of a bottle. Consequently sintered corundum is biocompatible but not bioactive.

Figure 10.67 shows the boundary zone between glass ceramic and bone. In order to achieve an intergrowth between the bone and implant, new apatite crystals must be formed (for which solution and ion diffusion processes in the implant surface are prerequisites). The solution processes must eventually stop at a certain depth because an intergrowth zone which is too broad can under continuous stress induce the implant to come loose. The maximum intergrowth zone for the new machinable and bioactive glass ceramic was measured to be 5–10 μm wide (see Fig. 10.67). With implants made from a foreign bioactive material, it was 100 to 150 μm wide (this does not imply that the reaction has come to an end by the time the measurement was taken).

Figure 10.68 shows the electron microprobe analysis result for the border layer between bone and implant. The X-ray intensity profiles of $Si_{K\alpha}$, $Ca_{K\alpha}$, $P_{K\alpha}$ and $K_{K\alpha}$ radiation allow the following conclusions to be drawn:

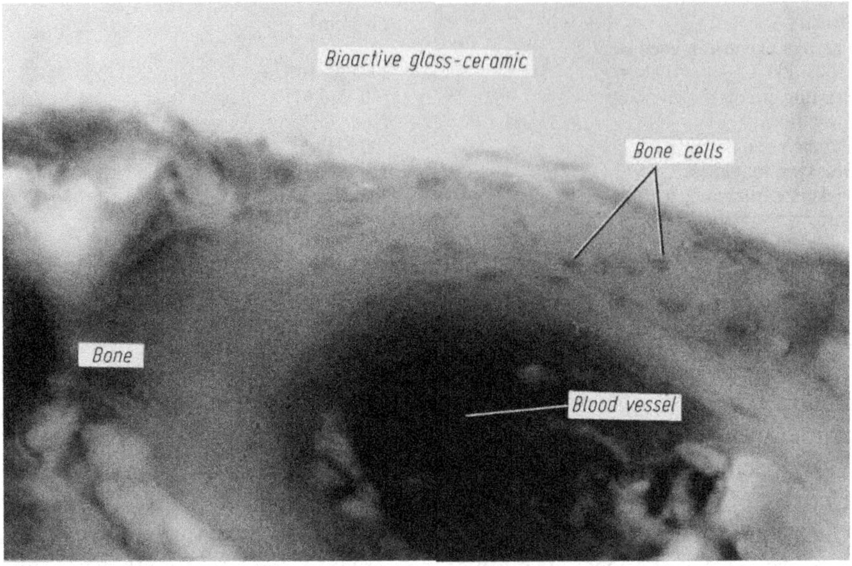

Fig. 10.65. Light micrograph of a bone-glass-ceramic intergrowth after 8 weeks in an animal bone dyed with alizarin complexon. Direct intergrowth is recognizable by the finger-shaped growth of the bone into the glass ceramic. More over-bone trabeculae and one blood vessel can be seen close to the glass ceramic

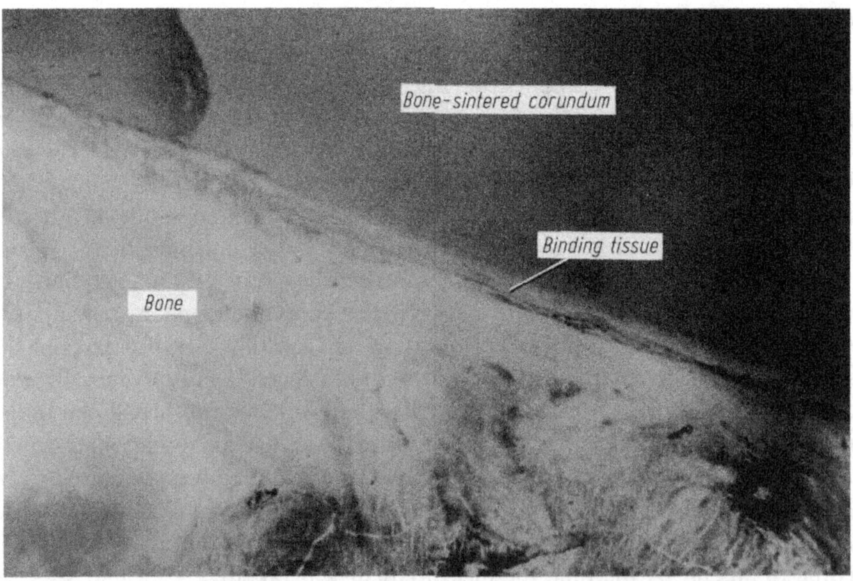

Fig. 10.66. Light micrograph of a bone-sintered corundum intergrowth after 16 weeks in an animal, bone dyed with alizarin complexon. The implant is encapsulated by a binding tissue (collagen layer and although it is biocompatible there is no direct intergrowth (Gummel [962])

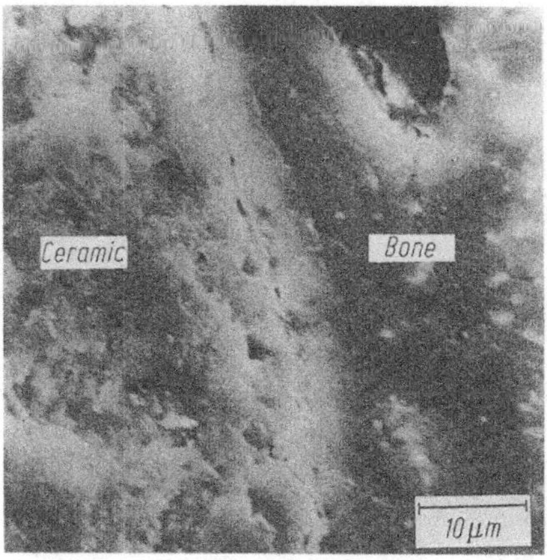

Fig. 10.67. Scanning electron micrograph of the boundary zone of bioglass ceramic and bone (of a guinea pig) after 16 weeks. A very good intergrowth zone of only ca. 5–10 µm is appearing

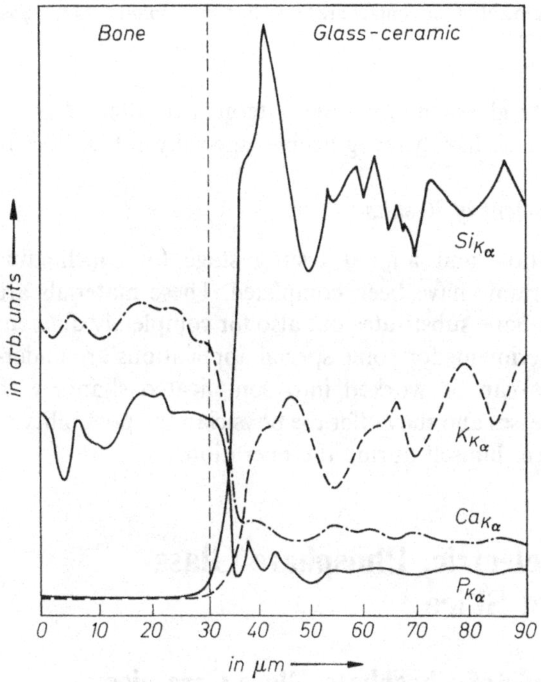

Fig. 10.68. Electron microprobe analysis of the intergrowth region (according to Fig. 10.67) X-ray intensity profiles in the border layer for $Si_{K\alpha}$, $K_{K\alpha}$, $Ca_{K\alpha}$ and $P_{K\alpha}$ radiation

- the glass matrix is insignificantly washed out of the glass ceramic surface. However SiO_2 does not diffuse into the bone.
- the potassium ions are washed out of the glass ceramic surface up to a depth of about 5 µm. The width of the intergrowth zone is therefore confirmed.
- the Ca and P content is higher in the intergrowth zone than in the glass ceramic. This suggests the formation of apatite (which leads to a firm intergrowth between the bioglass ceramic and the bone). A further growth of the bone crystals from the implant's substance removed from the surface can not be excluded.

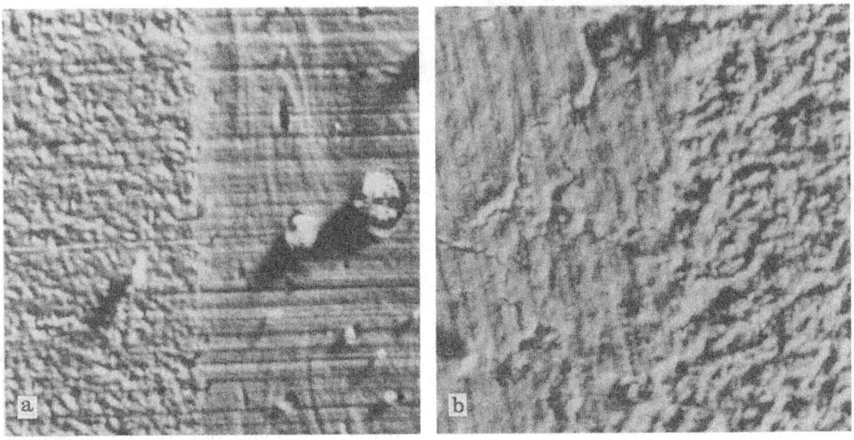

Fig. 10.69. Scanning electron micrograph of a glass-ceramic-bone intergrowth in an animal. **a** After 54 weeks; **b** after 71 weeks. No enlargement of the contact area by a further dissolution of the glass ceramic, but rather deeper intergrowth

Figure 10.69 a and b show the glass-ceramic-bone intergrowth after 54 or 71 weeks. The original contact zone is hardly recognizable especially in Fig. 10.69 b.

Interim Estimation and Assessment of Results

The laboratory experimentation and animal testing stage for machinable and highly bioactive glass ceramic have been completed. These materials are suitable as bone implants and bone substitutes but also for completely different applications in medicine. Experiments for some special applications are underway. The new glass ceramics can be worked into complicated shapes with normal metal machining processes and they offer the physician the possibility of changing them and fitting them himself during the operation.

10.5 Bioactive, Piezoelectric, Phosphate Glass Ceramics Free of Silica

10.5.1 Development Trends of Phosphate Glass Ceramics

Biocompatible or bioactive silicate bioglass ceramics have been successfully developed until now and have brought new healing possibilities and considerable progress in medicine. However the long-term reactions or interactions due to the contact of silicate compounds with human tissues are not yet fully known. For this reason, a pure biophosphate glass ceramic which would approximate more closely the chemical composition of the bone than all of the bioglass ceramics known until now, could be considerably more advantageous. There are preliminary experiments which point in this direction [964, 965], however they have not as yet been successful since almost all phosphate glasses have a relativcly homogeneous base glass structure. Consequently they do not phase separate which is a prerequisite for a controlled crystallization. An uncontrolled crystallization in the whole volume or at the surface will not yield the particular properties of the "vitrocerams".

The natural, structure-related barriers to the development of phosphate glass ceramics can in certain cases be overcome with the glass powder-sinter method [942]. A phosphate glass powder of a particular composition and grain size (which approximately corresponds to the size of the droplet shaped immiscibility regions in silicate glasses) is annealed, sintered and brought to crystallize. Crystallization will occur from the grain surface into the glass grain. By proper choice of the grain size, a regular crystallization and crystallite size can be achieved in the sintered product. The product is however porous and has a low mechanical strength when compared to massive glass ceramics. Such materials only find limited use in medicine. This applies also to sintered products from pure crystalline apatite [941].

The main goal of our work is therefore to find a way of transforming pure phosphate glasses into glass ceramics by means of a controlled crystallization in which apatite must be present to ensure bioactivity.

10.5.2 Structure and Crystallization Behavior of Phosphate Glasses

As a rule, silicate glasses contain a three-dimensional $[SiO_4]$ tetrahedron network. The introduction of network modifiers (oxides) leads to the break-up of oxygen bridges and to the deposition of the larger network modifier ions into the larger network voids formed (see Fig. 10.70 a). The network modifier ions are

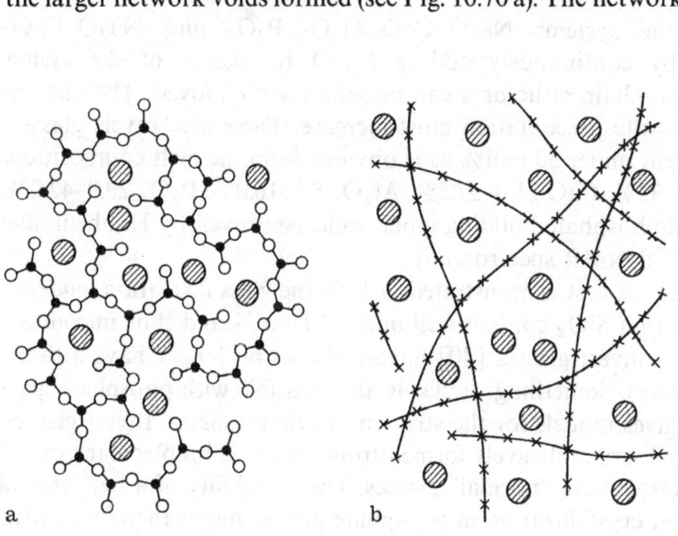

Fig. 10.70. Models for the structure of: **a** silicate glass with a network structure; ● = Si, ○ = 0, ⊘ = Na (Zachariasen and Warren); **b** a phosphate base glass with chain structure; x–x = PO_4–PO_4, ⊘ = Na

however not always statistically distributed over the three-dimensional network. Due to the different volume demands of the two different molecular building units, phase separation will occur.

The described phenomenon can be compared on one hand to the formation of a mixed crystal (in analogy to a homogeneous glass) and on the other hand to a eutectic crystal mixture (in analogy to a glass with microphase separation). Microphase separation in a glass is an important prerequisite for a controlled crystallization. Phosphate glasses have a different structure and therefore also a different crystallization behavior than silicate glasses.

The structure of acidic phosphate glasses consists of a network too. Only three corners of a $[PO_4]$ tetrahedron are however crosslinked. Basic phosphate glasses, in particular those with a ratio of network modifier to P_2O_5 greater than one, have a chain structure [400, 401, 404] (see Fig. 10.70 b). The spaces between the chains are also so large that another type of molecular building unit can find a place there. A trend towards phase separation is therefore not observed and the usual way of inducing a controlled crystallization cannot be used with these glasses. New methods must be found.

10.5.3 Development of Pure Biophosphate Glass Ceramics [966, 967]

Glasses of the System $CaO-Al_2O_3-P_2O_5$

Melts of the ternary system $CaO–Al_2O_3–P_2O_5$ [968] can be solidified as glasses in limited composition regions. ^{31}P-NMR analyses by Haubenreißer et al. [969] have clearly demonstrated the chain structure of such glasses. Heat treatment will induce surface crystallization (uncontrolled crystallization) in which the $AlPO_4$, $Ca(PO_3)_2$ and $Ca_2P_2O_7$ crystal phases are formed. Apatite is not precipitated.

Glasses of the systems $Na_2O–CaO–Al_2O_3–P_2O_5$ and $Na_2O–CaO–Al_2O_3–P_2O_5–F$ by continuously adding Na_2O to glasses of the system $CaO–Al_2O_3–P_2O_5$, chain structures can be selectively removed. The chances for an eventual apatite precipitation must increase. These are "invert glasses" with a P_2O_5 content under 50 mol% as is obvious from the melt compositions (mol%): Na_2O 11–32%, CaO 21.0–37.5%, Al_2O_3 8.3–16.0%, P_2O_5 29.9–42.0%. Only mono and diphosphate building units could be proven by Haubenreißer et al. [969] with ^{31}P-NMR spectroscopy.

Trapp and Stevels first demonstrated in 1959 the glass formation tendency of silicate melts with a SiO_2 content well under 50 mol% and thus introduced the new concept of invert glasses [40] (invert glasses no longer have a three-dimensional network). Something similar is also possible with phosphate glass melts. Fig. 10.71 gives models for the structure of these glasses. Invert glasses, that is glasses which are exclusively formed from very small molecular groups, crystallize much faster than "normal" glasses. The possibility of a new way of inducing controlled crystallization in phosphate glasses might therefore exist.

After a special thermal treatment, the following crystal phases can be precipitated from phosphate invert glasses (however only with an uncontrolled

Fig. 10.71. Models for the structure of a phosphate invert glass. triangles: $[PO_4]$ tetrahedra; circles: cations (Na^+, Ca^{2+}; Al^{3+})

crystallization): the cristobalite isotope $AlPO_4$ modification, the tridymite isotope $AlPO_4$ modification, β-$Ca_2P_2O_7$, a "complex phosphate" unknown until now, a diphosphate and apatite. The precipitation of an hydroxylapatite phase can be achieved.

The additional introduction of fluorine components in melts of the system Na_2O–CaO–Al_2O_3–P_2O_5 (mol%) (Na_2O 24.1–26.1%, CaO 21.7–25.5%, Al_2O_3 11.0–12.2%, P_2O_5 31.6–33.6%, F 6.1–10.9%) considerably changes the crystallization process. The following crystal phases will appear with an uncontrolled crystallization: fluoroapatite, low-temperature quartz isotopes $AlPO_4$ (Berlinite) [970] which demonstrates piezoelectric properties, and the complex phosphate phase. ^{27}Al-NMR analyses of Haubenreißer et al. [969] showed that the ratio of tetra- to hexa-coordinate Al^{3+} is 3:1. The structure suggested in Fig. 10.72 for the complex phosphate was deduced from these results, the ^{31}P-NMR analyses and from the chemical analysis of the glasses. Such a structure can in fact be formed totally free of stress. An exact structure determination remains necessary. Obtaining somehow fluoroapatite and piezoelectric berlinite crystal phases represent positive results even though at present no controlled crystallization is possible with such a base glass.

Glasses of the Systems Na_2O–CaO–Al_2O_3–P_2O_5–FeO/Fe_2O_3 and Na_2O–CaO–Al_2O_3–P_2O_5–F–FeO/Fe_2O_3

A controlled crystallization is not possible for glasses with a phosphate chain structure. Phosphate glasses with an invert glass structure (exclusively with mono or diphosphate building units), will on heat treatment nucleate and crystallize faster. In the corresponding melt diagrams they lie extremely close to the glass formation border. If such types of melts from the system Na_2O–CaO–Al_2O_3–P_2O_5 are oversaturated with FeO/Fe_2O_3, an effect is induced with cooling or subsequent heat treatment which resembles a phase separation but which has nothing to do with that. The chosen base glass compositions are (mol%): Na_2O 24.1–25.7%, CaO 24.9–31.2%, Al_2O_3 9.9–14.5%, P_2O_5 30.7–33.1%, FeO/Fe_2O_3 1.2–6.2%.

Figure 10.73 is an electron micrograph of a replica film of a chosen glass from the system Na_2O–CaO–Al_2O_3–P_2O_5–FeO/Fe_2O_3. A comparison of the

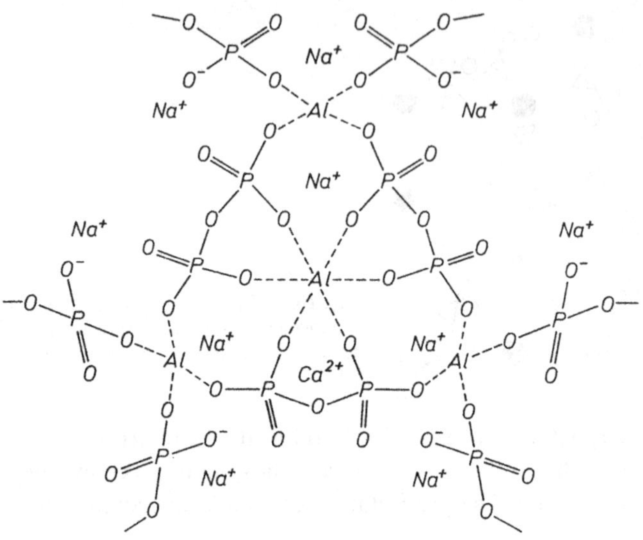

Fig. 10.72. Proposal for the structure of the crystalline "complex phosphate"

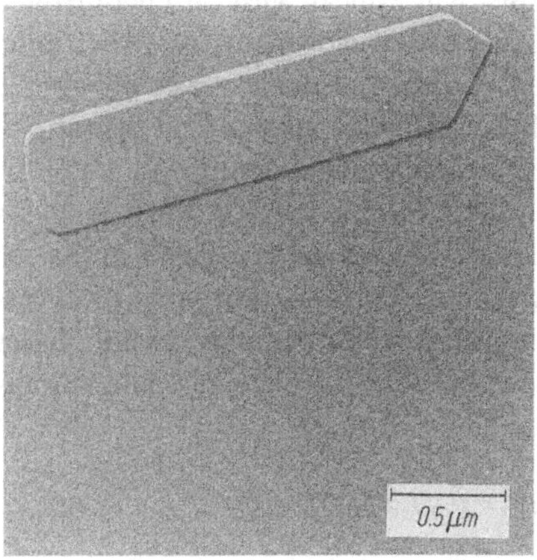

Fig. 10.73. Electron micrograph of a chosen glass from the system Na_2O–CaO–Al_2O_3–P_2O_5–FeO/Fe_2O_3 (replica method). The glass has a homogeneous structure after a rapid quenching; top: MoO_3 crystal test surface for comparison

sample fracture surface with the MoO_3 crystal reference surface demonstrates the clear homogeneous structure of the glass. Figure 10.74 was taken after a thermal treatment of the same glass: multiple, uniformly sized, in general ellipsoidal objects have appeared throughout the whole glass volume (similar to a phase separation). Furthermore each ellipsoidal region clearly contains a nucleus (see especially Fig. 10.75). The structures observed in Figs 10.74 and 10.75 are obtained after removal of the supersaturation of iron oxides in the base glass. The nucleus of the ellipsoidal regions is crystalline iron phosphate. As soon as this primary nucleus has reached the critical size, it grows further by epitaxial interactions with hydroxylapatite. If the lattice constants of apatite and from phosphates are compared, at least two constants of iron phosphates do not

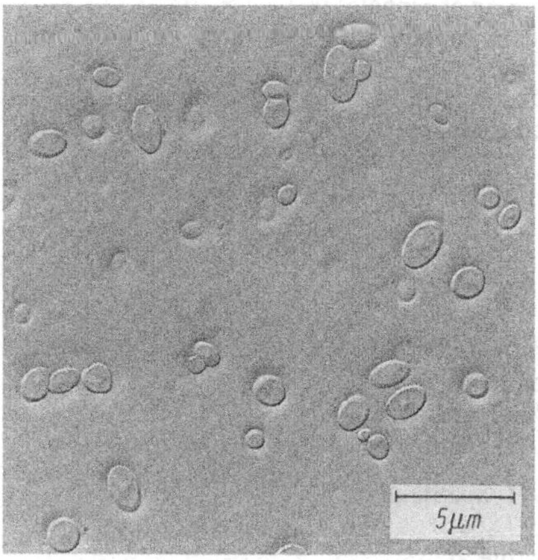

Fig. 10.74. (glass from Fig. 10.73) Oversaturation occurences predominate with annealing; spontaneous nucleation and crystallization occur evenly in the whole volume

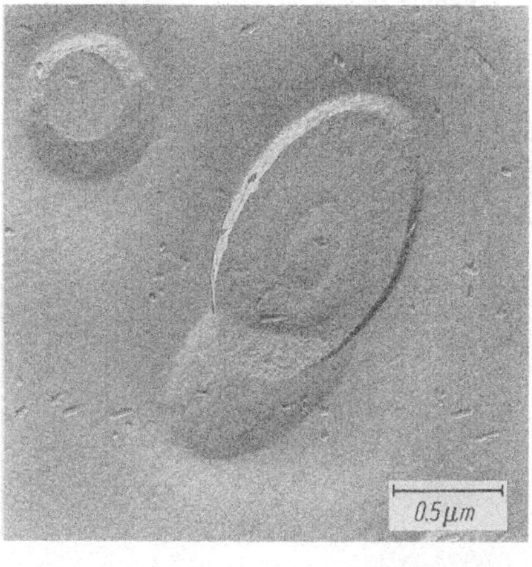

Fig. 10.75. Single crystallite from Fig. 10.74. An iron phosphate primary nucleus obviously grows by an epitaxial interaction with apatite

deviate by more than 15% from two constants of apatite (see Table 10.12). Epitaxial interactions are therefore possible on principle. Iron apatites might also be formed as intermediates.

An X-ray scanning photograph with $Ca_{K\alpha}$ radiation shows an enrichment in Ca in the ellipsoidal regions (see Fig. 10.76). X-ray diffraction patterns clearly demonstrate that the ellipsoidal regions consist of crystalline hydroxylapatite. Progressive heat treatment in such glasses provoke the precipitation of the crystalline complex phosphate as well as of tridymite or cristobalite $AlPO_4$ crystal phases. Figure 10.77 shows the tridymite $AlPO_4$ crystal phase. The extraordinarily large similarity with tridymite in silicate glasses is interesting. It can also be seen from the figure that, as the temperature increases, in the cristobalite modification of $AlPO_4$, the axes of the star-shaped crystal surround themselves with right-angled interdigitations. The large similarity with cristobalite in silicate glasses is surprising, too.

Table 10.12. Lattice parameters of a few iron-containing phosphates in comparison with the ones of fluoroapatite and hydroxylapatite

Crystal phase	Lattice parameter (Å)	Deviation compared to apatite
Fluoroapatite $Ca_5[F/(PO_4)_3]$	$a = 9.37$ $c = 6.88$	
Hydroxylapatite $Ca_5[OH/(PO_4)_3]$	$a = 9.43$ $c = 6.88$	
$Fe_3(PO_4)_3$	$a = 8.80$	− 6.1 to − 6.7% (1 · apatite a)
	$b = 11.50$	− 16.4% (2 · apatite c)
	$c = 6.25$	− 9.2% (1 · apatite c)
$NaFeP_2O_7$ I	$a = 7.11$	+ 3.3% (1 · apatite c)
	$b = 10.03$	+ 6.3 to 7.8% (1 · apatite a)
	$c = 8.09$	− 13.7 to − 14.3% (1 · apatite a)
$NaFeP_2O_7$ II	$a = 7.33$	+ 6.5% (1 · apatite c)
	$b = 7.90$	+ 14.8% (1 · apatite c)
	$c = 9.57$	+ 1.5 to 2.1% (1 · apatite a)
$Ca_xFe_y(PO_4)_6O_z$ [a]	$a = 9.39$	+ 0.2 to 0.5% (1 · apatite a)
	$c = 6.90$	+ 0.3% (1 · apatite c)

[a] x 9.5, y 0.5, z 1.25

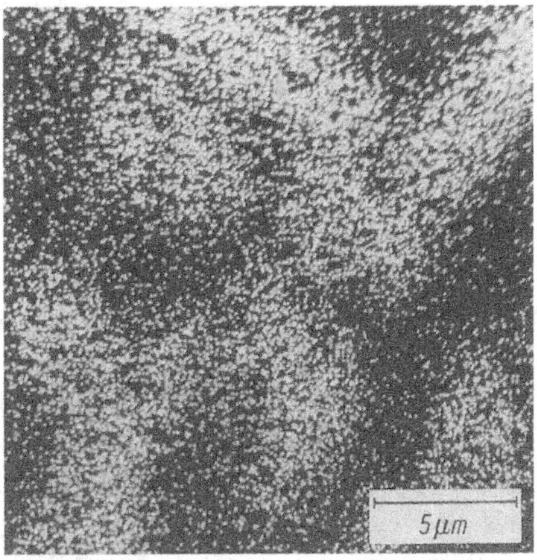

Fig. 10.76. $Ca_{K\alpha}$ scanning X ray of the glass in Fig. 10.74. The aggregation of the Ca^{2+} ions (white points) in the ellipsoidal regions is clearly shown

The precipitation of the low-temperature quartz $AlPO_4$ form (berlinite) with piezoelectric properties would be of considerably greater importance for the production of an optimal bioactivity of the phosphate glass ceramics than the precipitation of cristobalite or tridymite modifications [970]. Micro currents are known to enormously promote the healing of bone fractures [971, 972]. In

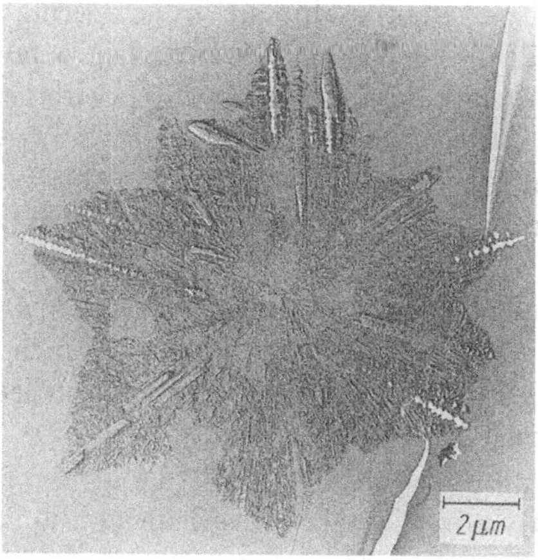

Fig. 10.77. Formation of the $AlPO_4$ crystal modification similar to tridymite in the matrix of the glass shown in Fig. 10.74 with further annealing. Transition to the $AlPO_4$ modification similar to cristobalite can be observed in the right-angled interdigitations at the ends of the main growth axes of the star-shaped crystals (scanning electron micrograph)

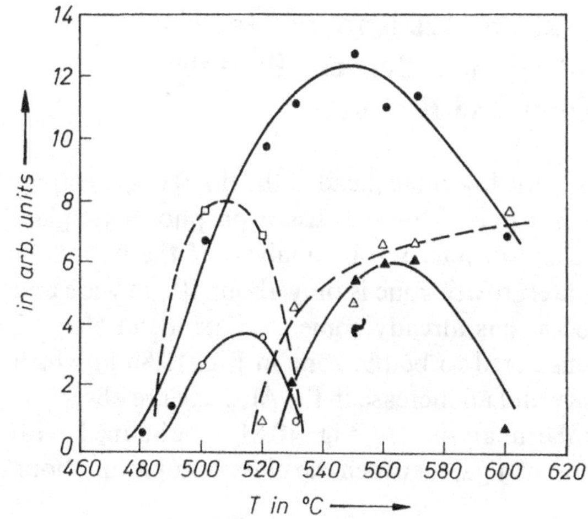

Fig. 10.78. Crystallization sequence of a glass from the system Na_2O–CaO–Al_2O_3–P_2O_5–F–FeO/Fe_2O_3 as a function of the temperature •–• apatite (211), o–o $AlPO_4$ similar to tridymite (1010), ▲–▲ berlinite (102), □–□ (Na, Ca) Fe_2 $(PO_4)_2$ crystal phase of the varulithe type, △–△ complex phosphate

the development of a bioactive, pure phosphate glass ceramic, the precipitation of apatite and the formation of the piezoelectric berlinite phase would complement each other very effectively with regard to bioactivity.

The presence of fluorine in melts from the system Na_2O–CaO–Al_2O_3–P_2O_5–F caused the precipitation of berlinite. This was also attempted with melts from the system Na_2O–CaO–Al_2O_3–P_2O_5–FeO/Fe_2O_3. The goal was in fact reached by adding fluorine in concentrations of 2.4 to 9.4 mol% [966, 967].

The crystallization process induced with annealing of glasses from the system Na_2O–CaO–Al_2O_3–P_2O_5–F–FeO/Fe_2O_3 is shown in Fig. 10.78. At lower temperatures the primary crystal phase consists of varulite type (Na, Ca) $Fe_2(PO_4)_2$ crystals and fluoroapatite as well as the tridymite form of $AlPO_4$. The latter however is decomposed at higher temperatures in favor of berlinite. The crystalline complex phosphate phase is simultaneously formed at about

525 °C. The optimal apatite and berlinite content is obtained by tempering at 560 °C. Consequently controlled crystallization has been achieved in a pure phosphate glass, with an invert glass structure, by supersaturating it with iron oxides. Fluoroapatite and the piezoelectric low-temperature quartz form of $AlPO_4$ lead to a higher bioactivity of the bioglass ceramics. It cannot be overlooked that the other crystalline phosphates contribute to an increase in the bioactivity of glass ceramics.

Later, it was demonstrated, that a fully analogous structure, nucleation and crystallization process can be obtained with this phosphate glass if the melt is supersaturated with either ZrO_2 or $FeO/Fe_2O_3 + ZrO_2$. The primary nucleus in the center of the ellipsoidal region consists of:
- a Na–Zr phosphate, $NaZr_2(PO_4)_3$ with an addition of ZrO_2
- a compound of the brianite type of $Na_2CaFe(PO_4)_2$ [973] with an addition of FeO/Fe_2O_3–ZrO_2.

10.5.4 Animal Experiments at the Academy of Medicine of Dresden on the Intergrowth Between Phosphate Glass-Ceramic Implants and Bones

Preliminary implants in the head of the lower leg (head of the tibia) of guineapigs show the growth of bone substance free of binding tissue on phosphate glass glass ceramics (Fig. 10.79). An electron microprobe analysis of the boundary zone (Fig. 10.80) shows that the intergrowth zone is only about 10 μm wide and that the required dissolving process has already come to a standstill after 12 weeks. The boundary zone is considered to be the zone in Fig. 10.80 in which a sudden drop in the $Ca_{K\alpha}$ intensity and an increase in the $Al_{K\alpha}$ one are observed. A completely different microprobe analysis (see Fig. 10.81) is obtained with a bad intergrowth or the formation of a gap between the glass ceramic and bone tissues.

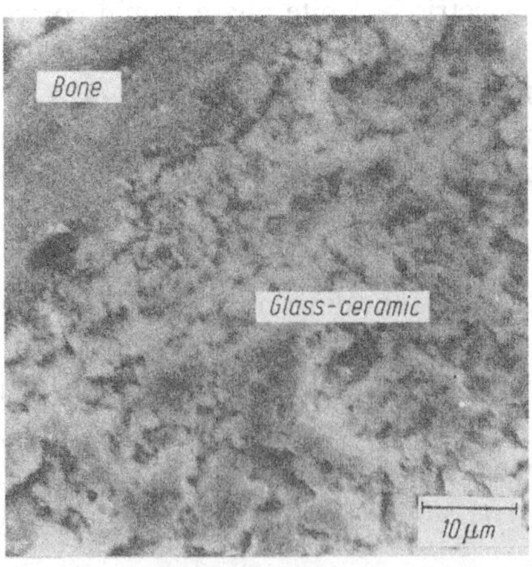

Fig. 10.79. Intergrowth of a pure phosphate glass ceramic and bone free of binding tissue in an animal (scanning electron micrograph)

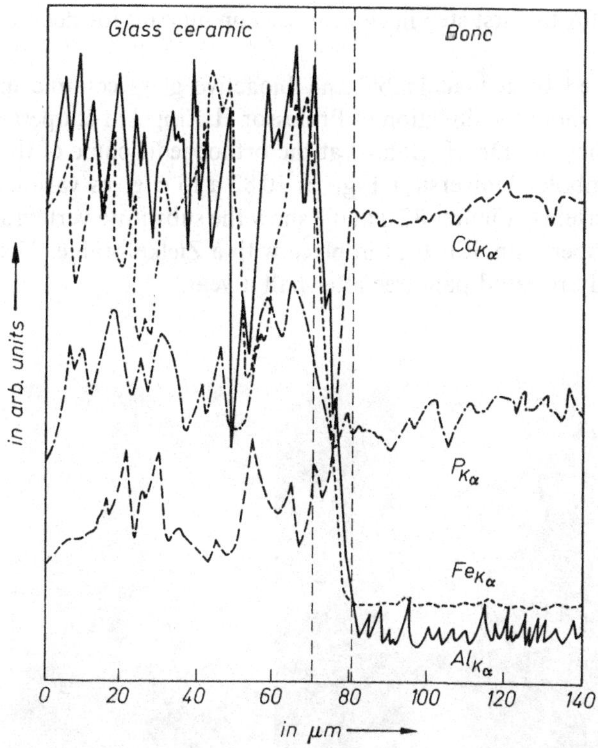

Fig. 10.80. Electron microprobe analysis of the intergrowth region from Fig. 10.79. The intergrowth region is not fundamentally wider in the pure phosphate glass ceramic than in silicate glass ceramics (compare to Fig. 10.68)

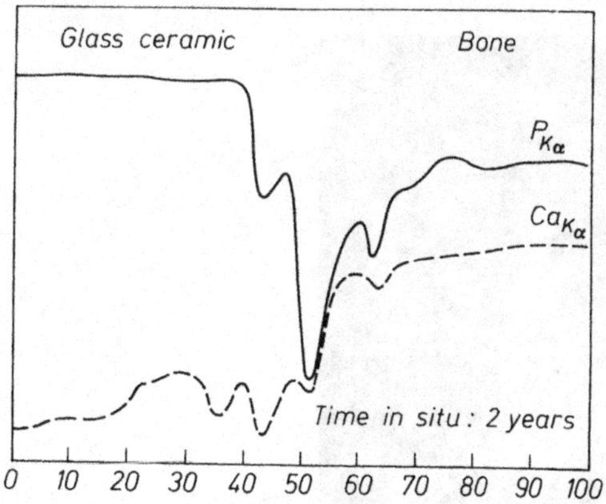

Fig. 10.81. Electron microprobe analysis of a glass-ceramic-bone contact. No direct intergrowth is seen in this case but rather formation of a gap (compare to Fig. 10.68 and 10.80)

10.5.5 Clinical Tests of the New Bioglass Ceramics on Humans

A series of the described bioglass ceramics has been successfully tested by our medical colleagues with animals. Other bioglass ceramics, which are designed for more unusual applications (e.g. in eyes) are still at that stage. Four clinics have been chosen for the first step in clinical tests on humans for definite cases.

- A vertebra was replaced by a machinable and bioactive glass ceramic in several tumor patients under the direction of Professor H. Zippel in cooperation with Dr. J. Gummel and Dr. H. Hähnel at the orthopaedic clinic of the Charite in Berlin (Humboldt University). Figures 10.82 a to c, gives various views of a destroyed vertebra. Figures 82 d and e show the substitute vertebra immediately after the operation, it is held in place with a Zielke bridge. The first patient was already released pain free after half a year.

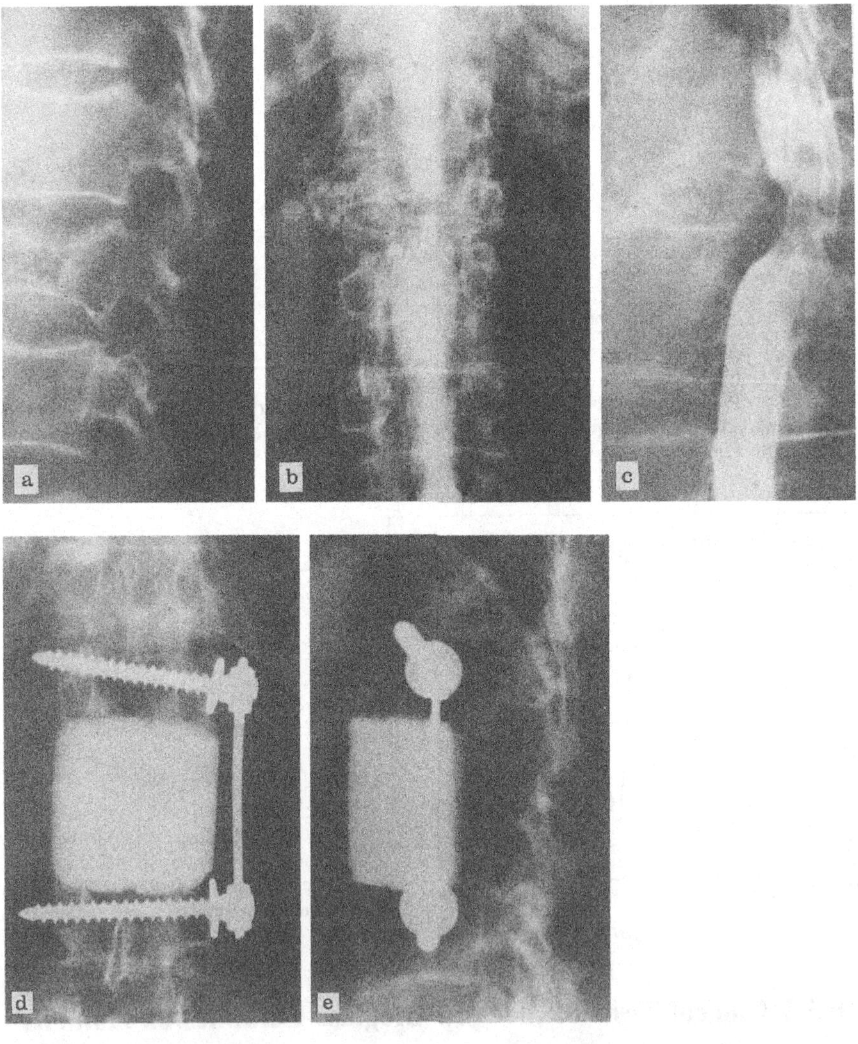

Fig. 10.82. Vertebra replacement by a machinable bioactive glass-ceramic in a patient with a tumor (Dr. J. Gummel, Humboldt University Berlin). **a–c** Several views of a damaged vertebra; **d, e** replaced vertebra (see text) (X rays)

- Machinable, biocompatible and bioactive glass ceramics were very successfully introduced in the middle ear as well as in the nose, jaw and whole skull region at the throat, nose, and ear clinic of the Friedrich Schiller University in Jena (operation by Dr. E. Beleites under the direction of Professor K.-H. Gramowski). The following operations have been carried out in the first stage of clinical tests.
  Restriction in tympanoplasty. (Operative correction of a damaged middle ear).
  Stapedectomy. (Removal of the stapes in whole or part and replacement with prostheses).
  Restoration of the internal auditory canal.
  Support for the base of the nose.
  Correction or repair of the eye socket.
  Restoration of the anterior wall of the frontal sinus.
  Rhinoplasty.

Figure 10.83 shows middle ear implants which were made by surgeons themselves. The hearing capacity could be restored. Animal testing has demonstrated that a biocompatible implant establishes itself completely free of irritation and is covered with epithelium. The intergrowth happened as if it was part of the body itself (see Fig. 10.84).

- Machinable bioactive implants were primarily used in a larger application range at the orthopaedic clinic of the Academy of Medicine of Dresden. The following operations have been done in the first stage of clinical tests under the supervision of Professor K.-J. Schulze with Dr. W. Purath and Dr. T. Schubert: Reconstruction of the socket for a dislocated hip joint at the stage of abnormal tissue development (Pericapsular osteotomy after Pemberton).

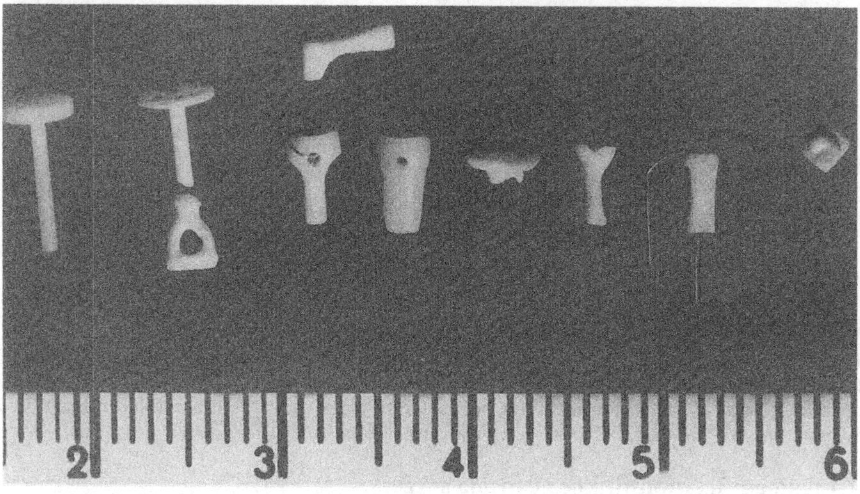

Fig. 10.83. Middle ear implants of machinable biocompatible and bioactive glass-ceramics which were made by the physician himself; scale: cm (Dr. E. Beleites, HNO Clinic of the Friedrich Schiller University in Jena)

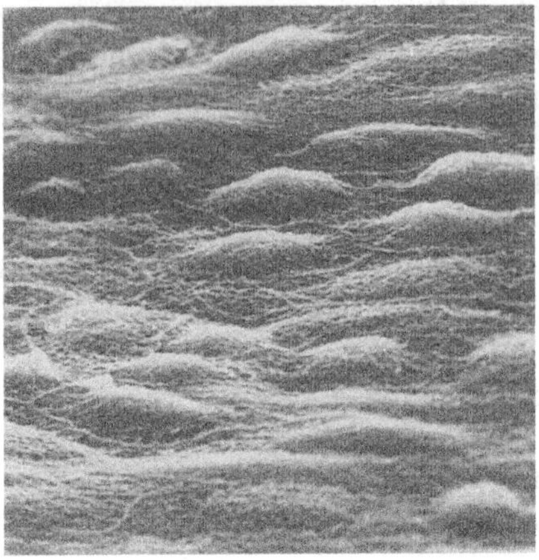

Fig. 10.84. Growing in of a machinable biocompatible or bioactive glass-ceramic and covered with epithelium. Animal test. (Dr. E. Beleites, HNO Clinic of the Friedrich-Schiller-University in Jena).

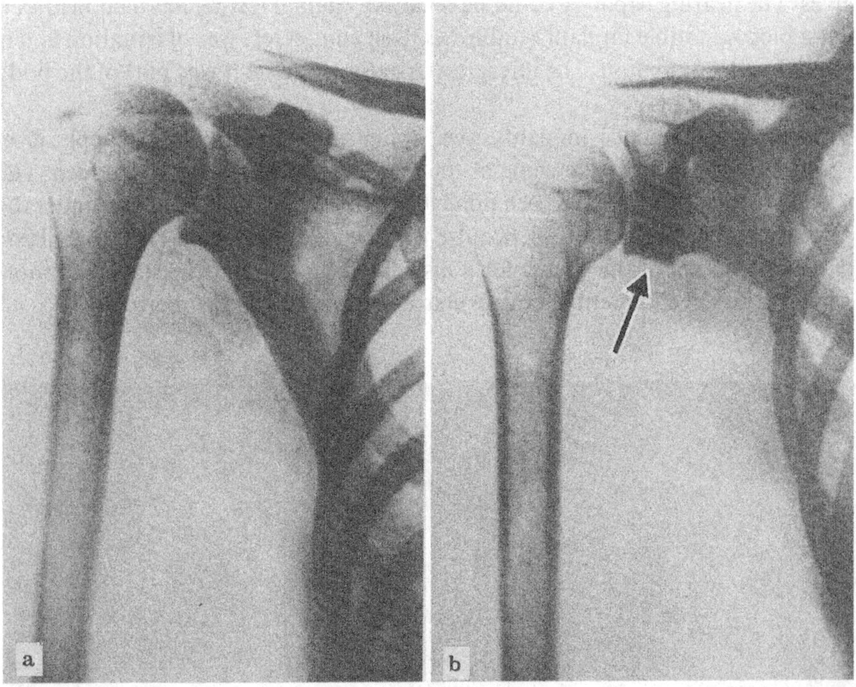

Fig. 10.85. Application of a wedge shaped implant made of machinable bioactive glass ceramic for treatment of a recidiving shoulder dislocation by the Eden Hybinette operation. **a** Prior to the operation; **b** 8 months after the operation, bone intergrowth in a good clinical result (Professor K.-J. Schulze, Dr. W. Purath, Dr. T. Schubert, Academy of Medicine in Dresden)

Ligament fixation in knee capsule surgery.
Tibia head bone surgery for the correction of tibia valgum (knock-knee) and tibia varum (bowleg).
Operation after Bandi.

Partial replacement of individual thoracic vertebrae.
Ventral spinal fusion of the cervical vertebrae after Robinson.
Shoulder joint surgery after Eden-Hybinette.
Distraction of joint surfaces to maintain clearance.
Filling in large bone cysts (glass-ceramic as filler).

Figures 10.85 and 10.86 give two examples of these many applications.

- Following successful experiments on pigs, roots of teeth were replaced by machinable, highly bioactive glass-ceramic implants by Professor Dr. R. Pinkert at the clinic and polyclinic for jaw-face surgery and the surgical section of the Academy of Medicine of Dresden. The first results are extraordinarily encouraging. Implants for the root of a tooth are shown in Fig. 10.87. They were produced by a spinning process from liquid glass and later made into ceramics. They were prepared according to a replica of the original one and their ceramic shape was obtained by spinning.
- Tests were successfully concluded on animals for the introduction of the aforementioned bioglass ceramics into the eye by Mr. M. Jütte at the Eye Clinic of the University of Jena. Clinical tests on humans are imminent.

The range of medical applications is growing continuously. The bioactive and biocompatible, machinable glass ceramics of the Bioverit-type were released from Jena in the spring of 1989 for general medical application in the

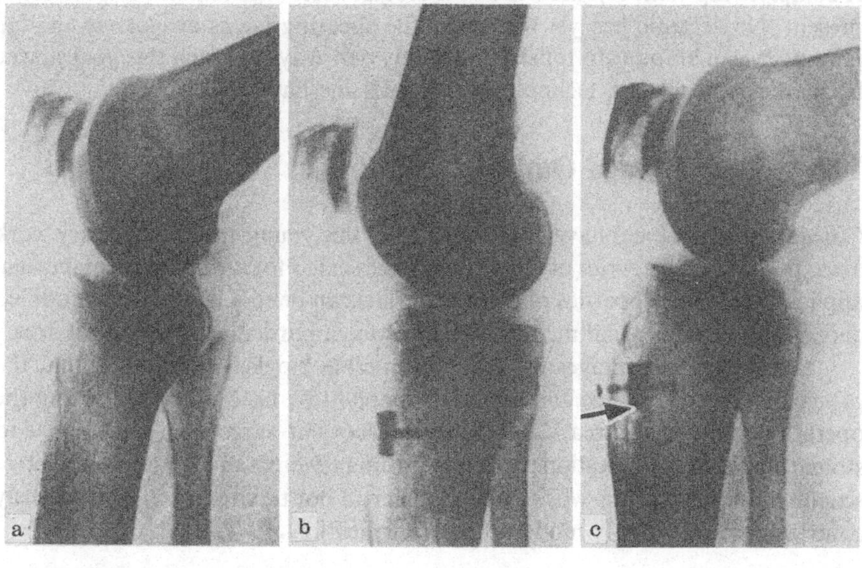

Fig. 10.86. Application of a crooked rectangular implant with a central cavity and an attaching pin made of machinable bioactive glass ceramic for the treatment of a patellofemoralarthose by the tuberositas-tibiae-ventralization according to Bandi. **a** Prior to the operation; **b** one week; **c** 8 months after the operation. Integration of the implant in the bone without a loss of correction (Professor K.-J. Schulze, Dr. W. Purath, Dr. T. Schubert, Academy of Medicine in Dresden)

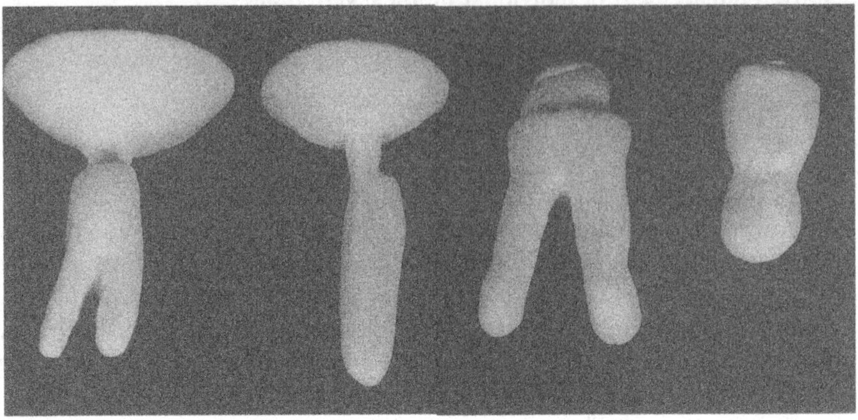

Fig. 10.87. Blank implant for a tooth root made of machinable highly bioactive glass ceramic, produced according to the centrifuge process (Professor Dr. R. Pinkert, Academy of Medicine in Dresden)

orthopaedic, traumatology as well as for the whole throat-head surgery field. Twenty one clinics and hospitals of the former DDR were involved in the three successful stages of clinical tests on humans. Until the end of 1990, 850 successful operations had been performed.

Figure 10.88 gives an informative overview as to which parts of the human skeleton have been successfully replaced by the bioglass ceramics of Jena. A complete hip-joint replacement with bioglass ceramics remains impossible at present. No ceramic has yet withstood the shearing forces at work in the hip joint (they can amount to tons). Completely new ways to reach this goal shared by the fields of science, technology and medicine have to be found.

10.5.6 Summary and Outlook

Considerable advances have been achieved in this young, interdisciplinary work area. Today a whole series of bioglass ceramics exists for a wide range of medical applications. The properties of these materials can be to a large extent modified depending on the special medical demands for a particular substitution area.

One of the advantages of these machinable bioglass ceramics is that the physician can still change and adjust the implant by his own means during the operation. This is of great importance in accident surgery, jaw surgery and stomatology. Many questions still remain unanswered and a large number of scientists and physicians world wide (which can not be cited here) continuously contribute to this highly modern field. (Applications [974–985].

→

Fig. 10.88. Applications of the bioglass ceramics from Jena as bone substitutes in humans. *1* Frontal sinus front wall, *2* middle ear, *3* ring around the eye, *4* nose, *5* lower jaw, *6* teeth (roots), *7* cervical vertebra, *8* thoracic vertebra, *9* shoulder, *10* dorsal vertebra, *11* filling of bone cysts, *12* hand (finger), *13* knee

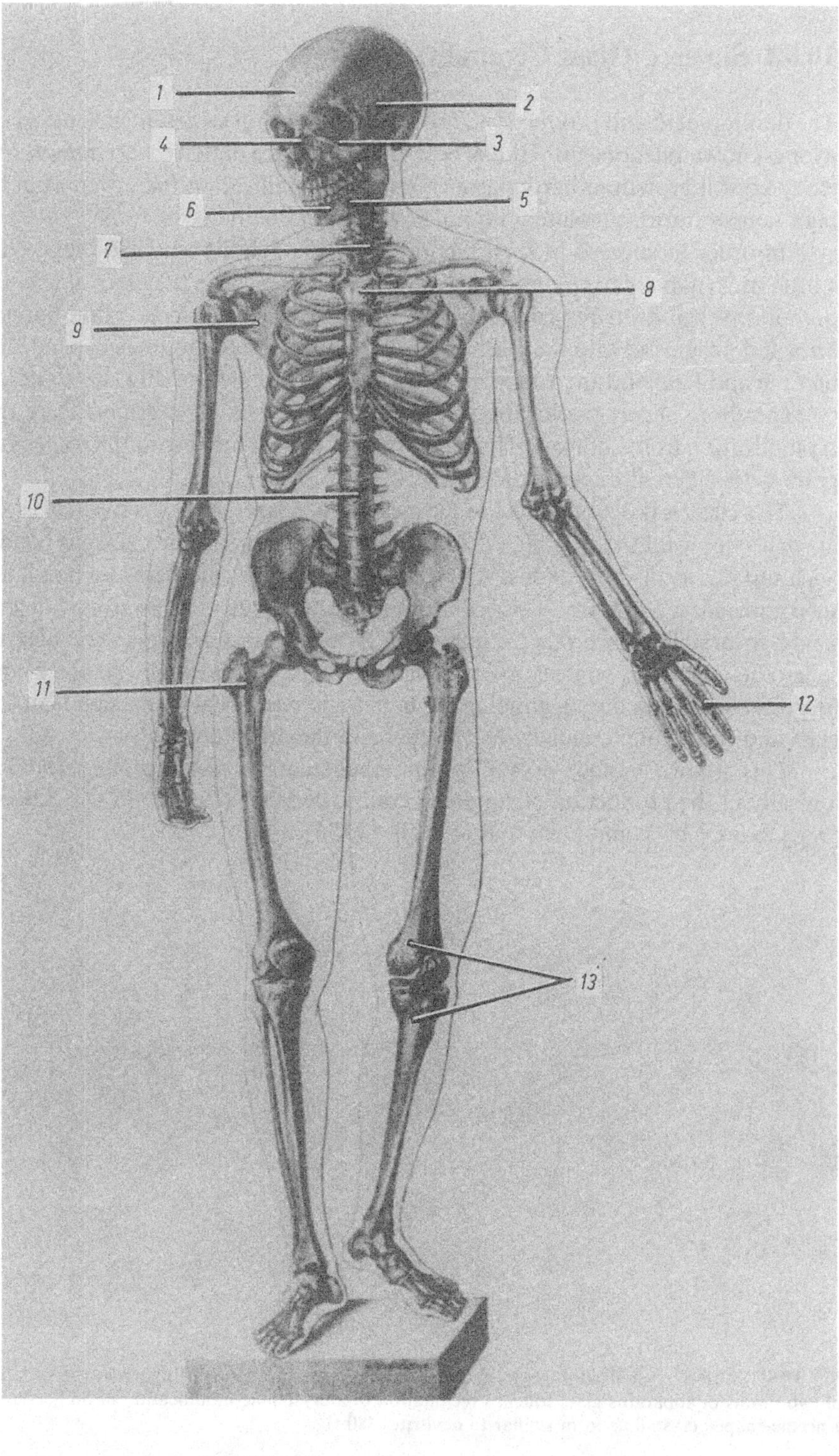
1
2
4
3
6
5
7
8
9
10
11
12
13

10.6 Sintered and Special Glass Ceramics

10.6.1 Sintered Glass Ceramics

The development and production of so-called sintered glass ceramics is based on a long-known phenomenon. It has been the experience of those who intended to study crystal growth in hard glasses resisting crystallization that, even at quite high temperatures, absolutely no nucleation could be induced.

In order to succeed in such laboratory work, Schönborn [986] suggested a little trick, namely to comminute, i.e., to coarsely pulverize the piece of glass in question. When heat-treated, a uniform crystal front starts at the grain boundaries and progresses into the glass grain. The shift of this front in its dependence on time and temperature can be measured easily. The picture (Fig. 10.89) taken by Schönborn shows the location of the contact of two glass grains. Here the crystallization fronts of two different crystalline phases move into the respective grain with different linear growth rates.

This success is doubtless due to the creation of a large, fresh, glass-surface area in connection with the very high demand for satisfaction of freshly severed bonds. Such surfaces avidly absorb dust – which might indeed be considered an inorganic solid mineralizer – or water. This is illustrated by the fact that freshly ground quartz sand is invariably absorbed in the human lung, causing silicosis, while street dust, in spite of its high SiO_2 content, is not. Solid mineralizers accidentally getting at the fresh fracture planes during grinding might enter into epitaxial interaction with the glass and contribute essentially to a decrease in the nucleation barrier.

This known surface crystallization phenomenon was exploited technologically in the production of sintered ceramic bodies by Sersale [987], Klause [988], Sack [989], and Sack and Scheidler [876].

Fig. 10.89. Surface crystallization (Schönborn). Uniform crystallization starting where the surfaces of two grains of apparatus glass touch. Precipitation of a crystalline modification of SiO_2 and of a needle-shaped crystalline form similar to devitrite (380:1)

The ceramic bodies described by Sack were produced by grinding certain glasses together with mineralizers, subsequently pressing them into shape, and heat-treating them between softening and working temperature. During heat treatment, a crystallization process guided by mineralizers starts at the fracture planes of the glass grains and progresses into the interior of the grain. Just as in the case of Pyroceram-type bodies, it is possible to obtain specific crystalline precipitations through the use of certain nucleants or mineralizers. Often these crystal types would not form under conventional conditions.

Comparison with production procedures of Pyroceram shows that the dimensions of the crystals formed are similar. In the case of Pyroceram, the crystallites start to grow in the center of a vitreous immiscibility region and progress to the phase boundary, attaining a mean size of 0.1 to 20 μm. In the case of production of ceramic bodies from glass powders of < 10 μm grain size, the crystallites grow from the surface of each grain toward the center.

Sack [989] lists as mineralizers compounds like magnesium oxide, magnesium silicate, lithium oxide, lithium silicate, lithium aluminate, etc. Reliable crystallizable glass compositions are given as follows (in mass%):

54–68 SiO_2	0–1 BaO
17–27 Al_2O_3	0–3 B_2O_3
4–12 MgO	0–3 TiO_2
0–10 C_2O	0.05 F

0.4 maximum alkali.

The following crystalline phases determine the properties of sintered glass ceramics: cordierite, anorthite, spinels, forsterite, diopside, spodumene, eucryptite, etc. Most of these phases and their properties are known from the general development of glass ceramics.

The firing shrinkage of sintered glass ceramics is (reproducibly) ca. 11–13%. Sintered glass ceramics are suitable for application in electronics, ceramic-metal combinations, antioxidation metal coatings, etc.

10.6.2 Special Glass Ceramics

Work on controlled crystallization and the development of glass ceramics continues worldwide in many directions – as was stated in the introductory paragraphs of this section. One example is the class of Solder Glass Ceramics. Modern technology can hardly be imagined without solder glass ceramics which have been adapted in many ways to specific properties of the elements to be joined, mostly in the electronics industry.

Another important example comprises the so-called "slag sitals" in the former Soviet Union. The designation "sital" taken from the names of the chief components of classical glass ceramics (Si–Ti–Al) is used for all glass ceramics in the former Soviet Union. Blast furnace slags with high remainders of Cu, Fe, Ni, Cr, etc. – which act as nucleants – happen to have compositions approaching the final composition of glass ceramics. Doped with certain additives, the continuous production of slag sitals was undertaken in the former Soviet Union and gained in importance, particularly for the chemical industry, by supplying acid- and alkali-resistant fittings, pump components, etc. It is indeed astounding that a continuous process could be developed, in spite of the natural compositional variations resulting from the smelting of various ores.

11 The Strength of Glass

11.1 Theoretical Strength

Glass is a brittle solid and, proverbially, its mechanical toughness is minimal. In the past, many experiments and developments aimed at improving the strength of glassware. Up to now these attempts were only partially successful, but they did open entirely new applications for glass on the basis of the improvements achieved.

Doubtless the strength of glass is primarily conditioned by the strength of the bonds between its constituents. Since these constituents vary greatly and since there are many secondary influences on the strength of glass, the question of how to strengthen glass has no simple answer.

An excellent "bibliographic study on the strength of glass and the possibilities to improve it" [990] critically evaluates more than 500 scientific papers published between 1920 and 1961. Recent reviews covering glass strength in detail have been compiled by Uhlmann and Kreidl [991], including one on improving strength by tempering by Gardon [902] and one on chemical strengthening by Bartholomew and Garfinkel [584].

Calculation of the strength of glass is founded on thermodynamic considerations and ionic structures. Both methods have one thing in common. Either material properties which are not sufficiently known enter the calculations, or crude approximations have to be used. Therefore, only approximate values can be expected for the theoretical strength of glass. However, it is interesting that results give values of the same order of magnitude: 9,810 to 29,420 MPa.

Polany [992] started from the fact that every fracture is accompanied by an increase in surface area and thus an increase in surface energy. An object could not be broken if this additional energy were not available in the form of the energy of elastic deformation and this over a layer whose thickness is equal to the intramolecular distance a.

Polany thus postulates:

$$\text{Theoretical strength} = \sqrt{4\sigma E/a}$$

where E = module of elasticity (68, 650 MPa); σ = surface tension, which after extrapolation to ambient temperature is $10^{-3}\ \text{N}\cdot\text{mm}^{-1}$: and a = intramolecular distance, 3.6×10^{-7} mm. The resulting value for the theoretical strength is 29,420 MPa.

Orowan [993] determined the work per surface area resulting from fracture. The work to form two new surfaces is equated to the change in tension as a function of intra-atomic distances. Orowan arrived at the relation:

$$\text{Theoretical strength} = \sqrt{\sigma E/a}$$

This value is only half that determined by Polany.

Finally, the calculation by Iglis [994] agrees with an assumption by Griffith [995, 996] – which will be discussed in more detail below – according to which microscopically small flaws are present in the glass. These flaws were postulated to act as stress centers. In fracture, the tension at the tip of such a flaw was supposed to equal the theoretical strength. Naray–Szabo and Ladik [997] started from the Coulomb forces between ions and arrived, for quartz, at a value of about 24,525 MPa. Weyl [998–1000] and Weyl and Marboe [1001] accepted this basis for calculation and held polarization effects responsible for the observed decrease in strength. This decrease is larger, the larger the polarizability of ions. Weyl's assumption was to make it plausible that the effective strength of glass is about one-tenth of that expected from the lattice energy and binding forces. In practice, surface flaws have a dominant influence on effective strength.

11.2 Effective Strength: Attempts at Theoretical and Practical Explanations

11.2.1 Theoretical Concepts Regarding the Strength of Glass

The effective strength of massive glass is not very high. In general it is around 50 MPa, compared to 390 MPa for organic fibers and 290 to 2,940 MPa for metals. However, if one reduces the dimension of the test samples – for instance using thin rods – a strength of about 690 MPa may be measured (Stockdale et al. [1002]). If one progresses to thin fibers (3 μm), a still higher strength is obtained (3,330 MPa). However, the difference from even the lowest estimate of theoretical strength (9,810 MPa) must be explained.

Calculations by Orowan [993, 1003], Condon [1004a], Polany [992], Weyl [998–1001] and Naray–Szabo [997] yielded theoretical strengths ranging from 9,810 to 29,420 MPa.

It is certainly remarkable that Griffith [995], long before modern explorations of the glass structure, arrived at the same value of 9,810 MPa by a completely different concept. He explained the small effective strength of glass by the presence of numerous, microscopically small, semielliptical flaws, now termed Griffith flaws. Extrapolating his relation to a zero flaw diameter, he arrived at the same theoretical strength (about 9,810 MPa). If the glass is loaded, fracture occurs, starting from these flaws, earlier than in a flawless piece. According to Griffith's first calculations, the effective strength would correspond to a flaw dimension of 5 μm. Observations on NaCl crystals can be interpreted by the Griffith flaw theory, according to Zwicky [1005]. Sodium chloride, with a proven low concentration of defects, has a strength of about 1569.6 MPa, a "normal" crystal only 4.91 MPa!

In the past, many attempts were made to reinforce the plausible Griffith flaw theory by experimental evidence and in some way make these flaws visible. For instance, Andrade and Tsien [1006] evaporated Na onto glass surfaces. Acloque et al. [1007] devised an ion exchange. Ernsberger [1008] connected ion exchange with the action of water on glass surfaces prepared in a special manner and carefully examined the formation of fissures. Although these attempts, as well as electron microscopy, unequivocally proved the presence of numerous surface flaws, they cannot be considered identical with Griffith's original model. Yet it was Griffith who, long before these investigations, suggested that the flaws related to lower effective strength might be predominantly associated with the surface. It is quite interesting that Littleton [1009] formulated this association in the sentence "We do not measure the actual strength of the glass, but the weakness of the surface." In other words, deficiencies in the glass surface are much more important than those in the bulk. Therefore, any attempts at a decisive improvement in the strength of glass must begin at the glass surface.

Such conclusions remain valid to this day, but also led toward underestimating the influence of composition and structure. True, as long as macroscopic flaws had such a predominant effect on the measured effective strength, workers like Gehlhoff and Thomas [1010] – were unable to find any relation between chemical composition and strength. It is only more recently that, succeeding in the exact elimination of surface influences, Watanabe et al. [1011], for example, could clearly demonstrate structural influences. Perhaps the most interesting further development of Griffith's theory was by Poncelet [1012, 1013]. As will be shown in later sections, his ideas about the "genesis of flaws" closely touch structural glass problems. Whereas Griffith later modified his original concept and located his flaws in the surface, Poncelet assumed that there are no flaws in the glass before loading – except for crude surface defects – and that they are forming only under load throughout the bulk. Poncelet's essential assumption is that the stress needed for flaw formation is only a part of the conventional fracture stress.

McAffee [1014] showed that thin glass membranes pass little or no helium. However, when subjected to a tensile stress, only one-half the fracture stress, helium permeability increases greatly. Apparently the origin or opening of flaws leads to increased gas permeability.

All further development in the theories of strength, like those of Cox [1015], Saibel [1016], Fisher [1017] and Gibbs and Cutler [1018] are built upon these concepts.

11.2.2 Experimental Investigations

11.2.2.1 A Demonstration of Griffith Flaws

It appeared plausible (Vogel [207]) to associate the Griffith flaws and their origin with the microheterogeneous structure, i.e., with the droplet-shaped immiscibility regions found in most glasses by numerous workers. For silicate glasses, the dimensions of these regions are of the same order as that of the Griffith flaws, which so far had not been made visible.

A silicate glass opacified by phosphates was prepared, with additives increasing the dimension of the droplet regions to the extent that these regions and their behaviour under tensile or compressive stress could be observed in the light microscope (Vogel [207]).

It is proposed that the semielliptical Griffith flaws, originating as postulated by Poncelet, are nothing else than the weakened microphase boundaries of droplet-shaped immiscibility regions or their ion cluster precursors.

Under tension, in an area containing a droplet-shaped immiscibility region, the primary fissure will first run vertically to the direction of tension, but will bend into a partial ellipse at the spherical inhomogeneity. This makes understandable the much-discussed dependence of the site of flaws upon the direction of tension (see Fig. 11.1).

Before subjecting the preparations with enlarged droplet regions to a load under microscopic observation, 1- to 2-mm thick glass foils were fabricated and polished. But – as shown in Figs. 11.2 and 11.3 – flaws had already formed around the droplet regions during the grinding and polishing process on the areas thus exposed to compressive and tensile stress. In both figures, small spherical shapes can be seen within the droplet region. They are bubbles which have no significance for the problem treated here. But it can be seen clearly that the formation of fissures occurred spatially in the shape of shells. The fissure has started to form in Fig. 11.2; in Fig. 11.3, in our opinion, a typical semielliptical Griffith flaw has been completed. It is particularly interesting that for very small droplets fissure starts right at the

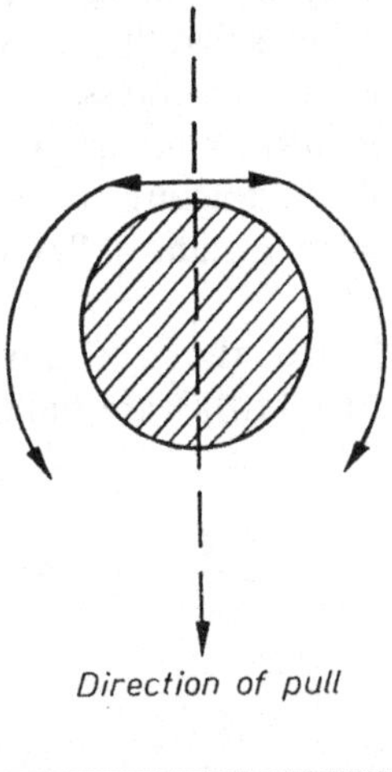

Fig. 11.1. Schematic of the origin of a Griffith flaw. If a glass region containing a droplet-shaped immiscibility is pulled, fracture on principle will start in a direction vertical to that of the pull. The fracture then will run around the droplet region in the shape of a semiellipse

Fig. 11.3. Droplet-shaped immiscibility region in a silicate glass opacified by phosphate. Completely developed semielliptical Griffith flaw. With increasing droplet size the fractures are developed at larger distance from the immiscibility regions (light micrograph. 200:1)

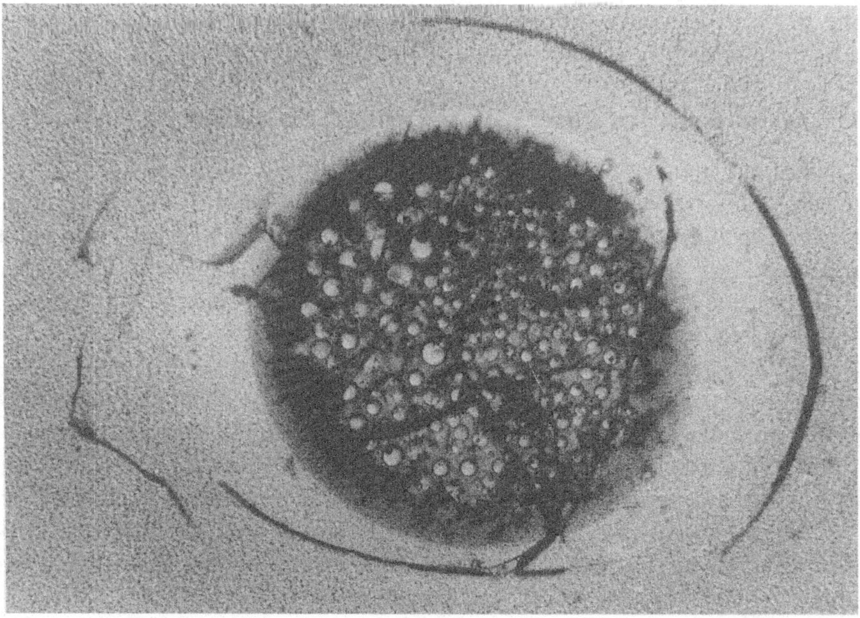

Fig. 11.2. Droplet-shaped immiscibility region in a silicate glass opacified by phosphate. Compression and tension caused by grinding and polishing will generate a dish-shaped fracture around the droplet-shaped immiscibility region. The spherical objects in the large droplet phase are bubbles (light micrograph. 640:1)

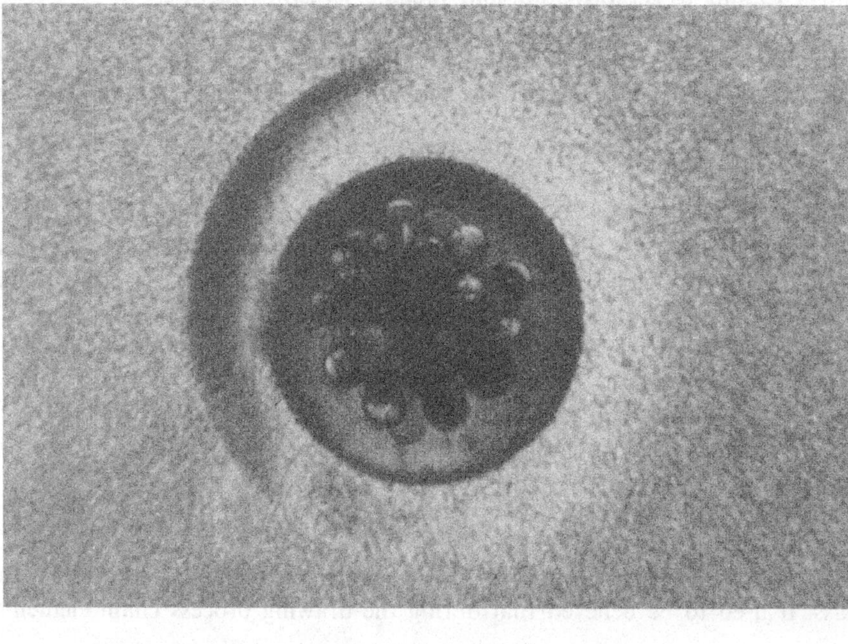

phase boundary. For larger droplets the fissure starts more and more outside the immiscibility region. Apparently the strain pattern around the droplet plays a decisive role.

11.2.2.2 Strength After Elimination of Crude Surface Defects

The normal engineering strength of bulk glass is only about 50 MPa. If such a glass is fire-polished, so that crude surface defects and fissures heal during a partial melting of the surface layer, the strength increases from 50 to about 200 MPa.

The same improvement is obtained if a surface layer is removed by etching with hydrofluoric acid, as long as the surface layer exceeds the depth of the fissures. The increase in strength now is from 50 to about 220 MPa. If one applies similar polishing and HF etching procedures to thin glass rods which exhibit normal engineering strength of about 690 MPa, the strength increases to 1,370 MPa after fire-polishing, and to 1,770–3,420 MPa after HF etching. This latter value may actually be considered to approach the strength of a glass fiber.

As new defects are formed, perhaps through friction of fire-polished or HF-etched glasses in contact with sand, paper, glass, or even with a hand, the strength immediately decreases to its original low value.

11.2.2.3 Fatigue

If a fresh glass surface is violated, e.g., by a diamond, and subsequently subjected to a small load of long duration in a normal atmosphere, a final strength test yeilds a noticeably smaller strength value than obtained without preloading. It must be assumed that the fissure is kept open under this small load and water enters, leading to gel formation and swelling, assisting fracture.

Fatigue is an important factor in predicting failure. Testing for reliable predictions requires the separation of bulk properties, flaw structure, and flaw reactions with the environment. The large new field of fracture mechanics deals with this problem in detail (see, e.g., Freiman [1019]).

11.2.2.4 Aging

If the glass violated by a diamond is subjected to a strength test 24 h later, the value is somewhat higher than for immediate testing. Apparently, at the extreme tip some chemical bonds reformed, leading to partial healing of the fissure.

If one waits much longer, however, the formation of a gel and swelling once more cause a decrease in strength. All these experiments demonstrate the enormous significance of surface defects.

11.2.3 The Strength of Glass Fibers

The strength of glass fibers far exceeds the normal engineering strength of bulk glass. It used to be believed that during the drawing process chain elements possibly present in the glass structure were oriented (Bartenev and Boykunenko [1020]; Hummel [861]). However, this assumption was abandoned when fracture strains in the direction and normal to the direction of the fiber axis had been determined experimentally and found to be equal. Nor could any anisotropy due to orientation be demonstrated.

The formation of a compression zone in the surface as a consequence of the fast cooling rate also had to be excluded because of the small fiber cross section.

According to Weyl [998–1000] and Weyl and Marboe [1001], the chemical bonding is generally stronger in the surface, since the screening of the central atom can be one-sided only. Since glass fibers have extremely large relative surface areas, this effect might be important for the interpretation of fiber strength.

In our opinion, the significant role played by immiscibility phenomena in glasses includes the formation of flaws. According to the present state of knowledge about microstructure and microphase separation, it must be assumed that a high cooling rate will suppress phase-separation phenomena, including the otherwise common formation of flaws.

Also, the number of flaws per volume unit must have an influence. Because of the small mass of fibers, this factor will cause an increase in strength. In fact, for very long fibers, i.e., an increase in volume, measured strength values are found to decrease.

Even though the structure of a fiber glass is expected to be much more homogeneous than that of bulk glass, Zarzycki and Mezard [803] unequivocally demonstrated immiscibility phenomena even in soda-lime glass fibers. This would explain the remaining difference between the strength of glass fibers (1,770 to 3,430 MPa) and the theoretical strength (9,810 MPa); and, also, the equality of fracture strain parallel and normal to the fiber axis. It has already been pointed out that the formation and direction of the flaws around droplet-shaped immiscibility regions will always depend on the direction of load.

One more fact supports our interpretation. In pure homogeneous SiO_2 glass no immiscibility phenomena can occur. And, indeed, fibers fabricated from this glass have practically theoretical strength.

11.3 Strengthening Methods in Practice

In practice, strengthening processes primarily prevent, or make inoperative, the influence of mostly macroscopic surface violations.

11.3.1 Tempering

In this manner the well-known process of tempering can be understood. A sufficiently thick-walled glass object is heated to just below the glass-transition temperature and is quenched severely by a blast of cold air or by contact with metal plates. During this procedure the surface glass will solidify very rapidly, while the glass below will continue to contract. In this way a strong compressive strain layer is formed in the surface while the interior experiences tensile stress, compensating surface compression (Fig. 10.20(a)). Fissures in the surface are compressed and prevented from opening i.e., initiating fracture under load. Obviously, the compressive layer will be fully-effective only if it is thick enough to exclude fissures from reaching into the interior zone of tensile strain. As a rule the increase in strength achieved is about 200 MPa. An extensive recent review of tempering is by Gardon [902].

11.3.2 Compound Glass

A simiilar process was found and applied as early as in 1891 by Otto Schott. A glass object is cased, during its formation from fluid glass, by a thin layer of another glass having a lower coefficient of expansion. After casing, the casing layer is under compression and causes an increase in strength in the way described before.

11.3.3 Silicon-Organic Coatings

Silvestrovich and Boguslavsky [1021] described a process in which a glass object is briefly immersed in a chlorosilane solution. Subsequent hydrolysis creates a thin SiO_2 surface film. Apart from the creation of a compressive layer, crude fissures are closely cemented. If HF treatment precedes the process, strength may increase from 50 to about 540 MPa. In general, antifriction coatings (Gaiser et al. [1022]) prolong the service life of containers by avoiding the formation of crude flaws by collision, for example, when bottles are crowded on conveyors. Silicone coatings have been widely replaced by coatings leading to deposition of TiO_2 (also SnO_2 and ZrO_2). (See, for example, the patent by Brocket [1023].) Applications are described by Southwick et al. [1024] and Budd [1025]. A recent extensive review of glass coatings including such coatings is by Dislich [1026].

Applied to glass fibers, coating with organic silicon compounds has multiple purposes. So-called "sizing" during the drawing process promotes gluing of the forming fiber bundle, thus decreasing the risk of tear. But these compounds also promote adhesion when the fibers are embedded in plastics. It is the inorganic character on one side and the organic character on the other side of a chlorosilane molecule that leads to good adhesion between glass and plastic.

11.3.4 Dealkalizing

Near the transition temperature, ion diffusion is accelerated and mobile ions migrate to the glass surface. If acidic gases (SO_2, HCl, $SOCl_2$), thionyl chloride, or SO_2Cl_2 (sulfurylchloride) are introduced, for example, during annealing, the diffusing ions, for example, Na, may be caught and prevented from diffusing back. The surface is deprived of alkali and enriched in SiO_2 and this surface layer will be under compression. The process took place inadvertedly in annealing furnaces and was studied extensively by Weyl. A recent study is by Schaeffer et al. [1027].

11.3.5 Ion Exchange: "Chemical Strengthening"

Ion exchange has gained increasing importance for strengthening mass-produced glassware by 50% or more (Philips Research Center [990], Kistler [1028], Acloque and Tachon [1029], Leibig [1030, 1031], and Bartholomew and Garfinkel [584]). This method is based on the immersion of a glass product in molten salts. In one process, larger ions (e.g., Na^+) were replaced in the surface by smaller ions (e.g. Na^+ by Li^+ from an Li_2SO_4 bath) (Hood and Stookey [1032]) or by Cu^+ from $CuCl_2$ vapor (Zijlstra and Burggraf [1033]) at high temperatures. However, at these temperatures glassware tends to be distorted, so the process is not suitable for commercial application (Bartholomew and Garfinkel [594]). In another process ("ion stuffing") smaller ions (e.g., Na^+) are replaced in the surface by larger ions (e.g., K^+ from a KNO_3 bath) at temperatures below those allowing the network to adjust to the larger ion introduced (Kistler [1028], Nordberg et al. [1034]). Aluminosilicate glasses are preferred in practical application because the rate of exchange is larger at lower temperatures, i.e., below the range of significant network relaxation. Later, more-sophisticated variations include double-ion exchange, combination with thermal tempering or acid etching, etc. The case of a combination with crystallization ("Chemcor®" process) was described earlier.

11.3.6 Multiple Layer Glass

A considerable increase in strength, especially for flat glass, has been attained by new techniques rather than by glass development.

Glaverbel, a Belgian glass company, developed a multilayer glass consisting of two thin panes strengthened by ion exchange and combined by organic interlayers (preferably polyvinyl butyral). This flat glass product (VHR = verre de haute resistance = high-strength glass) is distinguished by very high bending, impact resistance and torsion strength.

Corning Glass Works went beyond this stage in developing seven-layer household glasses produced in one production step. Above a slit-shaped main nozzle, three additional nozzles are provided (Fig. 11.4). The main nozzle is fed by a glass with high thermal expansion. Nozzles 2, 3, and 4 are fed with glasses of decreasing expansion. From the resulting ribbon, glassware can be cut or pressed, with a core enveloped by three casings of strongly decreasing α values.

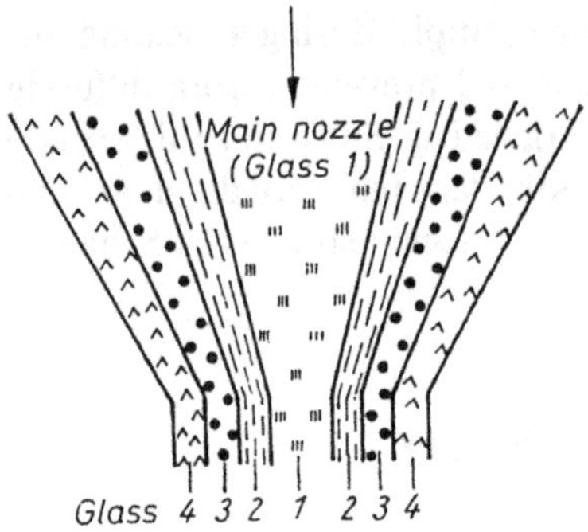

Fig. 11.4. Schematic at the drawing nozzle in the process of making high-strength multilayer glass, $\alpha_4 < \alpha_3 < \alpha_2 < \alpha_1$

Even if the outer layer is violated, the fracture will not propagate into the interior. This development represents a truly remarkable, yet logical step in the strengthening of commerical glassware.

12 Interaction Between High Energy Radiation and Glass

12.1 General Considerations

Glass and its properties are subject to a variety of changes under the influence of high energy radiation. The discussion of the effects of electron microscopy have provided, as a first example, the enormous changes in the glass structure by fast electrons. In general, high-energy radiation effects extend from the reduction of specific ions to metal (in lead glasses) to the collapse of the entire network.

The glass surface may be essentially removed using an ion beam. This phenomenon has been applied in the form of a special process called "ion etching" in electron microscopy (Bach [1035]). Conversely, bombarding the glass surface with specific ions alters the refractive index and dispersion of the glass surface layer ("ion implantation").

X and γ rays usually cause partial rupture of chemical bonds, partial destruction of the network, reduction of specific ions, introduction of defects, discoloration, or fluorescence.

Less-energetic UV radiation can cause different effects. Optical glasses of various composition, for instance, radiate various fluorescence colors under UV light. Photo effects, sometimes reversible, are observed; Nd_2O_3 glasses will emit coherent laser light. Even light of longer wavelength within the visible range of the spectrum has more or less recognizable effects on the transmission of glass. The discoloration of clear glass, e.g., in sunlight, is called solarization (from sol (Latin) = sun).

A few of these effects are selected for more detailed treatment in the following sections, including phenomena which can be utilized as well as phenomena which are to be suppressed in order to maintain the usefulness of the glass for particular purposes. Since the area of interaction of high energy and glass is extraordinarily large – including some phenomena (for example, of fluorescence) not yet sufficiently understood – the selection offered here is limited. Extensive general reviews are by Lell et al. [1036] and Kreidl [1037]; reviews on induced defects by Griscom [351].

12.2 Photosensitive Glasses Based on the Formation of Metal Colloids

The light sensitivity of some chemical compounds has been known for a long time and applied to classical photography. The photographic process is based essentially on photolysis, i.e., the splitting of a silver halide, preferentially bromide, under the influence of light with subsequent reduction of Ag to the metallic state:

$$AgBr + h\nu \rightarrow Ag^0 + 1/2Br_2$$

The gelatine, which at the same time protects the silver bromide colloid it contains, receives and ties up the nascent free bromine, thus preventing partial reversal. Around 1937, Dalton discovered similar effects in glass [1038], giving rise to a new, fast-spreading research activity in glass chemistry [1004b, 1039–1076].

The simplest type of photosensitive glasses (Table 12.1) is characterized by the conversion under neutral or weakly reducing conditions of a soda-lime glass batch containing small additions of Cu, Ag, Au, or Pd compounds as well as some CeO_2 and SnO_2 to an entirely clear, but photosensitive glass. Photosensitivity is conditioned on the simultaneous introduction of one of the forementioned noble metals and a cerium compound.

If such a glass is covered by, for example, a mask or a photo negative and subsequently irradiated by UV or X rays, an appropriate heat treatment (equivalent to photographic development) will create pattern or image in the glass. The formation of such an image is based on the precipitation, or, respectively, the formation of colloids of Cu, Ag, Au, or Pd on the irradiated locations.

The photo process involved is characterized by the equations:

$$Ce^{3+} + h\nu \rightarrow Ce^{4+} + e^-$$

$$Ag^+ + e^- \rightarrow Ag^0$$

X rays possess sufficient energy to release Compton electrons in the glass and to produce silver nuclei. Under UV irradiation, the detour via the redox system provided by the simultaneous addition of silver and cerium salts is required.

The addition of SnO_2 has the effect of a protective colloid just as in the case of colloidally colored (e.g., gold ruby) glasses. The images produced in photosensitive glasses of this type are irreversible and may appear yellow, orange,

Table 12.1. Example of the composition of a photosensitive glass based on the formation of metal colloids

Glass component	Concentration in mass %
Base glass:	
SiO_2	to 75
Na_2O or K_2O	to 20
CaO, PbO, ZnO, CdO	to 10
Al_2O_3	to 2
Doping	
Cu, Ag-, Au- or Pd-Compounds	0.05 . . . 0.3
CeO_2	to 0.05
SnO_2	to 0.2

blue, red, purple, or brown, depending on the size of the colloid particles. Thus they have certain features in common with true ruby glasses. The entire process should be conceived as photoinduced homogeneous nucleation.

12.3 Photosensitive Glasses Based on Partial Crystallization in Lithium and Barium Silicate Systems

12.3.1 Composition and Preparation

Since 1941 the patent literature (see especially [1038–1062, 1064–1067] has been reporting melt compositions for the production of photosensitive glasses based on the binaries Li_2O–SiO_2 and BaO–SiO_2, and the ternary Li_2O–BaO–SiO_2 (see Table 12.2), as well as the well-known nucleation by doping with Cu_2O, Ag_2O, or Au_2O with CeO_2. More recent is the use of so-called sensitizers in the form of halides or sulfates of Cu, Ag, Au, Li, or Ba. These melts, too, if carried out under weakly reducing, neutral, or weakly oxidizing conditions, solidify as perfectly clear photosensitive glasses.

12.3.2 Structure, Properties, and Microprocesses

As has been pointed out in earlier sections, binary Li_2O–SiO_2 and BaO–SiO_2 glasses have a pronounced tendency toward immiscibility. In certain compositional regions they will contain Li- or Ba-rich droplet regions in an SiO_2-rich

Table 12.2. Composition of the melts of photosensitive glasses based on the partial crystallization of lithium or barium silicate base glasses

Glass component	Concentration in mass %
Base glass:	
Li_2O	5 . . . 25
and/or BaO	3 . . . 45
SiO_2	70 . . . 85
Na_2O, K_2O, BeO, MgO	
CaO, SrO, Al_2O_3, B_2O_3	0 . . . small
Nucleation agent doping:	
Cu_2O, Ag_2O or Au_2O	0.001 . . . 0.3
CeO_2	0.005 . . . 0.05
Sensitizers:	
F^-	1 . . . 3
Cl^-	0.01 . . . 0.2
Br^-	0.02 . . . 0.4
J^-	0.03 . . . 0.6
SO_4^{2-}	0.05 . . . 0.1

matrix. Nucleation agents in conjunction with CeO_2 result in the enrichment in these ions also in the Li- or Ba-rich phases. Compared with the case of uniform distribution, this enrichment in additives in the microphases means an increase in nucleation rate, and thus in photosensitivity. Nor is it unimportant that the control of phase separation will influence the size of the crystallites which thus can be maintained uniformly small. This is particularly significant for the applications such as those to be described in the following section.

The halide or sulfate ions added as so-called sensitizers simply facilitate and influence phase-separation processes in the base glass, which, in turn, will increase photosensitivity.

It is known that F^- may replace O^{2-} in silicate glasses. Since, because of their univalence, they are unable to constitute network bridges, the mobility of mobile network-modifying ions is increased and the viscosity of the melt decreases. For different reasons, chlorine, bromine, iodine, or sulfate ions have the same effect. These ions cannot substitute for oxygen. However, their incorporation stretches the network because of their considerable size and so they, too, increase the mobility of network-modifying ions.

If these photosensitive glasses are irradiated by UV or X-ray light in the same manner as the photosensitive glasses based on the precipitation of noble metal colloids (described in the preceding section) the same primary photo-induced formation of Cu, Ag, or Au nuclei is obtained. However, the subsequent "thermal development process" now will not lead to an enlargement of these nuclei to colloidal dimensions. Rather, an epitaxial interaction occurs between the primary nucleus and the droplet phase, leading to the growth of an alien substance containing lithium or barium di- or metasilicate building groups since the lattice parameters of the metal nucleus and these silicates agree almost perfectly. This possibility was first recognized by Stookey [829]. Later an experimental proof was provided by Vogel and Gerth [189, 190], working with photosensitive model glasses. The phenomenon is an exemplary case of heterogeneous nucleation.

It should be mentioned that, on principle, photosensitivity may also be achieved in homogeneous lithium or barium silicate glasses (e.g., $Li_2O \cdot 2SiO_2$ glass) (Vogel [151, 206]). However, in the absence of phase separation no phase boundary stops crystal growth, so that rather large crystals may form (Fig. 12.1) which may be very disadvantageous for many applications.

12.3.3 Special Properties and Applications, Photoform, Photoceram

Glass described in the preceding sections may be partially transformed to crystalline regions by UV irradiation and subsequent "thermal development." In this manner scales, gradations, letterings, sketches, and images can be produced in glass. Scales or gradations used to be obtained either by letting HF act on a glass surface covered with paraffin removed in part by engraving, followed by filling the resulting grooves with paint; or by vapor deposition of metals. In both

Fig. 12.1. Uninhibited growth of a lithium disilicate crystallite ($Li_2O \cdot 2SiO_2$) in a base glass of equal stoichiometric composition on the basis of the homogeneous structure of the glass

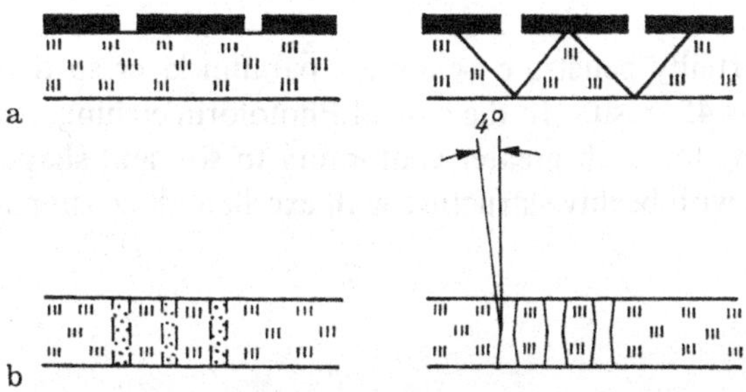

Fig. 12.2. Schematic of hydrofluoric acid etching of (a) a conventional glass after partial masking of the surface with paraffin, (etchings show inclines of 45°), and (b) of "Photoform" glass (incline only 4°)

cases the scales and gradations are only slightly resistant. Scales and gradations produced as crystalline regions in photosensitive glasses are free of these disadvantages.

12.3.3.1. Photoform

Some photosensitive glasses are, in addition, distinguished by significant differences in their solubility in diluted HF between the crystalline phase obtained and the remaining unirradiated clear glass. The illuminated crystallized phase is about 15–50 times more soluble in diluted HF than the remaining glass (Stookey [1050, 1053, 1055, 1065], Corning [1062]). This permits us to chisel reliefs and breaks in the glass, or to separate precisely dimensioned pieces by etching boundary lines. This special case may be termed photochemical punching; the entire process is called the "Photoform" process. It is of particular importance that the Photoform process allows slit widths as small as 0.01 mm, thus the production of gratings and sieves containing up to 50,000 holes per cm^2 and usable in TV technology. If the crystallites were large, as in Fig. 12.1, this achievement would be impossible.

Figure 12.2 is a schematic juxtaposition of HF etching in a conventional and a partially crystallized photosensitive glass. In the case of conventional etching,

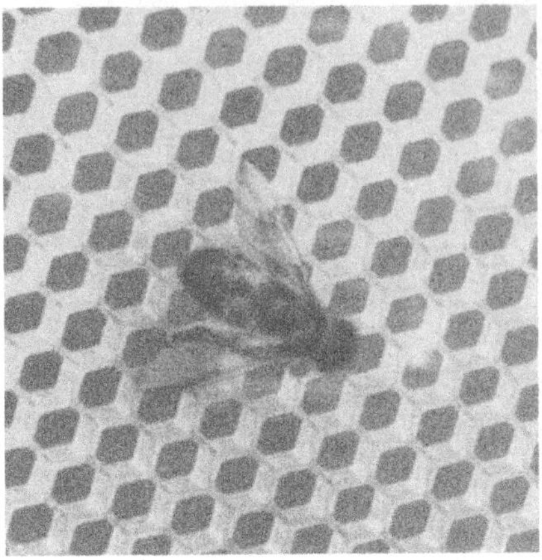

Fig. 12.3. Sieve plate with honeycomb structure produced by the "Photoform" process (Stookey [1066]. Beyer [1070])

the hydrofluoric acid will partially penetrate below the paraffin layer so that angles of the groove as high as 45° result. In the case of Photoform etching, the angle is only about 4°, leading to much greater conformity to size and shape. Figure 12.3 represents a sieve with beehive structure with excellent dimensional stability.

12.3.3.2 Photoceram

The Photoform process may be extended by reheating, at higher temperature, the piece obtained after "thermochemical development" and "photochemical punching." This reheating will lead to thermochemical reduction and formation of metal nuclei, followed by crystallization. The product so obtained is called Photoceram (Stookey [1065]). This ceramming is achieved with almost no shrinking. The Photoceram process permits production of the most complicated ceramic miniature elements of higher precision for modern electronics and electrical engineering. Conventional ceramic firing processes are incapable of such precision. Figure 12.4 exemplifies an electronic miniature element produced via the Photoceram process.

12.4 Dosimeter Glasses [1077–1082]

The exposure to X rays, particularly γ rays in the field of atomic energy techniques and related fields, encompasses the great danger of damaging the human organism. For this reason continuous monitoring of endangered persons by integrated, quantitative, radiation-dose measurement is necessary so that persons exposed to radiation doses exceeding a certain limit can be removed from the dangerous zone in time.

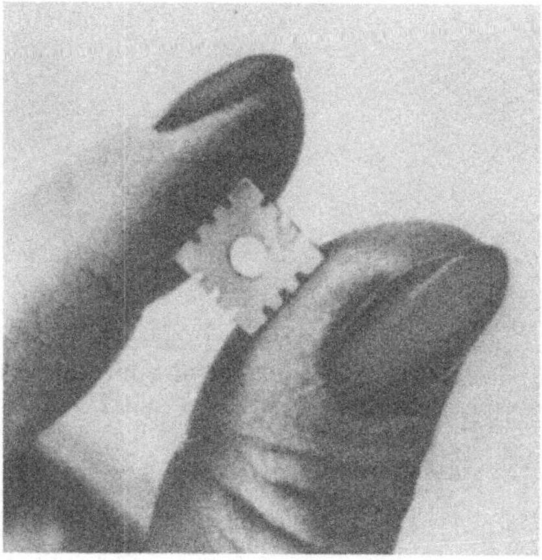

Fig. 12.4. Miniature part for electrotechnical and electronic applications separated from a glass pane according to the photochemical punching process, then ceramized (Henning [1075])

Moreover, exact dosimetry is necessary for all investigations of the effects of high-energy radiation on substances in the fields of physics, chemistry, and biology. An excellent possibility is offered by so-called glass dosimeters. An early unit of X- or γ-ray dose had been the Roentgen (r). One r is the radiation dose producing ions of both signs to the amount of one absolute electrostatic unit in $1\ \mathrm{cm}^3$ (i.e., 0.001 293 g) of air of 0 °C and 760 Torr. This amount equals 2.08×10^9 ion pairs of $1.61 \cdot 10^{12}$ ions/g air. The SI unit of the X or γ dose is $\mathrm{C \cdot kg^{-1}}$ ($1\mathrm{r} = 2.58 \cdot 10^{-7}\ \mathrm{C \cdot kg^{-1}}$).

Glasses exhibit quite different sensitivities to radiation doses of varying intensity. For this reason specific sensitive dosimeter glasses were developed for three γ-radiation regions (Jenaer Glaswerk [1082]).

γ-Radiation Dose Range $2.58 \cdot 10^{-4}$ to $2.58\ C \cdot kg^{-1}$ ($= 10^3$ to $10^7 r$)

Dosimeter glasses particularly sensitive to γ doses of $2.58 \cdot 10^{-4}$ to $2.58 \cdot \mathrm{C \cdot kg^{-1}}$ are derived from base glasses of the systems Na_2O–CaO–MgO–B_2O_3–SiO_2 colored green or blue by Cr_2O_3 or Co_2O_3. Under the influence of ionizing high-energy radiation, these colored ions trap electrons or holes corresponding to a change in valency measurable as a change in spectral transmission in the visible range.

Figure 12.5 exemplifies the change in transmission of a Cr-doped dosimeter glass after irradiation by varying ^{60}Co-γ-radiation doses [1082].

The color change reverts slowly in daylight ("fading"). For this reason dark storage is necessary for a dosimeter glass in use. Fading of a sample of this glass after irradiation by $2.58 \times 10^{-2} \cdot \mathrm{C \cdot kg^{-1}}$ is about 6% after one week and 10% after one month. Transmission changes are also affected by increases in temperature: for this reason, recalibration is necessary for use above 40 °C. Above 100 °C these dosimeter glasses cannot be used because of excessive fading.

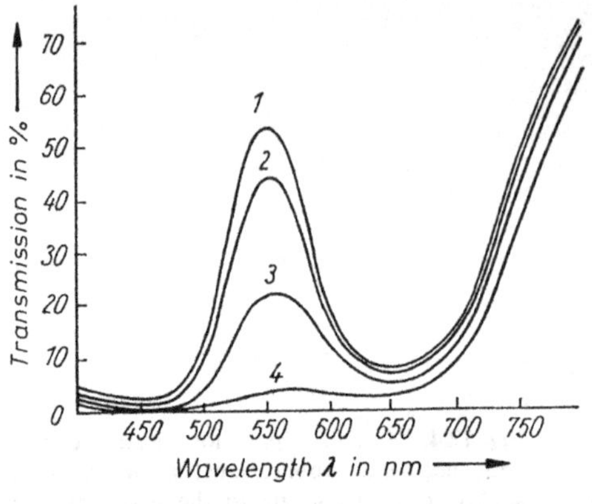

Fig. 12.5. Transmission curves of a Cr-doped dosimeter glass (4 mm thick) after (1) no, (2) $2.84 \cdot 10^{-3}\,C \cdot kg^{-1}$ (3) $2.84 \cdot 10^{-2}\,C \cdot kg^{-1}$, (4) $21.93 \cdot 10^{-2}\,C \cdot kg^{-1}$ ^{60}Co-γ irradiation

γ-Radiation-Dose Region $2.58 \cdot 10^{-3}$ to $2.58 \cdot 10^{-6}\,C \cdot kg^{-1}$ (10^{-4} to 10r)

Dosimeter glasses particularly sensitive to γ-radiation doses of about $2.58 \cdot 10^{-3}$ to $2.58 \cdot 10^{-6}\,C \cdot kg^{-1}$ consist of typically two-phase borosilicate glasses doped with AgBr and $Na_2\,(SiF_6)$ as sensitizers. Their behavior corresponds with that of photosensitive glasses based on the formation of metal colloids. For instance, measurement of the transmission of a glass colored brown by the precipitation of colloidal silver allows the irradiation dose to be determined. With incresing dose, the absorption edge of these glasses moves toward longer wave lengths. Again, fading is observed and immediate evaluation or dark storage are indicated.

γ-Radiation-Dose Region $1.29 \cdot 10^{-5}$ to $1.29 \cdot 10^{-7}\,C \cdot kg^{-1}$ (5 to 0.5 r) and Less

Dosimeter glasses sensitive to γ-radiation doses of $1.29 \cdot 10^{-5}$ to $1.29 \cdot 10^{-7}\,C \cdot kg^{-1}$ and less were first developed by Schulmann [1083] at the Naval Research Laboratory in Washington. Considering the replacement of an expensive Ag-containing crystalline dosimeter based on radiation-induced fluorescence by glass, Schulmann was aware of the poor solubility of Ag in conventional glasses. When questioned, Kreidl reported his synthesis, around 1940, of potassium-barium-aluminophosphate glasses doped with 2, 4, 8, and 16% silver salts. This base glass doped with 8% $AgNO_3$ became the first in a series of dosimeter glasses (Blair [1084]). Important refinements regarding higher sensitivity and less energy dependence followed (see, for example, Yokota and Nakajima [1085], Becker [1086], and the review by Jahn [1087]. The complicated mechanism of these radiation-induced changes remains to be explained in full detail, but involves electron or hole trappings at silver sites, as discussed, for example, by Yokota and Imagawa [1088], Yokota [1089] and Lell and Kreidl [1090].

12.5 Photochromic Systems and Glasses

12.5.1 Requirements for Photochromic Systems

The concept of photochromism, or phototropy, was coined by Marckwald [1091] for the phenomenon of a solid changing color when irradiated with light of a certain wavelength, but returning to its initial color in the dark (or on interruption of the active radiation). The process may be characterized by the equation:

$$A(\lambda_1) \frac{h\nu_1}{h\nu_2} B(\lambda_2)$$

The transition of the system A to the system B depends on wavelength.

Practice imposes two demands on photochromatic systems: (a) continuous regulation of light intensity and (b) as sudden as possible transmission change (thus ability to serve as glare-protective glazing).

The first demand originated particularly in the building sector. Quite early it became desirable to be able to produce glazes capable of automatically and continuously controlling solar irradiation, chiefly during the summer.

A satisfactory solution would be achieved if the irradiation medium became sufficiently reflective (assuming a pure white color). This would at the same time prevent heating of the system by absorption. Of course, the process would have to be reversible.

The second demand would be fulfilled satisfactorily if the system of illumination exhibited a short-term change in transmission (color).

12.5.2 Combination of Photochromic Organic Compounds and Glass

In the search for high reflectivity, it appears that around 1949/1950 organic compounds were found which unmix on heating and thus alter their scattering coefficients. The highly dispersed precipitate makes the originally homogeneous material heterogeneous and causes light reflection of above 50%. This effect (reversible unmixing dependent on temperature) was established in the following organic compounds or groups of compounds: polyvinyl methyl ether, certain polyglycoles, methyl cellulose, and acetalized or ketalyzed polyvinyl alcohols.

In 1949 Kühl and Grünwald [1092–1095] developed what was called a "Thermex" glass. It usually consists of two clear glass panes with an inserted organic layer which, with increasing temperature, changes its condition through thermocoagulation by splitting off reversibly hydrated particles. This causes a strong increase in reflection and a strong decrease in transmission. If solar irradiation ceases, the process is reversed and the system becomes clear once more. In this manner an automatic light curtain is obtained. The change in transmission τ and reflection α with temperature of a simple Thermex system

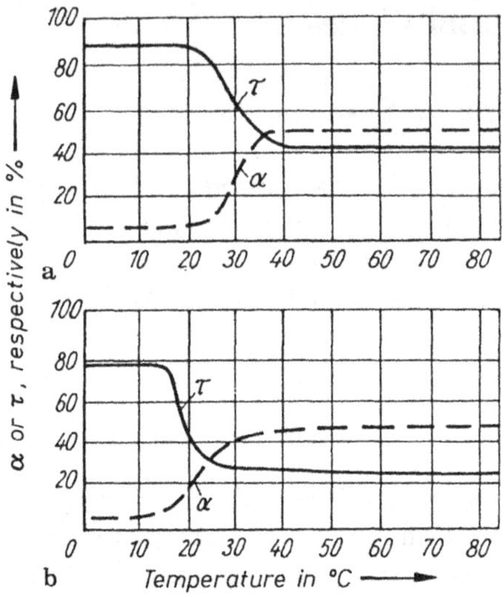

Fig. 12.6. Change in transmission τ and reflection α of "thermex glass" (two colorless glasses with an organic interlayer unmixing at a certain temperature) as a function of temperature (Reusch [1096]; (a) combination of two colorless glasses; (b) combination of a colorless and a heat-absorbing glass

(Reusch [1096]) is shown in Fig. 12.6 a. Here the change occurs between 20 and 30 °C. Figure 12.6 b shows the function of a Thermex system consisting of a clear glass pane and a heat-protecting glass pane with insertion of an organic plastic. We know that the heat-protecting glass will absorb the IR portion of the spectrum almost quantitatively while it is heating up itself. Under solar irradiation this heat is transferred to the organic system whose unmixing now causes the change in reflection and transmission at low temperatures (in Fig. 12.6 b below 20 °C).

If heat-absorbing substances are admixed to the organic plastic, the same effect is obtained and the temperature of change can be varied. Glass combinations with photochromic unmixing layers inserted have been used particularly in industrial large-area glazing.

Organic compounds which, on illumination, spontaneously and reversibly change from white to yellow, red, blue, or black comprise the following: spiropyrane, anilide ("Schiff's") bases, disulfoxide, hydrazone, osazone, semicarbazone, stilbene derivatives, succine anhydride, etc.

In combination with glass, the spiropyranes have proved to be by far the most suitable group of photochromic organic compounds. Wippler [1097] demonstrated various applications based on the combination of a colorless and a colored glass, permitting the selection of certain wavelengths of white light, with the insertion of spiropyrane compounds. Such organic photochromic systems react extremely fast.

But one important disadvantage must be registered for all organic phototropic media: The reversible color change is often lost after a quite limited number of cycles, especially if illumination is intense. The stability of the colored organic molecules has been impaired by various side reactions. Inorganic photochromic glasses are free of this disadvantage.

12.5.3 Inorganic Photochromic Glasses

12.5.3.1 Development and Application

"Fading," an effect interfering in the case of glass dosimetry, can, however, be advantageous for other technologies, particularly glare protection. In this application there is a demand for rapid darkening and as fast as possible clearing after cessation of irradiation. If the human eye is to be protected from a laser or nuclear explosion flash, darkening must be faster than closure of the eyelid. For sunglasses, too, fast as possible clearing after cessation of intensive illumination is indicated. Photochromic glasses are used, also, as display and storing media, image converts, light modulators, automatic polarizers, etc [1098–1107].

The existence of a photochromic effect in glasses after photolysis of a silver halide (or other compound) is no doubt connected with the particularities of glasses in their bonding and diffusion conditions. Although, after photolysis of a silver halide distributed in a gelatine layer the halogen is accepted or carried away by the gelatine matrix, this is on principle much less possible in glasses and the restitution of chemical bonds more probable. At the very beginning of the worldwide development of photochromic glasses, four directions involving certain demands of optimization became quite clear:

(1) Photochromic silicate glasses activated by rare earths (formation of color).
(2) Borosilicate glasses doped with silver halides.
(3) Borosilicate glasses doped with silver molybdate or tungstate.
(4) Borosilicate glasses doped with copper or cadmium halides.

The vast body of literature on photochromic glasses makes it impossible to present extensive citation and review. Within the context of this book, a short characteristic of each of these four glass groups will have to suffice. So far group (2) has found the most widespread technological application.

12.5.3.2 Photochromic Glasses Activated by Rare Earths

Cohen and Smith [1108] as early as 1962 reported on investigations of silicate glasses of variable transmission activated by rare earth ions. These investigations dealt with extremely pure silicate glasses molten under controlled reducing conditions. It should be stressed at this point that, unlike most photochromic glasses, the reversible color changes observed in these glasses are not caused by the precipitation of noble metals, but by the formation of color centers.

According to Swarts and Pressau [1109], a visible color center is formed at 570 nm when ionizing radiation is absorbed by a pure simple sodium silicate glass. Small additions of Ce, Eu, or Zr compounds favor color-center formation while others prevent it.

In the case of color-center enhancement by Ce and Eu ions, it may be assumed that Ce(III) ions, causing an absorption band at 315 nm and Eu(II) ions causing an absorption band at 332.5 nm in the glass set an electron trap near the

570 nm color center of the base glass, which will be filled under irradiation for example of 366 nm.

The center formed, appearing as an amethyst coloration, is unstable and decays rapidly a few seconds after cessation of irradiation. After prolonged irradiation, however, the intensity of the color-center band decreases, because all the Eu(II) or Ce(III) has been completely photooxidized. This decrease can be counteracted by irradiation with 253.7 nm (Hg) light. The forced quantitative photoreduction of Eu and Ce ions even increases the intenity of the 570 nm band, doubtless because in the melting process an equilibrium of valencies had been established rather than complete reduction of the Ce or Eu ions. Thus, more Eu(II) or Ce(III) ions are available as traps for color-center formation when irradiated with 366 nm light.

Solar irradiation acts similarly if additives are chosen which absorb energy in the range of solar light emission. Table 12.3 lists the base glass compositions investigated by Swarts and Pressau [1109] as to their photochromic effect. They concluded that acidic base glasses (see glass a in Table 12.3) are colored more intensively. While Ce and Eu additions of 0.1 to 0.0005 mass% increase color-center formation, Zr additions act quite differently. In an alkali-rich base glass, very small Zr additions increase phototropy, while in high-SiO_2 acidic glasses they have the opposite effect. On the other hand, small additions of Ti, V, or Fe compounds completely suppress phototropy. Impurities (heavy metals, transition elements) may impair phototropy either because of their high UV absorption or by competitive formation of electron traps and vacancies.

The position of the 570 nm absorption band in an Na_2O–SiO_2 glass is relatively constant for Na_2O contents between 8 and 30 mass%. But it shifts if Na_2O is replaced in part or entirely by other alkali or alkaline earth oxides. Thus the partial replacement of Na_2O by MgO or CaO causes a slight shift of the band toward shorter wavelengths.

The complete replacement of Na_2O by Li_2O or K_2O considerably weakens the phototropic behavior of the glass. The corresponding absorption bands lie at 494 and 717 nm, respectively.

Table 12.3. Base glasses doped with cerium or europium ions exhibiting the photochromic effect after color-center formation (Swarts et al. [1109])

Glass type	Na_2O-Content (Mass %)	CaO-Content (Mass %)	MgO-Content (Mass %)	SiO_2-Content (Mass %)
(a) Relatively basic base glass	31			69
(b) Acidic base glass	16			84
(c) Plate glass	14	12	2	72

Doping with 0.1 to 0.005 mass% Eu(II) or Ce(III) oxide.
Good color-center formation, especially on irradiation with UV light.

On principle, the phototropic behavior of glasses can be characterized by the phototropic equilibrium constant K^1 whose saturation value at uniform irrdiation is called K_0^1:

$$K^1 = \frac{1}{d} \log \frac{T_0}{T}$$

where d = thickness of the glass sample, T = percent transmission of colored glass at the maximum of the absorption band, and T_0 = percent transmission of colorless glass.

12.5.3.3 Borosilicate Glasses Doped with Silver Halides

Development and Preparation

Stookey and Schuler [1067] and Armistead and Stookey [1054] reported reversible photochromic glasses whose specific properties made them seem suitable for many practical applications. This group of new photochromic glasses was characterized by the formation of silver halide crystallites of colloidal dimensions in a borosilicate base glass during cooling or reheating. Depending on the particle size, such glasses may appear colorless, translucent, or opaque (white). Table 12.4 presents a survey of the composition of the borosilicate base glass, the doping with silver halides and sensitizers, and the spectral sensitivity of the glasses, dependent on the silver halide.

Table 12.4. Photochromic borosilicate glasses containing silver halide (Armistead et al. [1054]) and their spectral sensitivity

Glass components	Concentration in mass %
SiO_2	40 . . . 76
B_2O_3	4 . . . 26
Al_2O_3	4 . . . 26
Li_2O	2 . . . 8
and/or Na_2O	4 . . . 5
and/or K_2O	6 . . . 20
and/or Rb_2O	8 . . . 25
and/or Cs_2O	10 . . . 30
Doping with silver halide AgX	0.2 . . . 0.7 (transparent) 0.8 . . . 1.5 (opaque)
Halogen	0.2 . . . 0.4
Sensitizers	
As, Sn, Cu	0.005 . . . 0.5
Spectral sensitivity	
Doping with AgCl	300 . . . 400 nm
Doping with AgBr	300 . . . 500 nm
Doping with AgJ	300 . . . 650 nm

Glasses of this base composition may be melted and treated under oxidizing or conventional conditions. The most effective reheating temperature, which corresponds to the ripening process in a silver bromide gelatine emulsion, must be determined for each case.

A certain Al_2O_3 content is important for prevention of the undesirable precipitation of other crystalline phases or of excessive phase separation. But it is clear that here, too, controlled phase separation is a prerequisite for the enrichment of silver and halogen ions in a droplet phase (which is to be demonstrated), and the subsequent formation of silver halide molecules. Smith [1110] first pointed out the importance of phase separation in the base glass for the phototropy of a glass. Bach and Gliemeroth [1111] demonstrated experimentally the advantageous effect of connective structures on phototropy. They found, too, that the droplet-shaped glass phase containing the silver halide is the product of secondary phase separation not yet containing any crystallites.

Veit [1112] attaches no importance to immiscibility in the base glass. In contrast to this, extensive investigations by Bach, Gliemeroth and Mader [1111, 1113, 1114] as well as by Hoffmann [1115] have brought definite proof of immiscibility processes and suggestions of the importance of these for photochromism in glasses.

Formation of Silver Halide Crystals

According to Armistead and Stookey [1116], heavy metal halide crystallites form during cooling or reheating at a temperature higher than the glass transition temperature T_g of the glass. Since the melting point of the silver halide is below the T_g of the glass, the spherical microphases formed are liquid. Their formation occurs according to a nucleation and growth mechanism. While some silver halide is still present in the glass matrix, the diameter of the AgX phase increases proportionally to $t^{1/2}$. When the total precipitated quantity no longer changes, Ostwald maturation takes place: the relatively smaller AgX phases dissolve again and their material contributes to the growth of relatively larger AgX phases [1117]. The diameter of the microphases is then proportional to $t^{1/3}$. The diameter of the AgX phases can also be controlled by varying the conditions of the thermal treatment. Uniform size distribution of the AgX phases can be achieved by a hold at the temperature of nucleation (in general approximately 50 K above the T_g [1118].

The precipitated liquid AgX phases crystallize after cooling the glass below T_g. The microcrystals maintain their original spherical shape since the surrounding glass matrix is already solid. They remain under considerable tensile stresses and the photochromic properties can therefore be influenced by further heat treatments below T_g (reheating effect) [1119].

The intensity of the photochromic properties is determined by the concentrations of the AgX crystals (a minimum of 0.005 vol% is necessary) and the diameter of the microcrystals. Glasses with AgX crystals of a diameter under 5, nm hardly demonstrate any marked photochromic behavior. As the diameter of the crystal increases, the darkening intensity, that is the difference between the transmission before and after the radiation, increases. The tendency for light scattering also increases and consequently the glasses become opacified and no longer suitable for use as lenses for glasses. The optimal proportions are therefore AgX crystals between 10 and 15 nm large separated from each other by a distance of about 100 nm.

Spectral Sensitivity

Wavelengths of light causing darkening reach from the near-UV to about 650 nm, depending on the silver halide contained in the glass. (See Table 12.4.) Silver chloride glasses sensitive just to the UV would require direct UV or, at least, outdoor solar irradiation, because windows in buildings or cars absorb the efficient UV portion of sunlight. It has been noted that glasses sensitive only in the UV clear more rapidly when irradiated with longer-wave visible or near-IR light. Thus, they clear faster under visible light than in the dark.

It is important to stress at this point that it has become possible to develop borosilicate glasses containing silver halides with excellent photochromic efficiency. Darkening, however, is not directly proportional to light intensity. They react more strongly at high light intensities. Fading is an important consideration in the all-over characterization.

During the illumination of a photochromic glass, three processes occur simultaneously: (1) darkening (formation of color centers), (2) optical bleaching (fading), and (3) thermal bleaching.

The constant coloration attained under light impinging with constant intensity thus is the result of these three simultaneous processes, making it impossible to attain the maximal coloration expected from theory.

According to Smith [1110], the change in concentration of the absorbing color centers of a glass may be generally described by the following equation, assuming reactions of the first order:

$$\partial c/\partial t = K_d I_d A - (K_f I_f + K_t)c$$

where c = concentration of color centers, K_d, K_f, K_t = rate constants for darkening, fading (optical), and thermal bleaching; I_d = light intensity for darkening; I_f = light intensity for clearing; and A = concentration of sensitizable centers.

At equilibrium, $\partial c/\partial t = 0$. It follows that the color-center concentration is

$$c_s = K_d I_d A/(K_f I_f + K_t)$$

For very small K_t, for example, c_s is independent of the intensity of light, provided the relation I_d to I_f remains constant. If K_t is the determining equilibrium constant, c_s is proportional to the light intensity.

An interesting result of the investigations of Bach and Gliemeroth [1111] is that the formation of silver halide crystals of a certain size is not absolutely necessary for the photochromic effect. On the contrary, the largest rates of darkening and clearing are obtained if the precipitation of silver halide becomes stuck in the stage of a distribution of molecular dispersion within the vitreous microphases.

The principal reaction during irradiation of photochromic glasses containing silver halides may be described in the first approximation by the equation:

$$Ag^+Cl^- \overset{h\nu}{\rightleftarrows} Ag^0 + 1/2Cl_2^0$$

For the case of stabilization by copper the following additional reaction may take place:

$$Ag^+ + Cu^+ \overset{h\nu}{\rightleftarrows} Ag^0 + Cu^{2+}$$

All reaction products are metastable and may revert to their original condition.

While the rate of darkening is relatively insensitive to temperature, the rate of clearing depends strongly upon the temperature of the glass sample. With increasing temperature the rate of clearing increases.

Silver halide-containing borosilicate glasses are excellent for complete reversibility, i.e., they exhibit no fatigue phenomena in relation to time and number of darkening-clearing cycles.

During the past few years, many further developments have been carried out worldwide on the silver halide-containing photochromic borosilicate glasses – certainly the most important group among photochromic glasses (e.g., Gerth and Rehfeld [1120], Rehfeld and Rentsch [1121], Jenaer Glaswerk [1122] [1123], Kawamoto and Kikuchi [1124] (effect of γ irradiation). Araujo et al. [1125] (dichroic photochromic glasses), Morimoto and Mishima [1126] (slow-fading types), Kerko et al. [1127] (photochromic microsheets)).

Figure 12.7 shows typical darkening and clearing characteristics of a photochromic glass. The half-life time of darkening, $t_{D/2}$ here is 20 seconds, that of

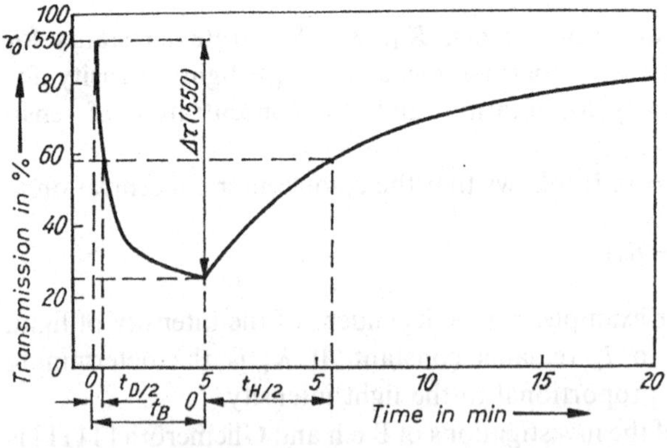

Fig. 12.7. Typical darkening and bleaching process in a photochromic glass (Heliovar I, Jenaer Glaswerk [1122]). 2-mm thick. at 25 °C; $t_{D/2}$ = half-life time of darkening, $t_{H/2}$ = half-life time of bleaching. $\Delta\tau$ (550) = darkening degree at λ = 550 nm

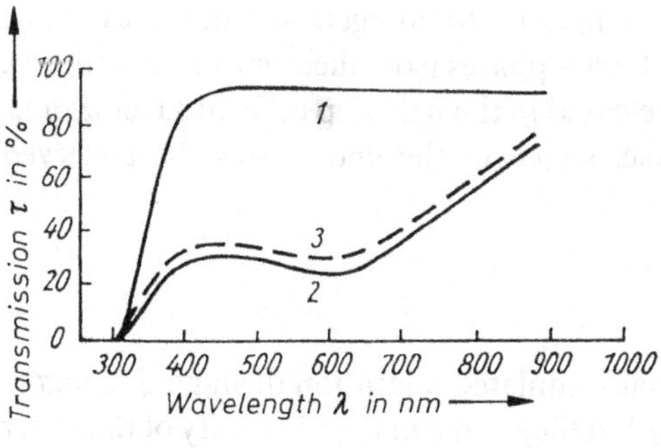

Fig. 12.8. Spectral transmission τ of a photochromic glass (Heliovar I, Jenaer Glaswerk [1122]) in bleached and darkened conditions (thickness 2 mm, temperature 25 °C); (1) bleached glass; (2) irradiated glass (mercury lamp of highest pressure type); (3) after illumination with sunlight

clearing, $t_{H/2}$ about 350 seconds. The degree of darkening $\Delta\tau$ at wave-length 550 nm is about 0.67.

Figure 12.8 shows spectral transmission of a photochromic glass in its clear condition (1) and after irradiation with a highest pressure mercury lamp (2) as well as with sunlight (3). Curves in Figs. 12.7 and 12.8 represent the state-of-the-art worldwide around 1978.

Mechanism of the Darkening and Clearing Reactions

Veit described a model for the reactions involved in the darkening and clearing of photochromic glasses in analogy to the chemistry and physics of the photographic process [1112, 1128]. The essential steps are presented here in a simplified way:

Darkening

A photoelectron of the AgX crystal is excited from the valence to the conduction band by absorbing an appropriate quantum of light ($h\nu\, E_{gap}$). In the valence band a missing electron ("hole") p^+ is left behind:

$$(AgX) \xrightarrow{h\nu} e^- + p^+$$

If these two charge carriers are not spatially separated, they recombine after a very short time so that no visible effect takes place. It is therefore necessary to momentarily fix one of the two charge carriers (the missing electron for reasons still to be explained) and to thus hinder recombination. This function can be undertaken by Cu^+ ions. Copper (I) ions are therefore called sensitizers and have a strong influence on the photochromic properties

$$Cu^+ + p^+ \rightarrow Cu^+p^+ (= Cu^{2+})$$

Whereas a copper (I) ion occupies a tetrahedral hole surrounded by halide ions (between lattice sites) in the AgX lattice, a copper (II) ion strives for a lattice site of a silver (I) ion due to the more favorable coordination proportions. It will actually displace the latter to a space between the lattice sites. These displaced silver ions are known to be capable of migration and can therefore move easily through the whole microcrystal. They migrate to energetically more favorable sites on the border of the crystal and glass phases (so-called sensitivity centers). There they trap the photoelectrons released in the first step to form silver atoms. They form a colloidal metal particle, which in the end causes the observed darkening.

Clearing

The copper (II) ions perform a thermal stimulated migration through the microcrystal (normally undirected diffusion). If they come in the proximity of the silver colloids, they virtually invert the process described above. The hole captured by the copper (I) ion withdraws an electron from a silver atom. It therefore contributes to the decomposition of the metal colloids. The more frequent copper (II) ions come into the vicinity of the colloids, the faster this decomposition takes place and consequently the faster the glass clears. It is therefore understandable that photochromic glasses which contain relatively small microcrystals always clear faster (and darken less intensely) than glasses with, in comparison, larger microcrystals, since the diffusion paths are shorter. The same effect is achieved with a temperature increase.

In contrast to the thermal fading, just described excitation of the silver colloids takes place under irradiation by light of the proper wavelengths. Electrons are also excited in the process and become available for recombination with the holes.

The explanations given here are very simplified. The original literature (for example Araujo [1129–1134] gives more exact and detailed explanations. This diffusion model for the reactions occurring in photochromic glasses containing AgX do not provide a sufficient explanation for all the phenomena observed. Alternative theories have therefore been developed [1135]. These however show definite weaknesses.

On principle, the effects known from the chemistry of photography for such ions as Cd^{2+}, S^{2-} and similar ones can be applied analogously to glasses containing AgX. These ions can always (in a definite way) stimulate or hinder partial reactions of the complex photochromic process. Thanks to an intensive development work, the most important properties of the photochromic glasses can be improved decisively with the use of such effects. Aside from the total absence of fatigue, the commercially available photochromic glasses distinguish themselves today by high reaction rates for the clearing process. The transitory change of the transmission (darkening and clearing) of a photochromic glass containing silver was measured at 550 nm and is shown in Fig. 12.7. The half time value for darkening, $t_{D/2}$, is currently between 10 and 15 seconds for typical glasses. The half time value for clearing, $t_{c/2}$, is in the range of 100 to 150 seconds. The darkening, $t_{c/2}$, is in the range of 100 to 150 seconds. The darkening intensity, $\Delta\tau$, (after 5 minutes radiation) should reach values of 0.60 with a transmission in the not-darkened state of 0.90.

The spectral-transmission degree of a photochromic glass in the cleared state and after irradiation with a mercury lamp of the highest pressure type or with sunlight is shown in Fig. 12.8. The clearly noticeable wide absorption band at 600 nm is caused by the silver colloids which form with the irradiation and later decompose.

The interrelations described here can just start to describe the problem since the number of the investigated base glasses and the combinations of activating agents (AgX) with sensitizers (Cu^+, Cd^{2+}, Tl^+, S^{2-}) is very large. For more detail the reader is referred to the original literature.

12.5.3.4 Borosilicate Glasses Doped with Silver Molybdate and Tungstate [1136]

Photochromic borosilicate glasses are highly sensitive toward UV and shorter-wave visible illumination as well as being very reversible. But this high sensitivity is also responsible for the fact that darkening is not proportional to intensity; at high intensities excessive darkening is observed. Better behavior was achieved by adding, to the same borosilicate base glass, silver molybdate or silver tungstate in concentrations of about 0.2 mass% Ag_2O and 2.5 to 10 mass% MoO_3, WO_3. Spectral sensitivity extended from 300 to 550 nm. The formation of silver molybdate or tungstate crystals is analogous to that of silver halide crystals. Their concentration should be 0.005 to 0.1 vol%. The most suitable size was found to be 4 to 20 nm (40 to 200 Å). The decisive advantage claimed for these glasses is their entirely proportional response to light intensity.

12.5.3.5 Borosilicate Glasses Doped with Copper or Cadmium Halides [846, 913, 1133, 1134]

An important step in the further development of photochromic borosilicate glasses is doping with copper or cadmium halides. These glasses show complete proportionality of darkening to light intensity while being free of silver, which may become scarcer in the future economy. The most advantageous doping levels were found to be:

Further developments lead to the system developed in former USSR:

0.2 to 1.5 CuO
0.5 to 2.5 Sb_2O_3
0.1 to 0.5 SnO
0.5 to 3.0 Cl^-

as well as the system of Morse [1137]

0.2 to 1.5 CuO
0.2 to 2.0 As_2O_3
0.1 to 0.5 WO_3 or MoO_3
0.5 to 3.0 Cl^-

Without going into detail it should be mentioned here that in all three cases a pronounced surface darkening of the glasses can be observed. This means that practically all the darkening occurs in a thin layer of about 0.2 to 0.3 mm thickness on the basis of the specific absorption properties of the copper(I) halide microcrystals [1138]. These silver-free photochromic glasses have until now been barely used in practice. This is due on one hand to the use of highly poisonous cadmium and on the other hand to the difficult technological control of this labile redox glass system. Their application can therefore only be promoted once this problem is overcome. There are certain advantages to the use of these silver-free glasses since they exhibit definitely different property combinations when compared to the silver-containing ones.

12.5.3.6 Thermally Darkening Photochromic Glasses ("TDPC")

A curious variation of AgCl photochromic glass is the thermally darkening photochromic glass (Randall and Seward [1139], Seward [1140]). This glass is dark in the stable state, darker at higher temperatures, and bleaches completely at a rate increasing with the intensity of visible light at room temperature, not at all above 300 °C. The base glass is a lanthanum borate glass. Optimum concentrations are 0.4–1.25 mass% Ag and a ratio of 1:1 to 4:1 for Cl(I, Br).

12.6 Laser Glasses

12.6.1 Introduction

The name laser is an abbreviation for *L*ight *A*mplification by *S*timulated *E*mission of *R*adiation. The laser is a light source producing intensive coherent electromagnetic radiation in the infra-red and optical spectral region. Maiman has first reported the laser properties of ruby crystals in 1960 [1141]. Immediately after that in 1961 Snitzer could prove laser activity in glasses doped with neodymium ions [1142]. Investigations on laser materials especially on laser glasses [1143–1155] show the frenzied development in this area in the last 30 years. The work of Weber [1149, 1156, 1157], Deutschbein [1158] as well as Neuroth [1159] stands out in the glass field. These authors have looked in particular into the influence of the base glass on the laser properties of Nd^{3+} doped laser glasses.

Of course the entire laser research, or application development can not be treated here, but the basic concepts and problems will be exposed.

12.6.2 Light Absorption and Light Emission in Solids

As already mentioned, stimulated emission was first observed in ruby. Therefore the typical absorption and emission processes in solids should also be more closely examined for this material.

Ruby is α-Al_2O_3 (rhombic) containing traces of Cr_2O_3 (between 0.01 and 1%, often 0.05). The introduced Cr^{3+} ions are the cause of the "ruby red" color as well as of the light emission (luminescence). Figure 12.9 shows the absorption spectrum of ruby in the visible spectral region [1160]. Both absorption bands at 410 and 560 nm are the main energy source for the luminescence. The energy

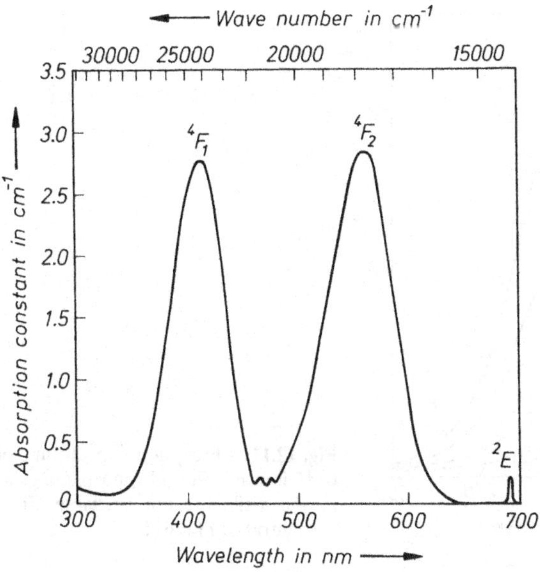

Fig. 12.9. Absorption spectrum of the ruby crystal in the visible region [1160] (light parallel to the crystal axes)

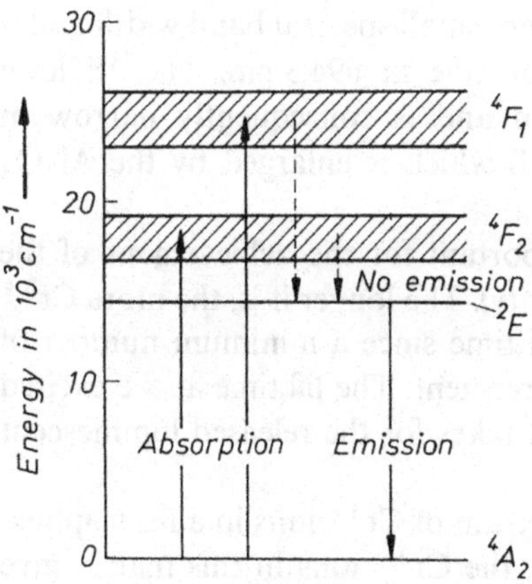

Fig. 12.10. Energy level diagram of Cr^{3+} (simplified with emission and absorption transitions [1142]

level diagram of the Cr^{3+} ion is given in Fig. 12.10. The transitions from ground state to the $^4F_1$ or $^4F_2$ level (hatched region) indicate the absorption bands. The Cr^{3+} ions therefore reach an excited state by absorbing light of a determined wavelength. The Cr^{3+} ions can only remain in this state for a very short time. After ca. 10^{-8} s a non-radiative transition to a lower energy state occurs. The energy is released as phonons and the whole substance warms up. This phenomenon is in principle common to all solids. For particular substances, such as ruby, this deactivation can be stopped at a so-called meta stable level (in ruby: ^2E level). The lifetime in this state is comparatively long (0.2–3 ms) since the transition is forbidden according to the selection rule. This ^2E level is the starting point for the emission of quanta of light. The emission is usually symbolized by an arrow which is oriented from a higher level to a lower level. An electron jumps from the ^2E level back to the ground level during the emission.

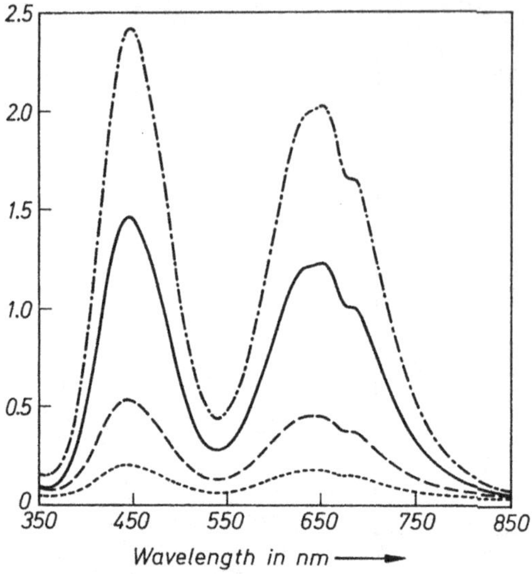

Fig. 12.11. Absorption spectrum of Cr^{3+} ions in a fluoride phosphate glass in the visible region (variable Cr^{3+} concentration) [1161]

The light emitted this way possesses a very small spectral band width and it can practically be described as an emission line at 694.3 nm. The ^2E level belongs to an inner shell of the Cr^{3+} ion and is consequently narrow in comparison to the F level of the outer shell which is enlarged by the Al_2O_3 ligand field.

The lifetime at the excited level is important for the achievement of the induced emission (prerequisite for laser activity). The longer it is, the more Cr^{3+} ions can be activated over a long period of time since a minimum number of excited laser ions is required for light reinforcement. The lifetime at the excited state is determined by measuring the time it takes for the released luminescent radiation to fade.

Figure 12.11 shows the absorption spectrum of Cr^{3+} ions in a fluorophosphate glass [1161]. In contrast to the ruby, the Cr^{3+} ions in this matrix give a green color due to a changed ligand field. This can be observed in the absorption spectrum (bands at 450 and 650 nm) as well as in the emission spectrum (very wide bands with a maximum at about 800 nm). Since the lifetime of the Cr^{3+} ions at the excited level is very short in the glass matrix, no laser activity could be achieved until now from Cr^{3+}. Nevertheless the Cr^{3+} ion could be introduced as a so-called "energy donor" for other laser ions in a glass matrix (see section of sensitization).

12.6.3 The Solid Laser

12.6.3.1 Laser Principle – Oscillator – Optical Pumps

The principle of the laser consists in achieving a synchronisation of the spontaneously emitting atomic centers. A population inversion is achieved with

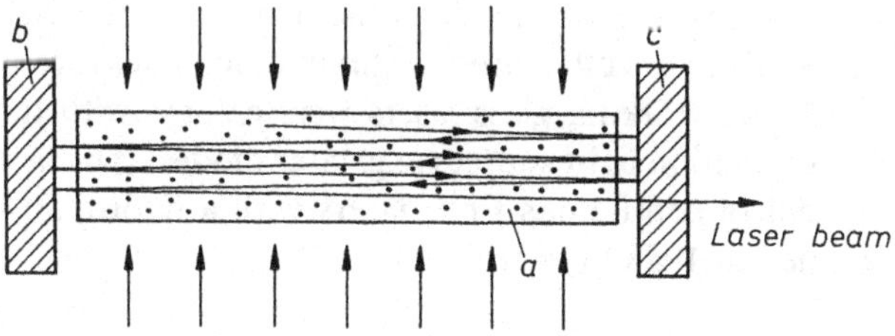

Fig. 12.12. Principle of the induced emission [1159] a. active substance b. mirror, totally reflecting c. mirror, partially transparent

a feedback of the obtained emission radiation in an oscillator. In order to obtain light reinforcement, it does not suffice to run the light wave just once through the active laser material. The light must be given the opportunity to run through the material as much as possible. The laser material is therefore placed between two mirrors, between which the light is reflected. Figure 12.12 shows such a mirror arrangement (oscillator) as well as the principle of the stimulated emission. The laser rod is usually polished flat at its edges and vacuum deposited with reflecting layers. The reflecting surfaces must be as flat and parallel as possible so as to minimize the deviation of the light ray due to multiple reflection. Such an arrangement of the components is named after the discoverers: the Fabry–Perot oscillator. The description of further oscillator types would exceed the scope of this book.

The efficiency of the laser is fundamentally determined by the diffraction loss in the oscillator, the reflection on the end mirrors as well as the scattering processes due to the optical inhomogeneities in the laser material. The light wave in the oscillator is therefore strengthened when diffraction, reflection and scattering losses are overcome. For this purpose a minimum number of excited laser ions which emit their energy into the light wave in the oscillator is necessary. Only very few ions however are in the excited state at room temperature according to the Boltzmann law. A population inversion can only be obtained with an irradiation of excited light of proper wavelengths – with the ruby, in the spectral region of 410 and 570 nm. More Cr^{3+} ions will therefore find themselves in the excited state than in the ground state and will contribute to the induced emission in the oscillator.

The pumping energy which is necessary for light reinforcement is called the threshold energy It is in an indirect way dependent on the transition probabilities in the energy level diagram of the laser ion, on the line width of the luminescence radiation, the diffraction, reflection and scattering losses in the oscillator.

12.6.3.2 Mode of Operation of Lasers

In the operation of lasers, two operation modes are distinguished principally: the discontinuous and continuous modes. The reason for this lies in the fact that the threshold limit of the stimulated emission (see Section 12.6.3.1) can only be obtained in a short time period. This is related among other reasons to the flash duration of (Xe, Hg) arc lamps, which are often used as stimulation light sources. New possibilities for continuously working glass lasers were created with the development of continuous stimulation light sources (e.g. laser diodes). The first successes with the development of continuous Er^{3+} lasers with a fluorophosphate base glass were published in 1988 [1162].

12.6.3.3 Properties of a Solid Laser Material

The requirements of a laser material are many and vary depending on the conditions of use. In laser glasses, the properties of the host glass as well as of the laser (active) ions must be considered (see Table 12.5 [1149]). These properties as well as their dependence on the glass composition were and are the object of a large number of publications [1150–1153, 1163, 1164].

Which glass is the most favorable material for a specific application will depend on the fulfilment of the following requirements:

- laser wavelength
- pulse duration
- spectrum and flash duration of the pump source
- optical configuration
- small/large signal reinforcement operation

The size of the glass laser as well as its surroundings during use, lead to questions regarding the economical production and chemical resistance. Finally the expense/use ratio is also decisive here and acts as limiting factor in the process of choosing the optimal laser glass. It is therefore not surprising that only 30 to 40 different laser glasses are available commercially (see Table 12.6).

Laser Ions

Until now all laser glasses use trivalent ions of the rare earths as active ions. The laser ions should absorb the pump light from flash lights and change it as effectively as possible into laser radiation. In other words: absorption and

Table 12.5. Fundamental properties for the characterization of a laser glass

a) host glass	b) laser ion
transperency region	absorption spectrum
homogeneity	probability for radiation emitting and non-emitting transitions
linear and non-linear refractive index	branching ratio of the fluorescence
thermo-optical and photoelastic properties	stimulated emission cross section
destructive threshold	fluorescent wavelength
chemical resistance	spectroscopic inhomogeneities
crystallization stability	
expansion coefficient	

Table 12.6. Selection of commercial Nd^{3+} laser glasses

Description	LG680	LG760	Q98	LHG80	GLS22
Producer	Schott	Schott	Kigre	Hoya	former USSR
Glass type	Silicate	Phosphate	Phosph.	Phosph.	Phosph.
Nd^{3+} concentration (10^{20} ions/cm^3)	1.4	1.4	3.3	3.1	2.0
effective cross section of stimulated emission, (10^{-20} cm)	2.9	4.3	4.3	4.8	3.9
Fluorescence wavelength λ (nm)	1061	1054	1053	1054	1055
Fluorescence half-value width $\Delta\lambda$ (nm) (width of the fluorescence peak at half-height)	27.8	20	26.3	20.2	19
Fluorescence lifetime τ (μs)	330	330	350	320	300
Non-linear refractive index $n_2(10^{-13}$ e.s.u.)	1.41	1.04	1.2	1.24	1.5
Thermo-optical constant $\Delta n/\Delta T$ (10^6/K)	8.0	0	0	1.8	0
Photoelastic constant B	21	19.4	?	17.7	?

quanta efficiency of the laser ion in the glass must be large. Figure 12.13 shows the energy level diagrams as well as the known laser transitions of important laser ions [1151].

The 1060 nm transition of Nd^{3+} was until now the transition used the most in research as well as in industry. This is a consequence of the comparatively good stimulation capacity of Nd^{3+} with flash lamps or pump lasers as well as of the unproblematic mode of operation at room temperature.

Laser ions which can work in the region of wavelengths greater than 1300 nm are gaining in importance. The production of radiation at 1550 nm, in the so-called window, has been successful with Er^{3+}. Lasers of these wavelengths can be used successfully for treatment of cornea sicknesses (opthalmology), as rangefinders and as amplifiers in data communication systems.

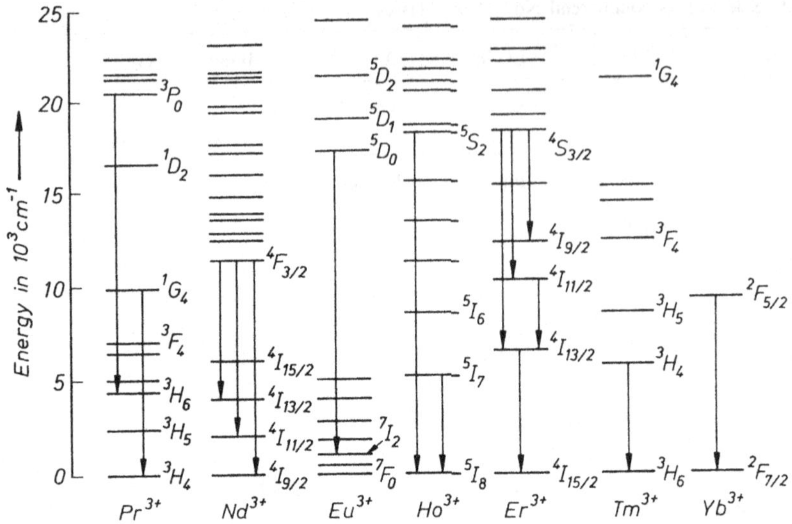

Fig. 12.13. Energy level diagram and laser transitions of chosen trivalent ions of the rare earths [1151]

Characteristic Parameters of Laser Glasses

The induced emission cross section σ is a measure of the reinforcement that a laser ray experiences in the path through the active material,

$$g = \sigma \cdot N$$

where g is the reinforcement and N is the number of laser ions per cm^3 of glass. The fluorescence lifetime τ is experimentally obtained from fluorescence decay curves as the time needed to reach 1/e of the initial intensity. The larger the lifetime, the more suitable the pump light would be for stimulation.

In particular with lasers of high efficiency, a low non-linear refractive index n_2 must be considered since otherwise the laser glass is destroyed by self-focusing (lens effect in active laser glass).

Of course the laser material can – even under very intensive pump radiation – show only a very low absorption at the laser wavelength. A good stability with regard to thermal shock in the stimulation process is also desirable in this context. This means the value $S(\mathrm{i} - \mathrm{u})/\alpha/E$ should be large (where S is the mechanical strength, u is Poisson's number, α is the coefficient of expansion and E is Young's modulus). One must strive for as small as possible values of the thermooptical and photoelastic constants $\Delta n/\Delta T$ or B since otherwise the radiation quality would drastically deteriorate.

The laser glass should possess a high destruction threshold towards strong laser radiation and be extremely resistant to the environment (e.g. humidity in the air).

12.6.4. Efficiency Increase by Sensitization

Sensitization of the Luminescence

Sensitization is one possibility, to increase the efficiency of a laser material. Sensitizers with comparatively intensive absorption bands should transfer their vibrational stimulation energy to the laser ion. "Sensitized luminescence" is understood to be a phenomenon in which particles emit radiation without having been directly stimulated by absorption. With regard to interionic relaxation, interactions appear by which a stimulated ion (donor) relaxes and another ion (acceptor, quencher) completely or partially absorbs the released energy. These energy transfer processes can be radiative or non-radiative. Theories about energy transfer were discussed by, among others, Förster [1165] and Dexter [1166] and have been continuously developed until the present.

Energy transfers between transition metal ions and ions of the rare earth metals were investigated for the first time by Murphy and coworkers [1167] in $LaAlO_3:Cr^{3+}:Nd^{3+}$. Melamed and coworkers [1168] demonstrated the sensitization of Nd^{3+} by UO_2^{2+} in a barium crown glass by measuring the lifetime of UO_2^{2+} while varying the concentration of the acceptor ions Nd^{3+}. It is however necessary to mention that the existence of an energy transfer does not guarantee an increase of the laser efficiency. The laser glass only exhibits all its properties during laser use.

In conclusion, the sensitization of the laser ion Er^{3+} in a special fluoride phosphate glass matrix will be described. Er^{3+} possesses relatively weak absorption bands (see Fig. 12.14). Pump light can therefore only be very ineffectively used for stimulation and consequently the degree of laser activity is very low. A successful energy transfer $Cr^{3+} - Yb^{3+} - Er^{3+}$ was obtained with a proper enrichment in Cr^{3+} and Yb^{3+}. This transfer occurs in two steps in which Yb^{3+} plays a central role (see Fig. 12.15). Figure 12.15 shows the energy level diagrams of Cr^{3+}, Yb^{3+} and Er^{3+} in a fluorophosphate glass including transfer steps and laser transitions for Er^{3+} at 1550 nm.

Figure 12.16 shows the spectral overlapping of the Cr^{3+} emission (curve 1) and the Yb^{3+} absorption (curve 2) in a fluorophosphate glass which is a prerequisite for a $Cr^{3+} - Yb^{3+}$ energy transfer.

The efficiency of the sensitization $Cr^{3+} - Yb^{3+} - Er^{3+}$ is demonstrated in Fig. 12.17: in the practical test, the laser wavelengths could be tuned from 1520 to 1600 nm in this laser glass at room temperature.

12.6.5 Applications of Lasers

Very high radiation power can be achieved over a short time for example with Nd^{3+} glass lasers. Moreover, the good coherence of the light rays led to the development of a series of technical applications. Very high temperatures could be obtained in a spatially restricted way. This can be used for melting, cutting, welding, drilling and vaporizing. The energy problems of humanity could

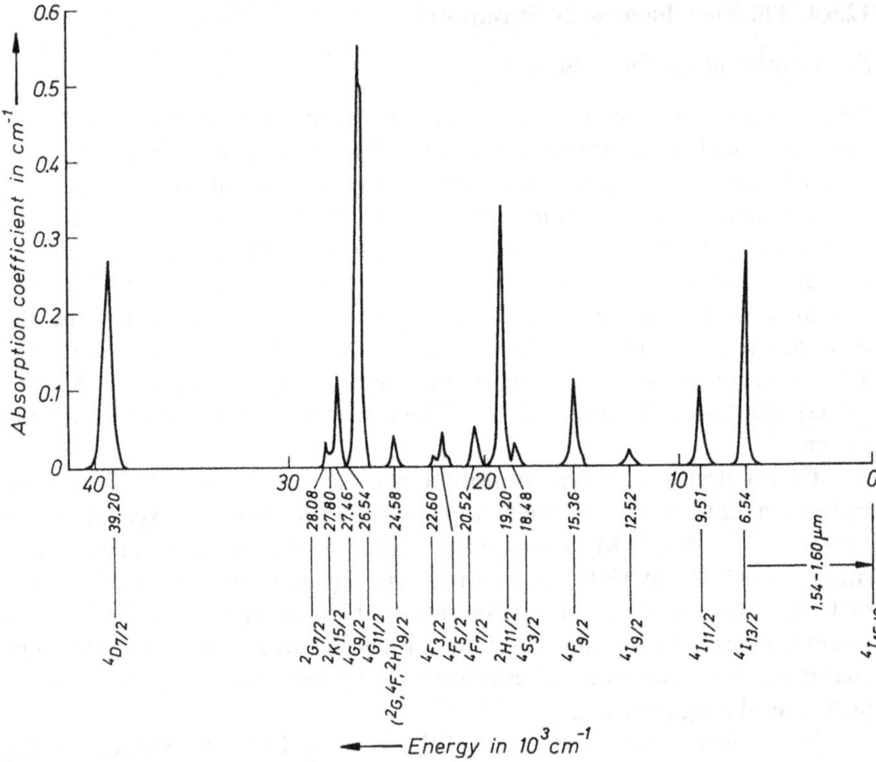

Fig. 12.14. Er^{3+} absorption spectrum ($n = 5 \cdot 10^{19}$ $Er^{3+} \cdot cm^{-3}$). Erbium in a fluoride phosphate glass matrix

eventually be solved by the use of giant impulse glass lasers for the initiation of controlled nuclear fusion.

The laser is of great importance in spectroscopy as a source of monochromatic radiation of high intensity (Raman spectroscopy)

Lasers are very important in communication technology because the light ray (10^{14} Hz) offers a larger bandwidth than radio waves (10^9 Hz) for modulation as a wave carrier.

The use of lasers for range finding is based on their good directional capacity.

The laser scalpel is firmly entrenched in biology and medicine. The focused laser ray is used here for the destruction of tissues. The point-like concentration of the ray is also of importance here. The operation occurs without physical contact and is therefore free of infection. The laser is an invaluable tool even the treatment of just injuries of the retina of the eye are considered.

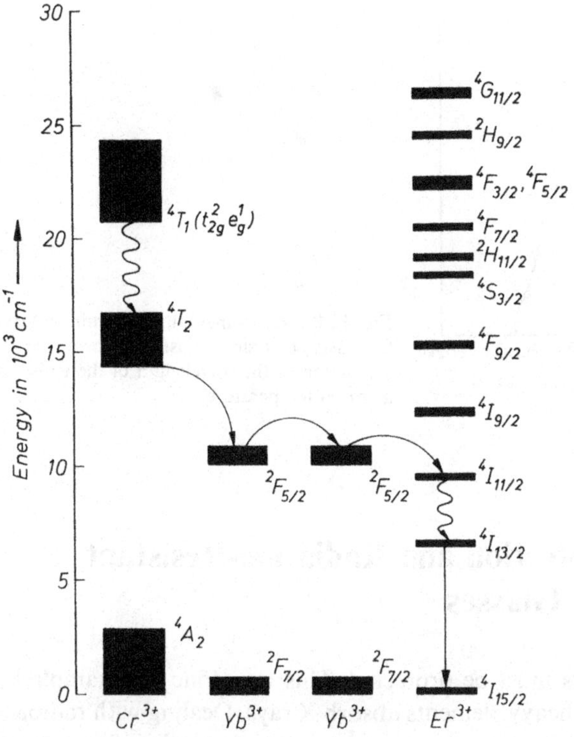

Fig. 12.15. Energy level diagram of Cr^{3+}, Yb^{3+} and Er^{3+} including transfer steps and Er^{3+} laser transitions at 1550 nm for a fluoride phosphate base glass

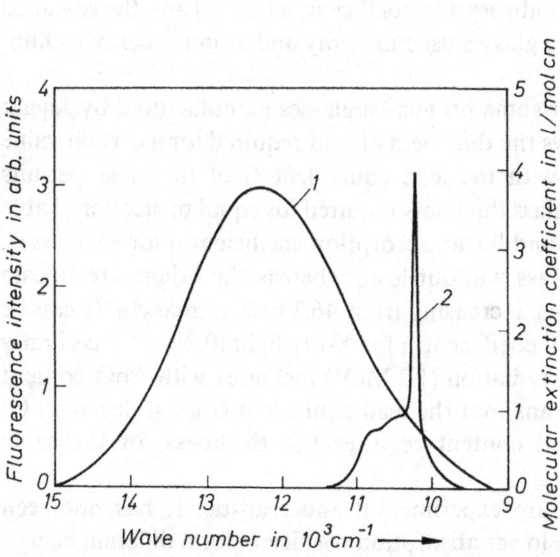

Fig. 12.16. Spectral overlap of the Cr^{3+}-(1) and the Yb^{3+}-(2) emission in a fluoride phosphate base glass as a prerequisite for a Cr^{3+}-Yb^{3+} energy transfer

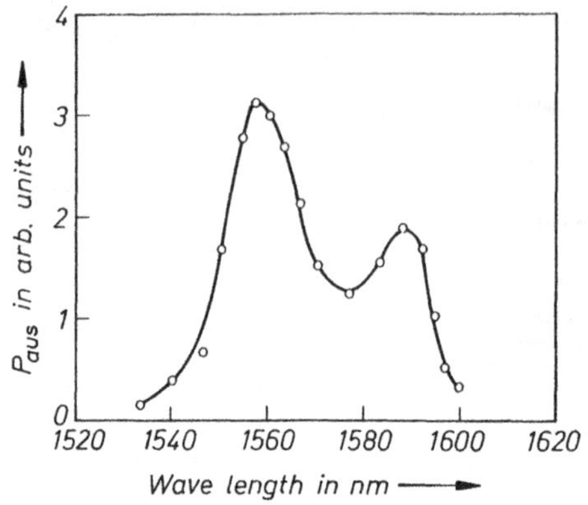

Fig. 12.17. Experimentally determined curve for laser emission. Laser performance as a function of the wavelength of the emission at room temperature

12.7 Radiation Protection and Radiation-Resistant ("Protected") Glasses

People dealing with X rays must be protected. This was done for example by wearing lead aprons, since heavy elements absorb X rays. Dealing with radioactive isotopes or strong γ-ray sources also requires secure protection. Since one is not permitted to enter space containing high-energy-radiation sources or radioactive substances and is forced to work through manipulators from far away, some visual connection becomes an unconditional necessity.

While the entire space is screened by thick lead concrete walls, the visual connection is screened by thick lead glass panes. The protective effect of lead glass is determined by the lead content (density) and the thickness of the glass. It is necessary to know the linear absorption coefficient μ (cm^{-1}) and the so-called lead equivalent G of protective glasses used in X-ray and atomic energy technology.

Table 12.7 lists values for some protective glasses manufactured by Jenaer Glaswerk [1088]. If one knows the thickness of lead required for a certain value of protection, mutiplication with the lead equivalent G of the glass permits immediate calculation of the glass thickness required for equal protection. Table 12.7 shows lead equivalent G and linear absorption coefficient μ for six glasses.

Glass 1 is a soda-lime glass without lead, whereas the others are silicate glasses with high lead contents, increasing from 46.3 to 71.3 mass%. It can be seen that the linear absorption coefficient μ for X-ray light (0.2 MeV), cesium-γ radiation (0.6 MeV), or Co-γ radiation (1.2 MeV) increases with PbO content (density) of the glass. This means that the lead equivalent G must decrease. Of course, a glass of higher lead content requires less thickness for the same protective effect.

The values in the table are experimental and realistic. It has not been explained why, in the case of a lower absorption coefficient for a different energy

Table 12.7. Properties of some radiation-protective glasses (Jenaer Glass Works [1088])

			A		B		C	
No	PbO-content of glass (mass %)	Density ρ ($g \cdot cm^{-3}$)	$E = 0.2$ MeV μ	G	$E = 0.6$ MeV μ	G	$E = 1.2$ MeV μ	G
			(in cm^{-1})		(in cm^{-1})		(in cm^{-1})	
1	—	2.52	0.31	32.7	0.20	5.51	0.14	3.97
2	46.3	3.68	1.66	6.09	0.32	3.45	0.19	2.93
3	51.2	3.86	1.89	5.35	0.34	3.24	0.20	2.78
4	55.5	4.08	2.12	4.78	0.36	3.05	0.21	2.65
5	64.0	4.60	2.67	3.79	0.42	2.66	0.24	2.33
6	71.3	5.13	3.24	3.13	0.47	2.36	0.26	2.14

ρ = density. G = lead equivalent. E = energy of irradiation in MeV. μ = linear coefficient of absorption.

of radiation, the lead equivalent also decreases. Perhaps this anomaly is connected with a completely different scattering behavior.

As has been mentioned before, X-ray or γ-ray irradiation causes many structural changes in the glass, from the detachment of electrons to the rupture of chemical bonds. This may often be recognized by the discoloration – from yellow to dark brown – suffered by the glasses. These effects caused by radiation-induced color centers could make the protective glass unusable. The glass must be stabilized against them, usually by the addition of about 1 to 2 mass% CeO_2. This stabilization is based on the function of Ce^{4+} as an electron trap. competing with the formation of the undesirable color centers.

Whereas in photosensitive glasses Ce^{3+} ions function as electron donors and condition photosensitivity in conjunction with silver ions:

$$Ce^{3+} + h\nu \xrightarrow{UV} Ce^{4+} + e^-$$

$$Ag^+ + e^- \rightarrow Ag^0$$

here Ce^{4+} ions function as electron acceptors (traps)

$$Ce^{4+} + e^- \rightarrow Ce^{3+}$$

causing considerable stabilization against discoloration.

This stabilization may also be achieved, and was first achieved in industrial practice (Kreidl [1170]. Kreidl and Hensler [1171]) for lead-free optical glasses used in installations for nuclear physics or nuclear technology. The stabilization is accompanied by a minor yellow coloration caused by the tail of the absorption band of Ce^{4+} reaching into the visible range. In the absence of Ce^{4+}, irradiation may even reduce ions which are otherwise not easily reduced to their elementary state. Phosphate glasses may become red under γ irradiation (Kreidl [1170]), perhaps due to reduction of P^{5+} to P^0. The red coloration can be removed by strongly oxidizing fuming HNO_3.

The effect of stabilization by cerium is considerable. Conventional optical glasses will show coloration clearly visible to the eye after doses of ca. $2.58 \times 10^{-4}\,C \cdot kg^{-1}$, and will become dark and useless after 2.58×10^{-2} to 2.58×10^{-1} irradiation. The so-called "protected" optical glasses stabilized by 1 to 2% CeO_2 show hardly any noticeable discoloration, even after $2.58 \times 10^{-1}\,C \cdot kg^{-1}$ doses of irradiation.

12.8 Transmission Changes of Colored Glasses under γ Irradiation [1172]

In 1955, Kreidl and Hensler [1171], comparing the effect of Ce^{4+} as an electron trap preventing discoloration of optical glasses with that of transition elements, found new induced colors in irradiated phosphate glasses containing iron, manganese, cobalt, nickel, vanadium, and copper. They attributed these changes to analogous competitive traps preventing the formation of the bands observed in the base glass. In 1958, Fanderlik systematically studied commercial color filter glasses using γ radiation up to $2.58 \cdot 10^{-1}\,C \cdot kg^{-1}$. According to Fanderlik's results, colored glasses can be divided into three groups, according to the response of their transmission behavior to γ irradiation.

(a) Colored glasses insensitive up to $1.032 \cdot 10^{-2}\,C \cdot kg^{-1}$ irradiation with no change in transmission behavior.
(b) Colored glasses sensitive up to $1.03 \cdot 10^{-2}\,C \cdot kg^{-1}$, with a considerable change in transmission in the visible region.
(c) Colored glasses, particularly cut-off and "striking" color glasses, whose absorption edge remains uninfluenced even up to a γ-irradiation dose of $2.58 \cdot 10^{-1}\,C \cdot kg^{-1}$, but whose transmission is reduced over the entire visible range. Group a involves practically all glasses containing Cu, Fe, Co, and Ni. Figure 12.18 demonstrates this finding for glasses NG5 (Fe + Co in borosilicate glass), BG7 (Cu in silicate glass), and UG1 (Ni in silicate glass) (but color changes in phosphate glasses observed by Kreidl and Hensler [1171] were considered at the higher dose of $2.58 \cdot 10^{-1}\,C \cdot kg^{-1}$ (10^6 r)).

Group b contains, above all, all glasses containing chromium. Figure 12.19 illustrates the considerable transmission change under γ irradiation attributed to some valence change. This behavior was used in the section on dosimetry.

The addition of a small amount of Cr also makes an otherwise resistant glass (e.g., one containing Cu) susceptible (see Fig. 12.19b and compare with Fig. 12.18b). Glass GG10 is a Cr-doped silicate glass, VG6 is a Cr + Cu-doped borosilicate glass.

Group c contains all "striking" color glasses whose steep absorption edge is conditioned by microcrystallites of the cadmium chalcogenides. Figure 12.20 shows that γ irradiation causes no change in the area from the UV to the cut-off edge. The horizontal part in the visible range, however, is shifted parallel toward

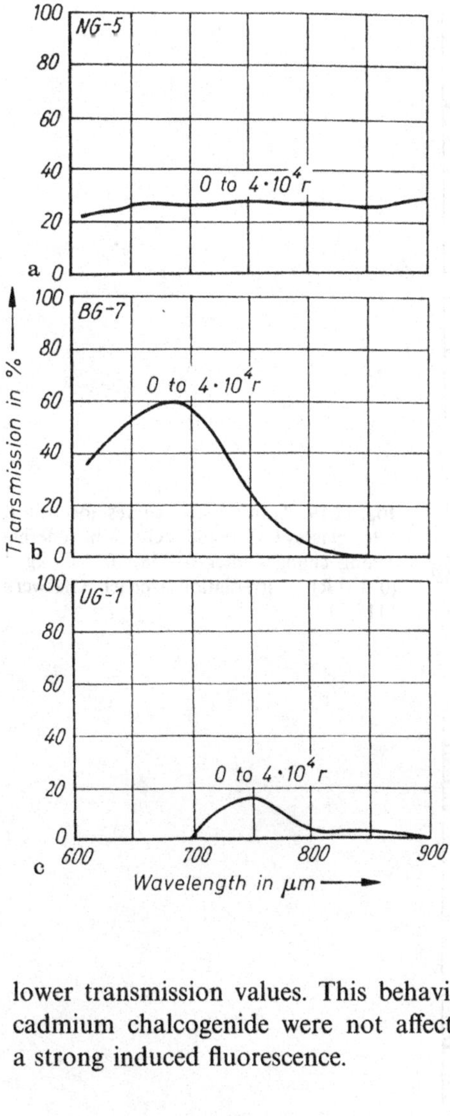

Fig. 12.18. Transmission curves for Schott filter glasses NB5. BG7, UG1 (a)–(c) which remain unaffected even after a γ dose of 0 to $10.32 \cdot 10^{-3}$ C·kg^{-1} (= 0 to $4 \cdot 10^4$ R) (Fanderlik [1172]

lower transmission values. This behavior indicates that the micro-crystals of cadmium chalcogenide were not affected by the γ dose applied, except for a strong induced fluorescence.

12.9 Solarization

Solarization is defined as a color change in glass under prolonged exposure to sunlight. Faraday [1173] observed and described the phenomenon for window glasses. Later it was established that the UV portion of sunlight causes the color and transmission changes, for example, in flat glass.

While fluorescence phenomena, also induced by UV irradiation, are based on the raising of electrons to higher levels by excitation and the emission of light of, on principle, longer wavelength on their return to the ground state, and while fluorescence stops immediately when the exciting UV irradiation stops, solarization effects are clearly related to permanent changes in structure

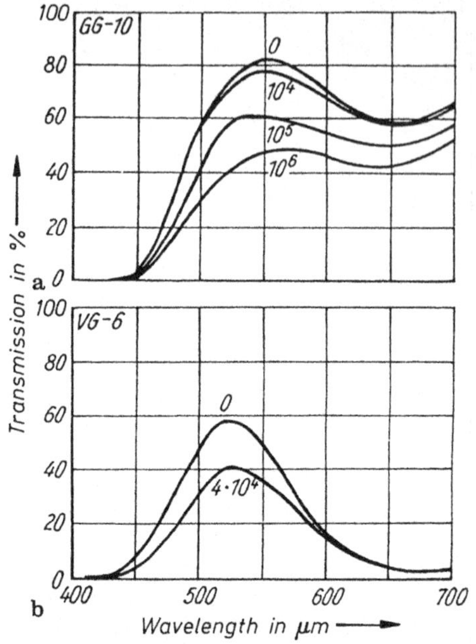

Fig. 12.19. Transmission curves for Schott filter glasses GG10 and VG6 (a–b) suffering strong changes after 0–2.58 · 10^{-1} C · kg^{-1} (0–10^6 R) γ irradiation (Jenaer Glaswerk [1172])

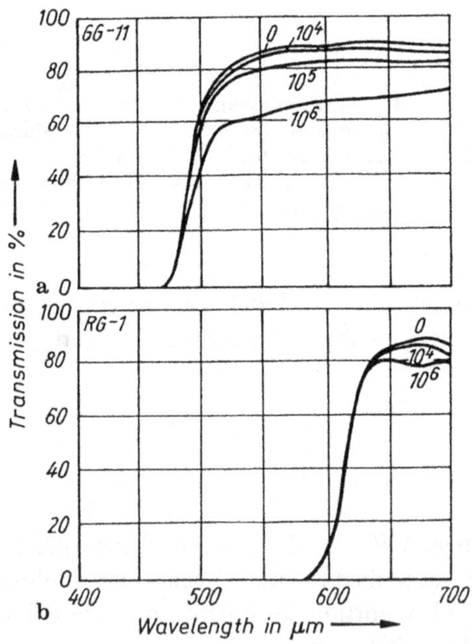

Fig. 12.20. Transmission curves for Schott filter glasses GG11 and RG1 (a–b) suffering no changes of the UV absorption edge but considerable changes in the horizontal profile after 0–2.58 · 10^{-1} C · kg^{-1} (0–10°) γ (Irradiation Jenaer Glaswerk [1172])

which can be reversed only by heating. Some solarization effects are reversed at relatively low reheating temperatures. Mathematical descriptions describing these transmission changes as functions of time and temperature are primarily due to Klemm and Berger [1174].

Almost unanimously, oxidation and reduction processes of heavy transition elements, which are present as trace impurities in almost all mass-produced technical glasses, are recognized as the main underlying cause of the solarization phenomena in glasses produced by the UV portion of sunlight. The manganese ions play a particular role in these processes. Long ago, small amounts of MnO_2 were added to many glass melts under the name of "glass-maker's soap." Its chief function was to minimize the bluish tint of glasses caused by Fe^{2+} ions by oxidation to Fe^{3+} with reduction of Mn^{4+} to Mn^{2+}. Fe^{3+} gives the glass a much weaker yellowish tint.

If such glasses are exposed to sunlight for a longer time, a bluish-violet coloration is observed, which is based on the photooxidation of Mn^{2+} to Mn^{3+}.

$$Mn^{2+} + h\nu \rightarrow Mn^{3-} + e^-$$

Mn_2O_3 is known to be violet. It appears that other ions contained in many glasses may function as electron acceptors and increase the solarization effect of Mn^{2+}.

According to Weyl [662], other redox equilibria may play a relatively significant role in the coloration by solarization of glasses, e.g., in addition to $4MnO + As_2O_5 \rightarrow 2Mn_2O_3 + As_2O_3$:

$$\begin{aligned}
&4FeO + As_2O_5 \rightarrow 2Fe_2O_3 + As_2O_3\\
&2Ce_2O_3 + As_2O_5 \rightarrow 4CeO_2 + As_2O_3\\
&Ce_2O_3 + V_2O_3 \rightarrow 2CeO_2 + 2VO\\
&Ce_2O_3 + Cu_2O \rightarrow 2CeO_2\\
&Ce_2O_3 + Cu_2O \rightarrow 2CeO_2 + 2Cu\\
&Ce_2O_3 + Ag_2O \rightarrow 2CeO_2 + 2Ag, \text{ etc.}
\end{aligned}$$

This may help in understanding the most varying solarization effects of glasses (see Table 12.8), as related to the kind and concentration of their heavy metal impurities.

Table 12.8. Solarization effects of glasses related to production (after Salmang [15])

Kind of glass	Color	
	Before exposure of sunlight	After exposure to sunlight
French colorless plate glass	bluish white	yellowish
German colorless plate glass	light green	light bluish
British colorless plate glass	light green	yellowish green
British crown glass	light green	light purple
Belgian window glass	brownish yellow	deep purple
British window glass	dark green	brownish green
American crystal glass	light bluish white	purplish white
American crystal glass	light bluish white	light yellowish green
American window glass	bluish green	no change

13 Survey of the Physical Basis of Some Glass Properties

13.1 Introduction

A mathematical presentation, or attempts at the calculation of glass properties, are as old as modern glass research or the systematic development of new glasses. In most cases, relations between composition and properties were established entirely on an empirical basis. One was forced frequently to introduce one, two, or more constants into equations describing these relations in order to obtain better agreement with experimental data.

It is for this reason that, with a few exceptions, such attempts at a mathematical description of glass properties and their relation were placed at the end of this book. Now, after the presentation of so many different glass structures, even within the same glass system, more understanding may be expected for mathematical solutions offering only limited satisfaction. It will be more easily understood why, in many cases, a linear dependence of properties on composition cannot even be expected and why certain equations will correctly reflect relations only within a limited area of concentrations. It is simply impossible to speak of a structural model for all glasses and to try, on this basis, to comprehend and calculate their properties. This certainly complicates these relations, but also makes them interesting.

In what follows, no comprehensive representation of the mathematical description of glass properties and property relations is intended. Rather, a limited selection of mathematical formulations, which have proved more or less reliable in practice and which have become indispensible in modern glass technology, is offered.

13.2 Refraction of Light, Dispersion and Abbe's Value

The most important characterizations of an optical glass are index of refraction, dispersion, and Abbe's number.

Up to the present time, the principal refractive index n_d referred to a wavelength of 587.1 nm (the d line of helium), but recently n_e, referring to a wavelength of 546.1 nm (the e line of mercury), where the maximum of the sensitivity of the human eye is observed, has been standardized as the principal index. In the United States standardization to n_e has not yet been accepted at the time of this writing.

The principal dispersion previously defined as $n_F - n_C$ is now generally accepted to be $n_{F'} - n_{C'}$.

The relative partial dispersions are defined as the relation of the various dispersions in the spectrum to the principal dispersion, that is, as:

$$\Delta n/(n_{F'} - n_{C'}) \text{ (formerly } \Delta n/(n_F - n_C))$$

The following Δn values are used in practice:

$n_{u2} - n_{u1}$, $n_{u1} - n_i$, $n_i - n_h$, $n_h - n_g$, $n_g - n_{F'}$, $n_g - n_F$, $n_F - n_e$, $n_{F'} - n_e$, $n_e - n_d$, $n_d - n_D$, $n_e - n_{C'}$, $n_e - n_C$, $n_{C'} - n_{i1}$, $n_{A'} - n_{i1}$, $n_{i1} - n_{i2}$, $n_{i2} - n_{i3}$, $n_{i3} - n_{i4}$, $n_{i4} - n_{i5}$, $n_{i5} - n_{i6}$, $n_{i6} - n_{i7}$, $n_{i7} - n_{i8}$, $n_{i8} - n_{i9}$. The chosen lines are listed in Table 13.1.

The Abbe number (also called ν value. ν value, constringence, reciprocal relative dispersion) is defined as

$$\nu_d = (n_d - 1)/(n_{F'} - n_{C'})$$

and more recently (except that the new standard is still debated in the United States:

$$\nu_e = (n_e - 1)/(n_{F'} - n_{C'})$$

If the relative partial dispersions for equal wavelengths

$$P_{1,2} = (n_1 - n_2)/(n_{F'} - n_{C'})$$

Table 13.1. Wavelengths of Chosen Spectral lines

Wavelength in nm	Symbol	Element	Spectral range
312.6	u2	Hg	↑
334.1	u1	Hg	UV
365.0	i	Hg	↓
404.7	h	Hg	violet
435.8	g	Hg	↑
480.0	F′	Cd	blue
486.1	F	H_2	↓
546.1	e	Hg	green
587.1	d	He	↑
			gold
589.3	D	Na	↓
643.8	C′	Cd	↑
656.3	C	H_2	red
768.2	A′	K	↓
852.1	i1	Cs	↑
1013.9	i2	Hg	
1128.6	i3	Hg	
1395.1	i4	Hg	
1529.6	i5	Hg	IR
1813.1	i6	Hg	
1970.1	i7	Hg	
2249.3	i8	Hg	
2325.4	i9	Hg	↓

are plotted vs. the Abbe number ν, one obtains approximately a straight line called the "normal" line. For all glasses lying on this line

$$P_{1,2} = a_{1,2} + b_{1,2}\nu_e$$

For any glass not conforming to this rule, the desirable deviation $\Delta P_{1,2}$ defined by $P_{1,2} = a_{1,2} + b_{1,2}\nu_e + \Delta P_{1,2}$ permits correction of the so-called secondary spectrum.

In practice, the use of partial dispersions requires high accuracy. To minimize small errors in measurements, it is advisable to smooth data by using a dispersion formula such as:

$$n^2 = A_0 + A_1\lambda^2 + A_2\lambda^{-2} + A_3\lambda^{-4} + A_4\lambda^{-6} + A_5\lambda^{-8}$$

where A_0 to A_5 are constants. The formula is derived by a series development of the theoretical formula, the term $A_1\lambda^2$ representing the influence of infrared absorption and the terms with negative exponents of λ representing that of the UV absorption regions on the refractive index [1175]. The refractive index, according to theory, increases rapidly toward infinity as it approaches the long wave side of any absorption band.

Since these conditions depend in a complex manner on the various resonating bonds between glass components and their electronic systems, as well as on density, additive factors for predicting refractive indices cannot be expected to give precise results. Nevertheless, empirical tables were developed, e.g., by Appen, [1176, 1177] for various constituents which gave reasonable results within conventional composition limits when inserted in a simple additive relation:

$$n = \frac{1}{100}\sum n_i p_i$$

where p = mole percent of constituent i and n_i = factor for constituent i. To minimize deviations, Gilard and Dubral [1178] gave the formula:

$$n = \frac{1}{100}\sum(n_i' p_i + n_i'' p_i 2)$$

using sets of two factors, n_i' and n_i'', in their table (p in mass%).

13.3 Density

Although density is not as important as the refractive index, its value is frequently required for calculating other parameters such as molar volume, molar refraction, etc.

Density, too, may be treated as an additive property over some compositional ranges. The first relation to composition, established empirically by Winkelman and Schott [1179], describes a linear dependence. In some cases this relation is applicable, in other cases it does not even cover the entire glass-formation region of a simple system. Just consider the crossing density curves of binary alkali silicate systems, or the turning points in the density curves of simple borate glasses.

Here, too, Gilard and Dubral [1178] tried to account for the deviations from linearity by the introduction of a quadratic term:

$$\frac{1}{\rho_{\text{glass}}} = \frac{1}{100}\sum[P_i/(\rho_i' + P_i^2/\rho_i'')]$$

Further modifications of such simple relations were, in part, successful in fitting experimental data quite well [1176, 1177, 1180–1183], but they never covered the entire range of concentrations.

13.4 Molar Refraction

In organic chemistry the molar refraction was mobilized successfully for clarification of the fine structure of compounds because it is a strictly additive combination of atomic increments for species such as C– , C= , C= , etc. These increments, which can be determined exactly, are available from tabulations.

According to Lorentz–Lorenz, the molar refraction (MR) is:

$$MR = \frac{n^2 - 1}{n^2 + 2}\frac{M}{\rho}\ (\text{cm}^3)$$

where n = refractive index, M = molecular weight, and ρ = density.

The molar refraction, MR, expressed in cm^3, may be conceived, in a first approximation, as a measure for the volume of the particles. The expression $(n^2 - 1)/(n^2 + 2)$ usually represents 68 to 79% of the molar refraction. The expression M/ρ is the molar volume (MV). The molar refraction is also known to be directly proportional to the polarizability α:

$$MR = \frac{4}{3}\pi N_{\text{L}}\alpha$$

where N_{L} is Loschmidt's number, i.e., the higher the polarizability of an ion, the more it contributes to the molar refraction.

For application to problems of glass chemistry, the classical Lorentz–Lorenz equation becomes:

$$MR_{\text{glass}} = \frac{n^2 - 1}{n^2 + 2}\frac{\bar{M}}{\rho}$$

where $\bar{M}$ is a mean molecular weight calculated from the glass composition. Originally it had been hoped that study of the molecular refraction in glasses could be as useful as in the case of organic chemistry, since in some cases excellent additivity and linearity of MR curves had been found. Later, however, it was recognized that even in simple binary glasses the polarizability of certain ions may vary considerably with the concentration of components, which leads to more or less pronounced deviations from linearity. Such deviations may be conceived as a direct measure of changed polarization and deformation effects.

Table 13.2 lists the ionic refractions of some important cations contained in glasses, according to Fajans and Kreidl [386]. Assuming that these values remain constant – an assumption certainly not valid for the larger cations in glasses – and calculating the molar refraction of oxygen R_O^{2-} by subtracting the values for the respective cations, one finds an enormous variation for $R_{o^{2-}}$: 3.05 to 7.97 cm^3. For SiO_2, $R_{o^{2-}}$ was calculated to be 3.67 cm^3. Certainly the anions are in some way contributing to the higher $R_{o^{2-}}$ values (Table 13.3), i.e., to the polarizability of at least the larger cations (Ba^{2+}, Pb^{2+}).

If one further considers that bridging and non-bridging oxygens must possess different increments of refractivity, and that in many glasses phase-separation phenomena occur which cause the enrichment of certain ions in

Table 13.2. Ionic Refractions (cm^3) of some cations occurring in glasses (Fajans and Kreidl)

Cation	Ionic refraction
Li^+	0.2
Na^+	0.5
K^+	2.2
Mg^{2+}	0.28
Ca^{2+}	1.33
Ba^{2+}	4.3
Pb^{2+}	3.1
B^{3+}	0.05
Al^{3+}	0.17
Si^{4+}	0.1
P^{5+}	0.07

Table 13.3. Ionic Refractions of oxygen calculated subtractively from the molar refractions of various compounds (Fajans and Kreidl [386]

Compound	$R_{O^{2-}}$ in cm^3
H_2O	3.05
Be_2SiO_4	3.35
SiO_2 (quartz)	3.55
SiO_2 (cristob.)	3.70
Mg_2SiO_4	3.83
$LiAl(SiO_4)$	4.01
$NaAl(SiO_4)$	4.03
Ca_2SiO_4	4.53
Sr_2SiO_4	4.67
Ba_2SiO_4	4.97
$BaSiO_3$	4.55
$Ba_2Si_3O_8$	4.30
$BaSi_2O_5$	4.10
BaO	7.97

microphases, it becomes understandable that in glasses strict linearity would be expected only in exceptional cases. Nevertheless, the study of molar refraction and the deviations from additivity have helped in achieving greater and deeper insight into the constitution of glasses.

13.5 Thermal Expansion

When a liquid glass melt solidifies, or a solid glass is heated above the glass-transformation range, the thermal expansion changes by a factor of about 3 because of changed amplitudes of vibration. The linear coefficient of expansion α is defined by:

$$\alpha_{\Delta T} = (1/l_0)(\Delta l/\Delta T)$$

where l_0 = initial length and Δl is change of length with change of temperature ΔT.

The cubic coefficient of expansion β is defined by

$$\beta_{\Delta T} = (1/V_0)(\Delta V/\Delta T)$$

where V_0 is the initial volume and ΔV is the change in volume with change of temperature ΔT.

Besides, generally

$$\beta \approx 3\alpha$$

In the glass sector and linear coefficient α is used predominantly; the range of temperature in which it has been determined is given as a subscript, e.g., $\alpha_{20\text{-}300}$. This is necessary because α is temperature-dependent.

The coefficient of expansion plays an important role in practice, e.g., for glass processing, sealing, annealing, tempering, strengthening, soldering with metals, shock resistance, etc.

It is generally customary to classify technical glasses according to their coefficients of expansion:

Low expansion: $\alpha < 60 \cdot 10^{-7}\,K^{-1}$
High expansion: $\alpha > 60 \cdot 10^{-7}\,K^{-1}$

Characteristic values are:

α "ULE" glass* ≈ 0
α SiO_2 glass $\approx 5.0 \cdot 10^{-7}\,K^{-1}$
α Pyrex type® $\approx 30 \cdot 10^{-7}\,K^{-1}$
α Conventional glass $\approx 90 \cdot 10^{-7}\,K^{-1}$
α Cs-borate glass $\approx 300 \cdot 10^{-7}\,K^{-1}$ (probably highest)

For sealing glass-to-glass without the risk that it will subsequently break off, the difference in expansion $\Delta\alpha$ should not exceed $3 \cdot 10^{-7}\,K^{-1}$.

If glasses with a larger difference must be combined, a series of "intermediate glasses" must be inserted, with each "intermediate glass" not differing by more than $\Delta\alpha = 3 \cdot 10^{-7}\,K^{-1}$ from the next one.

A first formula for calculating the linear coefficient of glass from its composition was again provided by Winkelman and Schott [1179]:

$$\alpha_{\text{glass}} = \alpha_1 P_1 + \alpha_2 P_2 + \alpha_3 P_3 \ldots + \alpha_n P_n$$

for which they gave factors $\alpha_1, \alpha_2 \ldots$ for components $1, 2, 3 \ldots$ present in concentrations given in mass percent $P_1, P_2, P_3 \ldots$. It is not surprising that different authors (Winkelman and Schott [1179], Gilard and Dubral [1178], English and Turner [1184], Hall [1185], Kumar [1186], Takahashi [1187] and Appen [1188]), frequently found different factors since they derived them from different systems. Again, Gilard and Dubral [1178] improved agreement of the additive formula by introducing a quadratic term:

$$\alpha = \sum (\alpha_i' Pi + \alpha_i'' P_i^2)$$

Naray-Szabo [1183] started from the assumption that network-formers contribute little to the total thermal expansion and accounted for them by a constant A.

$$\alpha = \sum \alpha_i P_i + A$$

Although the calculations attempted by Kumar [1186], Huggins [1180], Huggins and Sun [1181] and Demkina [1189] brought further improvements in special cases, a generally valid solution cannot be expected for the reasons presented.

13.6 Viscosity

In a melt, two layers can move against each other only by application of a certain force K. For this force the following relation is valid:

$$K = \eta q (\partial v / \partial x)$$

where K = force, η = a proportionality factor (viscosity), q = area of the layer, $\partial v / \partial x$ = the velocity gradient (x = thickness of the layer).

The proportionality factor in this relation of motion and force is called "viscosity". For an area $q = 1\ \text{cm}^2$ and a velocity gradient $\partial v / \partial x = 1$, K becomes η(in Pa · s, old unit: poise, 1 Pa · s = 10 P).

Viscosity is one of the most important properties for the entire glass technology. It plays an enormous role in all stirring processes, in the buoyancy of bubbles during the fining process, during forming, for nucleation and crystallization, and, last but not least, for the formation of microphases. Thus, for instance, multiple phase separation is the consequence of the impediment of diffusion processes by the high viscosity of the microphases.

Sometimes it is necessary in practice do deal with parameters derived from viscosity such as:

"Kinematic viscosity" $= \eta / \rho$ (viscosity/density)

or

"Fluidity" $= 1/\eta$

It is, in particular, the stirring process controlling the homogeneity of optical glasses by rod or wing stirrers which is usually related to "kinematic viscosity". If viscosity and density are not considered simultaneously, fracture or torsion of the stirrer might result.

Temperature Dependence of Viscosity

Frequent attempts have been made to derive a mathematical relation between viscosity and temperature permitting calculation of the viscosity for any given temperature, for such knowledge is an indispensible prerequisite for controlling many technological problems, especially those of glass-forming processes. As a rule, the temperature dependence of a kinetic process may be described by a Boltzmann relation. This was, indeed, done for the relation of viscosity and temperature:

$$\eta = K \exp E\eta/RT$$

where η = viscosity, K = constant, E_η = activation energy of viscosity, R = gas constant, and T = temperature. The Boltzmann relation is strictly valid only for fluids with spherical particles and nondirectional bonds. These conditions are certainly not fulfilled in glass melts. It may, therefore, not be expected that the activation energy of viscosity, E_η, will assume a constant value. Nevertheless, the application of this relation has promoted considerable progress in glass science.

The logarithmic form of the Boltzmann relation is:

$$\log \eta = K' + (E'_\eta/4.75\,T)$$

(the new constant K' contains the conversion factor to the decadic logarithm).

If one plots $\log \eta$ vs $1/T$, a straight line should result whose slope E_η represents the activation energy of viscosity. One would think that the change in viscosity with increasing temperature associated with the loosening or breaking of chemical bonds would yield insight into the bonding conditions in a glass. Many researchers have addressed this problem, as demonstrated in the critical evaluation by Scholze [70]. But the E_η values obtained have shown that such conclusions are not possible. Apparently, so many secondary influences (degrees of polymerization, specific groupings, etc.) exist that theoretical derivations regarding bonds remain speculative. E_η values for alkali borate and silicate melts at different temperature vary from 83.74×10^3 to $648.95 \times 10^3\,\mathrm{J \cdot mol^{-1}}$, and are, for a pure SiO_2 glass melt, as high as $711.76 \times 10^3 \cdot \mathrm{mol^{-1}}$. At any rate, these calculations were fruitful and demonstrated the variability of fine structures in glass.

Figure 13.1 shows a schematic representation of the $\log \eta - 1/T$ relation for various simple glasses and of the activation energies for viscosity (E_η) derived therefrom. In some cases no exact straight line will result. In this case, E_η for a certain temperature is obtained by applying a tangent.

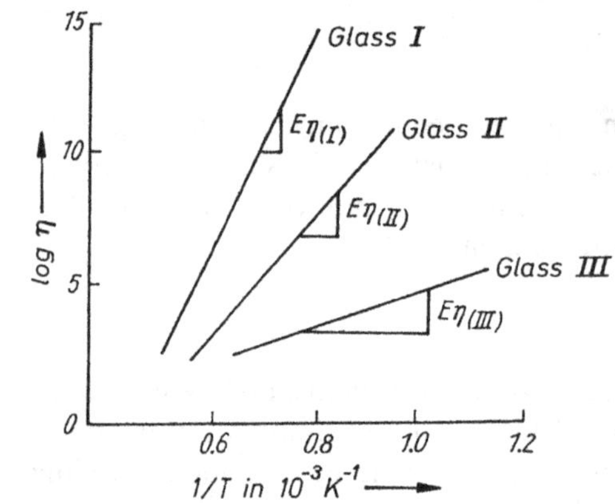

Fig. 13.1. Graphic evaluation of the activation energy of viscosity of a glass from the log $\eta - 10^3/T$ relation

Glasses with low values of E_η (flat $\log\eta - 1/T$ curve) (glass III in Fig. 13.1) are called "sweet". Glasses with high values of E_η (steep $\log\eta - 1/T$ curve, narrow softening range) are called "short" (glass 1 in Fig. 13.1).

As has been mentioned, application of the Boltzmann relation to the specific conditions of a glass melt cannot be expected to give a satisfactory agreement with experimental values for different temperatures. Numerous attempts have been made to modify the relation to make it more useful in practice. The best known and most successful is the so-called Vogel–Fulcher–Tammann (VFT) equation (Vogel [1190], Fulcher [1191] Tammann and Hess [1192]). A third constant, T_0, was introduced into the Boltzmann relation which represents a correlation of temperature T:

$$\eta = K\exp(E_\eta/T - T_0)$$

or in the more conventional logarithmic form:

$$\log\eta = A + B/(T - T_0)$$

where A, B, and T_0 are constants. The VFT equation is in quite good agreement with practice above the transformation region, but a theoretical interpretation of T_0 is missing.

An extensive discussion of theory and practice of viscosity-temperature as well as viscosity-composition relations will be found in a monograph on viscous flow and relaxation (Uhlmann and Kreidl [991]).

Dependence of Viscosity on Composition

There is no acceptable mathematical formulation for the relation between viscosity and glass composition, one more indication that a superposition of many influences determines the viscosity of a glass. Figure 13.2 shows the relation between viscosity and composition for binary alkali silicate glasses at various temperatures (Scholze [70]). The figure shows a break in the curve at 8–10 mol % alkali at 1400 °C. An extension of the curve for low alkali content into the region of high alkali content would lead to much lower viscosity values

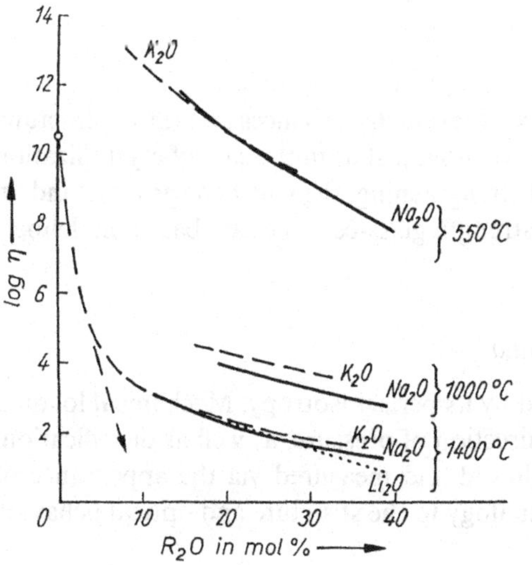

Fig. 13.2. Graphic representation of the relation between viscosity and composition of binary alkali silicate glasses for various temperatures. It is particularly the viscosity curves for 1400 °C which for compositions of 7–10 mol% M_2O show a strong change in direction which might be explained by beginning phase-separation processes

than those measured. This phenomenon may be explained plausibly. It has been shown by electron microscopy that at about 8–10 mol% alkali in a binary silicate glass, the first droplet-shaped alkali-rich regions are separated. Whereas between 0 and 8 mol% the alkali ions remain, after breaking bridges, statistically distributed in the SiO_4-tetrahedron network and induce a strong decrease of viscosity in the entire glass (steep slope 0–12% M_2O in Fig. 13.2), this is no longer true for higher concentrations of alkali. The aggregation of alkali ions in enclosed droplet regions means that they can contribute little to a decrease in viscosity. The viscosity of the glass will be determined predominantly by the matrix phase, much less by the droplet phase. Since in this case the matrix is an SiO_2-rich phase, the viscosity will decrease little if more alkali is added. The viscosity curves discussed in connection with the boron anomaly in binary alkali borates (Fig. 7.15) for various temperatures can be interpreted in a similar way. In the area of low concentrations of alkali and statistical distribution of alkali, some decrease of viscosity is observed. Then, however, beginning phase separation in connection with the boron anomaly even leads to an increase in viscosity after a minimum, and only at much higher alkali concentrations to a general decrease in viscosity.

At higher temperatures, where the boron anomaly is not observed and where the immiscibility region is no longer present, the same binary borate glasses exhibit just a continuous decrease in viscosity.

These few examples seem to show that in the future, too, mathematical relations of practical use for the dependence of viscosity on composition will be possible only for limited glass-composition ranges. An extensive collection of viscosity data for many compositions was presented by Scholze and Kreidl [1193].

13.7 Strain

Strain in glass may either be a source of severe disturbances and lead to fracture, or, if controlled, facilitate entirely novel uses, just as in the case of crystallization phenomena. Thermal and chemical strengthening, the Chemcor process, and, in part, the development of new high-strength glass-ceramics are based on deliberately produced strain in glass.

Strain Caused by Mechanical Loading

Homogeneous glass is characterized by its perfect isotropy. Mechanical loading leads to density differences in the direction of pressure, as well as densification, leading to strain which can be followed and measured via the appearance of birefringence. This provides some analogy to the structure and optical behavior of crystals.

A cubic crystal, e.g., an NaCl crystal, is optically isotropic on the basis of its structure and exhibits the same refraction in all directions. As in a glass, the optical indicator is a sphere. Any decrease in symmetry of the cubic system leads to a change in density in at least two of the three axis directions. The crystal will show birefringence.

Let us return to the application of a mechanical load to a body of glass. The birefringence caused by mechanical compression or tension can be described by the following relation:

$$D = KS$$

where D = birefringence in $nm \cdot cm^{-1}$, S = stress in Pa, K = stress-optical or photoelastic constant in $m^2 \cdot N^{-1}$ (old unit "brewster").

In 1902, Pockels [1194] made an important discovery while applying mechanical loads to glass. He found that, in general, glasses show an optically negative birefringence under load. Exceptions were glasses of high PbO content, unexplained for over 50 years. It was only in 1956 that Weyl [998] offered a theoretical explanation – to be discussed below – at a convention of German glass technologists. Pockels made, and proved experimentally, the logical conclusion that there must be a glass with a certain lead content which, under load, should exhibit no birefringence in monochromatic light.

The glasses confirming this conclusions were called "Pockels glasses" in the later literature. For a glass containing 1.5 mass% K_2O, Pockels determined that the decisive PbO content was about 75 mass % (and 22.8% SiO_2).

The theoretical interpretation by Weyl is illustrated in Fig. 13.3. If a cube of Na_2O–SiO_2 glass (case A, Fig. 13.3) is loaded, the large anions will be subjected to an enlarged polarization by the cations. The closer approach of the ion in the direction of pressure causes a deformation of the easily polarizable oxygen ions, i.e., to a shift in the electrons according to Fig. 13.3(A). If one now measures the refractive index in different directions, the resulting indicator is a rotation ellipsoid (Fig. 13.3(A), below), characterized by an optically negative character.

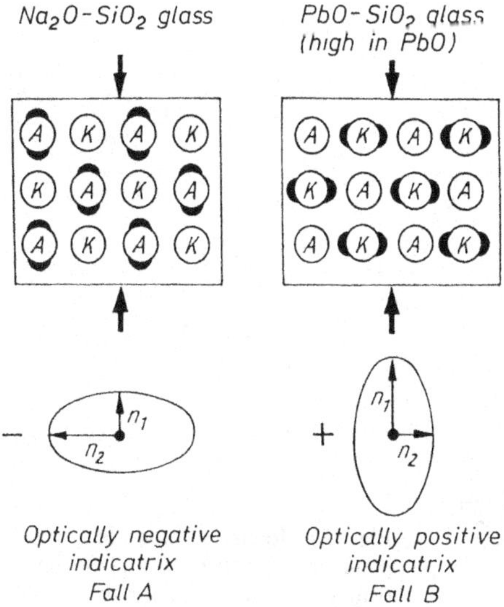

Fig. 13.3. Schematic of the entirely different strain-optical behavior of two glass cubes of different composition under load. An Na_2O–SiO_2 cube shows, like almost all glasses, an optically negative indicator; a silicate glass with a high content of PbO, however, shows an optically positive indicator

This is the normal case, easy to understand on the basis of a densification in the direction of pressure.

But, if a cube of a silicate glass of high PbO content is loaded in the same manner, the large polarizability by oxygen anions of the 6s electrons sitting on the outermost shell of the Pb^{2+} ion becomes effective (Fig. 13.3 (B)). This polarization behavior has been treated in connection with the structure of glasses of high PbO content (Chapter 7).

In this case of a polarization of lead cations by oxygen anions, the shift of electrons is just the reverse of the case for sodium silicate ions (compare A and B in Fig. 13.3). This, however, causes a higher refractive index in the direction of pressure: The indicator becomes optically positive (Fig. 13.3 (B) below).

Experimental investigations by Vogel in 1955 and 1956 led to the same results, but they had not been published when Weyl's publication appeared, and work was discontinued. Some of Vogel's results should be reported in this book since they fortify and supplement Weyl's theoretical conclusions.

If a glass cube is put under pressure from two opposite sides, and observed normal to the direction of pressure in convergent polarized white light in a polarization microscope, the picture is almost the same as for a uniaxial crystal. A dark cross is surrounded concentrically by many colored interference rings. The picture differs from that of a uniaxial crystal in that the sequence of colors decreases from center to periphery, but increases in the case of the glass cube because of the external pressure.

Using an accessory to the polarizing microscope (pressure gadget), it became possible to measure exactly on glass cubes, with 5-mm edges, the birefringence for monochromatic light of wavelength 486 nm (F), $\lambda = 589$ nm (D), and 656 nm (C) at pressures of 0–100 kP/25 mm^2 (room temperature).

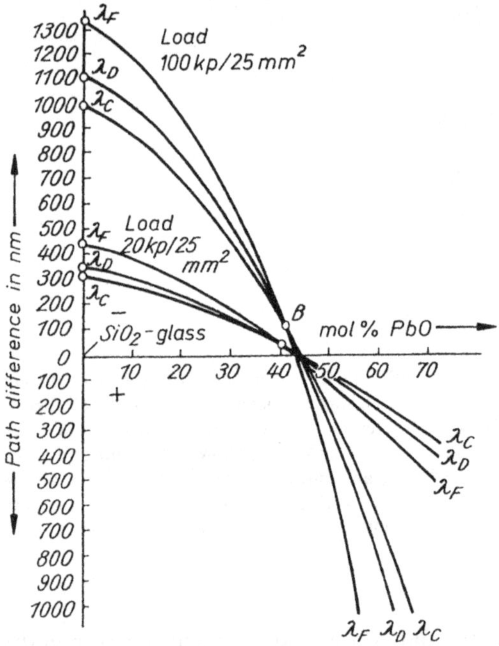

Fig. 13.4. Representation of the dependence of dispersion and of the optical character of birefringence under load of binary lead silicate glasses. Load 20 kP/25 mm^2 and 100 kP/25 mm^2. Determination of birefringence for the wavelengths of the *C*, *D*, and *F* lines

Part of the results for pure SiO_2 glass and a series of binary $PbO–SiO_2$ glasses are represented in Fig. 13.4. It can be seen that, with increasing PbO content, the negative birefringence first decreases to become zero for all three wavelengths at 42.5 mol% PbO.

This is true for all loads to 100 kP. In Fig. 13.4, results are represented only for two loads (20 kP, 100 kP) for clarity.

The dispersion of birefringence is turned around at the transition from the optically negative to the optically positive area. A binary lead silicate glass with 41% PbO exhibits no birefringence under any load. The intersection of the curves for different wavelengths, representing the absence of a dispersion of birefringence, lies in the negative area and migrates toward higher birefringences from A to B.

If one agrees with the interpretation presented for the behavior of the "Pockels" glasses, one should expect a similar behavior for glasses with a certain content of Tl_2O or Bi_2O_3, since the electronic structure of Tl^+ and Bi^{3+} is the same as for Pb^{2+} (Table 13.4). Glasses in the binary systems Tl_2O- and $Bi_2O_3–SiO_2$ were prepared in 1955 and confirmed this expectation. The behavior of the thallium silicate glasses under pressure could be studied quantitatively, like that of the lead silicate glasses. The behavior of the bismuth glasses could be studied only qualitatively because of their too-intensive coloration. Birefringence for D light vanished at 20.7 mol% Tl_2O in the system $Tl_2O–SiO_2$. Of course, Tl_2O introduces twice as many Tl^+ ions than PbO-introduced Pb^{2-} ions. And because of the difference in charge between Tl^+ and Pb^{2+}, polarization and deformation of Tl^+ by O^{2-} ions under load should be greater. Indeed,

Table 13.4. Number of electrons on the outermost shell of the elements and ions of thallium, lead and bismuth

Element	Tl	Pb	Bi
Atomic number	81	82	83
Number of electrons in *P* level	3	4	5
Number of electrons in 6s shell	Tl^{1+} 2	Pb^{2+} 2	Bi^{3+} 2

increased birefringence is observed. Whereas for glass cubes with 5-mm edges and for loads of 900 N and D light a path difference of +1150 nm was measured in the case of a glass with 67.3 mol% PbO, it was 3000 nm under the same conditions for a glass containing 42.5 mol% Tl_2O (the highest Tl_2O content achieved.)

Strain From Thermal Load

When heated, a glass body expands, whether more or less being dependent on the thermal coefficient of expansion. On relatively rapid subsequent cooling it shrinks and contracts. Because of the relatively low thermal conductivity of the glass in the exterior, a structure corresponding to a higher temperature is fixed, whereas the interior continues to contract, assuming a denser structure. Because of these differences in density, strain and birefringence will result. The thermally conditioned strain can be described by the following expression:

$$S = \alpha \Delta T E / 2(1 - \mu)$$

where S = strain, α = linear thermal coefficient of expansion, ΔT = temperature difference, E = modulus of elasticity, μ = Poisson's constant.

Comparing to the general expression

$$D = KS$$

the birefringence D corresponding to the above strain will be

$$D = K[\alpha \Delta T E / 2(1 - \mu)]$$

Brief reference should be made here to the importance of understanding these relations for the annealing process, especially of optical glass.

13.8 Surface Tension

The force field of a particle inside a liquid is generally screened and compensated completely by the partners surrounding it. The same particle placed in the surface of the liquid experiences only a one-sided saturation toward the inside. This causes one-sided attraction toward the inside, resulting in the usual contraction of the liquid in the shape of a droplet aiming at a minimum of surface energy. As has been pointed out before, Weyl therefore assumed that, on the basis of this one-sided compensation and the resulting one-sided attraction toward the inside, the chemical bonds must be reinforced at the surface. He explained the high mechanical strength of glass fibers, which have a very large surface, on this basis. The surface tension σ describing this phenomenon quantitatively has the dimension of work and is given in $N \cdot m^{-1}$.

Understanding the variations in surface tension is important for explaining and controlling technological phenomena. If, for instance, the surface tension of a liquid is measured in contact with a vacuum, or with gases containing molecules with dipoles of different strength, different values are obtained because such molecules attach to the liquid's surface, compensating the attraction to the inside – depending on the strength of their dipoles.

This may affect a glass melt under a changing atmosphere. Another important influence of surface tension is the mutual wetting of batch materials, for instance, that of sand grains by the primary melt. This influence will affect the entire rate of melting of a particular batch of raw materials. If the grain is well wetted by the primary melt, reactions of solution occur early and at low temperatures. However, if the surface tension of the primary melt is high, causing poor wetting of the sand grain, a large increase in temperature would be requested to initiate reactions of mutual solution. For this reason, many attempts were made to decrease the surface tension of the primary melt by small additives (JebsenMarwedel [1195–1200]. SO_2 accepted by the melt or introduced into the batch was found to decrease surface tension, resulting at times in positive effects.

For instance, the surface tension of a window glass of $\sigma = 0.30\,g \cdot N \cdot m^{-1}$ is reduced to $\sigma = 0.266\,N \cdot m^{-1}$ by 1% SO_3. We have had occasion to learn about the effect of certain trace elements when considering phase separation. This effect must be attributed to the change in the surface tension of the microphase in which the trace elements were enriched. In this connection, the fundamental studies of Dietzel [1201, 1202] on the effect of the most important constituent ions on surface tension were important. Figure 13.5 illustrates the relation of the factors Dietzel determined for various oxides to the value of r/Z (radius of the ion per charge unit). It is significant that the network-forming ions lie on the left, rising branch of the plot (surface tension increases with r/Z), while the network-modifiers lie on the right, descending branch (surface tension decreases with r/Z). Intermediate oxides like those of Zn, Mg, and Zr lie near the intersection of the branches.

Additive factors p_i, permitting an estimate of the melt surface tension ($\sigma = 1/100\, \Sigma \sigma_i p_i$), were given also by Lyon [1203], Appen [1177, 1204], Rubenstein [1205], and others. The factors deviate from each other, as would be expected if different base glasses with different structures were used in their derivation. In general, an increase in temperature, loosening the structure, decreases surface tension, generally by about $4 \cdot 10^{-3}\,N \cdot m^{-1}$ per 100 °C.

Among interesting exceptions are glasses high in PbO or B_2O_3. In the first case PbO, causing a lowering in surface tension, concentrates on the surface. The disorder brought about by high temperature decreases PbO content in the

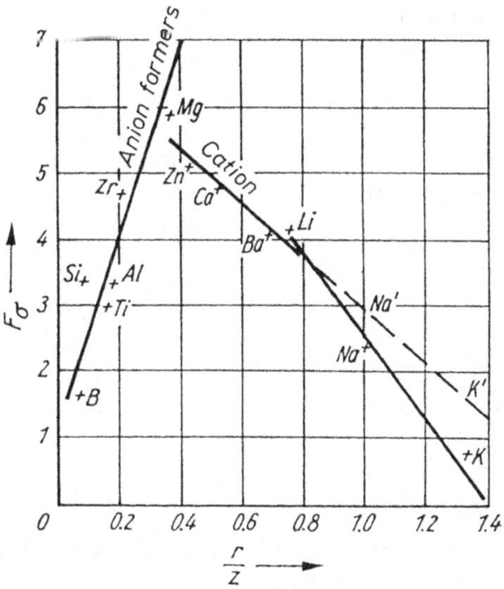

Fig. 13.5. Factors F_σ giving the effect of various oxides on the surface tension of a glass as a function of the r/Z value of these cations (Dietzel); r = ionic radius, Z = valence, F_σ = factor for 1 reduced mol% oxide

surface and increases surface tension at intermediate temperatures. Similarly, BO_3 groups enrich (in a leaf-like structure) at the surface. Again, disorder caused by an increase in temperature destroys this arrangement with a low surface tension.

13.9 Heat Conductivity, Specific Heat

Heat transfer and heat transport occur basically in two ways: by heat conduction and heat radiation. At low temperatures, conduction dominates, at high temperatures radiation. Here conduction is of primary interest.

The heat conductivity of a vitreous solid is defined by the heat content ∂Q flowing in time ∂t across the unit area F at a temperature gradient

$$\partial Q/\partial t = \Lambda_w F(\partial T/\partial x)$$

where Q = transported heat content in joules, t = time in seconds, F = area in cm^2, T = temperature in K, x = thickness of sample in cm, and Λ_w = heat conductivity in $W \cdot m^{-1} \cdot K^{-1}$. The heat conductivity is proportional to the specific heat. At absolute zero temperature $\Lambda_w = 0$, increases with temperature, and then, like the specific heat, approaches a constant value.

Knowledge of heat conductivity is essential. It is, for instance, an important prerequisite for the construction of furnaces, molds, and energy balances in glass technology. To a first approximation, it is possible to calculate the heat conductivity from additive factors (Ratcliffe [1206]).

$$\Lambda_w = 10^{-5} \sum \Lambda_i p_i$$

where Λ_i are the factors for component i and p_i their mass percentages.

A coarse approximation is obtained from density values ρ [1206].

$$\Lambda_w = (a/\rho) + b$$

where a and b are empirical constants.

Examples of Λ_w for three glasses at 0 °C are:

Component	Type	Λ_w
SiO_2	SiO_2	1.39
Na_2O–CaO–SiO_2	Soda-lime	0.91
K_2O–PbO–SiO_2	Lead glass	0.82

The specific heat has been discussed in connection with a description of the vitreous state. Its knowledge is as important as that of conductivity. Additive factors for approximate estimates were established very early by Winkelmann [1207] and refined later by other authors [1208–1210].

13.10 Electrical Conductivity

Among electrical properties, electrical conductivity is the property of greatest practical importance. On the one hand, it controls the use of the glass in electrical engineering applications at ambient temperature: on the other hand, high-temperature conductivity is important for contemporary methods of electrical melting of glass based on corrosion-resistant electrodes submerged in the melt and utilizing the electrical resistance of the melt itself.

Electrical conduction is based on the transport of electrons or ions. The "specific electrical conductivity" K is defined as the conductivity of a cylindrical body with a cross section of 1 cm^2 and a length of 1 cm. The "specific electrical resistance ρ" is defined as the reciprocal value of the electrical conductivity.

$$\rho = 1/K (K \text{ in } \Omega^{-1} \cdot \text{cm}^{-1})$$

At ambient temperature, conventional glasses have a very small electrical conductivity due to the structural features of the silicate network. A long time ago, Gehlhoff and Thomas [1211] characterized the conductivity by the temperature at which the specific conductivity has the value $K = 100 \cdot 10^{-10}$ ($\Omega^{-1} \cdot \text{cm}^{-1}$), the so-called "$T_{K-100}$ value." For a soda-lime (window) glass, T_{K-100} would be 150–200 °C.

This value was to be listed for technical glasses. Of great interest is the strong change in the specific electrical conductivity over large temperature ranges. For this reason, attempts at a mathematical formulation have been made, e.g., as early as 1908 by Rasch and Hinrichsen [1212].

$$\log K = A - (B/T)$$

where A and B are constants.

On a more theoretical basis, Stevels [37] obtained the similar relation

$$\ln K = A - (\phi/RT)$$

The value of A changes with both temperature and composition. The activation energy ϕ is a function of composition.

The activation energy is about $84 \cdot 10^3\,\mathrm{J \cdot mol^{-1}}$ for conventional glasses and is nearly constant between 50 °C and the glass-transition temperature T_g. Near T_g, however, the plot vs. $1/T$ shows a sharp break. This is the basis for the determination of T_g by measuring conductivity.

During the last 10 years there has been developing a vast body of inquiry regarding dielectric constant, dielectric loss, electronic conduction, and semiconduction phenomena. Here it is possible only to point out the existence of this specialized, fast-expanding field of glass research.

More recently, glasses have been found whose ionic conduction is several orders of magnitude higher than that of conventional silicate glasses (see for example, Tuller et al. [1213]). They are the subject of diligent and widespread study because of their potential use in solid electrolyte batteries, e.g., in electrically powered automobiles. The phenomenon has been described as superionic or fast-ion conduction. Typical ambient temperature conductivities are $\sigma \approx 10^{-2} - 10^{-3}\,\Omega^{-1} \cdot \mathrm{cm}^{-1}$ compared to ca. 10^{-13} for conventional glasses.

Typical compositions contain Li^+ or Ag^+ and halogens, particularly I. Anions can be oxides of B, Si, Ge, V, P, As, Cr, and Mo (see, for example, Minami [1214]). A glass 75 AgI–25$Ag_2(MoO_4)$, for example, has a conductivity of as much as $\sigma = 3.1 \cdot 10^{-2}\,\Omega^{-1} \cdot \mathrm{cm}^{-1}$.

References

1 Nölle G (1978) Technik der Glasherstellung. VEB Dtsch Verl Grundstoffind, Leipzig
2 Zschimmer E (1912) Die Glasindustrie in Jena. Diederichs, Jena
3 Kühnert H (1940) Otto Schott. Pott, Witten
4 Kühnert H (1946) Der Briefwechsel zwischen Otto Schott und Ernst Abbe über das optische Glas 1879-1881. Fischer, Jena
5 Kühnert H (1953 and 1957) Briefe und Dokumente zur Geschichte des VEB Optik Jenaer Glaswerk Schott u Gen, vols I and II. Fischer, Jena
6 Vogel W, Gerth K, Heindorf W (1965) Jenaer Rundschau 10:75–86
7 Morey GW (1936) Optical glass. US-Pats 2206081; GDR-Pat 691356; Brit Pat 462304; French Pat 810442; Swiss Pat 206664
8 Vogel W (1983) Wiss Z Friedrich-Schiller-Univ Jena Math Naturwiss Reihe 32:495–508
9 Heidenreich, E (1983) Wiss Zeitschr der Friedrich-Schiller-Univ Jena Math Naturwiss Reihe 32:527–539
10 Angell C (1988) Structural instability and relaxation in liquid and glassy phases near the fragile liquid limit. J Non-Cryst Solids 102:205–221
11 Tammann G (1903) Kristallisieren und Schmelzen. Barth, Leipzig
12 Tammann G (1933) Der Glaszustand. Voss, Leipzig
13 Tammann G (1923) Aggregatzustände, 2nd edn. Voss, Leipzig
14 Dietzel A (1948-1949) Glastech Ber 22:41–50, 81–86, 212–224
15 Salmang H (1957) Die physikalischen und chemischen Grundlagen der Glasfabrikation. Springer, Berlin Göttingen Heidelberg
16 Rötger H (1969) Silikattechnik 20:404–407
17 Rötger H (1970) Proc Symp Amorphous Materials, Sheffield, 1970, Soc Glass Techn, pp 125–132
18 Wong J, Angell C (1976) Glass structure by spectroscopy. Marcel Dekker, New York
19 Pye L, LaCourse W (1992) Sub-Tg phenomena. In: Proc 16th Int Congr Glass, Madrid, 1992. Bol Soc Esp Ceram Vidrio 31 C vol 2:15
20 Jebsen-Marwedel H (1959) Glastechnische Fabrikationsfehler. Springer Verlag, Berlin Göttingen Heidelberg
21 Dietzel A, Brückner R (1957) Glastech Ber 30:73–79
22 Günther G, Meister R (1951) Glastech Ber 24:1
23 Hinz W (1970) Silikate. VEB Verlag für Bauwesen, Berlin
24 Brill R, Grimm HG, Hermann C, Peters U (1942) Ann Phys 41:233–244
25 Pauling L (1945) The nature of the chemical bond. Cornell Univ, New York; German edn (1962) Verlag Chemie, Weinheim
26 Machatschki F (1953) Spezielle Mineralogie auf geochemischer Grundlage. Springer, Vienna
27 Warren BE, Bragg WL (1928) Z Kristallogr 69:168
28 Goldschmidt VM (1926) Geochemische Verteilungsgesetze der Elemente. Skr Nor Vidensk Akad Kl 1 Mat Naturvidensk Kl 8:7–156
29 Zachariasen WJ (1932) J Am Ceram Soc 54:3841
30 Warren BE (1933) Z Kristallogr Mineral Petrogr 86:349
31 Laves F (1939) Naturwissenschaften 27:65
32 Dietzel A (1942) Z Elektrochem 48:9–23
33 Dietzel A (1983) Phys Chem Glasses 24:172–180
34 Smekal A (1949) Glastech Ber 22:278–289
35 Mackenzie JD (1960 and 1962) Modern aspects of the vitreous state, vols 1 and 2. Butterworth, London
36 Weyl WA (1958 and 1959) Silicates Ind 23 and 24:16, 21
37 Stevels JM (1948) Progress in the theory of the physical properties of glass. Elsevier, Amsterdam
38 Stevels JM (1960-1961) Philips Tech Rdsch 22:337–349

39 Stevels JM (1954) Glass Ind 35:69
40 Trapp HJL, Stevels JM (1959) Glastech Ber 32:32–52
41 Huggins ML (1955) J Am Ceram Soc 38:172–175
42 Tilton LW (1957) J Res Natl Bur Stand Sect 59:139–154
43 Lebedev AA (1921) Trudy Cossud Opt Inst 2:57
44 Randall IT, Rooksby HP, Cooper BS (1930) Z Kristallogr 75:196–214
45 Valenkov NN, Porai-Koshits EA (1936) Z Kristallogr 95:195
46 Tudorovskaya NA (1955) In: Der Aufbau des Glases. Akad Wiss SSSR, Moskau
47 Schulz J, Hinz W, Silikattechnik, 6, 235 (1955)
48 Mackenzie ID (1956) Chem Rev 56:455–470
49 Turner WS (1934) Glastech Ber 12:409–413
50 Kusnetzov AI (1959) J Phys Chem (Moscow) 33:1492–1494
51 Botvinkin OK (1955) In: Der Aufbau des Glases. Akad Wiss SSSR, Moskau
52 Botvinkin OK (1960) In: Der glasartige Zustand. Akad Wiss SSSR, Moskau
53 Botvinkin OK (1940) Mitt Akad Wiss UdSSR Phys Ser 4:600
54 Hartleif G (1938) Z Anorg Chem 238:353–384
55 Müller RL, Markin BI (1957) In: Kitaigorodski JJ (ed) Technologie des Glases, VEB Verl Technik, Berlin; also: Oldenbourg, Munich
56 Evstrop'ev KS (1957) In: Kitaigorodski JJ (ed) Technologie des Glases, VEB Verl Technik, Berlin; also: Oldenbourg, Munich
57 Grebenstchikov JW (1957) In: Kitaigorodski JJ (ed) Technologie des Glases, VEB Verl Technik, Berlin; also: Oldenbourg, Munich
58 Molchanova OS (1957) Steklo i Keram 14:5–7
59 Landa L, Landa K (1992) In: Pye L, Lacourse W, Stevens H (eds) The physics of non-crystalline solids. Taylor & Francis, London, pp 376–380
60 Uhlmann D (1972) J Non-Cryst Solids 7:337
61 Uhlmann D, Kreidl N (1983) Treatise on glass, vol 1. Academic Press, New York
62 (a) Uhlmann D (1976) In: Kingery DW, Bowen H, Uhlmann D (eds) Introduction to ceramics. Wiley, New York; (b) idem (1983) In: Uhlmann D, Kreidl N (eds) Glass science and technology, vol 1. Academic Press, New York, p 1
63 Onorato P, Uhlmann D (1976) J Non-Cryst Solids 22:367
64 Van Der Sande J, Freed G (1982) Metal glasses. In: Uhlmann D, Kreidl N (eds) Treatise on glass, vol 1. Academic Press, New York, p 365
65 Haasen P (1983) J Non-Cryst Solids 56 [1/3]:191
66 Gupta P, Cooper A (1990) Topologically disordered networks of rigid polyropes. J Non-Cryst Solids 123:14–21
67 (a) Gupta P (1992) Rigidity, connectivity and glass forming ability. In: Pye L, Lacourse W, Stevens H (eds) The physics of non-crystalline solids. Taylor & Francis, London, pp 72–76; (b) idem (1992) Bol Soc Esp Ceram Vidrio 31 C vol 3:15–20
68 Wright A, Holme R, Grimley P, Sinclair R, Martin S, Price D, Galeener F (1991) The structure of some simple amorphous network solids revisited. J Non-Cryst Solids 129:213–32
69 Wright A, Bachra B, Clare A, Sinclair R, Hannon A (1991) The structure of alkali silicate glasses by n-diffraction. Am Ceram Soc Fall Meet 38 G 91
70 Scholze H (1965) Glas. Vieweg, Braunschweig; idem (1988) Glas, 3rd edn. Springer, Berlin, Heidelberg, New York
71 Trostel LT (1936) J Am Ceram Soc 19:271
72 Hummel FA, Tien TY, Kim KH (1959) J Am Ceram Soc 42:81–88
73 Hummel FA, Tien TY, Kim KH (1960) J Am Ceram Soc 43:192–197
74 Johnson DW, Hummel FA (1968) J Am Ceram Soc 51:196–201
75 Weyl WA (1962) The constitution of glasses, vol 1. Wiley, New York
76 Bernal JD (1959) Nature 183:141; ibid (1960) 185:68
77 Frenkel J (1946) Kinetic theory of liquids. Clarendon, Oxford
78 Stewart GW (1944) Am J Phys 12:321–324
79 Hoffmann LC, Statton WO (1955) Nature 4481:161–162
80 Porai-Koshits EA, Andre'ev NS (1958) Nature 4631:335–336
81 Porai-Koshits EA, Andre'ev NS (1959) J Soc Glass Technol 213:235–261
82 Zarzycki I, Naudin F (1967) J Phys Chem Glasses 1:11–18
83 Becherer G, Göcke W, Herms G (1971) Z Naturforsch 7:1177–1181
84 Patel IS, Schmidt PW, Ohlberg SM (1972) J Appl Phys 4:1636–1641
85 Urnes S (1971) J Phys Chem Glasses 3:82–84

86 Ruland W (1971) J Appl Cryst 1:70–73
87 Scholze H (1959) Glastech Ber 32:81–88
88 Florinskaya WA (1960) In: Der glasartige Zustand. Akad. Wiss. SSSR, Moskau
89 Bates T (1962) In: Mackenzie JD (ed) Modern aspects of the vitreous state, vol 2. Butterworth, London, pp 195–254
90 Kurkjian CR (1970) J Non-Cryst Solids 3:157–194
91 Hatton J, Rollin BV, Seymour EFW (1951) Phys Rev 83:672
92 Silver AH, Bray PJ (1958) J Chem Phys 29:984
93 Bray PJ, Silver AH (1960) In: Mackenzie JD (ed) Modern aspects of the vitreous state, vol 1. Butterworth, London, chapt 5, pp 92–119
94 Bray PJ, O'Keefe JG (1963) Phys Chem Glasses 4:47
95 Bray PJ, Leventhal M, Hooper HO (1963) Phys Chem Glasses 4:47
96 Bray PJ (1965) In: Proc 4th All-Union Conf Vitreous State, Leningrad, SSSR, 1964. Science Press, Leningrad, pp 237–251
97 Bray PJ (1965) In: Proc 7th Int Glass Congr, Brussels, Belgium, 1965. Inst Nat Verre, Charleroi, pp 40.1–40.9
98 Bray PJ, Kline D, Poch W (1966) Glastech Ber 39:175
99 Bray PJ (1967) In: Bishay, A (ed) Interaction of radiation with solids. Plenum, New York, pp 25–54
100 Bray PJ (1968) Silikattechnik 19:307–312; ibid 350–356
101 Bray PJ (1971) In: Proc 5th All-Union Conf Vitreous State, Leningrad, SSSR, 1969, Science Press, Leningrad, pp 191–193
102 Bray PJ (1970) Magnetic resonance. In: Proc Int Symp Electron and Nuclear Magnetic Resonance, Melbourne, Australia, 1969. Plenum, pp 11–40
103 Bray PJ (1975) Fizika i Khimiya Stekla (Leningrad) 1:490
104 Bray PJ (1974) Glass structure. In: Proc 10th Int Glass Congr, Kyoto, Japan, 1974. Jpn Ceram Soc, pp 13–1, 13–20
105 Bray PJ (1977) Investigation of glass structure by nuclear magnetic resonance techniques. In: Frischat GH (ed) The physics of non-crystalline solids. Trans Tech Publ, Clausthal–Zellerfeld, pp 65–80
106 Bray PJ, O'Keefe JG (1963) Phys Chem Glasses 4:37–46
107 Leventhal M, Bray PJ (1965) Phys Chem Glasses 6:113
108 Landsberger FR, Bray PJ (1970) J Chem Phys 53:2757
109 Kriz HM, Bray PJ (1971) J Magn Reson 4:69, 76
110 Kriz HM, Park MJ, Bray PJ (1961) Phys Chem Glasses 12:45
111 Kriz HM, Bishop SG, Bray PJ (1968) J Chem Phys 49:557
112 Kriz HM, Bray PJ (1971) J Non-Cryst Solids 6:27–36
113 Park MJ, Bray PJ (1972) Phys Chem Glasses 13:50–62
114 Kim KS, Bray PJ (1974) Phys Chem Glasses 15:47
115 Kim KS, Merrin S, Bray PJ (1976) J Chem Phys 64:4459
116 Hendrickson JR, Bray PJ (1974) J Chem Phys 61:2754
117 Müller-Warmuth W, Schulz GW, Neuroth N, Meyer F, Deeg E (1965) Z Naturforsch 20a:902
118 Müller-Warmuth W (1965) Glastech Ber 38:121, 405
119 Müller-Warmuth W, Poch W, Schulz GW (1966) Glastech Ber 39:415
120 Müller-Warmuth W, Poch W, Sieloff G (1970) Glastech Ber 43:5
121 Bishop SG, Bray PJ (1966) Phys Chem Glasses 7:73
122 Bishop SG, Bray PJ (1968) J Chem Phys 48:1709
123 Kramer F, Müller-Warmuth W, Scheerer J, Dutz H (1973) Z Naturforsch 28a:1338
124 Jellison GE Jr, Panek LW, Bray PJ, Rouse GB Jr (1977) J Chem Phys 66:802
125 Jellison GE Jr, Bray PJ (1978) unpublished data, Brown Univ, Providence
126 Kline D, Bray PJ (1966) Phys Chem Glasses 7:41
127 Rhee C (1971) J Korean Phys Soc 4:51
128 Rhee C, Bray PJ (1971) Phys Chem Glasses 12:156, 165
129 Rhee C, Bray PJ (1971) Phys Chem Glasses 12:165–174
130 Schulz GW, Müller-Warmuth W, Poch W, Scheerer (1968) J Glastech Ber 41:435
131 Weaver HE Jr (1953) Phys Rev 89:923
132 Holzmann GR, Anderson JH, Koth W, Lauterbur PC (1956) J Chem Phys 23:172
133 Mosel BD, Müller-Warmuth W, Dutz H (1974) Phys Chem Glasses 15:154
134 Lechus JA (1966) Master's Thesis, Brown Univ, Providence
135 Milberg ME, Otto K, Kushida T (1966) Phys Chem Glasses 7:14

136 Otto K, Milberg ME (1967) J Am Ceram Soc 50:513
137 Baugher JF, Bray PJ (1969) Phys Chem Glasses 10:77
138 Panek LW, Bray PJ (1977) J Chem Phys 66:3822
139 Van Vleck JH (1948) Phys Rev 74:1168
140 Lamb W (1941) Phys Rev 60:817
141 Ramsey NF (1950) Phys Rev 78:699
142 Krogh-Moe (1962) J Phys Chem Glasses 3:101
143 Krogh-Moe J (1965 Phys Chem Glasses 6:46
144 Greenblatt S, Bray PJ (1967) Phys Chem Glasses 8:213–217
145 Maurer RD (1956) J Chem Phys 6:1206–1209
146 Goldstein MJ (1963) J Appl Phys 7:1928–1934
147 Goldstein M (1965) J Am Ceram Soc 3:126–130
148 Hammel JJ (1972) Phys Electron Ceram 1972:963–983
149 Eitel W, Skaliks, W (1929) Z Anorg Allg Chem 183:263–286
150 Datta RK, Roy DM, Faile SP, Tuttle OF (1964) J Am Ceram Soc 47:153
151 Vogel W (1965) Struktur und Kristallisation der Gläser, 1st edn. (Appendix: Skatulla W, Elektronenmikroskopische Methoden in der Glasforschung) VEB Dtsch Verl Grundstoffind, Leipzig; (1971) 2nd edn.
152 Reimer L (1959) Elektronenmikroskopische Untersuchungs– und Präparationsmethoden. Springer, Berlin Göttingen Heidelberg
153 Picht J, Heydenreich J (1966) Einführung in die Elektronenmikroskopie. VEB Verl Technik, Berlin
154 Beeston BEP, Horne RW, Markham R (1973) Electron diffraction and optical diffraction techniques. In: Glauert AM (ed) Practical methods in electron microscopy. Elsevier, Amsterdam
155 Goodhew PJ (1973) Specimen preparation in materials science. In: Glauert AM (ed) Practical methods in electron microscopy. Elsevier, Amsterdam
156 Skatulla W, Horn L (1960) Exp Tech Phys 8:1–9
157 Skatulla W (1962) Silikattechnik 13:19–24
158 Müller H (1962) Präparation von techn.-phys. Objekten für die elektronenmikroskopische Untersuchung. Geest u. Portig, Leipzig
159 Wyckoff RWG (1949) Electron microscopy. Wiley–Interscience, New York
160 Borries BV (1949) Die Übermikroskopie. Saenger, Berlin
161 Kuhn P, Schimmel G (1957) Glastech Ber 30:463–470
162 Bradley DE (1954) Br J Appl Phys 5:55–60
163 Bradley DE (1958) Nature 181:875–877
164 Bradley DE (1960) Br J Appl Phys 11:506–509
165 Vogel W, Horn L, Völksch G (1982) J Non-Cryst Solids 49:221–240; idem (1982) Glass microstructure – surface and bulk. North-Holland, Amsterdam, pp 221–240
166 Schimmel G, Vogell W (1970-) Methodensammlung der Elektronenmikroskopie. Wiss Verl Ges, Stuttgart
167 Bach H (1973) Die Anwendbarkeit des Ionenstrahlätzens bei der Präparation für die Elektronenmikroskopie. In: Schimmel G, Vogel W (eds) Methodensammlung der Elektronenmikroskopie. Wiss Verl Ges, Stuttgart, chapt 2.4.2.1
168 Tögel K (1968) Siemens Z 42:550–556
169 Bethge H, Keller W (1960) Compt Rend Europ Reg Konf Elektr Mikr, Delft, Netherlands, 1960
170 Bethge H (1960) Phys Bl 1960; 223–227
171 Brümmer D (1978) Mikroanalyse mit Elektronen- und Ionensonden. VEB Dtsch Verl Grundstoffind, Leipzig
172 Bach H (1970) J Non-Cryst Solids 3:1–32
173 Völksch G, Reiß H, Horn L (1981) Silikattechnik 32:52–54
174 Zarzycki J (1990) In: Uhlmann D, Kreidl N (eds) Glass science and technology. Academic Press, New York, pp 253–271
175 Warren BE, Pincus AG (1940) J Am Ceram Soc 23:391–404
176 Markin BJ (1957) Technologie des Glases. In: Kitaigorodski JJ (ed) "Technologie des Glases", VEB Verl Technik, Berlin
177 Slayter G (1952) Am Ceram Soc Bull 31:276
178 Prebus AE, Michener GW (1954) Ind Eng Chem 16:147
179 Selyubsky WJ (1954) Steklo i Keram 11/10:19
180 Oberlies F (1956) Naturwissenschaften 43:224

181 Vogel W, Gerth K (1958) Glastech Ber 31: 15–28
182 Vogel W, Gerth K (1958) Silikattechnik 9: 353–358
183 Vogel W, Gerth K (1958) Silikattechnik 9: 495–501
184 Skatulla W, Vogel W, Wessel H (1958) Silikattechnik 9: 51
185 Vogel W (1958) Silikattechnik 9: 323
186 Vogel W (1958) Über Phasentrennung im Glas. In: Compt Rend Symp Fusion Verre, Brussels, 1958. Union Scientifique Continentale du Verre, p 741
187 Vogel W (1959) Silikattechnik 10: 241–250
188 Vogel W (1960) Über die Zellularstruktur des Glases. In: Der glasartige Zustand, Proc 3rd All-Union-Conf, Leningrad, SSSR, 1959. Akad Wiss UdSSR, Moscow, p 24; Engl Transl (1960) Consult Bur, New York
189 Vogel W, Gerth K (1962) Symp Nucleation Cryst Glasses Melts, Toronto, Canada, 1961. Amer Ceram Soc, p 11
190 Vogel W, Gerth K (1962) Z Chem 2: 261
191 Vogel W (1965) Angew Chem 77: 109
192 Vogel W (1966) Physica Status Solidi A 14: 255
193 Andre'ev NS, Boiko GG, Bokov NA (1970) J Non-Cryst Solids 1: 41–54
194 Andre'ev NS, Porai-Koshits EA (1970) Discuss Faraday Soc 1970: 135–144
195 Hammel JJ, Ohlberg SM (1965) J Appl Phys 4: 1442–1447
196 Vogel W, Rehfeld A, Ritschel H (1967) Silic Ind 5: 161–171
197 Vogel W, Schmidt W, Horn L (1969) Z Chem 9: 401–440
198 Vogel W, Schmidt W, Horn L (1971) In: Compt Rend 9th Int Congr Glass, Versailles, 1971. Inst Verre, pp 425–450
199 Vogel W, Schmidt W, Horn L (1972) Silikattechnik 4: 112–118
200 Vogel W (1975) Cent Glass Ceram Res Inst Bull 2-3: 95–107
201 Vogel W (1974) Wiss Z Friedrich-Schiller-Univ Jena Math Naturwiss Reihe 2: 341–357
202 Vogel W, Reiss H, Schrodt K (1967) Z Chem 7: 26
203 Vogel W, Reiss H, Wendler R, Seifert J (1974) In: Proc 10th Int Congr Glass, Kyoto 1974. Jpn Ceram Soc, pp 12/47–12/54
204 Vogel W, Gerth K (1967) Silikattechnik 18: 341–344
205 Vogel W (1963) Z Chem 3: 271–272
206 Vogel W (1963) Z Chem 3: 313–314
207 Vogel W (1964) Z Chem 4: 191–192
208 Vogel W (1964) Silikattechnik 15: 383–387
209 Vogel W, Ehrt D (1974) Z Chem 14: 396–404
210 Gerth K, Rehfeld A (1969) Silikattechnik 20: 227–228
211 Kühne K (1955) Z Phys Chem 204: 20–42
212 Kühne K, Skatulla W (1959) Silikattechnik 3: 105–119
213 Mazurin OV, Kluyev VP, Roskova GP (1970) Phys Chem Glasses 6: 192–195
214 Porai-Koshits EA, Averyanov VI (1968) J Non-Cryst Solids 1: 29–38
215 Ohlberg SM, Golob HR, Hollabaugh CM (1962) J Am Ceram Soc 1: 1–4
216 Ohlberg SM, Hammel JJ (1965) In: Compt Rend 7th Int Congr Glass, Brussels, 1965. Inst Nat Verre, pp 1–9
217 Moriya TA (1970) Bull Gov Ind Res Inst Osaka 1: 1–11
218 Moriya TA (1970) J Ceram Soc Jpn 6: 196–204
219 Seward TP, Uhlmann DR, Turnbull D (1968) J Am Ceram Soc 5: 278–285
220 Shaw RR, Uhlmann DR (1968) J Am Ceram Soc 7: 377–382
221 Srinivasan GR, Tweer J, Macedo PB, Sarkar A, Haller W (1971) J Non-Cryst Solids 3: 221–239
222 Tran TL (1965) Verres Refract 6: 416–428
223 Tran TL (1966) Verres Refract 1: 8–21
224 Watanabe M, Moriya TA (1961) Rev Electr Commum Lab 1-2: 50–71
225 Zarzycki J, Naudin F (1969) J Non-Cryst Solids 3: 215–234
226 Zarzycki J (1970) Discuss Faraday Soc London 50: 122–134
227 Zdanov SP, Koromaldi EV, Smirnova LG (1971) In: Compt Rend Trav 9th Congr Int Verre, Versailles, 1971. Inst Verre, pp 463–471
228 Charles RJ (1974) Wiss Z Friedrich-Schiller-Univ Jena Math Naturwiss Reihe 2: 341–348
229 Rindone GE (1958) J Am Ceram Soc 41: 41–42
230 Rindone GE, Ryder RJ (1957) Glass Ind 38: 29–31
231 Rötger H (1964) Silikattechnik 15: 71–78
232 Rötger H, Besen H (1963) Silikattechnik 14: 166–168

233 Hinz W, Knuth PO (1961) Glastech Ber 34:431–437
234 Johnson DW, Hummel FA (1969) Phasentrennungserscheinungen in Gläsern. In: Compt Rend Symp, Leningrad, 1968. Akad Wiss UdSSR, Leningrad
235 Johnson DW, Hummel FA (1974) Phasentrennungserscheinungen in Gläsern. Verl Akad Wiss UdSSR, Leningrad
236 Bondarev KT, Minakov VA (1960) Steklo i Keram 17:22–27
237 McDowell JF, Beall GH (1969) J Am Ceram Soc 52:17–25
238 Topping JA, Harrower JT, Murthy MK (1974) J Am Ceram Soc 57:209–212
239 Murthy MK (1961) J Am Ceram Soc 44:412–417
240 Kolomiets BT, Shilo VD (1966) Structure of Glass, vol 6. Consultants Bureau, New York, p 187
241 Krebs H, Welte H (1970) J Solid State Chem 2:182
242 Kreidl NJ, Ratzenboek W (1973) In: Proc 2nd Solid State Conf, Cairo, 1973
243 Feltz A, Voigt B (1974) Z Anorg Allgem Chem 403:61–71
244 Myers MB, Berkes JS (1972) J Non-Cryst Solids 8:804
245 Plumat ER (1968) J Am Ceram Soc 51:499
246 Kawamoto S, Tshuchihashi S (1969) Yogyo-Kyokaishi 77:328
247 Markova TP (1962) Vestn Leningr Univ 1962/17:96
248 Linke D (1975) In: Ber 8th Arb Tagung Elektronenmikrosk, Berlin, 1975. p 257
249 Bonceva-Nladenova Z, Linke D, Lippmann FJ (1972) In: Proc Conf Amorphous Liquid Vitr Semicond, Sofia, 1972, vol 1. pp 95–98
250 Kortüm G (1963) Einführung in die chemische Thermodynamik. Verlag Chemie, Weinheim
251 Vogel R (1959) Heterogene Gleichgewichte, 2nd edn. Geest und Portig, Leipzig
252 Brückner R, Poch W (1968) In: Compt Rend 8th Int Congr Glass, London, 1968. Soc Glass Tech
253 Cahn JW, Charles RJ (1965) Phys Chem Glasses 6:181–191
254 Volmer M (1939) Kinetik der Phasenbildung. Leipzig
255 Becker R, Döring W (1935) Ann Phys 24:719
256 Ohlberg SM (1965) In: Compt Rend 7th Int Congr Glass, Brussels, 1965. Inst Nat Verre
257 Hammel JJ (1965) In: Compt Rend 7th Int Congr Glass, Brussels, 1965. Inst Nat Verre
258 Burnett DG, Douglas RW (1970) Phys Chem Glasses 11:125–135
259 Neilson GF (1972) Phys Chem Glasses 13:70
260 Cahn JW (1965) J Chem Phys 42:93–96
261 Cahn, JW (1966) Acta Metall 14:1685–1692
262 Cahn, JW, Hilliard JE (1958) J Chem Phys 28:258–267; ibid (1961) 31:688–699
263 Haller W (1965) J Chem Phys 12:686–693
264 Tashiro N (1968) In: Compt Rend 8th Int Congr Glass, London, 1968. Soc Glass Tech
265 Ostwald W (1910) Kolloid Z 6:103–109
266 Filipovich VN (1967) Izv Akad Nauk SSSR Neorg Mater 3:993, 1192
267 Andre'ev NS, Mazurin OV, Porai-Koshits EA, Roskova GP, Filipovich VN (1974) Javlenie Liknacii v Steklach, Nauka, Leningrad
268 Kluge G, Wiedemann B (1974) Wiss Z Univ Jena 23:273
269 Zarzycki J, Naudin F (1971) J Non-Cryst Solids 5:415–425
270 Porai-Koshits EA (1975) Fizikai Chimija Stekla 1:385–394
271 Cook HE (1970) Acta Metall 18:297
272 Morral JE, Cahn JW (1971) Acta Metall 19:1037
273 Seward TP, Uhlmann DR (1968) J Am Ceram Soc 51:278–285
274 Schönborn H (1959) Silikattechnik 10:383
275 Vogel W, Byhan HG (1964) Silikattechnik 15:239
276 Vogel W, Höland W, Horn L, Völksch G (1985) 3rd Int Conf Struct Non-Cryst Mater. J Phys 46 C8:515–520
277 Wojciechovska J, Berak J, Trzebiatovski T (1956) Rocz Chem 30:750
278 Kashieva E, Dimitriev Y, priv commun
279 Kreidl NJ, Maklad MS (1969) J Am Ceram Soc 52:508–509
280 Maklad MS, Kreidl NJ (1971) In: Proc 9th Int Congr Glass, Paris, 1971. Inst Verre
281 Boulos, EN, Kreidl NJ (1972) Proc Canad Ceram Soc Annu Meet Febr 1972, Quebec. J Canad Ceram Soc
282 Rawson H (1967) Inorganic glass-forming systems. Academic Press, London
283 Volf MB (1961) Technical glasses. Pitman, London; idem, SNTL Publishers of Technical Literature, Prague
284 Rothenberg GB (1976) Glass technology. Recent developments. Noyes Data Corporation, Park Ridge, New Jersey

285 Doremus R (1973) Glass science. Wiley, New York
286 Kreidl N (1983) In: Uhlmann D, Kreidl N (eds) Glass forming systems, vol 1. Academic Press, New York. p 105 (with a section by Angell C)
287 Brückner R (1970) J Non-Cryst Solids 5:123–175, 177–216
288 Gaudin A (1839) Compt Rend Hebd Seances Acad Sci 8:678–679, 711
289 Porai-Koshits (1990) EA In: Uhlmann D, Kreidl N (eds) Glass science and technology, vol 4A: Structure, microstructure and properties. Academic Press, New York, pp 231–266
290 Uhlmann RD, Kreidl NJ (1980-1990) Glass science and technology, vols 1–5. Academic Press, New York
291 Rawson H (1967) Inorganic glass-forming systems. Academic Press, London, chapt 4, pp 45–68
292 Kitaigorodski JJ (1957) Technologie des Glases. VEB Verl Technik, Berlin; also: Oldenbourg, Munich
293 Jenaer Glaswerk (1980) brochures nos 1871, 1872, 1873, 1878
294 Clasen R (1988) Herstellung sehr reiner Kieselgläser durch Sintern submikroskopischer Glasteilchen. Habilitation, Univ Aachen
295 Osani H, Shioda T, Izawa T, Takata H (1976) Electron Lett 12:549–S50
296 Schultz PC (1974) J Am Ceram Soc 57:309–313
297 Geittner P, Küppers D, Lydtin H (1976) Appl Phys Lett 28:645–646
298 Bachmann P (1985) Pure Appl Chem 57:1299–1310
299 Nagel SR, MacChesney JB, Walker KL (1982) IEEE J Quantum Electron 18:450–476
300 Walker KL, Harvey JW, Geyling FT, Nagel SR (1980) J Am Ceram Soc 63:96–102
301 Irven J, Robinson A (1980) Phys Chem Glasses 21:47–52
302 Blankenship MG, Deneka CW (1982) IEEE J Quantum Electron QE-18:81–87
303 Izawa T, Kobayashi S, Sudo S, Hanawa F (1977) 3rd ECOC. pp 375–378
304 Wells J (1969) Encycl Chem Tech 18:134
305 Iler R (1979) The chemistry of silica. Wiley, New York
306 Moore H, Carey M (1951) J Soc Glass Technol 35:43–57 T
307 Imaoka M, Yamazaki T (1963) J Ceram Assoc Jpn 71:215–223
308 Dietzel A, Wickert H (1956) Glastech Ber 29:1–4
309 Kracek FC (1930) J Phys Chem 34:1583–1598
310 Brückner R, Chun HU, Goretzki H (1978) Glastech Ber 51 [1]:7
311 Greaves G, Raoux D (1983) The structure of non-crystalline materials. Taylor & Francis, London, p 55
312 Greaves G (1981) J Phys Coll C4:225
313 Greaves G. Effects of modes of formation on the structure of glass. To be published
314 Greaves G, Fontaine A, Largarde P, Raoux D, Gorman S (1981) Nature 293:611
315 Yasui I, Hasegawa H, Imaoka M (1983) Phys Chem Glasses 24 [3]:65
316 Imaoka M, Hasegawa H, Yasui I (1983) Phys Chem Glasses 24 [3]:72
317 Farnan L, Grandinetti P, Baltisberger J, Stebbins J, Werner U, Eastman M, Pines A (1992) Quantification of disorder in network-modified silicate glasses. Nature 358 [6381]:31–35
318 Bulkermann WA, Müller-Warmuth W (1992) A further 29Si MAS NMR study of binary alkali silicate glasses. Glastech Ber 65 [1]:18–21
319 Dietzel A, Sheybany HA (1948) Verres Refract 2:63–88
320 Dietzel A (1943) Naturwissenschaften 31:110–112
321 Uhlmann RD, Kreidl NJ (1983) Glass science and technology, vol 1. Academic Press, New York
322 (a) Vogel W (1979) Glaschemie, 1st edn. VEB Dtsch Verl Grundstoffind, Leipzig (2nd edn 1983); (b) idem (1985) Chemistry of glass (transl by Kreidl NJ). Am Ceram Soc, Columbus, Ohio; (c) idem (1985) Kemija Stakla (transl by Tkalcec E, Laslo R), SKTH Kemija u industriji, Zagreb, Yugoslavia
323 Dietzel A (1983) Glastech Ber 56:291–293
324 Day DE (1976) J Non-Cryst Solids 21:343–372
325 Weber P (1883) Sitzungsber Berliner Akad Wiss 1883 II:1233–1238
326 Schott O (1898) Z Instrumentenkd 11:330–337
327 Wiebe HF (1884) Sitzungsber K preuß Akad Wiss XXXVI:843–849; ibid (1885) XXXVII:1021–1028
328 Lengyel B, Boksay Z (1954) Z Phys Chem 203:93; ibid (1955) 209:157
329 Moynihan C, Lesikar A (1981) J Am Ceram Soc 64 [1]:40
330 Jain H, Peterson N, Downing H (1983) J Non-Cryst Solids 55:283

331 Rötger H (1958) Glastech Ber 31 [3]:54
332 Hendrickson J, Bray P (1972) Phys Chem Glasses 13 [2]:43; ibid [4]:107
333 (a) Ingram M (1980) J Am Ceram Soc 63 [5–6]:246; (b) idem (1983) Priv commun; (c) idem (1981) Solid State Commun 37:791
334 Ohlberg SM, Parsons JM (1965) Physics of non-crystalline solids. North-Holland, Amsterdam, pp 31–38
335 Biscoe J, Warren BE (1938) J Am Ceram Soc 21:287–295
336 Svanson SE, Forslind E, Krogh-Moe J (1962) J Phys Chem 66:174–175
337 Bray P (1958) J Chem Phys 29:984
338 Mozzi M, Warren BJ (1970) Appl Cryst 3:251
339 Richter H, Breitling G, Herre F (1954) Z Naturforschung 9A:390–402
340 Amini M, Mitra S, Hockney R (1981) J Phys Chem 14 [26]:3689
341 Bell R, Carnevale A (1981) Philos Mag B 43 [3]:389
342 Shartsis L, Capps W, Spinner S (1953) J Am Ceram Soc 36:319–326
343 Konijnendijk WL (1975) Philips Res Rep Suppl 1
344 Beekenkamp P (1966) Philips Res Rep Suppl 4
345 Kamitsos E, Chryssikos G, Patsis A (1992) Structure studies of single and mixed alkali borate glasses. Proc Int Conf Symp Solid State Ionics A-2. Adv Mater 1992:231–236
346 Bray P, Gravina S (1988) Glass 89, vol Ia. Nauka, Leningrad, pp 66–67
347 Bray P, Emerson J, Lee DH, Feller S, Bain D, Feil D (1990) J Non-Cryst Solids 129:240–248
348 Soppe W, Ebens W, Den Hartog HW (1988) Low frequency Raman spectroscopy of alkali borate glasses. J Non-Cryst Solids 105 [3]:251–257
349 Chryssikos G, Kamitsos E, Karakassides M (1990) Structure of borate glasses, pt 2. Phys Chem Glasses 31 [1]:109–16
350 Uhlmann D, Shaw R (1969) J Non-Cryst Solids 1:347
351 (a) Griscom J (1973) J Non-Cryst Solids 13:251; (b) ibid 31:241
352 Bray P (1978) In: Pye U, Frechette V, Kreidl N (eds) Borate glasses. Plenum New York, p 321
353 Rocket T, Foster W (1981) J Am Ceram Soc 64 [11]:C–148
354 Konijnendijk W, Stevels J (1976) J Non-Cryst Solids 20:193
355 Yun Y, Bray P (1978) J Non-Cryst Solids 27 [3]:363
356 Xiao S (1981) J Non-Cryst Solids 45:29
357 Feller S, Bray P (1979) J Non-Cryst Solids 33:273
358 Smets B, Lommen T (1981) Phys Chem Glasses 22 [6]:158
359 Zhong J, Wu X, Lium L, Bray P (1988) Structural modeling of borosilicate glasses via NMR studies. J Non-Cryst Solids 107 [1]:81–87
360 Martin S (1991) Review of the structure of phosphate glasses. Eur J Solid State Inorg Chem 28:163–205
361 Martin S, Bain D, Budhwani K, Feller S (1992) A 29MAS-NMR study of the short range order in lithium borosilicate glasses. J Am Ceram Soc 75 [5]:1117
362 Müller M, Reiss H, Vogel W (1984) Fiz Khim Stekla 10:247–249
363 Porai-Koshits EA, Zhdanov SP, Dobyschin DP, Kolykov GA (1955) In: Der Aufbau des Glases. Verl Akad Wiss UdSSR, Leningrad
364 Vogel W, Reiss H, Wendler R, Seifert J, Proc 10th Int Glass Congr, Tokyo, 1974. Ceram Soc Jpn, pp 12/47–12/54
365 Hood HP, Nordberg ME: US Pats 2215039 9/20 (1934); 2106744 2/1 (1938); 2286275 6/16 (1942); 2315329 3/30 (1943)
366 Nordberg ME (1944) J Am Ceram Soc 27:299–305
367 Nordberg ME (1956) In: Proc 1st Symp Art Glassblowing, Wilmington, Del, 1956. Am Scient Glassblowers Soc, Wilmington
368 Elmer TH, Nordberg ME (1961) In: Proc 6th Symp Art Glassblowing, Wilmington, Del, 1961. Am Scient Glassblowers Soc, Wilmington
369 Elmer TH, Chapman JD, Nordberg ME (1962) J Phys Chem 66:1517; ibid (1963) 67:2219–2222
370 Elmer TH, Nordberg ME (1958) J Am Ceram Soc 41:517–520
371 Taylor P, Owen D (1981) J Am Ceram Soc 64 [11]:C-158
372 Kalsing H (1957) Sprechsaal 90:220
373 Zagnazi H, Malassis M, Chaumont C, Bernier JC (1990) J Am Ceram Soc 73 [1]:163–169
374 Dimitriev YA, Kashtieva E, Ivanova I, Dzhambazov S (1986) J Mat Sci 28 [9]:3033
375 Elmer T (1991) Glastechn Ber 64[1]:1–8
376 Deppe J, Balkanski M, Wallis R, Massot M (1991) MRS Symp Proc 210:125–130
377 Beall G (1989) Rev Solid State Sci 3 [3/4]:333–359

378 Prabakar S, Rao K, Rao C (1992) An investigation of alkali aluminosilicate glasses by 29Si and 29Al magic angle spinning spectroscopy. Eur J Solid State Inorg Chem 29 [1]: 95–110
379 Jewell J, Spess M, Shelby J (1990) Effect of water concentration on the properties of commercial soda lime glasses. J Am Ceram Soc 73 [1] 132–135
380 Jewell J, Shelby J (1992) J Am Cer Soc 75 [4]: 178–183
381 Makishishima A, Hara T (1991) J Am Ceram Soc 74 [2]: 428–430
382 Shelby J, Minton S, Lord C, Tuzzolo M (1992) Formation and properties of yttrium aluminosilicate glasses. Phys Chem Glasses 33 [3]: 93–98
383 Applewhite A, Day P (1990) Properties of Y2O3-Al2O3-SiO2-MxOy glasses. J Non-Cryst Solids 120: 337–340
384 Kohli J, Shelby J (1992) A structural investigation of yttrium aluminosilicate glasses using 29Si and 29Al magic angle spinning nuclear magnetic resonance. Phys Chem Glasses 33 [3]: 73–78
385 Higby P, Merzbacher G, Aggarwal I, Friebele E (1990) Effect of small silica additions on the properties and structure of calcium aluminate glasses. SPIE 1327: 198–202
386 Fajans KN, Kreidl J (1948) J Am Ceram Soc 31: 105–114
387 Stanworth JE (1948) J Soc Glass Technol 32: 154–172 T
388 Bray PJ (1966) The Structure of Glass, pt II. In: Proc 4th All-Union Conf Glassy State, Leningrad, 1964. Consultants Bureau New York, pp 52–64
389 Mydlar M, Kreidl N, Hendren S, Clayton G (1970) Phys Chem Glasses 11 [6]: 196
390 Berger A (1932) Kolloid Z Beih 36: 1–42
391 Oberlies F (1964) Glastech Ber 37: 122–125
392 Vogel W, Heindorf W (1963) Z Chem 3: 394–395
393 Hahn-Weinheimer A (1954) Glastech Ber 27: 241
394 Biscoe J, Pincus AG, Smith J, Warren BE (1941) J Am Ceram Soc 24: 116–119
395 Kordes E, Becker H (1949) Z Anorg Allgem Chem 260: 185–207
396 Imaoka M (1962) In: Advances in glass technology, pt I. Plenum, New York, pp 149–164
397 Kordes E, Vogel W, Feterovsky R (1953) Z Elektrochem 57: 282–289
398 Kordes E (1941) Z Phys Chem 50: 194–212
399 Kordes E (1939) Z Anorg Allgem Chem 241: 1–38
400 Van Wazer JR (1950) J Am Chem Soc 72: 644
401 Van Wazer JR (1958) Phosphorus and its compounds, vol 1. Interscience, New York
402 Westmann AER (1960) Modern aspects of the vitreous state, vol 1. Butterworth, London
403 Westmann AER, Crowther J (1954) J Am Ceram Soc 37: 420–427
404 Westmann AER, Gartaganis PA (1957) J Am Ceram Soc 40: 293–297
405 Westmann AER, Murthy MK (1961) J Am Ceram Soc 44: 475–480
406 Westmann AER, Murthy MK (1962) Reser MK, Smith G, Inseley H (eds) Symp Nucleation Cryst Glasses Melts, Toronto, 1961. Am Ceramic Soc, pp 91–96
407 Bobovich L (1962) Opt Spectroscopy 13: 274
408 Müller KP (1969) Glastech Ber 42: 83
409 Brady GW (1958) J Chem Phys 28: 48
410 Wieker W, Thilo E (1961) Z Anorg Allgem Chem 313: 163
411 Kanazawa T, Ikeda M, Kawazoe H (1969) J Ceram Soc Jpn 77: 163
412 Huttenlocher H (1935) Z Kristallogr 90: 508
413 Wignall G, Rothon R, Longmann G, Woodward G (1977) J Mater Sci 12 [5]: 1039
414 Deutschbein O (1967) Rev Phys Appl 2: 29
415 Weber M (1976) Optical materials for neodymium fusion lasers. In: Critical material problems in energy production. Academic Press, New York
416 Weber M, Saroyan R, Ropp R (1981) J Non-Cryst Solids 44: 137
417 Weber M (1982) Wiss Z Friedrich-Schiller-Univ Math Naturwiss Reihe 32 [2/3]: 239
418 Alexe'ev N, Gapontjev V, Gromov A, Izyne'ev A, Kopylov Y, Kravchenko V (1979) Nov Neorg Stekla Mater Rasshir Zased, Riga, 1979. Rizh Politekh Inst, pp 6–8
419 Berzelius JJ (1934) Annal Physik Chem 32: 577
420 Lehners V, Wolensky E (1913) J Am Chem Soc 35: 718–733
421 Stanworth JE (1962) J Soc Glass Technol 36: 217
422 Stanworth JE (1952) Nature 169: 581–582
423 James JA, Stanworth JE (1954) J Soc Glass Technol 38: 421–435
424 Rawson H (1956) In: Proc 4th Int Congr Glass, Paris, 1956. Inst Verre, pp 62–69
425 Baynton PL, Rawson H, Stanworth JE In: Proc 4th Int Congr Glass, Paris, 1956. Inst Verre, pp 52–61
426 Baynton PL, Rawson H, Stanworth JE (1957) J Electrochem Soc 104: 237–240

427 Baynton PL, Rawson H, Stanworth JE (1957) Nature 179:434–435
428 Dale AE, Pegg EF, Stanworth JE (1954) Research 75:38–39
429 Dale AE, Stanworth JE (1949) J Soc Glass Technol 33:167–175 T
430 Imaoka M, Yamazaki T (1963) J Ceram Assoc Jpn 71:215–223
431 Vogel W, Bürger H, Müller B, Zerge G, Müller W, Forkel K (1974) Silikattechnik 25:205–209
432 Brady GW (1956) J Chem Phys 24:477
433 Brady GW (1957) J Chem Phys 27:303–304
434 Ito T, Sawada H (1940) Z Kristallogr 102:13–15
435 Yakhkind AK (1966) J Am Ceram Soc 49:670–675
436 Yakhkind AK (1969) In: Proc 5th All-Union Conf Struct Glass, Leningrad, 1969. Acad Wiss UdSSR
437 Zemann J (1968) Z Kristallogr 127:319–326
438 Zemann J (1964) Naturwissenschaften 51:334–335
439 Beyer H, Zemann J (1965) Naturwissenschaften 52:155
440 Leciejewicz J (1961) Z Kristallogr 116:345–353
441 Neuroth N (1968) Glastech Ber 41:243–253
442 Tanabe S, Hirao K, Kistermoto S, Soga N (1993) Adv Fusion Glass. Am Ceram Soc Ceram Trans 29:199–208
443 Tanabe S, Hirao K, Soga N (1990) J Non-Cryst Solids 122:79
444 Hirao K, Todoroki S, Soga N (1992) J Non-Cryst Solids 143:40
445 Hirao K (1992) Proc 16th Int Congr Glass, Madrid, 1992. Bol Soc Esp Ceram Vidrio 31 C vol 3:121–126
446 Hirao K, Todoroki S, Tanabe S, Soga N (1992) In: Pye L, Lacourse W, Stevens H (eds) The physics of non-crystalline solids. Taylor & Francis, London, pp 611–616
447 Yanagita H, Okada K, Miura K, Torstani H, Yamashita T (1991) Er3 + -doped fluorozircoaluminate glasses. Mater Sci Forum 67-68:521–526
448 Kozhukharov V, Buerger H, Neov S (1991) Glass forming properties and structure of halide tellurite glasses. Mater Sci Forum 67-68:143–148
449 O'Daniel H, Tscheiwilli L (1941) Z Kristallogr A 103:178
450 O'Daniel H, Tscheiwilli L (1942) Z Kristallogr A 104:124–141
451 O'Daniel H, Tscheiwilli L (1945-1948) Neues Jahrb Mineral Monatsh A pp 56–64
452 Thilo E, Lehmann HA (1949) Z Anorg Chem 258:332–355
453 Thilo E, Schröder H (1951) Z Physik Chem 197:39–62
454 Thilo E, Libau F (1952) Z Physik Chem 199:125–141
455 Jahn W, Thilo E (1953) Z Anorg Allgem Chem 273:72–80
456 Jahn W (1954) Z Anorg Allg Chem 276:113–127
457 Roy DM, Roy R, Osborn EF (1950) J Am Ceram Soc 33:85–90
458 Counts WE, Roy R, Osborn EF (1953) J Am Ceram Soc 36:12–17
459 Roy DM, Roy R, Osborn EF (1953) J Am Ceram Soc 36:185–190
460 Brandenberger E (1932) Schweiz Mineral Petrogr Mitt 12:243–246
461 Warren BE, Hill CF (1934) Z Kristallogr A 89:481–486
462 Heyne G (1933) Angew Chem 46:473–477
463 Izumitani T, Terai R (1952) Bull Gov Ind Res Inst Osaka 3:25–28
464 Imaoka M, Mizusawa S (1953) J Ceram Assoc Jpn 61:13–14
465 Imaoka M (1954) J Ceram Assoc Jpn 62:24–27
466 Löhmann R (1983) J Non-Cryst Solids 56 [1/3]:123
467 Vogel W, Gerth K (1958) Silikattechnik 9:539–543
468 Wright A, Etherington G, Erwin-Deoa J, Sinclair R (1982) J Phys Coll C9, suppl to 43 [12]
469 Sun K (1949) US Pat 2 466 509
470 Krylova L (1967) Steklo 1:101
471 Baldwin C, Almeida R, Mackenzie J (1981) J Non-Cryst Solids 44:309
472 Poulain M (1975) Mat Res Bull 10:243–246
473 (a) Lecoq A, Poulain M (1980) J Non-Cryst Solids 41:209; (b) idem (1980) Verres Refract 34 [3]:333
474 Takahashi S, Shibata S, Kanamori S, Mitachi S, Manabe T (1981) In: Bendow B, Mitra SS (eds) Advances in ceramics, vol 2. Am Ceram Soc, Columbus, Ohio
475 Drexhage M, Moynihan C, Boulos M (1980) Mater Res Bull 15 [12]:213
476 Bendow B, Banerjee P, Drexhage M, Goltman J, Mitra G, Moynihan C (1982) J Am Ceram Soc 65 [1]:C-8
477 2nd Int Symp Halide Glasses, Rensselaer Polytech Inst, 1983

478 (a) Krauss M, Erdmann C, Vogel W (1975) Mater Res Bull 10:243; (b) idem (1983) J Non-Cryst Solids 56:1; (c) idem (1981) Nature 293:279
479 Fonteneau G, Slim H, Lahaie F, Lucas J (1980) Mater Res Bull 15 [8]:1143; ibid [10]:1425
480 Slim H, Lucas J (1982) J Non-Cryst Solids 50:61
481 Drexhage M, El Bayoumi O (1982) Am Ceram Soc Bull 61[3]:369 (abstract)
482 Maze
483 Sigel GH (1977) In: Tomozawa M, Doremus R (eds) Treatise on materials science and technology of glass, vol I. Academic Press, New York, pp 58–59
484 Tran D (1983) In: Electronics Convention METSAC Sect IEEE, New York, 1983 (Available from authors at Naval Res Lab, Washington, DC, 20375)
485 Lucas J (1991) Halide glasses. Mater Sci Technol 9:453–491
486 Poulain Ma (1992) In: Pye L, Lacourse W, Stevens H (eds) The physics of non-crystalline solids. Taylor & Francis, London, pp 167–182
487 Poulain Ma, Poulain Mi, Messaddeq Y, Soufiane A (1992) Fluoroindate glass for active and passive fibers. In: Proc 16th Int Congr Glass, Madrid, 1992. Bol Soc Esp Ceram Vidrio 31 C vol 2:61–66
488 Izumitani T, Matsukawa K, Wang Y (1991) Light scattering of fluoride glasses. Mater Sci Forum 67-68:453–458
489 Poulain Mi, Poulain Ma (1991) Oxyfluoride glasses. Mater Sci Forum 67-68:129–36
490 Zhang X, Fonteneau G, Ma H, Lucas J (1991) Tellurium halide fibers. Mater Sci Forum 67-68:371–375
491 Stokovski S, Martin W, Yarema S (1980) Electrical magnetic and optical properties of glasses. In: Proc 5th Univ Conf Glass Science, Amsterdam, 1965. North-Holland, Amsterdam, p 481
492 Ehrt D, Krauss M, Erdmann C, Vogel W (1982) Z Chem 22 [8]:315
493 Erdmann H, Vogel W (1983) Z Chem 23 [1]:37
494 Vogel W (1983) Z Chem 23:111
495 Poulain M (1983) J Non-Cryst Solids 56:1–14
496 Poulain M (1975) Mater Res Bull 10:243–246
497 Lucas J (1986) J Non-Cryst Solids 80:83–91
498 Lucas J (1987) Wiss Z Friedrich-Schiller-Univ Jena: Naturwiss Reihe 37:719–733
499 Almeida RM, Mackenzie JD (1981) J Chem Phys 74:5954
500 Poulain M, Lucas J (1978) Verres Refract 32:505
501 Drexhage MG (1985) Treatise on materials science and technology, vol 26. Academic Press, New York
502 Tran DC, Sigel GH, Bendow B (1984) J Lightwave Technol LT 2, 5:566
503 Miyashita T, Manabe T (1982) IEEE J Quantum Electron QE-18:1432
504 Dennis LH, Laubengayer AW (1926) J Phys Chem 30:1519
505 Riebling EF (1963) J Chem Phys 39:1889–1895
506 Murthy KM, Aguayo J (1964) J Am Ceram Soc 47:444–447
507 Evstrop'ev K, Krupkin S, Galimov Y, Tarlakov Y, Shevyakov A (1970) Zh Prikl Spektrosk 13 [4]:655
508 Murthy KM, Scroggie B (1965) Phys Chem Glasses 6:162–167
509 Blinov V (1971) In: Stekloobraznye Sist Nov Stekla Ikh Osn Mater Vses Soveshch, Minsk, 1967. p 91
510 Blinov V (1971) Zh Khim 20B:507
511 Urnes S (1971) Phys Chem Glasses 12 [3]:82
512 Yoshimura T, Fukunaga J, Ihara M (1971) Yogyo-Kyokaishi 79 [915]:428
513 Sakka S (1977) Yogyo-Kyokaishi 85 [6]:299
514 Caslavska V, Strickler D (1969) J Am Ceram Soc 52 [3]:154
515 Zarzycki J (1957) Verres Refract 11:3–8
516 Ivanov AO, Evstrop'ev KS (1962) Dokl Akad Nauk SSSR 145:797–800
517 Riebling EF (1963) J Chem Phys 39:3022–3030
518 Murthy KM (1964) Nature 401:285–286
519 Schultz P (1981) In: Bendow B, Mitra SS (eds) Advances in ceramics, vol 2. Am Ceram Soc, Columbus, Ohio
520 Eysel W, Wolfe W, Newnham R (1973) J Am Ceram Soc 56 [4]:185
521 Heaton HM, Moore H (1957) J Soc Glass Technol 41:3–27 T
522 Van Uitert L, Grodkiewicz W (1981) J Am Ceram Soc 64 [12]:690
523 (a) Grodkiewicz W, O'Bryan H, Pressman L, Sineh G, Van Uitert L, Zydzyk G (1981) J Non-Cryst Solids 44:405; (b) idem (1981) Mater Res Bull 16:373

524 Karutz I, Stranski I (1957) Z Anorg Allgem Chem 292:330
525 Böttcher H, Plieth K, Reuber-Kürbs E, Stranski I (1951) Z Anorg Allgem Chem 266:302
526 Adams RV (1961) Phys Chem Glasses 2 No 4
527 Gmelin (1952) Handbuch der Anorg Chemie, Band Arsen. Verlag Chemie, Weinheim, p 249
528 Wells AF (1962) Structural inorganic chemistry. Clarendon, Oxford, pp 668–675
529 Pascal P (1958) Nouveau traité de chimie minérale, tom As-Sb-Bi. Masson, Paris, p 201
530 Winkler A (1967) Z Anorg Allgem Chem 350:320–325
531 Krebs H (1968) Grundzüge der anorganischen Kristallchemie. Enke, Stuttgart
532 Lin FC (1963) Glass Ind 44:19
533 Kordes E (1939) Z Phys Chem 43:173–190
534 Bishay A, Askalanie M (1968) In: Proc 7th Int Congr Glass, Brussels, 1968, pt 1 no 34. Inst Nat Verre
535 Hedden WA, King BW (1960) J Am Ceram Soc 43:387–388
536 Rao BVJ (1962) J Am Ceram Soc 45:555–563
537 Rao BVJ (1965) In: Proc 7th Int Congr Glass, Brussels, 1965, no 104. Inst Nat Verre
538 Brekhovskich SM (1959) Glastech Ber 32:437–442
539 Heynes MSR, Rawson H (1957) J Soc Glass Technol 41:794–802
540 Shabanova E (1967) Tr Gor'k Politekh Inst im AA Zhdanova 23 [4]:38
541 Gerhardt (1973) Rev Sci Instrum 44 [5]:657
542 White W (1965) J Am Ceram Soc 48 [2]:108
543 Marfels H (1969) Glastech Ber 42 [5]:161
544 Rao BVJ (1963) Phys Chem Glasses 4:455–463
545 Rao BVJ (1959) J Am Ceram Soc 42:487–490
546 Smoke E (1951) J Am Ceram Soc 39:87–90
547 Herczog A (1964) J Am Ceram Soc 47:107–15
548 Harrison DE, Hummel FA (1959) J Am Ceram Soc 42:487–490
549 Faulstich M/Jenaer Glaswerk Schott u Gen (1976) German Pat Appl 22 [59]:183
550 Searight C, Alexander M (1965) US Pats 3 198 641, 3 814 051, 3 294 558
551 Morey GW (1939) US Pat 2 150 694
552 Denton EP, Rawson H (1954) Nature 173:1030–1032
553 Roscoe HE (1968) Philos Trans R Soc London 158:1–27
554 Tammann G, Jenkel E (1929) Z Anorg Allg Chem 184:416–420
555 Rao BVJ (1965) J Am Ceram Soc 48:311–319
556 Adler D (1971) Amorphous semiconductors. CRC, Cleveland
557 Ohashi S, Matsuura T (1962) Bull Chem Soc Jpn 35:501–504
558 Rostkovski AP (1930) Zh Russ Fiz Khim Ova 62:2055–2059
559 Thilo E, Wieker C, Wieker W (1964) Silikattechnik 15:109–111
560 Stevels JM (1952) Philips Tech Rundsch 13:352
561 Dietzel A, Poegel HJ (1954) In: Atti 3rd Congr Int Vetro, Venice, 1953. Int Commiss Glass, Karlsruhe, Germany, p 219
562 Urnes S (1958) Glastech Ber 31:337
563 Van Uitert L, Bonner W, Grodiewicz W (1971) Mater Res Bull 6 [6]:283, 513
564 Vinogradov EE, Kirilenko JA, Yu J (1970) Inorg Mater (Engl Transl) 6:1678–1680
565 Angell C (1983) In: Uhlmann D, Kreidl N (eds) Glass science and technology, vol 1. Academic Press, New York, chapter 1
566 Tucker J (1973) Glass forming molten salt systems. In: Chem Pros Metallurgy, Richardson Conf (Imperial Coll Science), London, 1973
567 Hester R, Krishnan K (1968) J Chem Phys 49:4356
568 Furukawa T, Braiver S, White W (1977) J Chem Phys (submitted for publication)
569 Van Weckem H (1976) Thesis, Univ Amsterdam
570 Maier CA (1980) US Bur Mines Tech Progr Rep no 360
571 Schulz I (1957) Naturwissenschaften 20:536
572 Brehler B (1959) Naturwissenschaften 46:554
573 Imaoka M, Konagaya Y, Hasegawa H (1971) J Ceram Soc Jpn 79:97
574 Yamazaki T (1968) J Ceram Soc Jpn 76:100
575 Cooper E, Angell C (1983) J Non-Cryst Solids 56:75
576 Tick P (1986) Properties and structure of multicomponent phosphate and fluorine phosphate glasses. Collect Pap 14th Int Congr Glass, New Delhi, 1986, vol II. Indian Ceram Soc, pp 102–109
577 Mackenzie J, Zheng M (1992) Oxide- non oxide glasses. In: Proc 16th Int Congr Glass, Madrid, 1992. Bol Soc Esp Ceram Vidrio 31 C vol 1:1–24

578 Unuma H, Maekawa H, Kiyono H, Kawamura K, Maekawa T, Yokokawa T (1992) 29Si MAS NMR of Na-Si-O-N oxynitride glasses. J Ceram Soc Jpn 100 [11]: 1292–1296
579 Scholze H (1959) Glastech Ber 32: 142–152
580 Scholze H (1959) Glastech Ber 32: 278–281
581 Scholze H (1959) Glastech Ber 32: 314–320
582 Scholze H (1959) Glastech Ber 32: 381–386
583 Scholze H (1959) Glastech Ber 32: 421–426
584 (a) Bartholomew R, Garfinkel H (1980) Chemical strengthening of glass. In: Uhlmann D, Kreidl N (eds) Glass science and technology, vol 5. Academic Press, New York; (b) idem (1983) J Non-Cryst Solids 56 [1/3]: 331
585 Klement W, Willens RH, Duwez P (1960) Nature 187: 869–870
586 Güntherodt HJ (1977) Festkörperprobleme (Adv Solid State Phys) XVII: 25
587 Duwez P, Willens RH (1963) Trans Metall Soc AIME 227: 362
588 Beck H, Güntherodt HJ (1981/1983) Glassy metals, vols I/II. Springer, Berlin, Heidelberg, New York
589 Cahn RW (1982) J Phys 43, suppl 12, coll C9: 55–65
590 Chaudhari P, Turnbull D (1978) Science 199: 11–21
591 Davis HA (1976) Phys Chem Glasses 17: 159–173
592 Gaskell PH (1985) In: Wright-Dupuy A (ed) Glass. Nijhoff, Dordrecht, pp 54–71
593 Haasen P (1983) J Non-Cryst Solids 56: 191–199
594 Hillenbrand HG, Hornbogen E (1982) Z Werkstofftech 13: 407–415
595 Luborsky FE (1983) Amorphous metallic alloys. Butterworth, London
596 Takayama S (1976) J Mater Sci 11: 164–185
597 Van der Sande JB, Freed RL In: Uhlmann D, Kreidl N (eds) Glass science and technology, vols 1, 2, 3, 5. Academic Press
598 Wagner CNJ, Johnson WL (eds) (1984) Proc 5th Int Conf Liquids Amorph Metals. J Non-Cryst Solids 61: 427–431, 62: 1015–1020
599 Feltz A (1983) Amorphe und glasige anorganische Festkörper. VEB Akad Verl, Berlin
600 Egami T, Waseda Y (1984) J Non-Cryst Solids 64: 113–134
601 Warlimont H (1984) Z Metallkd 75: 679–685
602 Warlimont H (1984) Z Metallkd 75: 686–690
603 Perepezko JH, Galaup C, Cooper KP (1982) Mater Res Soc Symp Proc. Elsevier North Holland, New York, pp 1–11
604 Vind Nielsen HJ (1979) J Non-Cryst Solids 33: 285–289
605 Vind Nielsen HJ (1979) Z Metallkd 70: 606–608
606 Donald JW, Davies HA (1978) J Non-Cryst Solids 30: 77–85
607 Noda T, Inagaki M, Yamada S (1969) J Non-Cryst Solids 1: 285–302
608 Lersmacher B, Lydtin H, Knippenberg WF (1970) Chem Ing Tech 42: 659–669
609 Craievich AF (1976) Mater Res Bull 11: 1249–1256
610 Dübgen R, Popp G (1984) Z Werkstofftech 15: 331–338
611 Roy R (1969) J Am Ceram Soc 52: 344
612 Dislich H (1971) Glastech Ber 44: 1–8
613 Dislich H (1983) J Non-Cryst Solids 57: 371–388
614 Scherer GW (1987) J Ceram Soc Jpn 95: 21–44
615 Sakka S (1977, 1979, 1982) In Tomazova , Doremus R (eds) Treatise on materials science and technology, vol 22 (Glass). Academic Press, New York, pp 129–167
616 Zarzycki J (1985) In Wright A (ed) Glass – current issues. Nijhoff, Dordrecht, pp 203–223
617 Beier W, Meier M, Frischat GH (1985) Glastech Ber 58: 97–105
618 Jones WM, Fischbach BW (1988) J Non-Cryst Solids 101: 123
619 Konijnendijk WL, Duure H, Groenenij W (1973) Verres Refract 27: 11
620 Scholze H (1985) J Non-Cryst Solids 73: 669–680
621 Brinker C, Scherer G (1990) Sol-gel science. Academic Press, San Diego
622 Brinker C, Hurd A, Frye G, Schunk P, Ashley C (1992) Sol-gel thin-film formation. Chem Process Adv Mater 1992: 315–413
623 Uhlmann D, Boulton J, Tedwee G, Weisenbach L, Zelinski B (1990) Sol-gel synthesis of optical thin films and coatings. SPIE 1328: 270–295
624 Uhlmann D, Tedwee G, Boulton J, Denesuk M, Bommersbach W (1992) Wet chemical synthesis of optical films. Chem Process Adv Mater 1992: 483–496
625 Mackenzie J, Ulrich D (1990) Sol-gel optics, present status and future research. SPIE 1378 O1 12
626 Hench L, West J (1990) The Sol-gel process. Chem Rev 90 [1]: 33–72

627 Hench L, Vasconcelos W (1990) Gel-silica science. Annu Rev Mater Sci 20:269–298

628 (a) Schmidt H (1991) Sol-gel processing of ceramics. Ceram Trans 12:3–13; (b) idem (1992) Processing and optical properties of inorganic-organic films. Chem Process Adv Mater 1992:727–735

629 Reisfeld R, Jörgensen C (1992) Optical properties of colorants or luminescent species in sol-gel glasses. Struct Bonding (Berlin) 77:207–56; CA #54105r (1992) 117:346

630 Roy R (1992) Solution sol-gel technology and science: past, present and future. Chem Process Adv Mater 1992:997–1019, 1022–1033

631 Avnir D, Braun S, Ottolenshi M (1992) Encapsulation of organic molecules and enzymes in sol-gel glasses. Am Ceram Soc Symp 499:384–404; ACS Symp Ser 498:384–401; CA #176476h (1992) 117:313

632 Brinker CJ (1982) J Non-Cryst Solids 48:47–64; ibid (1984) 63:45–59

633 Brinker CJ, Scherer GW (1985) J Non-Cryst Solids 70:301–322

634 Yoldas BE (1984) J Non-Cryst Solids 63:145–154

635 Zarzycki J (1984) In: Hench LL, Ulrich DR (eds) Ultrastructure processing of ceramics, glasses, and composites. Wiley, New York, pp 27–42

636 Hench LL, Ulrich DR (1986) Science of ceramic chemical processing. Wiley, New York, pp 52–64

637 Schmidt H, Scholze H, Kaiser A (1984) J Non-Cryst Solids 63:1–11

638 Shoup R (1992) Properties of alkali silicate gels and their sintered fused silica glasses. In: Proc 4th Int Conf Ultrastruct Process Ceram Glasses Composites, 1989. Ultrastruct Process Adv Mater

639 Cheng YC, Hench L (1992) Sol-gel derived titania-silica gel glasses. In: Proc 4th Int Conf Ultrastruct Process Ceram Glasses Composites, 1989. Ultrastruct Process Adv Mater

640 Roncone R, Weller-Brophy L, Zelinski B (1992) Sol-gel synthesis of planar optical waveguides and integrated components. In: 4th Int Conf Ultrastruct Process Ceram Glasses Composites, 1989. Ultrastruct Process Adv Mater, pp 545–553

641 Zelinski B, Fabes B, Weisenbach L, Zauge T (1992) Sol-gel processing of passive components for integrated optics. Chem Process Adv Mater 1992:483–496

642 Chia T, West J, Hench L (1992) Fabrication of microlenses by laser densification on gel silica glass. Chem Process Adv Mater 1992:933–939

643 Yazawa T, Tanaka H, Nakamichi H, Yokoyama T (1991) Preparation of water and alkali durable porous glass membranes coated in porous alumina tubing by sol-gel method. J Membr Sci 60 [2B]:307–317

644 Arfsten NJ (1984) J Non-Cryst Solids 63:243–249

645 Beier W (1989) Habilitation, TH Clausthal-Zellerfeld

646 Schlichting J (1984) J Non-Cryst Solids 63:173–181

647 Höland W, Plumat E, Duvigneaud P (1982) J Non-Cryst Solids 48:205

648 Duvigneaud P, Höland W, Plumat E, Vogel W (1986) Ber Dtsch Keram Ges 63:523–526

649 Kamiya K, Ohya M (1986) J Non-Cryst Solids 83:208–222

650 Philipp G, Schmidt H (1984) J Non-Cryst Solids 63:283–292

651 Izumitani T, Nakagawa N (1965) In: Proc 7th Int Congr Glass, Bruxelles, 1965. Inst Nat Verre, pp 1–12

652 Käs HH (1972) Glastech Ber 45:1–9

653 Ehrt D, Jäger C, Vogel W (1987) Wiss Z Friedrich-Schiller-Univ Jena: Naturwiss Reihe 36:867–884

654 Ehrt D, Vogel W (1982) Feingerätetechnik 31:147–151

655 Vogel W, Gerth K, Heindorf W (1965) Jenaer Rundschau 10:75–86

656 Stachel D (1990) Habilitation, Otto-Schott-Inst Friedrich-Schiller-Univ, Jena

657 Reitmayer F, Schröder H (1975) Appl Opt 14:716

658 Koyama T, Ohtsuka S, Nagata H, Tanaka S (1992) J Cryst Growth 117:156

659 Stegeman G, Stolen R (1989) J Opt Soc Am B6:652

660 Koizumi K (1992) Non-linear optics, fiber lasers and microoptics. Bol Soc Esp Ceram Vidrio 31 C vol 1:45–60

661 Kordes E (1965) Glastech Ber 38:242–249

662 Weyl WA (1967) Coloured glasses, 2nd edn. Society of Glass Technology, Sheffield, England

663 Bamford C (1977) Color generation and control in glasses. Elsevier, Amsterdam

664 Kumar S (1959) Cent Glass Ceram Res Inst Bull 6:99–126

665 Bamford CR (1962) Phys Chem Glasses 3:189–202

666 Bates R (1961) In: Mackenzie, JD (ed) Modern aspects of the vitreous state, vol 2. Butterworth, London, pp 195–254

667 Juza R, Seidel H, Tiedemann J (1966) Angew Chem 78:41–51
668 Coenen M (1968) Glastech Ber 41:1–10
669 Gitter M (1970) In: Proc 3rd Wiss Tech Konf Glas-Feinkeram, Varna, Bulgaria, 1970. pp 91–105
670 Reinen D (1971) Angew Chem 83:991–999
671 Vogel W, Rehfeld A, Ritschel H (1969) In: Proc Symp Coloured Glasses, Prague, 1969. pp 114–127
672 Bunting EN (1930) J Am Ceram Soc 13:8
673 Igerson E, Morey GW, Tuttle OF (1948) Am J Sci 246:33
674 Kröger FA (1940) Physica 7:1–12
675 Dug E (1967) Glastech Ber 40:177–181
676 Dietzel A (1941) Glastech Ber 19:1–8
677 Hinrichs H (1928/29) Glastech Ber 4:51–54
678 Izumitani T, Matsuura T (1969) In: Proc Symp Coloured Glasses, Prague, 1969. pp 98–113
679 Karch Z (1964) Chim Ind (Paris) 92:218–219
680 Karch Z (1966) Sprechsaal 99:368–374, 396–400
681 Kuwabara G (1954) J Phys Soc Jpn 9:992–996
682 Löffler H (1938) Sprechsaal 71:406–408
683 Mie G (1908) Ann Physik 25:377–445
684 Sack W (1962) Glas Email Keramo Tech 13:126–136
685 Suckstorff GA (1930) Glastech Ber 8:270–275
686 Rehfeld A, Katzschmann R (1976) Dissertation, Bergakad Freiberg (VEB Jenaer Glaswerk Schott & Gen, Jena)
687 Zsigmondy R (1920) Kolloidchemie, 3rd edn. Leipzig
688 Dietzel A, Höfler W (1934) Glastech Ber 12:117–134, 297, 301; ibid (1936) 14:411–421
689 Dietzel A, Hirsch W (1935) Sprechsaal 68:16–20
690 Ahmed A, El Shamy T, Sharaf N (1980) J Am Ceram Soc 63:537
691 Kuhn A (1960) Kolloidchemisches Taschenbuch, 5th edn. Geest und Portig, Leipzig
692 Thiessen P (1942) Kolloid Z 101:241
693 Heymann E (1930) Kolloid Z 52:269
694 Raminskaya N (1980) Steklo i Keram 8:23
695 Berger E/Jenaer Glaswerk Schott & Gen, Jena (1934) DRP 604 146
696 Vogel W, Heintz H (1963) GDR W Pat 43 207
697 Schröder H, Neuroth N (1967) Optik 26:381–401
698 Sheperd ES, Rankin GA, Wright FE (1909) Am J Sci 28:293–330
699 Sun KH (1949) Glass Ind 30:199–200, 232
700 Hafner HC, Kreidl NJ, Weidel RA (1958) J Am Ceram Soc 41:315–323
701 Frerichs R (1950) Phys Rev 78:643; idem (1953) J Opt Soc Am 43:1153–1157
702 Fraser WA (1953) J Opt Soc Am 43:823
703 Tsugane S, Haradome M, Hioki R (1965) Jpn J Appl Phys 4:77–83
704 Feltz A, Lippmann FJ (1973) Z Anorg Allgem Chem 398:157–166
705 Heider F, Linke D (1973) Z Chem 13:418
706 Flaschen SS, Pearson AD, Northover WR (1959) J Am Ceram Soc 42:450; ibid (1960) 43:274–278
707 Borisova ZU, Panus VR, Obrazcov AA (1970) Vestn Leningr Univ Fiz Khim 22:121
708 Pearson AD, Northover WR (1962) In: Advances glass technology, pt 1. Plenum, New York, pp 357–365
709 Flaschen SS, Pearson AD, Northover WR (1960) J Appl Phys 31:219–220
710 Kawamoto I, Tsuchihashi S (1971) Yogyo-Kyokaishi 79:264–269
711 Asahara Y, Izumitani T (1974) Yogyo-Kyokaishi 82:583–586
712 Webber PJ, Sevage JA (1976) J Non-Cryst Solids 20:271–283
713 Linke D, Böckel I (1976) Z Anorg Allg Chem 419:97–107
714 Linke D, Heyder S (1976) Z Anorg Allgem Chem 425:155–168
715 Hilton AR, Jones CE, Brau M (1964) Infrared Phys 4:213–221
716 Linke D, Eberhardt G, Dietz Q (1975) In: Proc 6th Int Conf Amorphous Semicond, Leningrad, 1975. Acad Nauk SSSR, pp 251–255
717 Borisova ZU (1972) Khimiya Stekloobraznych polupro vodnikov. Leningrad Univ, p 107
718 Feltz A, Senf L (1975) Z Chem 15:119
719 Feltz A, Schlenzig E, Arnold D (1974) Z Anorg Allgem Chem 403:243–250
720 Feltz A, Thieme C (1974) Z Chem 14:32–33
721 Burkhardt W (1977) Dissertation, Univ Jena

722 Pazin AV, Obrazcov AA, Borisova ZU (1972) Izv Akad Nauk SSSR Neorg Mater 8:247
723 Hilton AR, Brau M (1963) Infrared Physics 3:69–76
724 Linke D, Hey H (1976) Z Chem 16:412–413
725 Egorova EA, Kokorina VF (1963) Opt Mekh Promst 1:33
726 Feltz A, Maul W, Schönfeld I (1973) Z Allgem Anorg Chem 396:103–107
727 Goryunova NA, Kolomiets BT (1955) Zh Tekh Fiz 25:984
728 Goryunova NA, Kolomiets BT (1956) Izv Akad Nauk SSSR 20:1496
729 Kolomiets BT (1964) Physica Status Solidi 7:359–372, 713–731
730 Ovshinsky SR (1966) US Pat 3 271 591
731 Ovshinsky SR, Fritzsche H (1971) Metall Trans 2:641–645
732 Bienenstock A, Betts F, Ovshinsky SR (1970) J Non-Cryst Solids 2:347
733 Betts F, Bienenstock A, Ovshinsky SR (1970) J Non-Cryst Solids 4:554–563
734 Betts F, Bienenstock A, Bates CW (1972) J Non-Cryst Solids 8–10:364–375
735 Betts F, Bienenstock A (1972) J Non-Cryst Solids 7:417
736 Bienenstock A (1973) J Non-Cryst Solids 11:447–458
737 Ovshinsky SR (1968) Phys Rev Lett 21:1450–1453
738 Fritzsche H, Ovshinsky SR (1970) J Non-Cryst Solids 2:393
739 Henisch H, Fagen EA, Ovshinsky SR (1970) J Non-Cryst Solids 4:538
740 Krebs H, Fischer P (1970) Ber Dtsch Keram Ges 47:518
741 Krebs H, Steffen R (1964) Z Anorg Allgem Chem 327:224
742 Krebs H, Ackermann F (1972) Glastech Ber 45:213–220
743 Feltz A, Büttner HJ (1972) Z Chem 12:392–393
744 Feltz A, Voigt B, Schlenzig E (1973) In: Proc Int Conf Amorphous Liquid Semicond, Garmisch Partenkirchen, 1973
745 Feltz A, Schirrmeister F (1972) Z Elektr Inform Energietechnik 2:203–205
746 Feltz A (1983) Amorphe und glasige anorganische Festkörper. Akad Verl, Berlin
747 Linke D (1972) In: Proc Konf amorph flüss Halbleiter, Sofia, 1972
748 Rawson H, Denton EP (1958) Brit Pat 801 570
749 Mackenzie JD (1966) US Pat 3 258 434
750 Kawamoto Y, Tsuchihashi S (1969) J Am Ceram Soc 52:626
751 Liu-Chun-Hua, Pashinkin AS (1962) Ber Akad Wiss UdSSR 146:1092
752 Cervinka L, Hruby A, Matyas M (1970) J Non-Cryst Solids 4:258–271
753 Stiegler H, Haberland DR (1972) J Non-Cryst Solids 11:147–152
754 Kenemann SA (1971) Appl Phys Lett 19:205–207
755 Igo T, Toyoshima Y (1973) J Non-Cryst Solids 11:304
756 Glaze FW, Blackburn DH (1957) J Res Natl Bur Stand 59:83–92
757 Specker H (1953) Angew Chem 65:299–303
758 Robinson RL, Scott WE (1933) Z Anorg Allgem Chem 210:57–66
759 Savage JA, Nielsen S (1964) Phys Chem Glasses 5:82–86
760 Kamamoto Y, Tsuchihashi S (1971) J Am Ceram Soc 54:131–135
761 Deeg EW (1962) In: Advances glass technology, pt 1. Plenum, New York, pp 348–355
762 Nikandrova GA, Orlova GM (1971) Zh Prikl Khim (Moscow) 44:1410; ibid (1970) 43:1210
763 Vinogradova GZ (1970) Zh Neorg Chim 15:1949
764 Bonceva-Mladenova Z, Linke D, Lippmann F-J (1972) In: Proc Conf Amorphous Liquid Semicond, Sofia, 1972. pp 95–98
765 Vaipolin AA, Porai-Koshits EA (1961) Sov Phys Solid State 2:1500–1508
766 Hilton AR (1973) J Electron Mater 2:211
767 Hilton AR, Hayes DJ, Rechtin MD (1975) J Non-Cryst Solids 17:319–338
768 Hilton AR (1966) Appl Opt 5:1877; idem (1968) Phys Chem Glasses 9 [5]:148
769 Ballard SS, Browder JS (1966) Appl Opt 5:1874
770 Anthonis HE, Kreidl NJ (1972) J Non-Cryst Solids 11:257
771 Ratzenboeck WH (1973/74) J Non-Cryst Solids 13:13
772 Cornet J (1975) Rev Phys Appl 10:409
773 Mott N (1967) Adv Phys 16:49
774 Emin D (1983) Jena Z Naturwiss 32 [2/3]:303
775 Schultz-Sellak C (1870) Ann Phys 139:162
776 Petz J, Kruh R, Amstutz N (1961) J Chem Phys 34 [5]:76
777 Leadbetter A (1974) J Non-Cryst Solids 15:250
778 Myers M, Felty E (1967) Mater Res Bull 2 [7]:715
779 Maruno S, Noda M (1972) J Non-Cryst Solids 7:1

780 Kastner M (1978) J Non-Cryst Solids 31: 223
781 Elliot S (1991) Chalcogenide glasses. Mater Sci Technol 9: 375–454
782 Fritzsche H (1992) The origin of photostructural changes in chalcogenide glasses. In: 16th Int Congr Glass, Madrid, 1992. Bol Soc Esp Ceram Vidrio 31 C vol 1: 61
783 (a) Kreidl N (1983) Inorganic glass forming systems. In: Uhlmann D, Kreidl N (eds) Glass science and technology, vol 1. Academic Press, New York, pp 107–300; (b) idem (1992) Oxide glasses free of classical network formers. Adv Fusion Glass Cer Transact 29: 175–189
784 Wei W, Corb B, Averbach B (1982) J Non-Cryst Solids 53 [1/2]: 19
785 Corb B, Wei W, Averbach B (1982) J Non-Cryst Solids 53 [1/2]: 29
786 Feltz A, Aust H, Blayer A (1983) J Non-Cryst Solids 55: 179
787 Malugani JP, Robert G (1980) Solid State Ionics 1 [5/6]: 519
788 Mercier R, Malugani JP, Fahys B, Robert G (1981) Solid State lonics 5: 663
789 Levasseur A, Brethous JC, Khala M, Hagenmuller P (1981) Solid State Ionics 5: 651
790 Cheng J, Tilloga G, Zarzycki J (1981) J Mater Sci 16 [9]: 2531
791 Ostwald W (1924) Licht und Farbe in Kolloiden. Dresden
792 Gehlhoff G, Thomas M (1928) Z Tech Physik 9: 172–175
793 Jochmann H (1956) Sprechsaal 89: 204–206
794 Springer L (1962) Sprechsaal 95: 511–513
795 Knizek I (1962) Sprechsaal 95: 569–585
796 Schönborn H (1929/30) Glastech Ber 8: 366
797 Inwald O (1899) Studien über die Zusammensetzung und die Eigenschaften von Phosphatgläsern. Dissertation, Berlin
798 Zschimmer E, Hesse K, Stoss L (1925) Sprechsaal 58: 513–517
799 Schönborn H (1959) Silikattechnik 10: 390–400
800 Andrews AJ, Clark GL, Alexander WH (1933) J Am Ceram Soc 16: 385–392
801 Weyl WA (1941) J Am Ceram Soc 24: 221–225
802 Trömel G, Harkort HJ, Hotop W (1948) Z Allgem Anorg Chem 256: 262
803 Zarzycki J, Mezard R (1962) Phys Chem Glasses 3: 163–166
804 Porai-Koshits EA, Goganov DA, Averyanov VJ (1964) In: Proc Int Conf Phys Non-Cryst Solids, Delft, 1964. p 117
805 Gerth K, Rehfeld A, Vogel W (1969) GDR W Pat 80526
806 Berger E, Klemm A (1936) Zeiß Nachr 2: 1–7
807 Christiansen P (1884) Wied Ann 23: 298; ibid (1885) 24: 439
808 Knudsen E (1934) Kolloid Z 66: 257–266
809 Knudsen E (1935) Kolloid Z 69: 35–43
810 Kleber W Silikattechnik (1962) 13: 5–10
811 Roy R (1960) J Am Ceram Soc 43: 670; idem (1962) Symp Nucleation Cryst Glasses Melts, Toronto, 1961. Am Ceram Soc, pp 39–45
812 Volmer M, Weber A (1925) J Phys Chem 119: 277
813 Turnbull D, Vonnegut B (1952) Ind Eng Chem 44: 1292–1298
814 Kossel W (1927) Nachr Ges Wiss Göttingen Math Phys Kl Fachgruppe, 135
815 Kossel W (1930) Naturwissenschaften 18: 901
816 Kossel W (1949) Das Molekül und der Aufbau der Materie. Vieweg, Braunschweig
817 Stranski IN (1928) Z Phys Chem Abt A 136: 259
818 Stranski IN (1929) Z Phys Chem Abt A 142: 453–466
819 Stranski IN (1944) Z Anorg Allgem Chem 252: 241
820 Stranski IN (1950) Naturwissenschaften 37: 289–296
821 Meyer K (1968) Physikalisch-Chemische Kristallographie. VEB Dtsch Verl Grundstoffind, Leipzig
822 Neuhaus A Fortschr Mineral (1950/51) 29/30: 136–296
823 Vogel W (1955) Silikattechnik 6: 510–517
824 De Vries RC, Roy R, Osborn EM (1954) Trans Br Ceram Soc 53: 531
825 Becker H (1913) German Pats 287394, 332578, 430387, 410351
826 Albrecht F, Eppler W, Maucher A (1957) German Pat 1007231
827 Albrecht F (1955) Beispiele angewandter Forschung. Fraunhofer Ges Förd Naturwiss, Munich, pp 19–22
828 Reaumur M (1739) Mem Acad Sci Inst Fr, 370
829 Stookey SD (1959) Glastechn Ber 32: 1–8 K
830 Stookey SD (1960) J Am Ceram Soc 43: 190
831 Stookey SD (1961) German Pat 844648, US Pat 3205079, German Pats 922733, 922734, 962110, German Pat Appls 1045056, 1090397

832 McMillan P (1979) Glass ceramics, 2nd edn. Academic Press, New York
833 Beall G, Montierth M, Smith G (1971) Chem Anlagen Verfahren 10:87
834 Duke D (1983) Glass ceramic technology. In: Uhlmann D, Kreidl N (eds) Glass science and technology. Academic Press, New York, p 403
835 Lungu SN, Popescu-Has D (1958) Ind Usoara 2:52–58; ibid (1959) 6:177–180
836 Lyng S (1967) Glastechn Ber 40:136–140; idem (1968) Glas Technol 9:179–184
837 Lyng S, Markali J (1970) Phys Chem Glasses 11:6–10
838 Sigayev WN (1975) Fiz Khim Stekla 1:403–406
839 Bobkova NM (1976) Izv Akad Nauk SSSR Neorg Mater 12:148–150
840 Pavlushkin NM, Sharafyev MS, Suliymenov ST (1968) Izv Akad Nauk SSSR Neorg Mater 4:635–637
841 Pavlushkin NM, Khodakovskaya RJ (1969) In: Proc Conf Glass Struct Leningrad, 1969. Akad Nauk SSSR; idem (1971) Izv Akad Nauk SSSR Neorg Mater 7:846–849
842 Varschal BG (1973) Izv Akad Nauk SSSR Neorg Mater 9:2206–2212
843 Khodakovskaya RJ (1968) Steklo 3:95–101
844 Khodakovskaya RJ, Pavlushkin NM (1967) Izv Akad Nauk SSSR Neorg Mater 3:1908–1915
845 Mukherhee SP, Rogers PS (1967) Phys Chem Glasses 8:81–87
846 Chyung CK, Beall GH, Grossmann DG (1974) In: Proc 10th Int Glass Congr, vol 14, Kyoto, 1974. Jpn Ceram Soc, pp 33–40
847 Beall GH, Montierth MR, Smith GP (1971) Glas Email Keramo Tech 22:409–419
848 Fluss FF (1958) Glas Instrum Tech 2:14–16
849 Weyl WA (1960) Nucleation, crystallisation and glass formation. Sprechsaal 93:128–136
850 Hinz W, Baiburt L (1960) Silikattechnik 11:455–459
851 Hinz W, Knuth P (1960) Silikattechnik 11:506–511
852 Vogel W, Gerth K (1962) Symposium on nucleation and crystallization in glasses und melts. In: Catalyzed crystallization in glass. Lect 63th Ann Congr Am Ceram Soc, Toronto, Canada, 1961. Am Ceram Soc, New York, pp 11–22
853 Vogel W, Gerth K (1962) Z Chem 2:261–274
854 Ziemba B (1961) Steklo i Keram 12:257–263
855 Nebrensky J, Voldan I (1961) Inf Prehl Statni Vyzk Ustav Sklarsky Hradci Kralove 1:3–22
856 Moriya T, Sakaino T, Endo M (1960) J Ceram Assoc Jpn 68:44–49
857 Baum W (1963) Glastech Ber 36:444–453, 468–481
858 Thakur RL (1963) Glass Cent Glass Ceram Res Inst Bull 10:51, 66
859 Marchesini L, Rappretti A (1961) Vetro Silic 5:5–11
860 Sakka S, Tashiro M (1961) J Ceram Assoc Jpn 69:109–118
861 Hummel FA (1950) J Am Ceram Soc 33:102; idem (1951) J Am Ceram Soc 34:235–239
862 Smoke EJ (1951) J Am Ceram Soc 34:87–90
863 Karkhanavala MP, Hummel FA (1953) J Am Ceram Soc 36:393–397
864 Prokopovicz TJ, Hummel FA (1955) J Am Ceram Soc 39:266–278
865 Levin EM, McMurdie HF, Hall FP (1956) Am Ceram Soc Bull 86:142
866 Mehmel M (1957) Sprechsaal 90:4, 5
867 Tashiro M, Wada M (1962) In: Proc 6th Int Glass Congr, Washington, 1962. Am Ceram Soc, pp 18–19
868 Petzoldt J (1967) Glastech Ber 40:385–395; ibid (1968) 41:181–189; ibid (1970) 43:127–137
869 Sack W, Scheidler H (1970) Glastech Ber 43:359–368
870 Müller G (1972) Glastech Ber 45:189–194
871 Pavlushkin NM (1967) Izv Akad Nauk SSSR Neorg Mater 3:733–736
872 Kruchin JD (1975) Izv Akad Nauk SSSR Neorg Mater 11:2055–2057
873 Lapin VV, Ivanov IT (1973) Izv Akad Nauk SSSR Neorg Mater 9:683–687
874 Pavlova VN (1975) Izv Akad Nauk SSSR Neorg Mater 10:391–392
875 Scheidler H, Sack W (1971) In: Proc 9th Int Glass Congr, Versailles, 1971. Inst Verre, pp 1069–1085
876 Sack W, Scheidler H (1966) Glastech Ber 39:126–130
877 Doenitz FD, Heidenreich E, Vogel W (1974) Silikattechnik 25:227–228
878 Heidenreich E, Ehrt R, Doenitz FD, Vogel W (1976) Silikattechnik 27:402–405
879 Heidenreich E, Doenitz FD, Ehrt R, Vogel W (1974) Silikattechnik 25:225
880 Heidenreich E, Ehrt R, Doenitz FD, Vogel W (1977) Silikattechnik 28:45–48
881 Hench LL, Freimann SW (1971) Advances in nucleation and crystallization in glasses. Am Ceram Soc
882 Heidenreich E, Doenitz FD, Vogel W (1974) Silikattechnik 25:229

883 Kumar S, Nag BB (1966) J Am Ceram Soc 49:10–14
884 Beall GH (1975) Austrian Pat 254431
885 Sack W (1964) BRD-AS 1596858
886 Sack W (1965) Chem Ing Tech 37:1154–1165
887 Daniels W, Moore R (1975) J Am Ceram Soc 58:217
888 Stookey S (1961) US Pat 3 325 265
889 Beall GH, Chyung CK, Watkins HJ (1974) US Pat 3 801 295
890 (a) Beall GH (1972) BRD OS 2224990, 3756838; (b) idem (1972) BRD OS 2133652
891 Grossmann DG (1974) US Pats 3 839 055, 3 839 056
892 Grossmann DG (1972) J Am Ceram Soc 55 [9]:446
893 Grossmann DG (1978) Technica 27 [12]:735
894 Vogel W, Heidenreich E, Ehrt R (1973) GDR W Pat 113885
895 Höland W, Zlateva K, Vogel W, Gutzow I (1982) Z Chem 22 [6]:197–202
896 Höland W, Vogel W, Mortier W, Duvigneaud P, Naessens G, Plumat E (1983) Glass Technol 24 [6]:318
897 Reif D, Greiner R (1974) GDR W Pat 111886
898 Matsushita T, Shiratori M, Tsunashima A, Kudaira K (1982) Yogyo-Kyokaishi 90 [6]:163
899 Höland W, Naumann W, Vogel K, Gummel J (1983) Wiss Z Friedrich-Schiller-Univ Jena Math Naturwiss Reihe 32:571–580
900 (a) Beall GH (1990) Tagungsband 4th Int Otto-Schott-Kolloq, Jena, 1990. Univ Jena, pp 234–251; (b) idem (1991) J Non-Cryst Solids, in preparation
901 Beall GH, Chyung CK, Stewart RL, Donaldson KY, Lee HL, Baskaran S, Hasselman DPH (1986) J Mater Science 21:2365–2372
902 Gardon R (1980) Thermal tempering of glass. In: Uhlmann D, Kreidl NJ (eds) Glass science and technology, vol 5. Academic Press, New York, pp 145–216
903 Stookey SD, Olcott JS (1961) US Pat 2 998 675; idem (1962) Ceram Abstr 45:58
904 Vogel W, Höland W (1982) Z Chem 22:429–437
905 Levin EM, Robbins CR, McMurdie HF (1964) Phase diagrams for ceramists. Am Ceram Soc, p 746
906 Vogel W, Doenitz FD, Dohndorf C (1974) GDR W Pat 110646
907 Doenitz FD, Vogel W (1980) Krist Tech 15:891
908 Müller G, Scheidler H (1974) Glastech Ber 47:10–13
909 Tucker RW, Stuart DR (1971) In: Proc 9th Int Glass Congr, vol 3, Versailles, 1971. Inst Verre, p 1119
910 (a) Vogel W, Heidenreich E, Weiss H (1976) GDR W Pat 194478; (b) idem (1976) GDR W Pat 194670
911 Heidenreich E, Doenitz FD, Erxleben H, Metz G, Ehrt R, Vogel W (1979) Wiss Z Friedrich-Schiller-Univ Jena Math Naturwiss Reihe, issue 2/3:432
912 Beall GH (1979) Wiss Z Friedrich-Schiller-Univ Jena Math Naturwiss Reihe, issue 2/3:415
913 Dalal K, Davis R (1977) Am Ceram Soc Bull 56 [11]:991, 1076
914 Takusagawa N, Skito H (1973) Yogyo-Kyokaishi 81 [10]:424
915 Lyng S, Markali J (1970) Phys Chem Glasses 11:6–10
916 Vogel W, Höland W, Naumann K (1980) GDR WP 223925
917 Höland W, Naumann K, Seiferth HG, Vogel W (1981) Z Chem 21:108
918 Vogel W, Heidenreich E (1975) GDR WP 122062, Int Kl C03c 3/22
919 Beall GH, Reade FR (1975) US-Pat 4 140645, Int Kl C03c 3/22
920 Andrus RL (1975) US-Pat 4084973, Int K1 C03c 3/22
921 Reade FR (1975) US-Pat 4083709, Int K1 C03b 32/00
922 Bras E (1976) DE-OS 2633744, Int K1 C03c 3/22
923 Roger PS, Williamson J (1969) Glass Technol 10:128
924 Pavlushkin NM, Sarkisov PD, Levina VS, Rogninskaya MS (1973) Neorg Mater 10:2198
925 Höland W, Heidenreich E, Vogel W (1982) GDR WP 237729
926 Höland W, Anh Dung N, Heidenreich E, Tkalcec E, Vogel W (1982) Glastech Ber 55:41
927 Höland W, Anh Dung N, Heidenreich E, Tkalcec E, Vogel W (1982) Glastech Ber 55:70
928 Höland W, Wange P, Carl G, Vogel W, Heidenreich E, Erxleben H (1984) Silikattech 35:181–184
929 Doenitz ED, Koch K, Vogel W (1982) Silikattech 33:18–21
930 Höland W, Zlateva K, Vogel W, Gutzow J (1982) Z Chem 22:197–202
931 Höland W, Vogel W, Heidenreich E (1985) Silic Ind 1–2:3–6
932 Höland W, Anh Dung N, Heidenreich E, Tkalcec E, Vogel W (1982) Glastech Ber I 55:41–49

933 Höland W, Anh Dung N, Heidenreich E, Tkalcec E, Vogel W (1982) Glastech Ber II 55:70–74
934 Vogel W, Höland W (1987) Angew Chem 99:541–558; idem (1987) Angew Chem Int Ed Engl 26:527–544
935 Pantano CG, Clark EA, Hench LL (1974) J Am Ceram Soc 57:412
936 Ogino M, Ohuchi F, Hench LL (1980) J Biomed Mater Res 14:55
937 Ogino M, Hench LL (1980) J Non-Cryst Solids 38/39:673
938 Köhler S, Retemeyer K, Berger G, Carow R, Meyer R (1980) Dtsch Gesundheitswes 35:1343
939 Hentrich RL, Graves GA, Stein HG (1971) J Biomed Mater Res 5:25
940 Köster K, Heide H (1977) Z Orthop 115:604
941 Akao M, Aoki H, Kato K (1981) J Mater Sci 16:809
942 Zhu Peinan, Cheng Zongquion, Huang Jianong, Sim Quinlian, Feng Meizhen, Chien Pufan, Yang Quinghung, Cai Tidong (1982) J Non-Cryst Solids 52:503
943 Pernot F, Zarzycky J, Bonnel F, Rabischong P, Baldet R (1979) J Mater Sci 14:1694
944 Kokubo T, Shigematsu M, Nagashima Y (1982) Bull Inst Chem Res Kyoto Univ 60:260
945 Kokubo T, Ito S, Shigematsu M, Nakamura T, Yamamuro T, Higashi S (1983) 13th Int Glaskongr, Hamburg, 1983. Dtsch Glastech Ges
946 Hench LL (1975) J Non-Cryst Solids 19:27
947 Hench LL, Paschall HA (1973) J Biomed Mater Res Symp 4:25
948 Hench LL, Splinter RJ, Greenlec TR, Allen WC (1971) J Biomed Mater Res Symp 2:1117–1147
949 Käs H (1973) DAS 2349859
950 Brömer H, Käs H, Pfeil E (1973) DBP 2326100
951 Wihsmann F, Berger G, Kirsch M (1983) Wiss Z Friedrich-Schiller-Univ Jena Math Naturwiss Reihe 32:553
952 Thieme V, Hofmann H, Heiner H, Berger G (1982) Z Exp Chir 15:310
953 Köhler S, Retemeyer K, Berger G, Kunth R (1984) Stomatol DDR 34:557
954 Vogel W, Höland W, Naumann K (1982) GDR Pat WP CP 3 B/237 728/2
955 Grossmann, DG (1972) DOS 2208236
956 Vogel W, Höland W (1982) Adv Ceram 4:125
957 Vogel W, Höland W (1982) Z Chem 22:429
958 Höland W, Naumann K, Seyferth HG, Vogel W (1981) Z Chem 21:108
959 Vogel W, Höland W, Naumann K (1980) GDR WP 0153108
960 Höland W, Vogel W, Mortier WJ, Duvigneaud PH, Naessens G, Plumat E (1983) Glass Technol 24:318
961 Trojer F, O'Conner PG, Tannenberg H (1975) DOS 2606540
962 Vogel W, Höland W, Naumann K, Gummel J (1986) J Non-Cryst Solids 80:34
963 Gummel J (1983) Habilitation, Academy of Medicine, Dresden
964 Abe Y, Nagoya A (1981) DBP 3 142813
965 Abe Y, Hosoe M, Kasuga T, Ishikawa H, Shiukai N, Suzuki Y, Nakayama J (1982) J Am Ceram Soc 65:189
966 Vogel J, Höland W, Vogel W (1984) GDR WP 259561/1
967 Vogel W, Vogel J, Höland W, Wange P (1987) 3rd Int Otto-Schott-Kolloquium, Jena, 1986. Wiss Z Friedrich-Schiller-Univ Jena: Naturwiss Reihe 37:841–854
968 Stone PE, Egan EP, Lehr JR (1956) J Am Ceram Soc 39:96
969 Haubenreißer U, Vogel J, Höland W, Vogel W (1987) 3rd Int Otto-Schott-Kolloq, Jena, 1986. Wiss Z Friedrich-Schiller-Univ Jena: Naturwiss Reihe 37:763–776
970 Landolt-Börnstein (1969) Neue Serie Gruppe III, vol 2. Springer, Berlin
971 Nixon J (1985) Br Med J 290:490
972 Schubert T (1981) Dissertation, Med Akad, Dresden
973 Wange P, Vogel J, Vogel W, Höland W, Götz W (1990) German Pat DE 4004196A1
974 Schulze KJ, Schubert T (1988) Knorpel-Knochen Transplantationen. Thieme, Stuttgart, pp 139–144
975 Schubert T, Purath W, Liebscher P, Schulze KJ (1988) Beitr Orthop Traumatolog (Berlin) 35:7–16
976 Purath W, Dörrschmidt V, Schubert T, Liebscher P (1987) Beitr Orthop Traumatolog (Berlin) 34:547
977 Beleites E, Gudziol H, Höland W (1986) HNO-Praxis 11:169–176
978 Neuperth G, Vogel W (1984) Exp Pathol 26:113–116
979 Beleites E, Neupert G, Augsten G, Vogel W, Schubert H (1985) Laryngol Rhinol Otol 64:217–220

980 Pinkert R, Naumann K, Vogel W (1987) Philips J Res 6:339–342
981 Pinkert R (1991) Philips J Res 8:43–45
982 Pinkert R (1990) Zahn-Mund-Kiefer-Heilkunde 78:411–416
983 Donath K, Sitzmann F, Bauer G (1986) Zahnärztl Implantol II:117–180
984 Hench L (1986) Adv Ceram Mat 1:306–310, 324
985 Yamamuro T, Hench L, Wilson J (1990) Handbook of bioactive ceramics, vol I. (Bioact Glasses and Glass-Ceramics) CRC, Boca Raton
986 Schönborn H (1962) Silikattechnik 13:419–424
987 Sersale R (1953) Ric Sci 23:1993–2008
988 Klause K (1956) InterCeram 5:14–16
989 Sack W (1959) Beiträge zur angewandten Glasforschung. Schott, Mainz, pp 111–120
990 Philips Research Centre, Eindhoven (1961) Bibliographische Studie über die mechanische Festigkeit von Glas und die Möglichkeiten, sie zu verbessern. In: Symp Union Scient Cont Verre, Florence, 1961
991 Uhlmann D, Kreidl N (1985) Viscous flow and relaxation. In: Uhlmann D, Kreidl N (eds) Glass science and technology, vol 3. Academic Press, New York
992 Polany M (1921) Z Physik 7:323–327
993 Orowan E (1934) Z Kristallogr A 89:327–343
994 Iglis CE (1913) In: Proc Inst Naval Archit, pt 1, 55:219
995 Griffith AA (1920) Philos Trans R Soc London Ser A 221:163–198
996 Griffith AA (1924) In: Proc 1st Int Congr Appl Mech, Delft. pp 55–63
997 Naray-Szabo I, Ladik J (1960) Nature 188:226–227
998 Weyl WA (1957) Glastech Ber 30:269–282
999 Weyl WA (1957) Bull Centr Glass Ceram Res Inst 4:121–140
1000 Weyl WA (1958) Bull Centr Glass Ceram Res Inst 5:104–108
1001 Weyl WA, Marboe EC (1959) Glastech Ber 32 K VI:1–18
1002 Stockdale GF, Tooley FV, Ying CW (1951) J Am Ceram Soc 34:116–121
1003 Orowan E (1945/46) Trans Inst Eng Shipbuild Scotl 89:65–212
1004 (a) Condon EU (1954) Am J Phys 22:224–232; (b) idem 22:310–317
1005 Zwicky F (1929) Proc Nat Acad Sci, 253
1006 Andrade EN, LC Tsien (1937) Proc R Soc London Ser A 159:346–354
1007 Acloque P, Le Clerc P, Ehrmann P (1959) Colloque USCV sur la Nature des Surfaces Vitreuses Polies, Paris, 1959
1008 Ernsberger FM (1960) Proc R Soc London Ser A 257:213–223
1009 Littleton JT (1923) Phys Rev 22:510–516
1010 Gehlhoff G, Thomas M (1926) Z Tech Phys 7:105–126
1011 Watanabe M, Caporali RV, Mould RE (1961) Phys Chem Glasses 2:12–23
1012 Poncelet EF (1944) Metals Technol 11:1684
1013 Poncelet EF (1948) Am Soc Met (Cleveland) 201–227
1014 McAffee KB (1958) J Chem Phys 28:218–229
1015 Cox SM (1948) J Soc Glass Technol 32:127–146 T
1016 Saibel E (1948) Am Soc Met (Cleveland) 275–281
1017 Fisher JC (1948) J Appl Phys 19:1062–1067
1018 Gibbs P, Cutler IB (1951) J Am Ceram Soc 34:200–206
1019 Freiman S (1980) Fracture mechanics of glass. In: Uhlmann D, Kreidl N (eds) Glass science and technology, vol 5. Academic Press, New York, pp 21–78
1020 Bartenev GM, Bovkunenko AM (1956) Tech Phys SSSR 26:2508–2516
1021 Silvestrovich CI, Boguslavsky IA (1960) Steklo i Keram 17:7–12
1022 Gaiser R, Lyon K, Scholes A (1965) Ceram Ind (Sevres Fr) April
1023 Brocket W (1964) US Pat 3 130 071
1024 Southwick R, Wasylyk J, Smay G, Kepple J, Smith E, Augustson B (1981) Thin Solid Films 77 [1–3]:41
1025 Budd S (1981) Thin Solid Films 77 [1–3]:23
1026 Dislich H (1985) Glass coatings. In: Uhlmann D, Kreidl NJ (eds) Glass science and technology, vol 2. Academic Press, New York
1027 Schaeffer H, Mecha J, Freude E, Weisskopf K, Knödler H (1981) Glastech Ber 54 [8]:267
1028 (a) Kistler S (1962) J Am Ceram Soc 45 [2]:59; (b) idem (1958) Brit Pat 917 388
1029 Acloque P, Tachon J (1962) Symp Mech Strength Glass, Charleroi, Belgium, 1962. Union Scientifique Continentale du Verre
1030 Leibig EC (1944) US Pat 2 075 466

1031 Leibig FC (1944) US Pat 2 198 733
1032 Hood H, Stookley S (1961) US Pat 2 998 675
1033 Zijlstra A, Burggraf F (1968) J Non-Cryst Solids 1:49
1034 Nordberg M, Mochel E, Garfinkel H, Olcott J (1964) J Am Ceram Soc 47:215
1035 Bach H (1972) Int J Mass Spektrom Ion Phys 9:247–252
1036 Lell E, Kreidl N, Hensler J (1966) Radiation effects in quartz, silica, and glasses. In: Burke J (ed) Progress in ceramic science, vol 4. Pergamon, New York, pp 1–96
1037 Kreidl N (1971) In: Kriegel W, Palmour H III (eds) Ceramics in severe environments. Plenum, New York, p 521
1038 Dalton RH (1947) US Pats 2 422 472, 2 326 012
1039 Armistead WH (1950) Canad Pat 442272
1040 Armistead WH (1950) US Pat 2515 936
1041 Stookey SD (1950) US Pat 2515 275
1042 Stookey SD (1950) US Pat 2515 937
1043 Stookey SD (1950) US Pat 2515 938
1044 Stookey SD (1950) US Pat 2515 939
1045 Stookey SD (1950) US Pat 2515 940
1046 Stookey SD (1950) US Pat 2515 941
1047 Stookey SD (1950) US Pat 2515 942
1048 Stookey SD (1950) US Pat 2515 943
1049 Stookey SD (1954) US Pat 2684 911
1050 Stookey SD (1954) US Pats 2628 160, 2 651 146, 2 682 134
1051 Stookey SD (1954) US Pat 2651 145
1052 Stookey SD (1955) US Pat 2732 298
1053 Stookey SD (1955) US Pat 2971 853
1054 Armistead WH, Stookey SD (1961) German Pat 809847
1055 Stookey SD (1962) German Pat 844648
1056 Stookey SD (1966) German Pat 922733
1057 Stookey SD (1966) German Pat 922734
1058 Stookey SD (1947) Engl Pat 635649
1059 Stookey SD (1947) Engl Pat 636151
1060 Stookey SD (1947) Engl Pat 668767
1061 Corning Glass Works (1952) Engl Pat 769090
1062 Corning Glass Works (1947) Engl Pats 636152, 654740, 684134, 699898
1063 Stookey SD (1948) Canad Pat 442273
1064 Stookey SD (1949) Ind Eng Chem 41:856–861
1065 Stookey SD (1953) Ind Eng Chem 45:115–118
1066 Stookey SD (1954) Ind Eng Chem 46:174–176
1067 Stookey SD, Schuler FW (1956) In: Proc 4th Int Glass Congr, Paris, 1956. Inst Verre
1068 Otley OO, Weyl WA (1953) US Pat 2 722 519
1069 Reinhart F (1956) Glas Email Keramo Tech 7:153–156
1070 Beyer M (1956) Mater Methods 43:143–147
1071 Bondarev KT, Borodai FJ (1960) Steklo i Keram 17:1–4
1072 Tashiro M, Sakka S (1959) J Ceram Soc Jpn 67:263–268
1073 Hall AJC, Hayes JG (1958) Res Appl Ind 11:461–465
1074 Henning W (1961) Glas Instrum Tech 5:3–5
1075 Henning W (1961) Glas Email Keramo Tech 12:319–321
1076 Barth K (1960) Silikattechnik 11:101–105
1077 Fischer R, Ahlendorf W, Schabinski G (1959) Z f ärztl Fortb 53:1225–1228
1078 Nagli G, Prosch U, Vormum G (1960) Isotopentechnik 1:43–46
1079 Fischer R, Ahlendorf W, Schwabe F (1961) Radiobiol Radiother 2:359–364
1080 Suckow W (1961) Kernenergie 4:830–834
1081 Fischer R, Schwabe F, Unverricht A, Ahlendorf W (1962) Isotopentech 2:125–128
1082 VEB Jenaer Glaswerk Schott u Gen, Jena (1976) Druckschrift 067
1083 Schulman J, Ginther R, Klick C (1951) J Appl Phys, 1479
1084 Blair (1966) J Am Ceram Soc 43:426
1085 Yokota R, Nakayima S (1965) Health Phys 11:241
1086 Becker K (1967) Personnel Radiation Dosimetry, 1st Health Phys Soc Midyear Topical Symp, Chicago, 1967
1087 Jahn W (1969) Glastech Ber 42 [5]:176
1088 Yokota R, Imagawa H (1966) J Phys Soc Jpn 23 [5]:

1089 Yokota R (1967) J Phys Soc Jpn 23 [1]: 129
1090 Lell E, Kreidl N (1967) In: Bishay N (ed) Interaction of radiation with solids. Plenum, New York, p 199
1091 Marckwald W (1899) Z Phys Chem 30: 140
1092 Kühl G, Grünwald WD (1966) German Pat 939595
1093 Kühl G, Grünwald WD (1966) German Pat 953193
1094 Kühl G, Grünwald WD (1967) German Pat 1115451
1095 Kühl G, Grünwald WD (1967) German Pat 1023752
1096 Reusch G (1965) In: Proc 7th Int Glass Congr, Brussels, 1965. Inst Nat Verre, pp 110/1–10
1097 Wippler C (1965) In: Proc 7th Int Glass Congr, Brussels, 1965. Inst Nat Verre, pp 109/1–8
1098 Bertelson RC (1971) Photochroism – Techniques of chemistry, vol III. Wiley Interscience, New York
1099 Araujo RJ (1971) In: Bertelson RC (ed) Photochroism – Techniques of chemistry, vol III. Wiley Interscience, New York
1100 Robinson CC, Snitzer E (1964) German Pat 1596724
1101 Smith GP (1967) J Mater Sci 2: 139
1102 Aschentov Y, Sukhanov V (1971) Opt Spectros (Engl Trans) 30: 612
1103 Kiss ZH (1970) Phys Today 1: 42
1104 Stookey SD (1972) US Pat 3 252 394
1105 Eichelberger WE, Megla GK (1973) US Pat 3 541 330
1106 Maurer RD (1972) US Pat 3 365 678
1107 Gliemeroth G (1970) Umschau, 210
1108 Cohen AJ, Smith HL (1962) Science 134: 981
1109 Swarts EL, Pressau JP (1965) J Am Ceram Soc 48: 333
1110 Smith GP (1965) In: Proc 7th Int Glass Congr, Brussels, 1965. Inst Nat Verre
1111 Bach H, Gliemeroth G (1971) Glastech Ber 44: 35–44
1112 Veit M (1983) Dissertation, Friedrich-Schiller-Univ, Jena
1113 Gliemeroth G, Mader KH (1970) Angew Chem 82: 421–433
1114 Gliemeroth G (1968) Glas Email Keramo Techn 19: 269–271
1115 Hoffmann HJ (1990) In: Dürr H, Bonas, Laurent (eds) Photochroism: Molecules and systems. Elsevier, Amsterdam
1116 Armistead WH, Stookey SD (1964) Science 144: 150–158
1117 Moriya J, Tokunaga S, Kawai T (1972) Yogyo-Kyokaishi 112: 121
1118 Kriltz A, Müller M (1990) Silic Ind 55: 343
1119 Besen H (1978) Proc 10th Int Glass Congr, Prague, 1978, vol III
1120 Gerth K, Rehfeld A (1974) German Pat 2149568; GDR Pat 97121
1121 Rehfeld A, Rentsch J (1975) German Pats 2260879, 2256775; GDR Pat 101375
1122 VEB Jenaer Glaswerk Schott u Gen, Jena (1980) Druckschrift 0 107/1
1123 VEB Jenaer Glaswerk Schott u Gen, Jena (1972) Druckschrift 0 31 G
1124 Kawamoto T, Kikuchi R (1978) Yogyo-Kyokaishi 86 [8]: 374
1125 Araujo R, Borelli N, Chodak J, Hakes G, Meiling G (1978) Dichroic photochromic glass, German Pat 2 747 919
1126 Morimoto S, Mishima M (1981) Yogyo-Kyokaishi 59 [7]: 55, 129
1127 Kerko D, Odile JP, Quinn C, Tick P (1979) Photochromic microsheets, US Pat 4 168 339
1128 Müller M, Gitter M, Vogel W, Veit M (1985) Poster volume: 100 Jahre Jenaer Glas. VEB Jenaer Glaswerk Schott u Gen, Jena
1129 Araujo RJ (1968) Appl Opt 7: 781
1130 Araujo RJ, Borelli NF (1976) J Appl Phys 47: 1370
1131 Araujo RJ, Borelli NF, Nolan DA (1979) Philos Mag B 40: 279
1132 Araujo RJ, Borelli NF, Nolan DA (1981) Philos Mag B 44: 453
1133 Araujo RJ (1972) US Pat 3 325 299; German Pat 1494093
1134 Araujo RJ (1972) German Pat 1496091
1135 Ferley L, Mattern T, Lehmann G (1987) J Non-Cryst Solids 92: 107
1136 Sawchuk LG, Stookey SD (1972) US Pat 3 293 052; German Pat 1496082
1137 Morse DL (1981) Inorg Chem 20: 777
1138 Müller M, Voigt J, Kriltz A, Vogel W (1988) Silikattech 39: 221
1139 Randall J, Seward T III (1973) Thermally darkening photochromic glass, US Pat 3 734 754
1140 Seward III T (1975) J Appl Phys 46 [12]: 689
1141 Maiman TH (1960) Stimulated optical radiation in ruby. Nature 187: 493–494
1142 Snitzer E (1961) Optical maser action of Nd^{3+} in a barium crown glass. Phys Rev Lett 7: 444–446

1143 Deeg EW, Faulstich M, Neuroth N (1963) Phys Verh 3:204
1144 Maurer RD (1963) Appl Opt 2:87–88
1145 Gandy HW, Ginther RJ (1962) Appl Phys Lett 1:25–27
1146 Etzel HW, Gandy HW, Ginther RJ (1962) Appl Opt 1:534
1147 Patek K (1970) Glass lasers. Butterworth, London
1148 Brown DC (1981) High-peak-power Nd-glass laser systems. Springer, Berlin
1149 Weber MJ (1987) Recent developments in glass for lasers. 3rd Otto-Schott–Kolloq, Jena, 1986. Wiss Z Friedrich-Schiller-Univ Jena: Naturwiss Reihe 37:689–698
1150 Alexe'ev NE, Gapontjev VP et al (1980) Lasernie Phosphatnie Stekla, Nauka, Moskau
1151 Reisfeld R, Jorgensen CK (1977) Lasers and excited states of rare earths. Springer Heidelberg
1152 Gan F (1985) Chin Phys Lett 5:145
1153 Krupke WF (1974) IEEE J Quantum Electron QE-10:450
1154 Brierley MC, France PW, Millar CA (1988) Lasing at 2.08 m and 1.38 m in a holmium doped fluoro-zirconate fibre laser. Electron Lett 24:539
1155 Seeber W, Ehrt D (1990) Silikattechik 41:230–233
1156 Weber MJ (1977) Laser Program Annual Report, Lawrence Livermore Lab UCRL-50021-76; idem (1978) UCRL–50021–77
1157 Weber MJ (1982) Glass lasers
1158 Deutschbein O, Pautrat C, Svirchevsky IM (1967) Rev Phys Appl 1:29
1159 Neuroth N (1964) Chem Ing Tech 36:947–956
1160 Maiman TH, Hoskins RH, D'Haenens IJ, Asawa CK, Evtuhov V (1986) Stimulated emission in fluorescent solids, vol II: Spectroscopy and stimulated emission in ruby.
1161 Seeber W, Ehrt D (1991) priv commun
1162 Heumann E, Ledig M, Ehrt D, Seeber W, Duczynski E, v.d.Heide HJ, Huber G (1988) Appl Phys Lett 52:255
1163 Wybourne BG (1965) Spectroscopic properties of rare earth. Interscience, New York
1164 Judd BR (1962) Phys Rev 127:3
1165 Förster T (1949) Z Naturforschung 4a:321
1166 Dater DL (1953) J Chem Phys 21:836
1167 Murphy J, Ohlmann RC, Mazelsky R (1964) Phys Rev Lett 13:135
1168 Melamed NT, Hirayama C, French RW (1965) Appl Phys Lett 7:170
1169 Ledig M, Heumann E, Ehrt D, Seeber W (1990) Opt Quantum Electron 22:107–122
1170 Kreidl N (1953) In: Garzanti (ed) 3rd Int Congr, Venice, 1953. Stabilimento Graphico Romo
1171 Kreidl N, Hensler J (1955) J Am Ceram Soc, 423
1172 Fanderlik M (1958) Veda Vyzk Prum Sklarskem V:29–45
1173 Faraday M (1825) Ann Chim Phys 25:99
1174 Klemm A, Berger E (1935) Glastech Ber 13:349–368
1175 VEB Jenaer Glaswerk, Jena (1950) catalog no 0 115 (Optisches Glas)
1176 Appen AA (1949) Ber Akad Wiss UdSSR 69:841–844
1177 Appen AA (1956) In: Compt Rend 4th Int Glass Congr, Paris, 1956. Inst Verre, pp 36–40
1178 Gilard P, Dubral L (1938) Verre Silic Ind 9:25–28, 37–39, 50–52
1179 Winkelmann A, Schott O (1894) Ann Phys 51:730–746
1180 Huggins ML (1940) J Opt Soc Am 30:420–430
1181 Huggins ML, Sun KH (1943) J Am Ceram Soc 26:4–11
1182 Stevels JM (1946) J Soc Glass Technol 30:173–191, 192–197, 303–305, 306–309
1183 Naray-Szabo I (1959) Glastech Ber 32:185–189
1184 English S, Turner WES (1927) J Am Ceram Soc 10:551–560; ibid (1929) 12:760
1185 Hall FP (1930) J Am Ceram Soc 13:182–199
1186 Kumar S (1959) Glastech Ber 32 K V:26–35
1187 Takahashi K (1953) J Soc Glass Technol 37:3N–7N
1188 Appen AA (1954) Silikattechnik 5:113–114
1189 Demkina LI (1960) Steklo i Keram 17:474–476
1190 Vogel H (1921) Phys Z 22:645–646
1191 Fulcher GS (1925) J Am Ceram Soc 8:339–366, 789–794
1192 Tammann G, Hesse W (1926) Z Anorg Allg Chem 156:245–257
1193 Scholze H, Kreidl N (1984) Viscosity of inorganic glasses. In: Uhlmann D, Kreidl N (eds) Glass science and technology, vol 3. Academic Press, New York
1194 Pockels F (1902) Ann Phys 7:745–771
1195 Jebsen-Marwedel H (1956) Glastech Ber 29:223–238
1196 Jebsen-Marwedel H (1954) Kolloid Z 137:118–120
1197 Jebsen-Marwedel H (1958) Naturwissenschaften 45:260

1198 Jebsen-Marwedel H (1958) In: Compt Rend Symp Fusion Verre, Brussels, 1958. Union Scientifique Continentale du Verre
1199 Jebsen-Marwedel H (1957) Kolloid Z 150:137–140
1200 Jebsen-Marwedel H (1958) Glastech Ber 31:431–438
1201 Dietzel A (1942) Kolloid Z 100:368–380
1202 Dietzel A (1942) Sprechsaal 75:82–85
1203 Lyon KC (1944) J Am Ceram Soc 27:186–189
1204 Appen AA (1954) Silikattechnik 5:11–12
1205 Rubenstein C (1964) Glass Technol 5:36–40
1206 Ratcliffe EH Glass Technol 4:113–128
1207 Winkelmann A (1893) Ann Phys 49:401–420
1208 Sharp DE, Ginther LB (1951) J Am Ceram Soc 34:260–271
1209 Moore J, Sharp DE (1958) J Am Ceram Soc 41:461–463
1210 Schwiete HE, Ziegler G (1955) Glastech Ber 28:137–146
1211 Gehlhoff G, Thomas M (1925) Z Tech Phys 6:544–554
1212 Rasch E, Hinrichsen FW (1908) Z Elektrochem 14:41
1213 Tuller H, Button D, Uhlmann D (1980) J Non-Cryst Solids 40:93
1214 Minami T (1983) J Non-Cryst Solids 56:15

Index

Abbe, E. 2
Abbe microscope 71
Abbe ratio ν_e 15
Abbe's diagram 18
Abbe's number 208
Abbe's value 409
absorbable glasses and glass ceramics 331
absorption bands 211
–, Ag ^{o} at O.34 211
–, Cr^{6+} at O.36 211
–, Fe^{3+} at O.38 211
–, Ti^{4+} at O.37 211
–, V^{5+} at O.32 211
absorption centers 211
absorption edge of SiO_2 glass 225
absorption effect of Fe^{3+} and Fe^{2+} ions 251
absorption of colorless base glass 224
absorption spectrum of Cr^{3+}ions 395
accident surgery 358
acid etching 372
acid treatment 121
acid-extraction process 151
additives 305
aggregation processes 54
aging 369
$Al(PO_3)_3$ crystal 290
Al_2O_3–SiO_2 glass 96, 296, 298
alkali borate glasses 138
alkali silicate glasses 128
alkali-alkaline earth silicate glasses 135
aluminate glasses 257
aluminium titanate (Al_2TiO_5) 299
aluminoborate glasses 157
aluminosilicate glasses 157
amber-glass 246
amorphous semiconductors 260
amphibole group 302
animal experiments 352
animal or clinic testing stages
–, replacement in eyes 330
–, replacement in head area 330
–, replacement in leg 330
–, replacement in nose 330
–, replacement in roots of teeth 330
–, replacement in shoulder 330
–, replacement in the middle ear 330
–, replacement in throat 330
–, replacement of vertebrae 330
annealing of optical glass 31
annealing point 30
anomalous BeO, MgO, ZnO, Al_2O_3–P_2O_5 systems 166
anomalous partial dispersion 184
anomaly line 148
anorthite 361
apochromatic condition 208
application of the bioglass ceramics 330
application of the bioglass ceramics from Jena as bone substitutes in humans 358
applications of lasers 400
arsenic oxide 189
– glasses 190
–, BaO–As_2O_3 190
–, binary systems 190
–, CaO–As_2O_3 190
–, fining agent 189
–, glass formation, Li_2O–As_2O_3 190
–, glass formation, Na_2O-As_2O_3 190
–, glass formation, properties of As_2O_3 glasses 190
–, glass formation, properties of As_2O_5/As_2O_3 glasses 191
–, glass formation, SrO–As_2O_3 190
–, glass formation, structure and properties of glasses fo high arsenic oxide content 192
–, glass formation in K_2O–, PbO–, Sb_2O_3–, MoO_3–, SeO_2–, TeO_2– and V_2O_5–As_2O_3-systems 190
arsenic sulfide glasses 260
–, transmission curves of two As_2S_3-glasses 260
artifical marble 302
asborption spectrum of the ruby crystal 394
astigmatism 6
asymmetrical immiscibility dome 148
athermal optical glasses 216
atomic energy techniques 379
atomic refraction 7
avoid contamination 127

B–O–B bond angles 69
B_2O_3 glass 138
B_2O_3–SiO_2 matrix 118
^{11}B NMR spectrum 67
$Ba_{L\alpha}$ X-ray intensity curve 118
bad intergrowth 352
barium borate glasses 157

barium borosilicate 83
barriers to the development of phosphate glass ceramics 345
basic building elements 43
–, As_2S_3 43
–, in BeF_2 43
BeF_2 glasses 93
beryllium fluoride glasses
–, density 177
–, phase separation 180
–, properties 176
–, ranges of glass formation 176
–, refractive index 178
binodal curve 109
bioactive and biocompatible, machineable glass ceramics of the bioverit-type 357
bioactivity 330, 340
biochemical intergrowth 332
biocompatibility 330, 340
bioglass ceramics 114
– into eyes 357
biotite $K(Mg_{1-x}Fe^{II})_3[(Al_{1-y}Fe_y{}^{III})(Si_3O_{10}F_2)]$ 327
birefringence 32
blackberry-type aggregation of droplets 116
BO_2 chain in calcium metaborate 68
BO_3 triangles 137
BO_4 tetrahedra 138
body fluids 332
bone intergrowth 341
bone-glass-ceramic intergrowth 341
bone-sintered corundum boundary zone 341
borate glasses 137
borid acid anomaly 188
borofluorophlogopite $(KMg_3(BSi_3O_{10}F_2))$ 302
boron anomaly 138
boron oxide anomaly 137
borosilicate glass opacified by phosphate 271
borosilicate glasses 10, 137, 145
borotellurite glasses 116
boroxol rings 68, 69, 137, 139, 140
boundary zone 331
brianite type of $Na_2CaFe(PO_4)_2$ crystals 352
bridging oxygen 130
– ions 49
brucite layer 318
bubbles 1
building substance of the bone 335

cabal $(CaO–Al_2O_3–B_2O_3)$ glasses 157
cabbage-like crystal aggregats 323
cadmium borate 157
calcium metaphosphate phase 115
calcium orthophosphate phase 115
canasite $(Ca_5Na_4K_2Si_{12}O_{30}F_4$ 302
β-$Ca_2P_2O_7$ 347
carbon
– electrode 76
– film 75
– replica method 75
– vacuum deposition 87
carbonate glasses 199
cassius gold purple 248
catalytic influence of water vapor 306
Cd-S-Se ruby glass 242
cell compatibility tests 332
cell cultures 341
ceramic crucible 15
ceravital 331
chain structures 163, 346
chain structures of pure phosphate glasses 161
chalcogenide glass 80, 97
Chemcor process 302, 372
chemical durability 175
chemical etching 84
chemical etching technique 85
–, acetone 84
–, alcohol 84
–, water 84
chemical resistance 125, 136, 150
chemical shift 62
chemical strengthening 372
chlorosilane solution 371
Christiansen effect 276
Christiansen filter 275
chromatic aberration 2
chromophore concentration 229
chrysotile crystals 324
classical biomaterials 330
classical theories of glass structure 41
–, Goldschmidt 41
–, Huggins 50
–, indermediate 46
–, kinetic theorie of Uhlmann 54
–, Lebedev's crystallite theory 50
–, Tammann 41
–, Tilton 50
–, Vitron theory 50
–, Weyl's screening theory 49
–, Zachariasen-Warren network theory 49
–, –, intermediates 43
–, –, low-order 3-dimensional network 42
–, –, network former 43
–, –, network modifier 43
cleavage 318
clinical tests 353
cluster formation 54
clustering 45
Co^{III} complexes
–, croceo-type 227
–, flavo-type 227
–, luteo-type 227
–, melano-type 227
–, praseo-type 227
–, purpureo-type 227
–, roseo-type 227
–, violeo-type 227
Co-γ-radiation doses 379

coagulation 248
coefficient of expansion 19, 25, 138
coexisting phases 102
coherent laser light 374
collagen layer 342
colloidal dispersoids 246
colloidal silver 250
color centers 384
color defect 10
color distortion 208, 212
color indicator 117
color-center band 385
color-center formation 384
coloration by solarization of glasses 408
coloration of high-lead silicate glasses 161
coloration of silver and nickel halides 228
colored glass types 226
colored glasses 223
–, absorption and emission 223
–, Fresnel factor 224
–, internal transmittance 223
–, reflection losses 224
–, transmittance 223
– under γ irradiation 404
colorless glasses 123
complex phosphate 347, 348
composite material 330
composition of microphases 116, 291
compound glass 371
compression layer 307
compression zone 370
compressive stress 366
Compton electrons 375
conduction band 210
conductors 255
connective structures 107, 110
construction of furnaces 424
continuous change of properties 44
control of phase separation 120
controlled crystallization 238, 290, 304
–, controlled microphase separation 291
–, criteria of controlled crystallization 291
–, prinziples of controlled crystallization 290
controlled heat treatment 120
controlled microphase separation 304
controlled surface crystallization 303
conventional implants 330
cooling conditions 129
cooling curve 32
cooling rate 26, 370
coordination groups 49
–, BeF_4 49
–, BO_3 49
–, PO_4 49
–, SiO_4 49
coordination of Fe^{2+} or Fe^{3+} 229
cordierite 361
cordierite crystal 335
cordierite field 311, 316
correction or repair of the eye socket 355
corrosion-resistant electrodes 425
Coulomb's law 284
counterpolarization 159, 197
counterpolarizing effect 257
Cr-doped dosimeter glass 380
cristobalite 161
– (SiO_2) 299
– isotope $AlPO_4$ modification 347
critical nuclear dimension 291
critical point 103
critical radius 282
critical size 348
critical temperature 146
crude surface defects 369
crystal glasses 157
crystal growth 129, 280, 283
– rate 129
crystalline
– $AlPO_4$ 165
– apatite 331
– Fluorspar 212
– GeO_2 188
– semiconductors 255
– solid 25
– solid state 280
– substrate 291
crystallite 288
– of $Ca_3(PO_4)_2$ 271
– of cristobalite 289
crystallization 22
– as a defect 291
– as a defect in glass 287
– behavier of phosphate glasses 345
– kinetics 316
– of glasses 280
– sequence in an NiO-doped base glass 308
– sequence of a TiO_2-doped $MgO–Al_2O_3–SiO_2$ base glass 312
– temperature 293, 295
crystals as an optical medium 212
culture cell test 340
curve of supersaturation 280
curved mica crystals 333
cut-off filters 236

darkening intensity 391
dealkalizing 371
Debye T^3 relation 24
degree of undercooling 305
dendrites 288
denningite (Mg, Ca, Zn) Te_2O_5 173
density 411
– fluctuations 44, 124
– in binary alkali silicate glasses 131
dental protheses 334
destroyed vertebra 354
development of bioglass ceramics 331
–, biocompatible and machinable glass ceramics 332
–, biophosphate glass ceramics 346

–, machinable glass ceramics with mica and cordierite crystals 333
–, bioactive glass ceramics 335
–, phosphate glass ceramics 344
development trends of phosphate glass ceramics 354
devitrite 331, 360
diameter of the crystallite 52
diborate groups 69, 140
dichroism 264
dielectric constant 426
dielectric loss 132, 426
diffraction 57
diffusion 264
– coefficient 134
– enthalpy 291
– zone 113
dilatometric softening point 30
diopside 361
dipolar interaction 61
dipole character 158
dipole formation 159
direct penetration 73
discoloration 374
discontinuous property changes 53
dispersion 8, 409
– behavior of glasses 208
– light filters 275
–, partial dispersion 10
–, relative dispersion 10
disphosphate building units 346
distraction of joint surfaces to maintain clearance 357
distribution of heavy metal ions 116
divalent (anion) glasses 176
dosimeter glasses 163, 379
dosimetry 380
double-ion exchange 372
doubly weakened models 174
drawing nozzle 373
drawing process 369
droplet agglomeration 114
droplet-shaped regions of immiscibility 75, 79
dust in the optical glasses 288

early stages of phase separation 109
Eden Hybinette operation 356
edges of crystals 289
effective strength 365
elastic and inelastic scattering of electrons 72
elastic deformation 364
elastic friction 134
electric electron lenses 71
electric resistance 82
–, optical glasses 82
–, technical glasses 82
electrical conductivity 44, 132, 425
–, specific electrical conductivity 425
–, T_{K-100} value 425
electrical heating 17
electrical resistance 25
electrical resistivity 134
electromagnetic radiation 60
electron
– absorption 87
– acceptore (traps) 404
– backscattering micrograph 114
– beam 74
– beam image 76
– beam microanalyzer 89
– microprobe 85
– microprobe, Auger electrons 85
– microprobe, backscattered electrons 85
– microprobe, cathode luminescence 85
– microprobe, high energy electrons 86
– microprobe, secondary electrons 85
– microprobe, secondary signal types 86
– microprobe, X-rays 85
– microscopy 57, 71
– scanning micrograph 89
– source 72
electronegativity 46
electronic conduction 426
electronic structure of silicon 34
electroporcelain 302
ellipsoidal regions 348, 349
energy dispersion analysis 87
energy levels 60, 255
energy of formation for a molecule 285
enstatite ($MgSiO_3$) 302
epitaxial interaction 283, 291, 360, 377
– with apatite 349
epithelium 355
equilibrium Fe^{2+}/Fe^{3+} 253
etching of the fresh fractured surface 82
β-eucryptite $LiAl[SiO_4]$ 300, 361
eutectic gap 305
extraction 150

Fabry-Perot oscillator 396
fading 380, 384
fast-ion conduction 426
fatigue 369
Fe^{2+}/Fe^{3+} ratios 325
$(Fe^{III}O_4)$ complexes 251
$(Fe^{III}O_6)$ complexes 251
ferrimagnetic properties 329
ferrimagnetic spinel $Fe^{III}(MgFe^{III}O_4)$ 327, 328
ferroelctric glass ceramics 189
ferroelectric germanatae glasses 189
fiber strength 370
field strength 46
filling in large bone cysts 357
fining 162
– reactions 162
fire-polishing 369
firing shrinkage of sintered glas ceramics 361
flat glass 304

flaw theory 365
flaws 365
flint glasses 157
flocculation 248
flow point 30
fluidity 416
fluorescence 374
– colors 374
fluoride glasses free of beryllium 181
fluorine opal glasses 272
fluoroapatite 347
fluorophosphate glasses 183
fluorophosphate laser glass 184
fluorotitanosilicate glass 278
formation
– of colloids of Cu, Ag, Au or Pd 375
– of different habits 291
– of fissures 366
– of shells (halos) 112
– of silver halide crystals 387
forsterite 361
fracture flags 79, 293
fracture mechanics 369
fracture velocity 75
free enthalpy of mixing 106
free enthalpy of nucleation 283
free enthalpy of the interface 281
free mixing enthalphy 103
free volume enthalpy 281
fructose 139
fulgurit 124
functional change of microphases 109

γ irradiation discolors 162
gel glasses 203
germanate glass anomaly 188
germanate glasses 96, 187
–, $BaO–GeO_2$ 187
–, $Bi_2O_3–GeO_2$ 187
–, $CaO–GeO_2$ 187
–, $K_2O–GeO_2$ 187
–, $Li_2O–GeO_2$ 187
–, $M_2O–Al_2O_3–GeO_2$ 187
–, $M_2O–B_2O_3–GeO_2$ 187
–, $Na_2O–GeO_2$ 187
–, $PbO–GeO_2$ 187
–, properties 187
–, refractive indices 189
–, $SrO–GeO_2$ 187
–, structure 188
–, $Tl_2O–GeO_2$ 187
–, $ZnO–GeO_2$ 187
–, ternaries $M_2O–Al_2O_3–GeO_2$ 187
glass (transition) temperature 25
glass ceramics 290, 295
–, bioactive glass ceramics 335
–, bioglass ceramics 330
–, chain silicate glass ceramics 302
–, ferrimagnetic glass ceramics 324
–, high-strength glass ceramics 307
–, machinable glass ceramics 300
–, mica-containing glass ceramics 302
–, phosphate glass ceramics 344
–, sintered glass ceramics 360
–, special glass ceramics 361
–, with minimal coefficients of thermal expansion 295
glass colored by ions, depencence of absorption on network-former 227
glass dosimetry 380, 384
glass fiber 369
glass formation 22
–, tendency to glass formation 24
glass processing 295
–, automatic machine forming 295
–, blowing 295
–, drawing 295
–, tubing 295
glass properties 409
–, calculation 409
–, linear dependence of properties on composition 409
–, mathematical description of glass properties 409
glass structure 365
glass system, SiO_2-B_2O_3 47
glass system $SiO_2–Na_2O–CaO–Ca_3(PO_4)_2–MgO–K_2O–CaF_2$ 331
glass systems 18, 47
–, B_2O_3-P_2O_5 47
–, SiO_2-P_2O_5 47
glass thermometer problem 134
Glass Works Schott & Gen. 7
glass-ceramic implants 341
– and bones 352
glass-ceramics 52
glass-formation regions in binary phosphate systems 163
glass-formers SiO_2, B_2O_3, and P_2O_5 48, 95
glass-maker's soap 408
glasses based on $ZnCl_2$ 199
glasses colored by ions 226
–, BG blue 233
–, characterize of color glass families of Schott catalog 233
–, chromophores for the production of ionically colored glass 232
–, dependence of absorption on modifiers 228
–, dependence of absorption on the valency of the chromophore 229
–, dependence of absorption on the coordination number of the chromophore 230
–, GG yellow 233
–, NG neutral (grey) 233
–, OG orange 233
–, RG red 233
–, SG protective 233
–, UG violet (either IR-transmitting or not) 233
–, VG green 233
glasses colored by metal colloids 246

–, absorption effect of colloidal gold 247
–, antimony ruby grayish-brown 249
–, bismuth ruby grayish-brown 249
–, cobalt ruby brown 249
–, lead ruby grayish-brown 249
–, platinum ruby grayish-brown 249
–, precipitation of other colloidal particles, 249
–, tin ruby brown 249
glasses containing antimomy oxide 193
–, glass formation 193
–, high expansion coeffinciet 194
–, infrared wind 194
–, polarizability of Sb 193
–, properties 193
–, structure 194
glasses containing Bismuth oxide 194
–, Bi_2O_3–B_2O_3 194
–, Bi_2O_3–GeO_2 194
–, Bi_2O_3–P_2O_5 194
–, Bi_2O_3–SiO_2 194
–, glass formation 194
glasses of high lead content 157
glasses with other chromophores 245
–, Ag_2S (heavy metal ruby) 245
–, CuS (heavy metal ruby) 245
–, FeS (iron ruby, or amber glass) 245
–, Mo_2S_3 (heavy metal ruby) 245
–, NiS (heavy metal ruby) 245
–, PbS (heavy metal ruby) 245
–, Sb_2S_3 (antimony ruby) 245
–, W_2S_3 (heavy metal ruby) 245
glassiness 129
glassy carbon 330
glucose 139
glycerol 139
gold ruby glasses 242
grain boundaries 360
granularity of the deposits 79
Griffith flaws 366
Griffith's theory 366
growth of crystals 22
growth stages of a crystal lattice 287

habit changes 289
halide glasses 182
head of tibia 341
heat conductivity 424
heat-absorbing glasses 163, 250
–, system (MgO–ZnO–Al_2O_3–P_2O_5)–FeO 252
–, system (Na_2O–CaO–SiO_2–P_2O_5–Al_2O_3)–FeO 251
–, system Na_2O–CaO–SiO_2 doped with FeO 251
heavy cation fluoride glasses 183
heterogeneous nucleation 283, 292
HF etching procedures 369
HF-resistant glasses 163
high cross section (phosphate glass) 183
high-H_2O glasses 200
high-lead germanate glasses 189
high-quartz lattice 316
high-quartz mixed crystals 306, 308, 310
high-quartz solid solution phase 316
high-temperature catalysts 150
hip-joint replacment 358
homogeneity 6
homogeneous nucleation 281
household dishes 302
human body fluids 340
human skeleton 358
hydrolytic grade 152
hydrolytic resistance 19, 137
hydroxylapatite phase 347
hygroscopic glass 163
hysteresis curve of the glass containing magnetite 330

ice-point depression 134, 20
ilmenite ($FeTiO_3$) 246
immiscibility behavior of binary lithium silicate glasses 297
immiscibility behavior of tellurite glasses 173
immiscibility
– gap 96, 103
– phenomena 93
– processes 133
– region 367
incidental fluctuations 117
indirect (replica) preparations 75
induced emission 396
inert gas atmosphere 258
infinite chains 50
infinitesimal fluctuations 106
influence of F content on droplet dimension 273
influence of impurities 289
influence on the formation of a crystalline habit 290
infrared absorption 411
infrared optics materials 181
inorganic photochromic glasses 384
–, borosilicate glasses doped with silver halides 384
–, borosilicate glasses doped with silver moybdate or tungstate 384
–, borosilicate glasses doped with copper or cadmium halides 384
–, photochromic silicate glasses activated by rate earth 384
instability 106
INT reduction test 340
intensity of scattering 263
interaction between high energy radiation and glass 374
interatomic forces 283
interatomic interaction 283
interface tension 120
intergrowth 331

– between phosphate 352
– region 353
– zone 352
interionic relaxation 400
intermediate glasses 414
internal damping 132
internal friction 134
intrusion of sulfur 250
invert glass structure 137
invert glasses 50, 129, 184, 346
ion diffusion 371
ion etchin 374
ion exchange 372
ion exchange process 332
ion implantation 374
ionic aggregation 133
ionic chain 285
ionic conduction 426
ionic etching technique 85
ionic refraction 413
ionic structures 363
IR absorption of Fe^{2+} 229
IR and UV absorption 209
IR optics 182
IR resonance peak 210
IR spectroscopy 57
IR transmission of germanate, tellurite, and aluminate glasses 257
IR-transmitting chalcogenide glasses 258
–, binary and ternary chalcogenide systems 259
–, other chalcogenide systems 261
IR-transmitting glasses 254
–, crystalline semiconductors 256
–, eigenvibration of particles 255
–, high-pressure hot-pressed materials 255
–, high-temperature resistant IR materials 256
–, IR-transmitting materials 256
–, lattice vibrations 255
–, materials hot-pressed at high pressures 256
iron apatites 349
iron phosphate 348
iron-clad lines 11
iron-containing phosphates 350
γ-irradiation 404
irradiation of photochromic glasses 389

jaw surgery 358
Jena Apparatus Glass 19, 20
Jena Duran 20
Jena Glass 19
Jena method 10
Jena Normal Glass 16^{III} 20
Jena Normal Glass 20
Jena Rasothern 20
Jena Thermometer Glass 59^{III} 20

K_2O–PbO–SiO_2-system 246
K_2O–SiO_2 system 130
kaolin 250
kinematic viscosity 415
kinetic hindrance 23
kinetic theories of glass formation 45
kinetics of immiscibility in glasses 105
kinetics of the immiscibility processes 337
Knudsen glasses 276

labile equilibrium 28
laboratory glass 151
Lambert's law 249
lamp cylinders 19
–, cylinder glass 19
lanthanum borate glasses 157
Larmor frequency 66
laser glasses 393
–, coherent electromagnetic radiation 393
–, energy level diagrams 400
–, fluorescence half-value 398
–, fluorescence lifetime 398
–, fluorescence wavelength 398
–, laser ions 397
–, mode of operation of lasers 396
–, Nd^{3+} doped laser glasses 393
–, non-linear-refractive 398
–, photoelastic constant 398
–, sensitization of the luminescence 400
–, thermo-optical constant 398
laser materials 165
laser principle 395
laser wavelength 397
lattice absorption of crystalline CdSe 243
lattice energy 284, 365
lattice parameters 291
lattice plane 285
law of the iron line 10
layer packets 318
LDH release test 340
lead equivalent 403
lead glass rule of Zschimmer 161
lead phosphate glasses 161
lead-containing systems 157
Li_2O–SiO_2 system 130
ligament fixation in knee capsule surgery 356
ligands 226
light microscope 71
light scattering 160
light scattering and color of microdisperse two-phase glass 275
light-transmitting fibres 126
linear expansivity 150
liquid melt 280
lithium orthoclase $LiAl[Si_3O_8]$ 300
lithium salt Li_2SO_4 melt 303
lithium silicate glass 83
location of nucleation 291
low-temperature quartz isotopes $AIPO_4$ (Berlinite) 347

Macor 301
magnesium titanate ($MgTiO_3$) 299

magnetic electron lenses 72
magnetite (Fe_3O_4) 326
magnetite crystallization 326
mannite 139
material contrast 87
matrix function 119
matrix phase 113
mechanical bending strength 340
mechanical ejection tests 341
mechanical properties of the machinable bioactive glass ceramics 341
melt relics 123
membrane for gas diffusion 150
meta stable level 394
metaphosphate glasses 164
metal colloids 242
metal glasses 56, 200
–, composition of metal glasses 201
–, production of metallic glasses 200
–, properties 202
–, super conducting 202
metal nuclei 379
metallic colloids 246
metastability 103, 106
metastable equilibrium 26
metastable subliquidus miscibility gap 105
methods of production of silicate glass 125
–, MCVD (modified chemical vapor deposition) method 125, 127
–, OVD (outside vapor deposition) method 125, 128
–, PCVD (plasma-activated chemical vapor-deposition) method 125, 128
–, VAD (vapor axial deposition) method 125, 128
Mg-titanate $MgO \cdot TiO_2$ 309
micro currents 350
microcrystalline entities 51
microcrystallites 51
microheterogeneity 52, 82
microheterogeneous structure 54, 366
microinhomogeneities 93
middle ear implants 355
mie scattering effects 242
miscibility gaps 101
mixed alkali effect 48
mixed crystals of the Mg, $Fe^{II}(Fe_x{}^{III}Al_{1-x})_2O_4$ type 326
mixed-alkali effect 132
mixed-ion effect 48
mixed-oxide effect 133
model glasses 174
model liquid 340
model properties 174
model structure 93
models of liquids 57
module of elasticity 364
molar refraction 412
molecular building units 134
molecular theories 284
monovalent (anion) glasses 176
MoO_3 crystal test plane 317
MoO_3 crystal test surface 77
Mössbauer spectroscopy 57
Mössbauer spectra 325
mullite $3Al_2O_3 \cdot 2SiO_2$ 309
mullite field 313, 325
multilayer glass 373
multiple layer glass 372
multiple phase separation 111, 112
multivalent alcohol 139

n_d–v_d diagram 13
Na_2O–CaO–Al_2O_3–P_2O_5–F 346
Na_2O–SiO_2 system 130
natural crystalline silicates 36
–, chain silicates 40
–, cyclosilicates 36
–, group silicates 40
–, inosilicates 36
–, island silicates 40
–, layer silicates 40
–, nesosilicates 36
–, network silicates 40
–, phyllosilicates 36
–, ring silicates 40
–, sorosilicates 36
–, tectosilicates 36
neglection factor 280
network theory by Dietzel 45
$Ni_{K\alpha}$ X-ray intensity curve 118
nickel spinel 308
nickel spinel $(Ni, Mg)Al_2O_4$ 309, 310
NiO-doped base glass 309
nitrate glasses 197
–, $[M^{2+}(NO_3)]^{2-}$ complexes 197
–, glass formation in ternaries 198
–, $M^I NO_3 - M^{II}(NO_3)_2$-$M^{III}(NO_3)_3$ 198
–, structure of nitrate glasses 198
–, systems 197
NMR spectra of ^{12}O 68
NMR spectrum 61
NMR spectrum of ^{11}B 61
NMR theory 60
non-bridging oxygens 130
non-crystalline (glassy) solid 24
non-linear refraction 221
non-linear refractive index (fluoride glasses) 183
norbergite ($2MgO \cdot SiO_2 \cdot MgF_2$) 317
normal line 20, 410
nose curve 55
nose temperature 55
nuclear magnetic dipole 60
nuclear magnetic resonance 57
nuclear magnetic resonance method 59
nuclear magneton 60
nucleation 22, 280
– barrier 283, 360
– process 281
– temperature 293, 295

numerical aperture 71

obsidian 1
opacified glasses 263
–, classification of opacified glasses 266
–, mechanism of opacification 263
–, opacification by diffraction 263
–, opacification by reflection 263
–, opacification by refraction 263
–, opalescence 263
–, particle size 263
opacifiers 274
opal glasses based on SnO_2, TiO_2, ZrO_2, CeO_2, ZnO 274
opalescence 92, 264
opalescent glass 150
opaque glass 129
open immiscibility gaps 292
operation after Bandi 356
optical absorption losses 127
optical attenuation 186
optical attenuation loss 127
optical glass 2
optical glass families 10
–, barium crowns (BaK) 11
–, barium flints (BaF) 11
–, beryllium fluorocrowns (BeFK) 180
–, borosilicate crowns (BK) 11
–, crown glass (K) 7
–, dense and extra dense crowns (SK and SSK) 11
–, dense barium flints (BaSF) 11
–, dense flints (SF) 11
–, flint glass (F) 7
–, fluor crowns (FK) 11
–, fluorophosphate crowns (FPK) 183
–, fluorophosphate heavy crowns (FPSK) 183
–, heavy phosphate crowns (PSK) 11
–, light barium flints (BaLF) 11
–, phosphate crowns (PK) 11
optical glass systems 18
–, $BaO–B_2O_3–P_2O_3–Al_2O_5$ 18
–, $BaO–B_2O_3–P_2O_5–(Al_2O_3)$ 18
–, $BaO–B_2O_3–SiO_2$ 18
–, $BaO–B_2O_3–SiO_2–La_2O_3–ThO_2$ 18
–, $CaF_2–AlF_3–MgF_2–SrF_2–BaF_2–Me^{II}(PO_3)_2–Me^{III}(PO_3)_3$ 18
–, $CaF_2–BaO–B_2O_3$ 18
–, $CaF_2–La_2O_3–B_2O_3$ 18
–, $KF–AlF_3–BeF_2–CaF_2–SrF_2–BaF_2–CdF_2–LaF_3–CeF_3–ThF_4$ 18
–, $KF–AlF_3–K_2TiF_6–Sb_2O_3–PbO–B_2O_3–SiO_2$ 18
–, $KF–PbF_2–AlF_3–NaPO_3–Mg(PO_3)_2–Al(PO_3)_3$ 18
–, $La_2O_3–Nb_2O_3–Ta_2O_5–ThO_2–B_2O_3$ 18
–, $La_2O_3–ThO_2–BaO–CaO–MgO–B_2O_3$ 18
–, $La_2O_3–ThO_2–CdF_2–B_2O_3$ 18
–, $La_2O_3–ThO_2–Ta_2O_5–B_2O_3$ 18
–, $La_2O_3–TiO_2–B_2O_3$ 18
–, $La_2O_3–TiO_2–Ta_2O_5–B_2O_3$ 18
–, $La_2O_3–TiO_2–ZrO_2–Wo_3–B_2O_3$ 18
–, $Li_2O–ZrO_2–La_2O_3–Ta_5O_3–B_2O_3–(SrO, CaO, BaO)$ 18
–, $LiF–NaF–KF–ZnF_2–MgF_2–BaF_2–PbF_2–AlF_3–LaF_3–ThF_4–Al(PO_3)_3$ 18
–, $Na_2O–B_2O_3–SiO_2$ 18
–, $Na_2O–B_2O_3–SiO_2–Al_2O_3–F$ 18
–, $Na_2O–BaO–B_2O_3–SiO_2$ 18
–, $Na_2O–BaO–PbO–B_2O_3–SiO_2$ 18
–, $Na_2O–BaO–ZnO–PbO–SiO_2$ 18
–, $Na_2O–K_2O–CaO–SiO_2$ 18
–, $Na_2O–PbO–SiO_2$ 18
–, $NaF–BaF_2–BeF_2–NaAlF_6–Al_2(SiF_6)_3–NaPO_3$ 18
–, $NaF–BaF_2–BeF_2–NaPO_3$ 18
–, $NaF–KF–PbF_2–BeF_2–Pb(PO_3)_2$ 18
–, $NaF–LaF_3–CeF_3–BeF_2–Be(PO_3)_2–CePO_4–La(PO_3)_3$ 18
–, $NaF–TiO_2–GeO_2$ 18
–, $NaPO_3–Sr(PO_3)_2–K_2TiF_6$ 18
–, $PbO–CdO–ZnO–B_2O_3$ 18
–, $PbO–SiO_2$ 18
–, $Sb_2O_3–Al_2O_3–SiO_2$ 18
–, $TeO_2–Al_2O_3–MgO–SrO–BaO–ZnO$ 18
–, $TeO_2–Ba(PO_3)_2–Pb(PO_3)_2$ 18
–, $TeO_2–La_2O_3$ 18
–, $TeO_2–PbBr_2$ 18
–, $TeO_2–PbO–PbOCl_2$ 18
–, $ThO_2–Gd_2O_3–Ta_2O_5–Nb_2O_5–B_2O_3$ 18
–, $ZnO–MoO_3–WO_3–CdO–B_2O_3$ 18
optical pumps 394
optical silica glasses 124
optimal bioactivity 350
optimization of the machinability 321
organic compounds 22
organic high polymers 330
organic interlayers 372
oriented deposition 291
oscillator 395
Ostwald-Miers region 280
outside deposition methods 128
oversaturation 292
oxycarbonate glasses 200
oxyhalide glasses 199
oxyhydrogen flame 128
oxynitride glasses 199

P–O–P angle 163
paratellurite (α-TeO_2)
partial dispersions 208
partial replacement of individual thoracic vertebrae 357
penetration (einsink) point 31
pentaborate group 140
pericapsular osteotomy after Pemberton 355
petalite $LiAl[Si_4O_{10}]$ 300
phase seperation in glasses 92
phase-separation phenomena 370
phlogopite structure 319

phonons 394
phospate glasses, phase separation 166
phosphate base glass with chain structure 345
phosphate glass ceramics 352
phosphate glass powder 331
phosphate glasses 159, 163
–, structure of phosphate glasses 163
phosphate opal glasses 266
phospho-tellurite glasses 173
photoceram 377
photochemical punching 377
photochromatic glasses 118
photochromic organic compounds 382
photochromic organic compounds
–, anilide (Schiff's) bases 383
–, disulfoxide 383
–, hydrazone 383
–, methyl cellulose 382
–, osazone 383
–, polyglycoles 382
–, polyvinyl alcohol 382
–, polyvinyl methyl ether 382
–, reversible color change 383
–, semicarbazone 383
–, spiropyrane 383
–, stilbene 383
–, succine anhydride 383
photochromic processes 388
–, darkening (formation of color center) 388
–, optical bleaching (fading) 388
–, thermal bleaching 388
photochromic systems 382
photochromism or phototropy 382
photoform 377
photographic process 374
photolysis 375
– of a silver halide 384
photooxidizion 385
photoreduction 385
photosensitive glasses 374
– based on partial crystallization in lithium and barium silicate systems 376
piezoelctric 344
– berlinite crystal phases 347
– berlinite phase 351
– properties 350
planes of maximum rate of moving 289
platinium crucible 2
platinum melt crucible 17
Pockels glasses 419, 421
Poisson's constant 422
polarizability 158
polarization coefficient 216, 218
polishing 369
polymerization 22
polyphosphate glasses 164
polyvinyl butyral 371
porcelain for household use 302
pore diameter 150
pore size of a Vycor-type glass 122
porous (thirsty) Vycor glass 150
porous SiO_2 glass 122
potassium phlogopite ($KMg_3(AlSi_3O_{10}F_2)$) 301
precipitation of apatite 269
precipitation of mica phases 300
precrystalline condition 291
principal dispersion 410
properties of classical pyroceram 297
properties of glass ceramics 296
–, abrasion resistance 296
–, chemical durability 296
–, electrical properties 296
–, low porosity 296
–, machinable drilled 300
–, machinable milled 300
–, machinable threaded 300
–, machinable turned 300
–, mechanical strenght 296
–, negative coefficient of expansion 296
–, transparency 300
–, zero expansion 296
properties of machinable glass ceramics of corning glass works 301
properties of the machinable mica (phlogopite) cordierite glass ceramics 335
properties of the machinable glass ceramics 333
proportionality of darkening 392
protected glasses 403
protective colloid 375
pseudo-binary system $CaO \cdot SiO_2$–$3CaO \cdot P_2O_5$ 117
Pt–Ir–C mixed layer 78
pulse duration 397
pumping energy 396
Pyrex 155
– Glass 20
–, electron microscopic 155
– -type glasses 151
pyroceram 295

quadrupole interaction 60, 64
quartz (SiO_2) 299
quartz glass 123

γ-radiation dose 379
radiation protection 403
radiation-dose measurement 379
radiation-protective glasses 404
Rasotherm 155
–, glass 151
–, investigation 155
raw material for the production of optical glasses 222
Rayleigh scattering 265
Rayleigh's relation 263
reciprocal relative dispersion 410
refraction of light 409
refractive index in binary alkali silicate glasses 132

refractive indexes 9
regenerative gas furnace 10
regions of immiscibility 101
reheating effect 387
relative partial dispersions 410
replacement of oxygen by fluorine ions 315
replica process 75
resonance absorption 226
restoration of the anterior wall of the frontal sinus 355
restoration of the internal auditory canal 355
restriction in tympanoplasty 355
rhinoplastsy 355
richterite ($KNaCaMg_5Si_8O_{22}F_2$) 302
Ringer solution 340
roots of teeth 357
Ruby 393
ruby glasses 246
ruby red color 393
rupture of chemical bonds 374

$(Si_2O_5)^{2-}$ layers 318
"säkulärer Anstieg" 134
saturated solution 288
saturation curve 281
scanning electron microscope 85
scattered X rays 29
scattering absorption 72
Schott, Otto 2
screw dislocation 286, 287
second moment 61
secondary phase separation 292
secondary separation process 113
secondary spectrum 6, 8, 209, 411
selenium-pink ruby glass 246
self-diffusion 132, 284
sellaite (MgF_2) 317
semiconducting materials 258
semiconduction phenomena 426
semiconductor crystals 221
–, CdSe 221
–, CdTe 221
semiconductors 25
semielliptical flaws 365
sensitivity centers 391
sensitizers 376, 390
shadowing with metals or WO_3 77
shear strength of the implant 341
sheet-shaped mica crystals 332
shells around microphases 112
shock resistance 414
short glasses 417
shoulder joint surgery after Eden-Hybinette 357
shrinkage 26
silica glass 123
silicate glass powder 331
silicon-organic coatings 371
silicone coatings 371
silicophosphate glasses 253
silicosis 360
silver diffusion 250
silver halide crystallites 386
silver iodide borate glasses 157
silver ions 250
silver nuclei 375
silver stain 249
silver-free photochromic glasses 392
sintered corundum 330
SiO_2 shell 113
^{29}Si-NMR spectra 62
skeletons 288
slag sitals 361
slanted shadowing 78
small-angle X-ray diffraction 57
small-angle X-ray scattering 94
sodium aluminoborate glasses 157
sodium borate glass 95
sodium borosilicate glass 108
sodium phlogopite ($Na_{=0,5-1}Mg_3(AlSi_3O_{10}F_2)$) 316
sodium silicate glass 97
softening interval 24
softening point 30
sol-gel method 128
sol-gel method for production of glasses and glass ceramics 203
–, the alkoxide sol-gel method 202
–, the silica hydrosol method 203
–, the Ormocer method 203
–, schematic of the production of silica glass according to the alkoxide-gel method 204
–, schematic of the silica-hydrosol process 205
–, molecular building units of ormocers 205
–, application of gel glasses 206
–, physical properties 206
solar irradiation 385
solarization 406
solid solution of SiO_2 in β-eucryptite 300
specific heat 424
spectral transmission of a photochromic glass 390
sphene ($CaTi(OSiO_4)$) 299
spherical aberrations 6
spherical lamella arangement of the phlogopite crystals 322
spherulites 288
spherulitic growth 293
spin vector of magnitude 60
spinels 361
spinodal 103
– area 147
– immiscibility 105
– mechanism 105
spiral growth 286
splitting of a silver halide 375
β-spodumene $LiAl[Si_2O_6]$ 300, 361
spontaneous nucleation 349

stability 103, 106
stabilization against discoloration 404
stable chemical compound 291
stapedectomy 355
statistic fluctuations 95
statistical distribution of alkali 417
statistical distribution of the network-modifier 54
statistical fluctuations 93, 106
stepwise phase-separation 111
stepwise separation 114
steric hindrance 174
sterical hindrance 134
stimulated emission 396
stirrer 17
stirring methods 1
stirring processes 415
stomatology 358
Stookey's basic patents 295
Stookey's fundamental studies 295
strain 419
–, caused by mechanical loading 419
–, dispersion of birefringence 421
– distribution 303
–, from thermal load 421
–, negative birefringence under load 419
–, photoelastic constant 419
– point 30
strength of glass 364
strengthened by controlled crystallization 303
strengthened model glasses 175/76
strengthening methods 370
strengthening of a special glass 302
stress centers 288
striae 6
striking glasses 235
–, base glass with microheterogeneous structure 238
–, chromophores (CdS, CdSe, CdTe) 235
–, coloring mechanism in striking glasses 236, 241
–, lattice absorption of crystalline CdSe 243
–, phase-separation effects in the base glass 244
–, shift of the absorption edge with heat treatment 244
–, shift of UV absorption edges 236
–, transmission curves of Schott striking color glasses 237
– ,colors between yellow, orange, and red (GG, OG,RG) 235
strontian glass 2
structural elements of silicates 34
–, bond angles 34
–, bond distance 34
–, edge length 34
–, SiO_4 Tetrahedron 34
structural inhomogeneities 77
structural units in crystalline arsenic and phosphorus oxides 193
structural units of Na_2O–P_2O_5 glasses 164
structure elements 74
structure of a crystallite 52
structure of acidic phosphate glasses 347
structure of crystalline ZrF_4 184
structure of liquids 57, 58
–, Bernals's model 58
–, Frenkel's model 58
–, Stewart's model 58
structure of pure B_2O_3 137
structure of pure Se glass 262
structures of $Zr(Hf)F_4$ glassses 183
subliquidus immiscibility region 110
substitute vertebra 354
supercritical fluctuations 109
supersaturation of iron oxides 348
support for the base of the nose 355
Supremex Glass 2950^{III} 20
surface compression 304
surface crystallization 360
surface energy 364
surface roughness 334, 340
surface tension 364, 422
swarm-like aggregates 92
sweet glasses 417
system BaO–B_2O_3–SiO_2 111
system BaO–SiO_2 136
system CaO–Al_2O_3 257
system CaO–Al_2O_3–P_2O_5 346
system CaO–P_2O_5–SiO_2 115
system K_2O–CaO–SiO_2 246
system K_2O–Na_2O–CaO–MgO–SiO_2–F 302
system K_2O–ZnO–SiO_2 239
system K_2O–ZnO–SiO_2–CdS 239
system Li_2O–Al_2O_3–SiO_2 299
system Li_2O–Al_2O_3–SiO_2–TiO_2 295
system MgO–Al_2O_3–SiO_2 305
system Na_2O–B_2O_3–SiO_2 145
system Na_2O–CaO–Al_2O_3–P_2O_5 346
system Na_2O–CaO–Al_2O_3–P_2O_5–FeO/Fe_2O_3 and Na_2O–CaO–$Al_2O_3P_2O_5$-F-FeO/Fe_2O_3 347
system Na_2O–CaO–Al_2O_3–P_2O_5–F–FeO/Fe_2O_3 351
system Na_2O–K_2O–CaO–SiO_2–F 302
system Na_2O–MgO–Al_2O_3–SiO_2–F 316
system NaF/KF–BeF_2–PbF_2–Pb $(PO_3)_2$ 214
system TiO_2–SiO_2 296
system ZnO–SiO_2 298
system based on $ZrF_4(HfF_4)$ and ThF_4 181
system ZrO_2–Li_2O–Al_2O_3MgO–SiO_2 302
system ZrO_2–MgO–SiO_2 302

^{125}Te-NMR spectrum 64
tank 152, 156
technical glass 19
tellurim halide glasses 183
tellurite (β-TeO_2)
tellurite glass 80, 97, 163
–, glass-formation range 166
–, halogen and sulfate containing system 167

–, optical properties 166
–, structural unit of tellurite glasses 173
–, structure 172
temperature coefficient of the refractive index 216
temperature dependence of dispersion 275
temperature dependence of the borid acid anomaly 139
tempering 370
– processes 28
TeO_4-polyhedra 173
ternary system K_2O–ZnO–SiO_2 238
tetraborate groups 69, 139, 140
tetragonal cell of crystalline PbO 159
tetrasilicofluoromicas ($K_3Mg_3(Si_4O_{10}F_2)$) 302
tetravalent (anion) glasses 176
theoretical strength 364
thermal (glow) electron emission 72
thermal development process 377
thermal expansion 218, 413
–, cubic coefficient of expansion 414
–, linear coefficient of expansion 413
thermal fading 391
thermal light wave aberration 216
thermal reistance 150
thermal tempering 372
thermally darkening 393
thermex glass 382
thermochemical development 379
thermochemical reduction 379
thermodynamic immiscibility gaps 96
thermodynamics of irreversible processes 109
thermodynamics of phase separation 98
–, equilibrium 98
–, free energy 98
–, free enthalpy 99
–, stability 98
–, stability of mixed phases 99
– ,standard and rest potentials 99
thermometer 20
– glass 132
thermooptical constant 216, 218
three dimensional SiO_4 tetrahedron network 123
three-layer packets 320
three-phase glass 149
threshold energy 396
throat-head surgery field 358
tibia head bone surgery 356
tin opal glasses 274
tissue layer 341
titanate glasses 195
–, Cs_2O–TiO_2 195
–, K_2O–TiO_2 195
–, P_2O_5–TiO_2 195
tooth root 358
topographical contrast 87
toxic effect 330
toxicity 332
trace impurities 124
transition interval 26
transition temperature 25
transmission
– curves of a dispersion filter 276
– electron microscopy 93
– in the infrared beyond 10 262
transparent silica glass 125
transport mechanism 132
triborate group 140
tridymite 161
– isotope $AlPO_4$ modification 347
tris-buffer solution (2-amino-2 hydroxymethyl-1,3 propanediol-HCl–H_2O mixture) 340
trivalent (anion) glasses 176
Turnbull's equation 291
Tyndall effect 94, 277
Tyndall scattering 181, 288

ultraphosphate glasses 164
undercooled liquid 25
undercooling 23, 103, 282
–, magnitude of undercooling 23
unusual partial dispersion 212
–, in alum (KAL $(SO_4)_2 \cdot 12H_2O$) 215
UV absorption 411
– edge 210, 225
UV irradiation 375
UV permeability 125
UV radiation 374
UV resonance peak 210
UV transmitance 125

vandate glasses 196
–, As_2O_3–V_2O_5 system 196
–, BaO–V_2O_5 system 196
–, electrical properties 196
–, electronic semiconductors 196
–, GeO_2–V_2O_5 system 196
–, P_2O_5–V_2O_5 system 196
–, PbO–V_2O_5 system 196
–, polyhposphate vanadate chain 197
–, TeO_2–V_2O_5 system 196
–, trimetaphosphate vanadate ring 197
varulite type (Na, Ca) $Fe_2(PO_4)_2$ crystals 351
ventral spinal fusion of the cervical after Robinson 357
vertrebra replacement by a machinable bioactive glass-ceramic 354
viscosity 415
–, temperatures 30
–, activation energy of viscosity 416
–, dependence of viscosity on composition 417
–, temperature dependence of viscosity 416
–, viscosity and composition of binary alkali silicate glasses 418
viscosity-temperature relation 22
vitreous carbon 202

–, densities of the three solig carbon forms 202
–, solid pyrolisis of highly crosslinked aromatic polymers 202
vitroceram 290
Volmer's relation 282
volume strength 307
Vycor
– glass 122
– process 122, 149
– -type glasses 146, 149

water glasses 129
water resistance of the bioglass ceramic 339
wave conductors 128
wave guides in modern communication 186
waveguide glasses 189
wavelength dispersion analysis 87
weakened model 93, 174
willemite (Zn_2SiO_4) 299
WO_3 films 76
wollastonite 331
– ($Ca_3(Si_3O_9)$) 299
working point 30

X-ray
– diffraction 28
– intensity curve 88, 117
– low angle scattering 124
– patterns of glass 29
– scanning micrograph 89

Z_1 temperature 30
Zachariasen-Warren network theory 42
Zeiss Foundation 11
zinc tellurite $Zn_2Te_3O_8$ 173
zirconium fluoride glasses 184, 222
–, glass formation 184
–, properties 184
–, structure 184
–, system ZrF_4–BaF_2 184
–, system ZrF_4–BaF_2–LaF_3 184
Zrf_x network 183

Springer-Verlag and the Environment

We at Springer-Verlag firmly believe that an international science publisher has a special obligation to the environment, and our corporate policies consistently reflect this conviction.

We also expect our business partners – paper mills, printers, packaging manufacturers, etc. – to commit themselves to using environmentally friendly materials and production processes.

The paper in this book is made from low- or no-chlorine pulp and is acid free, in conformance with international standards for paper permanency.

Made in the USA
Monee, IL
13 September 2026

62973768R00267